Chemical Metallurgy

256

Chemical Metallurgy

Second Edition

J.J. MOORE, PhD, BSc, CEng, MIM
Professor and Head, Department of Metallurgical and Materials Engineering,
Colorado School of Mines, Golden, Colorado, USA

Co-authors

E.A. Boyce, CChem, FRSC
M.J. Brooks, BSc
B. Perry, PhD, BSc, CEng, MIM
P.J. Sheridan, PhD, BSc, CChem, MRSC
Lecturing staff in the Materials Technology Division at
Sandwell College of Further and Higher Education

BUTTERWORTH
HEINEMANN

Butterworth-Heinemann Ltd
Linacre House, Jordan Hill, Oxford OX2 8DP

 PART OF REED INTERNATIONAL BOOKS

OXFORD   LONDON   BOSTON
MUNICH   NEW DELHI   SINGAPORE   SYDNEY
TOKYO   TORONTO   WELLINGTON

First published 1981
Second edition 1990
First published as a paperback edition 1993

British Library Cataloguing in Publication Data
Moore, J. J. (John J)
  Chemical metallurgy – 2nd ed.
  1. Chemical metallurgy
  I. Title
  668.9

ISBN 0 7506 1646 6

Library of Congress Cataloging in Publication Data
Chemical metallurgy/edited by J. J. Moore: co-authors
E. A. Boyce . . . et al. – 2nd ed.
p. cm.
Includes bibliographical references
ISBN 0 7506 1646 6
1. Chemistry, metallurgic I. Moore, J. J. (John
Jeremy), *1944*– II. Boyce, E. A.
QD132.C45   1990
669'.9–dc 20   90–1336

Printed and bound in Great Britain by Hartnolls Ltd,
Bodmin, Cornwall

Preface

Chemical metallurgy encompasses the extraction and refining of metals, liquid metal treatments, and the corrosion protection and surface treatment of metals. A further important area that has received recent attention is the chemical synthesis and processing of metals and materials – especially the high technology materials, ceramics and microelectronic materials. Chemical metallurgy processing principles are also very much applicable to these materials as they are to the extraction and refining of base metals, Cu, Zn, Fe etc. A study of each of these topics demands an understanding of the principles of thermodynamics, reaction kinetics and electrochemistry. Previous chemical metallurgy texts have concentrated mainly on the physical chemistry and technology associated with extractive metallurgy and, in some cases, metallic corrosion. This book has gone a stage further by including a chapter on bonding and periodicity. This is felt to be important since the thermodynamics and kinetics of metallurgical reactions are largely governed by the breaking and forming of bonds between elements and compounds which in turn are determined by atomic structure. In addition, owing to the growing importance of recycling of metallic wastes and the requirement of materials students to understand the general principles associated with metal melting procedures, a brief chapter on metal melting and recycling is also included.

Each contributing author has had extensive experience in teaching his particular speciality to materials students and it is therefore hoped this experience has been used to suitably highlight the main problem areas encountered by students.

The assumption has been made that the reader has an understanding of mathematics and chemistry between GCE 'O' and 'A' level or TEC levels II to III in the UK, or has completed freshman and sophomore chemistry and physics in a North American university. It is intended that this book should act as an introductory text for materials students studying for first degrees, TEC higher diplomas and certificates and Graduateship of the Institution of Metallurgists.

This book should also be of use to scientists and engineers entering employment in the metallurgical and materials processing and metal finishing industries or the teaching profession.

It is the intention of this book to avoid subjecting the reader to a highly statistical approach but rather to provide an understanding of the *fundamental chemical principles* and to demonstrate the *application* of these principles to process metallurgy, materials synthesis and processing, and corrosion protection.

The first five chapters emphasise the fundamental chemical principles involved in metallurgical reactions. Since it is felt that the understanding of quantitative thermodynamics and its application to process metallurgy often prove to be a major problem area for students, example calculations and exercises are included for each chapter. An additional chapter on slag chemistry has also been added in this second edition in order to provide a more thorough understanding of slag–metal reactions. The final three chapters emphasise the applications of the chemical principles to the extraction and refining of metals, metal melting and recycling and metallic corrosion. Each of the chapters is supported by the most significant references or further reading list, whichever is thought to be the more useful, so that the reader may extend his or her knowledge of that particular subject area.

The editor wishes to thank the various authors and publishers who have given permission to use certain figures and data. Acknowledgements to some of these are included at the appropriate places within the text while others include:

John Murray Publishers Limited, for permission to use data from the *Chemistry Data Book*, by J.G. Stark and H.G. Wallace.
Portcullis Press Limited, for permission to use data from *Fundamentals of Foundry Technology*, edited by P.D. Webster, to construct *Table 8.1*.
International Metals Reviews, for permission to use data published in *Review No.* 234, 'Recycling of non-ferrous metals', by J.J. Moore, in Chapter 8.
Longman, for permission to use physical chemistry and thermodynamic data and illustrations from *Physical Chemistry of Iron and Steel Manufacture*, by C. Bodsworth and H.B. Bell.

The undermentioned for permission to use appropriate data from their texts to construct $\Delta G^{\ominus} - T$ diagrams (*Figs 2.10, 2.11, 2.12, 2.13, 7.6, 7.10, 7.20*):

D.W. Hopkins, *Physical Chemistry and Metal Extraction*, published by J. Garnett Miller, 1954.
The Chemical Society, *Principles of Extraction of Metals*, by D.J.G. Ives.
Pergamon Press Limited, *Metallurgical Thermochemistry*, by O. Kubaschewski and Ll. Evans, 1958, and *Fundamentals of Metallurgical Processes*, by L. Foudurier, D.W. Hopkins and I. Wilkomirsky, 1978.
Academic Press, *Physical Chemistry of Melt in Metallurgy*, by F.D. Richardson, 1974.
British Iron and Steel Research Association—now The Corporate Laboratories, British Steel Corporation, Sheffield.

The editor is indebted to Dr A.O. McDougall and Mr P. Groves (Chapters 3 and 6), Professor C. Bodsworth (Chapters 2, 4, 7), Dr P. Boden (Chapter 9) for their advice and expert guidance in minimising the number of errors in the text. Any inaccuracies which may have crept through are, of course, due to the fallibility of the editor.

Acknowledgement is also given to Mrs J.A. Moore and Mr J. Plumb, for help in the graphics and artwork, and to Mrs E. Chambers for typing the manuscript.

J.J. Moore

Contents

Units and Symbols

SI units are used throughout the text with the exception of the unit atmosphere for pressure in thermodynamic applications. This is allowable, however, since the currently used chemical contents which relate to pressure do so with reference to 1 atm rather than 1.01325×10^5 N m^{-2}.

As there is no universally accepted set of symbols for Chemical Metallurgy data, the most generally accepted symbols are used. In certain cases the same symbol is used for several functions but this should not present any confusion to the reader since where this is done the functions are unconnected. The list of symbols used is given below.

A	mass number, area, pre-exponential factor, constant in Debye–Hückel and Onsager equations
A_m	mass number in oxide film calculation
a	activity, constant in Langmuir and Tafel equations, ion size parameter (Debye–Hückel equation)
atm	atmosphere (pressure)
B	constant in Debye–Hückel and Onsager equations, basicity ratio
b	constant in Langmuir and Tafel equations
C	heat capacity
C_p	heat capacity at constant pressure
C_V	heat capacity at constant volume
c	speed of light, concentration
D	diffusivity, diffusion coefficient
d	electron sub-shell
E	energy, electrode potential, emf
E_A, E_B	activation energy
e	electron
e	electron charge, Henrian interaction coefficient
ⓔ	free electrons in oxide film
F	Faraday constant, force
f	electron sub-shell
f	fugacity, frequency, Henrian activity coefficient on molar scale
G	free energy
G^*	free energy of activation
g	acceleration due to gravity
H	enthalpy
h	Planck constant
I	ionic strength, total current
i	current density
i_0	exchange current density

$i_-,\ i_+$	current carried by anions and cations respectively
$\overrightarrow{i}\ \overleftarrow{i}$	rates of anodic and cathodic reactions respectively
i_{pol}	anodic dissolution rate in electropolishing
J	flux (rate of mass flow per unit area)
K	electron shell
K	equilibrium constant
K_1 to K_{12}	temperature dependent constants in oxide film growth
K_c	concentration equilibrium constant
K_p	pressure equilibrium constant
K_w	ionic product (equilibrium constant) for water
K_{dist}	distribution coefficient or ratio
k	rate constant, Sievert's law constant, Boltzmann constant
k_a	rate constant for adsorption
k_d	rate constant for desorption
k_m	mass transfer coefficient
L	electron shell
L_e	latent heat of evaporation (or vaporisation)
L_f	latent heat of fusion
l	subsidiary quantum number, distance
ln	Naperian logarithms ($\log_e$)
M	electron shell
M	molar mass (molecular weight), molality
m	magnetic quantum number, mass
N	electron shell
N	number of neutrons, number of radioactive nuclei, Avogadro's number, mole fraction
N_0	number of radioactive nuclei at time 0
n	neutron
n	principal quantum number, number of gram molecules (moles), order of reaction
P	total pressure
p	electron sub-shell, proton
p	partial or vapour pressure, order of reaction
p_0	saturated vapour pressure
$\overset{\circ}{p}$	vapour pressure of pure component
Q	partition function, charge
q	quantity of heat
R	Rydberg constant for hydrogen, gas constant, electrical resistance, corrosion rate with inhibitor
R_0	corrosion rate without inhibitor
r	rate, radius
S	entropy
s	electron sub-shell
s	spin quantum number
T	absolute temperature (Kelvin)
T_f	melting point (fusion)
t	time
$t_{1/2}$	half-life
$t_-,\ t_+$	transport numbers of anions and cations respectively
U	internal energy
V	volume
VPN	Vickers Pyramid Number
v	velocity
W	atomic mass, probability
w	weight, work done
x	amount adsorbed, order of reaction, distance

y	order of reaction, oxide film thickness, Henrian activity coefficient on the molar scale
Z	atomic number, collision frequency, chemical (or electrochemical) equivalent, charge on ion
z_-, z_+	charges on anion and cation respectively

Greek symbols

α	alpha particles, single-phase solution
β	beta particles, single-phase solution
β^+	beta ray (positive electron)
β^-	beta ray (negative electron)
Γ	surface excess concentration
γ	gamma ray, interatomic distance, surface tension (interfacial and surface free energy), Raoultian activity coefficient
δ	effective boundary layer, thickness of double layer at electrode–electrolyte interface
ϵ	permittivity, Raoultian interaction coefficient
η	polarisation (overvoltage), viscosity
θ	fraction of surface covered
κ	electrical conductivity
Λ	molar conductivity
Λ_0	molar conductivity at zero concentration
λ	wavelength, decay constant, molar conductivity of ions, e.g. λ_{A^+}
μ	chemical potential (partial molar free energy), reduced mass
π	constant (3.142), pi bond
ρ	resistivity, density
σ	molecular diameter, sigma bond
ϕ	absolute electrode potential

Abbreviations

AE	auxiliary electrode, e.g. platinum wire
HER	hydrogen electrode reaction, i.e. $H^+ + e = \frac{1}{2}H_2$
IHP	inner Helmholtz plane
KLA	Knife line attack
OHP	outer Helmholtz plane
PZC	potential of zero charge
RDS	rate determining step
RE	reference electrode, e.g. calomel
SFE	surface free energy
SCC	stress corrosion cracking
VCI	volatile corrosion inhibitor
VHI	Van't Hoff isotherm
VPI	vapour phase inhibitor
WE	working electrode, i.e. metal studied
[ ]	concentration
[ ]$_0$	initial concentration
□	ionic vacancy, e.g. $Ni^{2+}\square$
⊕	positive hole, e.g. Ni^{3+}

Prefixes

Δ	change in a function, i.e. difference in a function between reactants and products, e.g. ΔG, ΔH, ΔS, ΔU, ΔV
d	very small change or increment in a function, e.g. dG, dT, dP, differential

Suffixes

The following symbols are used to specify physical states. They are placed in parentheses after the formula of the substance to which they refer.

g	gaseous, e.g. $H_2(g)$
l	liquid, e.g. $H_2O(l)$
s	solid, e.g. $FeO(s)$
aq	dissolved at effectively infinite dilution in water (aqueous), e.g. $M^{n+}(aq)$

Subscripts

A	atomisation, e.g. ΔH_A
a	anode, e.g. i_a, E_a, η_a
app	applied, e.g. E_{app}
b	breakdown, e.g. E_b
c	cathode, e.g. i_c, E_c, η_c; combustion, e.g. ΔH_c
cell	cell, e.g. E_{cell}, ΔG_{cell}
cont	contact, e.g. E_{cont}
corr	corrosion, e.g. I_{corr}
crit	critical, e.g. i_{crit}, r_{crit}
D	dissociation, e.g. ΔH_D
dil	dilution, e.g. ΔH_{dil}
elec	electronic, e.g. U_{elec}
E	electron affinity, e.g. ΔH_E
F	Flade, e.g. E_F; forward reaction, e.g. r_F
f	formation, e.g. ΔG_f
fus	fusion (applied to entropy and enthalpy), e.g. ΔS_{fus}
I	ionisation, e.g. ΔH_I
i	interstitial ion, e.g. Zn_i^{2+}
L	limiting, e.g. i_L
lat	lattice, e.g. ΔH_{lat}
M	metal, e.g. E_M
m	integral molar function, e.g. ΔG_m, ΔV_m
neut	neutralisation, e.g. ΔH_{neut}
nuc	nuclear, e.g. U_{nuc}
nucleation	nucleation, e.g. $\Delta G_{nucleation}$
ox	oxidation, e.g. I_{ox}; oxide, e.g. V_{ox}
p	polarised, e.g. E_p
pass	passivating, e.g. i_{pass}
prods	products, e.g. G_{prods}
R	reverse reaction, e.g. r_R
Res	electrical resistance (in external circuit), e.g. E_{Res}
r	thermodynamic reversible, e.g. E_r
reacts	reactants, e.g. G_{reacts}
red	reduction, e.g. I_{red}
rot	rotational, e.g. U_{rot}
sol	solution, e.g. ΔH_{sol}; electrolyte solution resistance, e.g. E_{sol}
trans	transformation, e.g. ΔH_{trans}
total	total, e.g. U_{total}
vib	vibratory, e.g. U_{vib}
vol	volume, e.g. ΔG_{vol}, ΔH_{vol}

Superscripts

A	activation, e.g. η^A
C	concentration, e.g. η^C

E	excess thermodynamic function, e.g. G^E
id	ideal thermodynamic function, e.g. G^{id}
m	mixing, e.g. ΔH^m
P	products, e.g. C_p^P
R	reactants, e.g. C_p^R; resistance, e.g. η^R
$\circ$	pressure thermodynamic standard state, e.g. ΔG°
$\ominus$	arbitrary standard thermodynamic state, e.g. $\Delta G^{\ominus}$
$-$	partial molar function, e.g. $\bar{G}_{Zn}$
$\ddagger$	transition or activated state

1

Bonding and Periodicity

The properties of all materials are directly related to the types of atoms present in the material, the nature of the bonds between these atoms and their resulting distribution. It follows, therefore, that to produce materials with specific properties a particular chemical structure must be formed. Many materials result from natural chemical processes but the demands of modern technology have led to the manufacture of increasing numbers of materials for particular applications. Where chemical processes are involved, the bonds between atoms are broken in some substances (reactants) and new bonds are formed to produce the final products.

The breaking and making of chemical bonds involves energy changes and hence the study of thermodynamics is extremely valuable in the understanding of these processes. The rates at which these changes occur are very variable so that a knowledge of chemical kinetics is essential in the understanding and control of chemical reactions. A modern area of study involves the mechanisms whereby these changes occur.

It must be realised that all materials have their limitations so that they must be chosen carefully for a particular application. The present chapter endeavours to explain the various chemical structures that are present in materials and how these structures control the chemical and physical properties. Chapter 2 discusses the thermodynamics of reactions—the energies involved with changing the structures of reactants to form those of the products—while Chapter 3 discusses the reaction kinetics of these reactions—the rates at which the reactant structures can be changed to give those of the desired products.

1.1 The atomic structure of elements

Physicists have now discovered several hundred fundamental atomic particles but only the electron, proton and neutron are of direct interest to the metallurgist or materials engineer.

Protons and neutrons have unit mass whereas the electron has approximately 1/1840 unit mass. Neutrons are electrically neutral, protons have a positive charge while electrons have an equal but negative charge compared with the proton. Most of the atom is empty space with a very small but positively charged nucleus (about 10^{-15} of the atomic volume) and a corresponding number of electrons moving around the nucleus, making the atom as a whole electrically neutral.

1.1.1 The nucleus

The number of protons in an atomic nucleus, the atomic number, determines the fundamental nature of an atom. It is found, however, that the number of protons (Z) in a nucleus may be accompanied by varying numbers of neutrons (N), resulting in atoms with the same atomic number but of different atomic mass (protons + neutrons). Such atoms are known as isotopes. A general name for a nucleus with a particular number of protons and neutrons is a nuclide. Owing to electrostatic repulsion of the positive protons there must be some very powerful nuclear forces binding the nuclear particles.

These specific nuclear forces are not yet fully understood as they are not magnetic, gravitational or electrostatic and they act over a very limited distance ($\sim 10^{-14}$ m).

For a nucleus to be stable it must be less than a certain maximum size as determined by the opposing binding and repelling forces. This gives rise to an optimum ratio of charged (protons) to uncharged (neutrons) particles. Nuclei with an excess or deficiency of neutrons compared with the optimum ratio have too much energy which has to be released in the process of radioactive decay. All nuclei with masses greater than 209 are unstable to some extent and no nuclei with masses greater than 238 are found in nature. The number of disintegrations occurring in the sample per unit time is termed the activity of the radioactive material. The unit of activity is the curie (a disintegration rate of 37×10^9 s^{-1}).

1.1.2   Atomic spectra

When electrons in an atom are energised by heat or electric spark or discharge, their distance from the nucleus is increased and the atom is said to be excited. As the electrons revert to the original, or *ground* state, electromagnetic radiation is emitted which, on refraction using a spectrometer, produces a spectrum.

Study of atomic spectra has led to important developments in the understanding of electron behaviour in atoms. Atomic spectra consist of lines corresponding to definite frequencies. The frequency is directly related to the energy emitted by an electron during a transition within the atom. This was interpreted by Bohr on the basis of the quantum theory developed by Planck. The quantum theory states that a body can only change its energy by a definite *whole number* multiple of energy units or quanta (i.e. it cannot change its energy continuously).

The value of a quantum is related to the frequency of the radiation emitted, by the expression:

$$E = hf$$

or

$$E = hc/\lambda$$

where

E is the value of a quantum (in joules)
h is Planck's constant ($6.626\,176 \times 10^{-34}$ J s)
f is the frequency of the radiation (c/λ)
c is the velocity of light
λ is the wavelength of the radiation

Spectral lines for a particular element appear in series (*Fig. 1.1*). It was found that the frequencies of these spectral lines were related by the formula:

$$\frac{1}{\lambda} = R\left(\frac{1}{n_2^2} - \frac{1}{n_1^2}\right)$$

where R is the Rydberg constant for hydrogen, i.e. $1.097\,373\,8 \times 10^7$ m^{-1}, and n_1 and n_2 are integers ($n_1 > n_2$).

1.1.2.1   Bohr theory

On the basis of the quantum theory, Bohr suggested the presence of definite energy levels within the atom and that electrons could move only in certain selected circular orbits (stationary states) in which they did not radiate energy. Each stationary state corresponded with a definite energy level; emission of energy was caused by movement

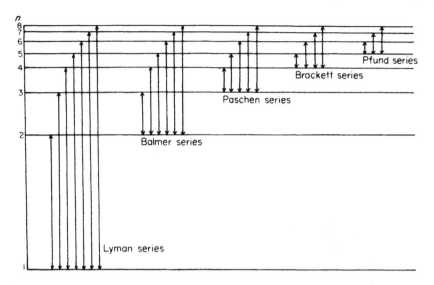

Figure 1.1 Energy levels for some of the electron transitions relating to the hydrogen spectrum

of an electron to a lower stationary state and absorption of energy by movement to a higher state, i.e.

$$E_1 - E_2 = hf$$

where E_1 is the energy of one stationary state and E_2 the energy of a lower stationary state.

The Bohr theory appeared to account accurately for the spectral lines of the hydrogen spectrum but it was found that atomic spectra under high resolution possess a 'fine structure', i.e. they consist of two or more lines very close together. The Bohr theory did not account for this fine structure which was explained by Sommerfeld in 1915 who suggested that some of the orbits are elliptical as well as circular. It was also found that many single lines split when the source of radiation is placed in a magnetic field (Zeeman effect). This fine structure indicates that the electron can exist in an increased number of orbits. The term *quantum number* is used to describe the various energy states and four such numbers are necessary to label each electron.

1.1.2.2 Quantum numbers

The principal quantum number (n) corresponds to the Bohr stationary states. The orbit nearest the nucleus has a value of 1, the second 2, and so on. Letters have also been used to designate these orbits originating from Moseley's work on X-ray spectra, the letter K corresponding with quantum number 1, L with 2, M with 3 and N with 4. The maximum number of electrons with the same quantum number is given by the expression $2n^2$. Hence, in shell 1 the maximum is 2, in shell 2 it is 8, in shell 3 it is 18 and in shell 4 it is 32.

The subsidiary quantum number (l) describes the differing energy levels of electrons within a particular shell and is related to their spatial distributions. Electrons do not really travel in circular or elliptical orbits and it is more accurate to refer to electron *orbitals* which represent the volume of space where there is a high probability of finding the electron. Within a particular shell there are n possible values for l which range from 0 to $(n-1)$. The letters s, p, d, f are used to designate these sub-shells where s = 0, p = 1, d = 2 and f = 3. The letters correspond to the words sharp, principal, diffuse and fundamental which were used to describe the related spectral lines. The s-orbitals are spherical but the p-, d- and f-orbitals are directional as represented in *Fig. 1.2*.

The magnetic quantum number (m) describes the differing spatial distributions of orbitals with the same shape and is determined by the way in which the lines of the

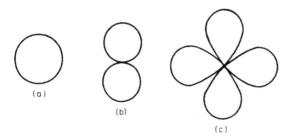

Figure 1.2 Surface limits of: (a) s-orbital; (b) p-orbital; and (c) d-orbital

atomic spectra split up under the influence of a magnetic field (Zeeman effect). There are $(2l + 1)$ possible values for m which are between $-l$, 0, $+l$. Thus, for p-electrons $(l = 1)$ the values of m would be -1, 0, $+1$. This indicates that the p-electrons have three possible spatial distributions which are, in fact, at right angles to each other. The d-electrons have values -2, -1, 0, $+1$, $+2$, i.e. five possible distributions, and the f-electrons have seven $(-3, -2, -1, 0, +1, +2, +3)$.

The Pauli exclusion principle states that no two electrons in any atom may have all their quantum numbers the same. This concept involves the idea of electron spin which necessitates a further quantum number (s) which has two values $(+)$ and $(-)$, related to the direction of spin.* Hence, an orbital can hold two electrons provided that each electron has opposed spins, when they are said to be 'paired'. The maximum number of electrons in the first four shells is indicated in *Table 1.1*.

1.1.2.3   Electronic configuration of the elements

The arrangement of electrons in an atom in its ground state is that which gives the most stable arrangement as described within the limitations of allowable quantum numbers. Hydrogen (atomic number 1) has one electron in the 1s-level and helium (atomic number 2) has two in the same level as the spins of the two electrons are opposed. Lithium (atomic number 3) has two electrons in the 1s-level and one electron in the next lowest level, the 2s, so that its structure may be written as $1s^2 2s^1$. Beryllium (atomic number 4) has the structure $1s^2 2s^2$. Boron (atomic number 5) must have the fifth electron in the 2p-level as the 2s sub-shell is full. Carbon (atomic number 6) and nitrogen (atomic number 7) have two and three electrons respectively in the 2p-level but none of these are paired. Hund's rule states that the number of unpaired electrons in a given energy level must be a maximum because the negatively charged electrons tend to repel each other. For elements with higher atomic numbers than nitrogen the electrons pair in the 2p-levels, e.g. oxygen, fluorine and neon. The electronic structures of the first twelve elements are given in *Table 1.2* (the arrows pointing up or down indicate

Table 1.1 NUMBERS OF ELECTRONS IN THE FIRST FOUR SHELLS

Values of quantum numbers	First shell	Second shell	Third shell	Fourth shell
n: (1 →)	1	2	3	4
l: (0 → ($n-1$))	0	0 | 1	0 | 1 | 2	0 | 1 | 2 | 3
m: (+l → −l)	0	0 | −1 0 +1	0 | −1 0 +1 | −2 −1 0 +1 +2	0 | −1 0 +1 | −2 −1 0 +1 +2 | −3 −2 −1 0 +1 +2 +3
s: (±)	±	± | ± ± ±	± | ± ± ± | ± ± ± ± ±	± | ± ± ± | ± ± ± ± ± | ± ± ± ± ± ± ±
No. of electrons in (i) sub-shells	2s	2s 6p	2s 6p 10d	2s 6p 10d 14f
(ii) shells	2	8	18	32

differing electron spins). The p-orbitals are arranged at right angles to each other along the x, y and z axes. The sequence in which the energy levels are filled is known as the 'Aufbau' principle and is shown in *Fig. 1.3*. It can be seen that the order of filling is:

1s, 2s, 2p, 3s, 3p, 4s, 3d, 4p, 5s, 4d, 5p, 6s, 4f, 5d, 6p, 7s, 5f, 6d

The arrangement of the electrons differs slightly from the above for a few elements.

Table 1.2   ELECTRON STRUCTURES OF THE FIRST TWELVE ELEMENTS, H to Mg

Sub-shells \ Elements	H	He	Li	Be	B	C	N	O	F	Ne	Na	Mg
3s											↑	↑↓
2p$_z$							↑	↑	↑	↑↓	↑↓	↑↓
2p$_y$						↑	↑	↑	↑↓	↑↓	↑↓	↑↓
2p$_x$					↑	↑	↑	↑↓	↑↓	↑↓	↑↓	↑↓
2s			↑	↑↓	↑↓	↑↓	↑↓	↑↓	↑↓	↑↓	↑↓	↑↓
1s	↑	↑↓	↑↓	↑↓	↑↓	↑↓	↑↓	↑↓	↑↓	↑↓	↑↓	↑↓

*Spin quantum number (s) is not to be confused with the s sub-shell in subsidiary quantum number.

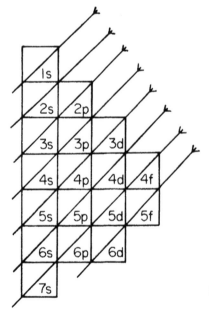

Figure 1.3 Sequence in which energy levels are filled

These deviations involve placing one or two electrons in the $(n-1)$ d-level. Examples are chromium with an outer structure of $3d^5 4s^1$ rather than $3d^4 4s^2$ and copper with $3d^{10} 4s^1$ rather than $3d^9 4s^2$. These electronic structures are more stable due to the stability of a half-filled or full d sub-shell. Electronic configurations of all the elements are given in Appendix 1.

1.2 The periodic table of the elements

Elements in the periodic table (*Table 1.3*) are arranged in horizontal rows (periods) in order of increasing atomic number so that each element has one more orbital electron than the preceding element. The number of elements in each period is determined by the sequence in which the energy levels are filled. As the physical and chemical properties of an element are decided by the number and arrangement of the orbital electrons, the periodic table produces a remarkable arrangement of the elements with related electronic configurations. The order in which the energy levels are filled determines the number of elements in each period as shown in *Table 1.4*. Elements in the vertical arrangements of the table (groups) have similar outermost electronic configurations and hence, similar properties. Those with one outer s-electron are in group I and those with two outer s-electrons are in group II. The two groups comprise the *s-block* of elements which are very electropositive metals, i.e. they react chemically by the loss of electrons from the outer shells to produce positively charged ions (cations), e.g. Na^+, Mg^{2+}.

Groups III to VII contain elements with outer electronic configurations of two s-electrons and one to five p-electrons respectively. Those with eight outermost electrons (except for helium which has two) are the noble gases of group 0.

Elements with the outermost p-orbitals filling from one to six are known as *p-block* elements. This block consists of both metals and non-metals which display a gradual increase in electronegativity (gain of electrons to form negatively charged ions or anions, e.g. Cl^-, Br^-) on moving from group III to group VII. Similarly, elements where d-orbitals are filling are known as *d-block* or *transition elements*. There are several series of these elements in which one to ten d-electrons are filling in the 3d-, 4d- and 5d-levels respectively. These elements exhibit both electropositive (loss of electrons to form positively charged ions) and electronegative characteristics (e.g. Mn forms Mn^{2+} and MnO_4^-).

Finally, elements where f-orbitals are filling are known as *f-block* elements. There are two series of these elements in which 4f- and 5f-orbitals fill from 1 to 14. The first series contains the lanthanoids and the second the actinoids. The characteristics of these elements are similar to those of the transition elements and, indeed, have often been described as *inner transition* elements.

The periodic table is thus now based on the electronic configurations of elements but it must be noted that the table was originally evolved by Mendeléef purely on observed behaviour of the elements when arranged in order of increasing atomic masses. Characteristic properties of the s-, d- and p-block elements and their important compounds are discussed in Section 1.4.

The bonding between elements in forming compounds is dependent on the electronic configuration of the constituent elements and therefore largely determines the properties of the resulting compound.

1.3 Chemical bonds

Noble gases are particularly unreactive elements with outermost electronic structures of eight electrons (two in the case of helium). This lack of reactivity reflects the great

Table 1.3   PERIODIC TABLE OF THE ELEMENTS

	s-block																		p-block						
Period \ Group	I	II														III	IV	V	VI	VII	0
1	1 H																				2 He
2	3 Li	4 Be														5 B	6 C	7 N	8 O	9 F	10 Ne
3	11 Na	12 Mg					d-block									13 Al	14 Si	15 P	16 S	17 Cl	18 Ar
4	19 K	20 Ca	21 Sc	22 Ti	23 V	24 Cr	25 Mn	26 Fe	27 Co	28 Ni	29 Cu	30 Zn				31 Ga	32 Ge	33 As	34 Se	35 Br	36 Kr
5	37 Rb	38 Sr	39 Y	40 Zr	41 Nb	42 Mo	43 Tc	44 Ru	45 Rh	46 Pd	47 Ag	48 Cd				49 In	50 Sn	51 Sb	52 Te	53 I	54 Xe
6	55 Cs	56 Ba	57* La	72 Hf	73 Ta	74 W	75 Re	76 Os	77 Ir	78 Pt	79 Au	80 Hg				81 Tl	82 Pb	83 Bi	84 Po	85 At	86 Ru
7	87 Fr	88 Ra	89† Ac	104 −	105 −	106 −															

		f-block													
*LANTHANOIDS		58 Ce	59 Pr	60 Nd	61 Pm	62 Sm	63 Eu	64 Gd	65 Tb	66 Dy	67 Ho	68 Er	69 Tm	70 Yb	71 Lu
†ACTINOIDS		90 Th	91 Pa	92 U	93 Np	94 Pu	95 Am	96 Cm	97 Bk	98 Cf	99 Es	100 Fm	101 Md	102 No	103 Lw

Table 1.4   ORDER IN WHICH ENERGY LEVELS ARE FILLED RELATED TO THE PERIODIC TABLE

	Period	Sub-shells filled				No. of elements in each period
Short periods	1	1s				2
	2	2s			2p	8
	3	3s			3p	8
Long periods	4	4s		3d	4p	18
	5	5s		4d	5p	18
	6	6s	4f	5d	6p	32
	7					This period is incomplete but appears to follow the same pattern as period 6

stability of these atoms which cannot be improved by compound formation. This led to the first modern theory regarding the mechanism of bond formation. The idea was that atoms combined in such a way as to attain noble gas type electronic configurations thus resulting in greater stability.

Ways in which atoms attain a stable electronic configuration are by losing, gaining or sharing electrons. Ionic bonding results by the combination of an electropositive element with an electronegative element. Covalent bonding results when the combining elements are electronegative and metallic bonding when the elements are electropositive. These different bond types were originally thought to be completely distinct but it is important to realise that these distinctions are now a matter of convenience and most actual bonds are intermediate in type.

1.3.1 The ionic bond (electrovalency)

Ionic bonding commonly occurs between elements positioned just before the noble gases in the periodic table, e.g. groups VI, VII and groups I or II. These elements readily obtain a noble gas type structure by transfer of a small number of electrons to form positive and negative ions which then bond by electrostatic attraction. Consider the case of sodium chloride. The electronic configuration of Na is $1s^2 2s^2 2p^6 3s^1$ and that of chlorine $1s^2 2s^2 2p^6 3s^2 3p^5$. It is obvious that both these elements could attain a more stable electronic configuration by the loss of the 3s-electron from sodium and the gain of this electron by the chlorine atom. The resulting sodium ion is positive due to the loss of an electron and has the configuration $1s^2 2s^2 2p^6$ while the chloride ion is negative due to electron gain, $1s^2 2s^2 2p^6 3s^2 3p^6$. Similarly, in the case of calcium chloride, chlorine atoms gain electrons to form Cl^- but calcium, with the configuration $1s^2 2s^2 2p^6 3s^2 3p^6 4s^2$, must lose two 4s-electrons to form the stable argon configuration $1s^2 2s^2 2p^6 3s^2 3p^6$ with a resultant double positive charge, Ca^{2+}. Hence, for each Ca^{2+} there must be two Cl^- ions in calcium chloride. Other simple ionic compounds may be formed containing group I and II metal ions (cations) and group VI and VII anions.

All the bonds in ionic solids are strong and non-directional as there are no separate molecules: they are giant structures. As a result, these compounds have high melting and boiling points and are very hard, they are usually soluble in polar solvents such as water and are electrolytes. Ionic reactions are usually very rapid. Before studying ionic lattices it is useful to understand the following characteristics of atoms and ions.

1.3.1.1 Size of atoms

Atomic size decreases with increasing atomic number across the same period due to increased nuclear charge (i.e. protons) which pulls electrons of the same and inner shells closer to the nucleus. When extra shells of electrons are added to atoms (moving from a lower to higher period) the size of the atom increases. Thus, on descending a group in the periodic table, there is a marked increase in the size of the atoms.

Formation of ions also causes changes in size. A positive ion formed by *removal* of one or more electrons from the outermost shell often results in complete removal of this shell with a dramatic decrease in size. As a result of electron removal the ratio of positive charges on the nucleus (protons) to the number of electrons is increased hence the effective nuclear charge is increased and the electrons are pulled closer to the nucleus. Negative ions are formed when one or more electrons are *added* to the atom resulting in a fall in the effective nuclear charge and hence an expansion of the electron 'cloud', resulting in an increase in size. (Examples of actual sizes of atoms and ions are given in *Tables 1.9, 1.10, 1.11*.)

A knowledge of size is important in the understanding of bond type and the geometry of crystal structures.

1.3.1.2 Ionisation energy

Formation of positive ions depends upon the ease of removal of electrons from the neutral atom. The minimum energy to remove an electron from an isolated gaseous atom of an element is known as the ionisation energy. It represents the energy absorbed in the process,

$$M_{(g)} \longrightarrow M^+_{(g)} + e$$

This is the *first* ionisation energy. As further electrons are removed values of second, third, fourth, etc. ionisation energies are obtained. These values are obtained from spectra and expressed in kJ mol^{-1}.

Several interrelated factors influence the value of ionisation energy such as the size of the atom, the nuclear charge, type of electron and the screening effect of inner electrons. As electrons are less tightly held in larger atoms, ionisation energy values will decrease with increased size (see *Table 1.5*).

Table 1.5 FIRST AND SECOND IONISATION ENERGIES OF SELECTED ELEMENTS (kJ mol^{-1})

	First	Second		First	Second
H	1316		*p-elements*		
s-elements			B	805	2426
Li	525	7306	Tl	596	1976
Cs	382	2426			
			C	1096	2356
Be	906	1766	Pb	722	1456
Ba	508	972			
			N	1406	2866
d-elements			Bi	780	1616
Ti	667	1316			
Cr	659	1596	O	1316	3396
Fe	768	1566	Te	876	1806
Cu	751	1966			
			F	1686	3376
			I	1016	1846

Table 1.5 shows that it is easier to form an ion of low charge than one of higher charge, that is, it is more difficult to remove successive electrons due to the increase in effective nuclear charge as more electrons are removed. Due to orbital distributions, the s-electron is nearer to the nucleus than the p-, d- or f-electron and is, therefore, more tightly held so that the ionisation energy decreases in the order s, p, d, f—other factors being equal.

First and second ionisation energies of the elements hydrogen to nitrogen are plotted in *Fig. 1.4* which also indicates the great stability of a noble gas type structure and the extra stability of a full s-level in Be and B$^+$ and a half-filled p-level in N and O$^+$.

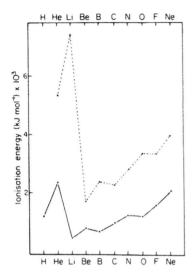

Figure 1.4 First (–) and second (- - -) ionisation energies of the elements H–Ne

1.3.1.3 Electron affinity

Electron affinity is the energy evolved when an isolated gaseous atom gains an electron forming an anion. Adding further electrons requires *expenditure* of energy and, therefore, electron affinities after the first are positive. Some actual values ($\Delta H/kJ\ mol^{-1}$) are: F, -354; Cl, -370; Br, -348; I, -320; O, -148; O$^-$, 850; S, -206; S$^-$, 538; N, -9; H, -78.

Electron affinities cannot be measured very easily and indirect methods such as the application of the Born–Haber cycle may be used. The factors that determine whether

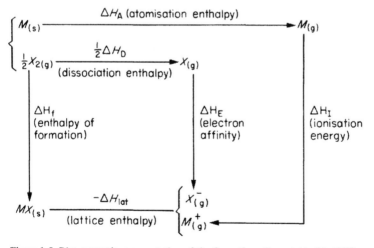

Figure 1.5 Diagrammatic representation of the formation of a metal halide (*MX*)

given elements will combine to form an ionic solid may be found by considering the energy charges involved. Consider the formation of an alkali metal halide MX from solid metal M and gaseous halogen X_2 as illustrated in *Fig. 1.5.* Known as the Born–Haber cycle, this relates the lattice enthalpy of an ionic solid to other thermochemical data. The formation of MX involves converting the elements in their standard states, i.e. pure M and X_2 at 1 atm pressure into gaseous states (ΔH_A, ΔH_D) then to ions (ΔH_E, ΔH_I) and finally into crystal lattice (ΔH_{lat}).

Applying Hess's law, the enthalpy of formation, ΔH_f, will be equal to the algebraic sum of the energy terms involved in the above route:

$$\Delta H_f = \Delta H_A + \tfrac{1}{2}\Delta H_D + \Delta H_E + \Delta H_I - \Delta H_{lat}$$

Thus,

$$\Delta H_E = \Delta H_f - \Delta H_A - \tfrac{1}{2}\Delta H_D - \Delta H_I + \Delta H_{lat}$$

When electron affinities are known the cycle may be used to calculate lattice enthalpies of unknown substances. The concept of lattice enthalpy is important as it measures the energy to break up a lattice, i.e. it relates to the process $MX(s) \rightarrow M^+(g) + X^-(g)$ in which the ions are separated by an infinite distance. If the lattice enthalpy is greater than the energy of solvation of a substance then the substance will probably be insoluble.

1.3.1.4   Electronegativity value

Electronegativity is defined as the power of an atom to attract electrons to itself when it is combined in a compound. Small atoms are the most electronegative, i.e. those at the top right (p-block) of the periodic table, e.g. F and O. The least electronegative elements are those at the left of the periodic table in groups I and II (s-block). Electronegativity increases across a period and decreases down a group in the periodic table.

The bond between two atoms A and B is usually intermediate between pure covalent (Section 1.3.2) $A–B$ and pure ionic $A^+ - B^-$. The bond enthalpy is increased as a result of its partial ionic character. This energy increase was used by Pauling to calculate electronegativity values. Hence electronegativity values can be used to find the approximate ionic character of a bond.

Pauling calculated that two elements with an electronegativity difference of 1.7 would form a bond of 50 per cent ionic character. Thus, elements with an electronegativity difference greater than 1.7 would form mainly ionic bonding and those with an electronegativity difference of less than 1.7 would form mainly covalent bonding. Selected electronegativity values are given in *Table 1.6.*

Table 1.6 SELECTED ELECTRONEGATIVITY VALUES (PAULING)

H	2.1	p-elements	
s-elements		B	2.0
Li	1.0	Tl	1.8
Cs	0.7		
		C	2.5
Be	1.5	Pb	1.8
Ba	0.9		
		N	3.0
d-elements		Bi	1.9
Ti	1.5		
Cr	1.6	O	3.5
Fe	1.8	Te	2.1
Cu	1.9		
		F	4.0
		I	2.5

Other factors affecting ionic character may be summarised in Fajan's rules which state that ionic bonding is favoured if:

(i)   the charge on *either* ion is *small*;
(ii) the *negative* ion is *small*;
(iii) the *positive* ion is *large*.

1.3.1.5   Structures of ionic solids

Each ion in an ionic solid is surrounded by oppositely charged ions. In a crystal consisting of A^+ and B^- ions, the number of B^- ions around an A^+ ion is termed the coordination number of A^+. Similarly, the coordination number of B^- is the number of A^+ ions surrounding an ion of B^-.

Coordination numbers of A^+ and B^- in the compound AB are the same but in compound AB_2 the coordination number of A^{2+} is twice that of B^-. The value of the coordination number is related to the ratio of the radius of the positive ion (r^+) to that of the negative ion (r^-) i.e. r^+/r^-. This ratio is the limiting value for preventing the anions touching and can be obtained from simple trigonometrical calculations. Coordination numbers and their limiting radius ratios are summarised in *Table 1.7*.

Table 1.7 RELATIONSHIP BETWEEN COORDINATION NUMBER, STRUCTURE AND RATIO

Coordination numbers	*Structure*	*Limiting radius ratio,* r^+/r^-
2	linear	0–0.155
3	triangular	0.155–0.225
4	tetrahedral	0.225–0.414
6	octahedral	0.414–0.732
8	body-centred cubic	0.732–1
12	close packing	1

When $r^+/r^- = 1$, ions of the same size are involved. Some metals have a coordination number of 12.

Examples of important crystal structures will now be considered.

(i) Crystal structures of *AB* type ionic compounds

(a) Sodium chloride   The radius ratio (Na^+/Cl^-) of 0.52 suggests an octahedral structure with coordination number of 6, which is indeed found to be the case. The structure is illustrated diagrammatically in *Fig. 1.6*. Other compounds with this type of structure include the halides of the alkali metals (except the chloride, bromide and iodide of caesium), the oxides and sulphides of magnesium, calcium, strontium and barium.

(b) Caesium chloride   As the caesium ion is larger than that of sodium, the radius ratio (Cs^+/Cl^-) is 0.93 and the resulting structure is body-centred cubic with a coordination number of 8 as in *Fig. 1.7*. The other halides of caesium have the same structure except the fluoride.

(ii) Crystal structures of *AB₂* type ionic compounds

The two common structures of this type are CaF_2 (fluorite) and TiO_2 (rutile). Fluorite has a radius ratio of 0.73 with coordination numbers of 8 (Ca^{2+}) and 4 (F^-). Since there are twice as many fluoride ions as calcium ions each Ca^{2+} ion is surrounded by eight F^- and each F^- ion is surrounded tetrahedrally by four Ca^{2+} ions as shown in *Fig. 1.8*. In effect, this structure is very similar to that of CsCl, but since there are only half the number of Ca^{2+} as F^-, the Ca^{2+} occupy only four of the eight possible cubic

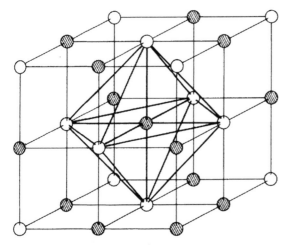

Figure 1.6 Sodium chloride structure showing the octahedral arrangement of ions

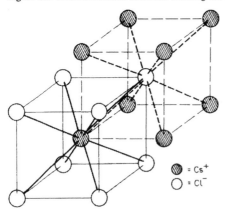

= Cs⁺
= Cl⁻

Figure 1.7 Caesium chloride structure showing the body-centred arrangement of both caesium and chloride ions

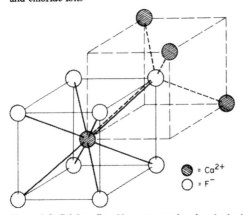

= Ca²⁺
= F⁻

Figure 1.8 Calcium fluoride structure showing the body-centred arrangement of fluoride ions and the tetrahedral arrangement of calcium ions

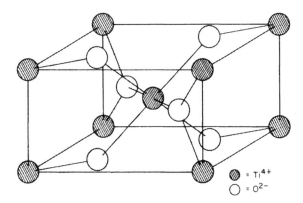

Figure 1.9 Rutile (TiO_2) structure showing the octahedral arrangement of the O^{2-} ions and the trigonal arrangement of Ti^{4+} ions

positions producing a tetrahedral pattern. The radius ratio for rutile (TiO_2) is 0.49 and the resulting structure has coordination numbers of 6 and 3, each Ti^{4+} is surrounded octahedrally by six O^{2-} and each O^{2-} has three Ti^{4+} in a triangular arrangement (*Fig. 1.9*).

The radius ratio rules are useful but cannot be rigorously applied because ionic radii cannot be measured accurately and other factors may influence crystal structure.

1.3.1.6 Applications of ionic compounds

These compounds are used widely in industry. Of particular importance in metallurgy are electrodeposition, electrowinning and electrorefining of metals (see Sections 6.3, 6.4, and 7.4). The above processes are essentially:

$$M^{n+} + ne \longrightarrow M$$

where M is a metal, e an electron. Other examples include:

(i)   Corrosion studies (see Chapter 9) where $M \longrightarrow M^{n+} + ne$ represents the reaction at the corroding metal.

(ii)   Formation of slags (see Chapter 5), e.g.
$2CaO + SiO_2 = 2Ca^{2+} + SiO_4^{4-}$ or Ca_2SiO_4.

(iii)   Study of ionic refractories which are related to high lattice energies in compounds such as MgO due to the high charge and small size of the ions.

1.3.2 The covalent bond

In 1916 G.N. Lewis proposed the idea that a covalent bond is formed when there is a mutual sharing of a pair of electrons between two atoms. The sharing of the electrons constitutes the bond and the shared pair can be considered as contributing to the electronic configurations of both atoms. An example of this type of bond is given by the chlorine molecule (Cl_2). Chlorine atoms have the outer electronic configuration $(3s)^2(3p)^5$ so that if two such atoms each contribute one electron to the bond and also share both bonding electrons they will both acquire the stable argon configuration $(3s)^2(3p)^6$. The chlorine molecule may be represented diagrammatically as:

$$: \overset{\bullet\bullet}{\underset{\bullet\bullet}{Cl}} \times \overset{\times\times}{\underset{\times\times}{Cl}} \times$$

where $\cdot$ and $\times$ denote outer electrons. A single line is often used to denote a covalent bond so that the chlorine molecule may be represented thus: Cl—Cl.

The tetrachloromethane molecule (CCl_4) is made up of one carbon atom and four chlorine atoms. As the carbon atom has four outer electrons $(2s)^2(2p)^2$ it needs to share four electrons to attain a noble gas configuration. Chlorine atoms, however, are only one electron short so four chlorine atoms each bond with one carbon atom. Thus, CCl_4 may be represented as:

$$\ddot{\text{C}}\text{l}$$
$$: \overset{..}{\text{C}}\text{l} \times \text{C} \times \overset{..}{\text{C}}\text{l} : \qquad \text{or} \qquad \text{Cl}-\underset{|}{\overset{|}{\text{C}}}-\text{Cl}$$
$$: \overset{..}{\text{C}}\text{l} : \qquad\qquad\qquad \text{Cl}$$

Other simple examples are water and ammonia:

$$:\overset{..}{\text{O}}\overset{.}{\text{x}} \quad \text{or} \quad \text{H}-\text{O}-\text{H} \qquad\qquad \text{H} \; \overset{..}{\underset{.x}{\text{N}}}\overset{x}{} \; \text{H} \quad \text{or} \quad \text{H}-\text{N}-\text{H}$$
$$\overset{x\cdot}{\text{H}} \qquad\qquad\qquad\qquad\qquad \text{H} \qquad\qquad\qquad \text{H}$$

water   ammonia

Both oxygen and nitrogen attain the stable configuration $(2s)^2(2p)^6$ and hydrogen attains the helium configuration $(1s)^2$. It is also possible for the electron pair to be supplied by one atom but shared by both. Such bonds are called coordinate or dative covalent bonds. As an example, it will be noted that the ammonia molecule previously described has a lone or non-bonding pair of electrons and, as a result, it can react with a hydrogen ion to form the ammonium ion NH_4^+:

$$\text{H} \; \overset{x\cdot}{\underset{\cdot x}{\text{N}}}: \quad + \; [\text{H}]^+ \longrightarrow \left[\text{H} \; \overset{x\cdot}{\underset{\cdot x}{\text{N}}}: \text{H} \right]^+$$

A coordinate bond is represented by an arrow pointing from the donating atom. Hence the ammonium ion may be depicted:

$$\left[\text{H}-\text{N} \rightarrow \text{H} \right]^+$$

1.3.2.1   Shapes of covalent molecules

As electron orbitals possess directional properties (e.g. s, spherical and p, mutually at right angles) covalently bonded atoms will occupy definite relative spatial positions. Since an orbital cannot contain more than two electrons, it follows that an orbital containing one electron from one atom will overlap with a similar electron from the other atom to form an electron pair which is the basis of covalency. Consider the formation of the hydrogen chloride molecule HCl. The outer electronic configuration of chlorine is $(3s)^2(3p)^5$ and hence there is one unpaired electron in the $3p$-level. If this electron overlaps with the $1s$-orbital of the hydrogen atom with the spins of the two electrons opposed, the two waves fuse forming a bond between the hydrogen and chlorine atoms (*Fig. 1.10*).

Simple bonds of this type are known as σ(sigma)-bonds. The strength of the bond is related to the degree of overlap of the orbitals and change of orbital distribution often occurs in order to obtain the maximum degree of overlap.

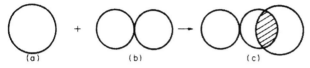

Figure 1.10 Bonding in hydrogen chloride: (a) s-orbital; (b) p-orbital; (c) overlap of s- and p-orbitals to form a σ-bond between H and Cl

In the Sidgwick–Powell approach to stereochemistry, molecular shape is related to the number of electron pairs; both bonding and lone pairs in the outer shell of the central atom. These pairs of electrons repel each other and are, therefore, oriented in space at the greatest distance apart. As a result, it is possible to predict molecular shapes as indicated in *Table 1.8*.

Table 1.8 RELATIONSHIP BETWEEN NUMBER OF ELECTRON PAIRS, SHAPE AND BOND ANGLES OF COVALENT COMPOUNDS

No. of electron pairs	*Shape*	*Bond angles*
2	linear	180°
3	trigonal planar	120°
4	tetrahedral	109° 28'
5	trigonal bipyramidal	120° and 90°
6	octahedral	90°

Carbon has a valency of four thus in carbon compounds of the type CX_4, all the bonds are the same and are directed tetrahedrally. As carbon has the electronic configuration $(1s)^2(2s)^2(2p)^2$ there are only two unpaired electrons and hence it would appear that the valency should be two. However, it is possible to uncouple the 2s-electrons and promote one of them to the empty 2p-level thus producing four unpaired electrons. If these orbitals were used for bonding in this form we should expect different types of bonds as three bonds would be derived from the p-orbitals and one from the s-orbital. The fact that all the bonds are equal is explained by hybridisation. The energy of the s- and p-orbitals is combined and four sp^3 hybrid orbitals of the same shape and energy are formed. These orbitals have one lobe much larger than the other and repulsion between the four orbitals is at a minimum if they point to the corners of a regular tetrahedron with a bond angle of 109° 28' (*Fig. 1.11*).

Lone pairs of electrons may occupy one or more of the tetrahedral positions, e.g. in molecules such as ammonia and water. The lone pairs repel bonding pairs more

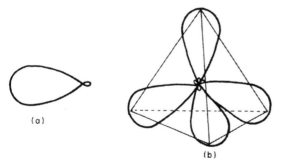

Figure 1.11   (a) Hybridised orbital; (b) tetrahedral distribution of four sp^3 hybridised orbitals

strongly, thus reducing the bond angle of ammonia to $106°\ 45'$ and that of water to $104°\ 22'$.

Other types of hybridisation occur with the mixing of one s- and two p-orbitals (sp^2) and also with one s- and one p-orbital (sp). The orbital shapes of sp^2 and sp are similar to those of sp^3-orbitals and their spatial distributions are $120°$ (trigonal) and $180°$ (linear) respectively.

1.3.2.2   Multiple bonds

Multiple bonding occurs when there are two or three bonds between two atoms. In these cases one bond will be a σ-bond surrounded by one π-bond in the case of double bonding and two π-bonds in the case of a triple bond. Whereas *end* overlap of orbitals

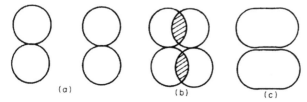

Figure 1.12 Formation of a π-bond: (a) two parallel p-orbitals; (b) merging of orbitals to form (c) a π-bond

produces σ-bonds, *lateral* overlap produces π-bonds (*Fig. 1.12*). Where more than two orbitals overlap laterally the resulting bonding is described as *delocalised* (see graphite structure).

1.3.2.3   Intermolecular forces

Some covalent compounds have continuous covalent bonding throughout and are known as giant structures, but most are made up of separate molecules which vary in size from very small molecules such as hydrogen to extremely large polymeric structures. Intermolecular (van der Waals) forces operate between such molecules and are relatively weak. Hence these substances are either gases, liquids or low-melting-point solids. When covalent molecules are formed from atoms of differing electronegativities the bonds will have partial polar character ($\delta+$ or $\delta-$), e.g.

$$\overset{\delta+\ \ \ \delta-}{\text{H}-\text{Cl}}$$

This will result in dipole–dipole attraction between correctly oriented molecules providing a major contribution to the van der Waals forces. These are known as Keeson forces. Debye forces, caused by the interaction of a permanent dipole with a bond system capable of being polarised, will further increase intermolecular forces. Another large contribution to van der Waals forces is caused by London forces which result from the polarisation of one molecule by another due to the oscillation of electron clouds.

The variation in the strength of intermolecular forces is enormous, from substances such as the noble gases with very low forces to large polymeric molecules containing high dipolar forces. Some of these polymers are powerful adhesives as a result of their extremely high van der Waals forces.

1.3.2.4   Covalent giant structures of industrial importance

(i) Diamond and graphite

An excellent example of the effect of bond type on the properties of materials is the case of the element carbon which exists in two allotropic modifications: graphite and

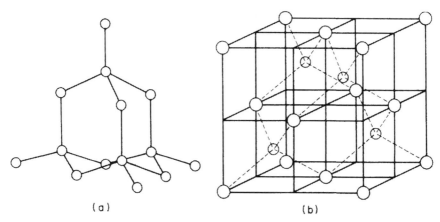

Figure 1.13 Structure of diamond: (a) tetrahedral arrangement of carbon atoms in diamond; (b) unit cell

diamond. Graphite is a good electrical conductor and is used as a lubricant whereas diamond is a non-conductor and an abrasive, yet both are made of carbon atoms only.

Each carbon atom in diamond utilises sp^3-orbitals to form four tetrahedrally directed bonds to each of four other carbon atoms (*Fig. 1.13*). The diamond crystal can, therefore, be regarded as one giant molecule (a macromolecule). The melting (subl.) point of diamond is above 3500 °C due to the fact that breakdown of the lattice would involve the rupture of many extremely stable carbon–carbon bonds. Such structures produce substances which are very hard.

Graphite forms a two-dimensional layer structure in which each carbon is linked to three other carbon atoms in a trigonal plane (see *Fig. 1.14*). The atoms are sp^2 hybridised with the unhybridised p-orbitals overlapping sideways to form delocalised π-bonding. Delocalised π-electrons are mobile and hence graphite is an electrical conductor. The layers of graphite are held by very weak van der Waals forces and easily slide over each other, imparting lubricating properties to the material. Graphite is the thermodynamically stable form of carbon.

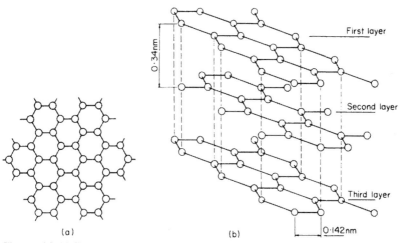

Figure 1.14 (a) Trigonal arrangement of carbon atoms in graphite resulting in a layer of interlocking hexagons; (b) arrangement of layers in graphite

(ii) Silicon carbide

This material exists in a number of crystalline forms with related structures. In each case carbon is surrounded tetrahedrally by four silicon atoms and each silicon is surrounded tetrahedrally by four carbon atoms. Owing to the similarity to the diamond structure, silicon carbide is a very hard, refractory and brittle material.

(iii) Silica (silicon (IV) oxide)

Carbon can form double bonds so that when it forms CO_2 a discrete molecule is formed. Silicon, however, cannot form double bonds due to its size and hence, SiO_2 forms a giant structure. Whereas CO_2 is a gas, SiO_2 is a high-melting-point solid existing as quartz, tridymite and cristobalite. In all of these forms silicon is bonded tetrahedrally to four oxygen atoms and each oxygen is bonded to two silicon atoms from two tetrahedra. The essential difference between these structures is the arrangement of the SiO_4 tetrahedra (see Section 5.3).

(iv) Silicates

A large proportion of the earth's crust consists of silicates. Fusion of silica with alkali produces the silicate ion (SiO_4^{4-}) together with more complicated ions. The way in which SiO_4^{4-} tetrahedra are linked provides a useful basis for classification:

(a) Discrete anions, in which simple SiO_4^{4-} tetrahedra are present (as in ortho-silicates or a limited number of tetrahedra) are combined, e.g. two units as in pyrosilicates $(Si_2O_7)^{6-}$ or cyclic structures such as $(Si_3O_9)^{6-}$ or $(Si_6O_{18})^{12-}$

(b) Extended anions, in which tetrahedra are: (i) linked into chains of indefinite length, e.g. $(SiO_3)_n^{2n-}$, $(Si_4O_{11})_n^{6n-}$; or (ii) sheets of indeterminate area, e.g. $(Si_2O_5)_n^{2n-}$

(c) Three-dimensional silicates, in which the four oxygens of a SiO_4 unit are shared with other tetrahedra to form a three-dimensional network.

1.3.3 The metallic bond

A simple approach in modern terms would be to describe metallic bonding as delocalised σ-bonding. Since the number of outer or valence electrons in metals is small it is impossible to form individual σ-bonds from one atom to all its neighbours, hence all the valence electrons of the atoms in a crystal are shared (delocalised) throughout the structure. These electrons are therefore mobile, thus accounting for the high electrical and thermal conductance in metals as there are no underlying directed bonds. This continuous σ-bonding also accounts for the strength and high melting points of metals.

A more comprehensive account of this type of bonding must involve *band theories*. Consider the electronic structure of an aluminium atom in the ground state, which is $1s^2 2s^2 2p^6 3s^2 3p^1$, comprising differing and discrete energy levels associated with each quantum state. When aluminium atoms combine, the 3s and 3p valence electrons are used not simply to form a few discrete bonding orbitals with neighbouring atoms, but an enormous number of *delocalised* orbitals with all the atoms in the crystal. As the 3s-electrons in the atoms have the same quantum state, they will try to form molecular orbitals of exactly the same energy. This, however, is not possible and hence the orbitals form energy states extremely close to each other since they are derived from the same atomic quantum level. Even though the energy increment from one value to the next of these orbitals is inappreciable, the energy range, due to the massive numbers involved, will be significant and is termed an energy band. Each band, therefore, corresponds in principle to a quantum level of an isolated atom. Similar comments apply to the 3p-electrons of aluminium which form a 3p-band. The energy spread of the 3s- and 3p-bands is such that they overlap, as shown in *Fig. 1.15*.

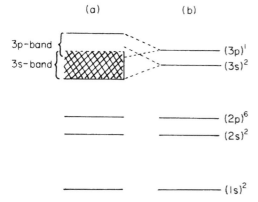

Figure 1.15 Energy levels in: (a) aluminium metal formed by the bonding of (b) aluminium atoms

This, however, does not illustrate the number of electrons in any portion of a band and hence energy is often plotted against $N(E)$, the number of available energy levels per energy increment, i.e. between E and $E + \delta E$. *Figure 1.16* shows a partly filled band in which electrons move extremely easily as the energy difference between E and $E + \delta E$ is imperceptible. However, there is a large energy gap between the partly filled band and a higher unfilled band so that for an electron to move across this gap a great deal of energy would be required. A similar representation for a non-metal is given in *Fig. 1.17*.

In this case, all the energy levels in the lower band are *completely* filled and would need considerable energy to be supplied for an electron to be able to jump to the unfilled band. This results in poor electron mobility and electrical conductivity; non-metals are usually insulators rather than conductors. Metals are conductors because

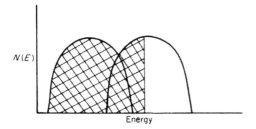

Figure 1.16 Overlapping energy bands in a metal which are only partly filled

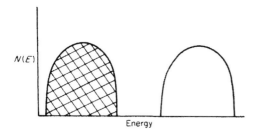

Figure 1.17 Separate energy bands in a non-metal showing one completely filled band (hatched) and one empty band separated by a considerable energy gap

electrons can readily move within the unfilled bands. As metals change from one structure to another relatively easily they are malleable and ductile because the lower parts of the bands are relatively insensitive to structural changes.

1.4 A more detailed study of the periodic table of the elements

1.4.1 The s-block elements

The s-block elements comprise group I (lithium group) and group II (beryllium group). All these elements have their outermost electrons in the s-level. Physical characteristics of these elements are listed in *Table 1.9*.

The chemistry of these elements is dominated by the relative ease with which s-electrons are lost and a stable noble gas type configuration is attained. This tendency is shown by the low ionisation energies of the elements which are related to the large size of the atoms. These metals, therefore, are very powerful reducing agents combining vigorously with non-metals to produce ionic compounds with M^+ cations by group I and M^{2+} by group II. Increasing reactivity occurs in each group with increase in mass and size as will be noted from the decreasing ionisation energies and electronegativity values. Group I metals are more reactive than those in group II and hence the heavier members of group I are the most reactive of all metals. Most compounds of group I elements are soluble but those of group II are much less so as the cations of the latter have a double positive charge leading to higher lattice enthalpies (see *Table 1.9*).

The first members of the two groups, as in other groups of the periodic table, are somewhat anomalous. Lithium shows many characteristics in common with its diagonal neighbour magnesium in group II and beryllium, the compounds of which are covalent, shows similar diagonal relationships with aluminium in group III.

It is interesting to note that both sodium and magnesium are used industrially to reduce titanium (IV) chloride in the production of titanium metal.

1.4.1.1 Oxides

All the s-elements form oxides containing the O^{2-} anion producing oxides of formulae M_2O (group I) and MO (group II). In addition, peroxides containing the anion O_2^{2-} are known, M_2O_2 for group I, MO_2 for group II, and the larger elements potassium, rubidium and caesium form superoxides, containing the anion O_2^-, of formula MO_2.

Each of the group I oxides is extremely basic and dissolves in water producing hydroxide MOH. Those of group II are also strongly basic but significantly less so than those of group I. Beryllium is exceptional in that BeO is covalent and the compound is amphoteric thus displaying acidic properties as well as basic.

1.4.1.2 Chlorides

Chlorides of the s-elements are predominantly ionic with the exception of beryllium chloride. The ionic chlorides have high melting points, are soluble and stable in water and are electrolytes. The anhydrous chlorides of magnesium and beryllium as a consequence of their greater covalent character are hydrolysed by water. The elements are all prepared by the electrolysis of fused chlorides.

1.4.2 Transition elements (d-block)

Elements in which the penultimate shell of electrons is expanded from eight to eighteen by the addition of d-electrons are termed d-block or transition elements. Many physical and chemical properties of these elements are similar. They are all metals, good conductors of heat and electricity, ductile and form alloys with other metals. The melting points and boiling points of these metals are generally high and are related to the enthalpies of atomisation as illustrated for the first transition series in *Fig. 1.18*. The densities of the elements are high due to the filling of penultimate electrons resulting

Table 1.9 PHYSICAL CHARACTERISTICS OF THE s-BLOCK ELEMENTS

Group I

	Electron configuration	Metallic radius (nm)	M^+ ionic radius (nm)	First ionisation energy (kJ mol⁻¹)	Electronegativity	Lattice enthalpy of chloride (kJ mol⁻¹)	Melting point (°C)	Boiling point (°C)	Density (g cm⁻³)	Principal oxidation state
Li	(He)2s¹	0.152	0.060	525	1.0	846	180	1330	0.53	+1
Na	(Ne)3s¹	0.186	0.095	500	0.9	771	97.8	890	0.97	+1
K	(Ar)4s¹	0.231	0.133	424	0.8	701	63.7	774	0.86	+1
Rb	(Kr)5s¹	0.244	0.148	408	0.8	675	38.9	688	1.53	+1
Cs	(Xe)6s¹	0.262	0.169	382	0.7	645	28.7	690	1.90	+1

Group II

	Electron configuration	Metallic radius (nm)	M^+ ionic radius (nm)	Ionisation energy (kJ mol⁻¹)		Electronegativity	Lattice enthalpies (kJ mol⁻¹)			Melting point (°C)	Boiling point (°C)	Density (g cm⁻³)	Principal oxidation state
				1st	2nd		Chloride MCl_2	Oxide MO	Sulphide MS				
Be	(He)2s²	0.112	0.031	906	1766	1.5	3006	3889	3238	1280	2477	1.85	+2
Mg	(Ne)3s²	0.160	0.065	742	1456	1.2	2493	3513	2966	650	1110	1.74	+2
Ca	(Ar)4s²	0.197	0.099	596	1156	1.0	2237	3310	2779	850	1487	1.54	+2
Sr	(Kr)5s²	0.215	0.113	554	1066	1.0	2112			768	1380	2.62	+2
Ba	(Xe)6s²	0.217	0.135	508	972	0.9	2018	3152	2643	714	1640	3.51	+2

Table 1.10 PHYSICAL CHARACTERISTICS OF THE d-BLOCK ELEMENTS Sc–Zn

	Electron configuration (with respect to Ar)	Metallic radius (nm)	Ionic radii (nm)	Ionisation energy (kJ mol⁻¹)		Electronegativity	Melting point (°C)	Boiling point (°C)	Density (g cm⁻³)	Principal oxidation states
				1st	2nd					
Sc	$(Ar)3d^1 4s^2$	0.160	0.081(+3)	638	1246	1.3	1540	2730	2.99	3
Ti	$(Ar)3d^2 4s^2$	0.146	0.090(+2) 0.068(+4)	667	1316	1.5	1675	3260	4.54	2,3,4
V	$(Ar)3d^3 4s^2$	0.131	0.074(+3) 0.059(+5)	654	1376	1.6	1900	3000	5.96	2,3,4,5
Cr	$(Ar)3d^5 4s^1$	0.125	0.069(+3) 0.052(+6)	659	1596	1.6	1890	2482	7.19	2,3,6
Mn	$(Ar)3d^5 4s^2$	0.129	0.080(+2) 0.046(+7)	722	1516	1.5	1240	2100	7.20	2,3,4,6,7
Fe	$(Ar)3d^6 4s^2$	0.126	0.076(+2) 0.063(+3)	768	1566	1.8	1535	3000	7.86	2,3,6
Co	$(Ar)3d^7 4s^2$	0.125	0.078(+2) 0.063(+3)	763	1646	1.8	1492	2900	8.90	2,3
Ni	$(Ar)3d^8 4s^2$	0.124	0.078(+2) 0.062(+3)	742	1756	1.8	1453	2730	8.90	2,3
Cu	$(Ar)3d^{10} 4s^1$	0.128	0.096(+1) 0.069(+2)	751	1966	1.9	1083	2595	8.92	1,2
Zn	$(Ar)3d^{10} 4s^2$	0.133	0.074(+2)	914	1736	1.6	420	907	7.14	2

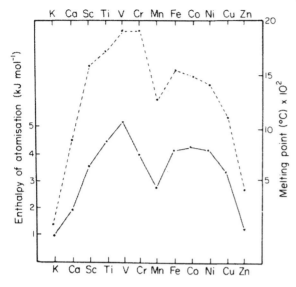

Figure 1.18 Relationship between enthalpies of atomisation (–) and boiling points (- - -) for the elements K–Zn

in increased nuclear charge which pulls in the electrons. Increasing nuclear charge also makes it more difficult to remove electrons resulting in increased ionisation energies across a series causing a decreasing basic character. Ionisation energies are intermediate between those of the s- and p-blocks of elements and hence they are less reactive than the s-block elements. This means that they will tend to exhibit covalency as well as electrovalency with the higher oxidation states being covalent. Physical characteristics of these elements are listed in *Table 1.10*.

1.4.2.1 Magnetic properties

Most substances when placed in a powerful magnetic field are repelled. These substances are described as being diamagnetic. Some substances, however, are weakly attracted in the field and are termed paramagnetic. A few substances, e.g. iron, cobalt and nickel, exhibit a high degree of paramagnetism and are said to be ferromagnetic.

Table 1.11 RELATIONSHIP BETWEEN NUMBER OF UNPAIRED ELECTRONS AND COLOUR OF AN ION OF THE TRANSITION ELEMENTS

No. of 3d-electrons	No. of unpaired electrons	Simple ions and their colours
0	0	$Sc^{3+}Ti^{4+}$ (colourless)
1	1	V^{4+} (blue)
2	2	V^{3+} (green)
3	3	V^{2+} (violet) Cr^{3+} (green)
4	4	Cr^{2+} (blue) Mn^{3+} (violet)
5	5	Mn^{2+} (pink) Fe^{3+} (yellow)
6	4	Fe^{2+} (green)
7	3	Co^{2+} (pink)
8	2	Ni^{2+} (green)
9	1	Cu^{2+} (blue)
10	0	$Cu^{+}Zn^{2+}$ (colourless)

Most simple ions of the transition elements are paramagnetic as they contain unpaired d-orbitals.

1.4.2.2 Colour

The presence of unpaired d-orbitals is also responsible for colour in transition element ions. Colour is due to absorption of radiation in the visible region of the spectrum which can occur with these elements due to electron transitions within the incomplete d-orbitals. The s- and p-block elements usually are not coloured as the energy to promote electrons to an outer shell is often high (e.g. ultraviolet) and hence does not produce colour. There is some correlation between colour and the number of unpaired electrons in an ion as illustrated in *Table 1.11*.

1.4.2.3 Catalytic properties

The common catalysts (see Section 3.5.5) are usually d-block elements and their compounds, e.g.

iron	(manufacture of ammonia)
vanadium (V) oxide	(manufacture of sulphuric acid)
nickel	(hydrogenation processes)
platinum	(oxidation of ammonia for nitric acid production)

The large number of outer s- and penultimate d-orbitals of the metals may form unstable intermediates with the reactants on the surface of the catalyst which would accelerate the reaction.

1.4.2.4 Interstitial compound formation

The transition metals appear to be unique in the ability to fit small atoms such as hydrogen, boron, carbon and nitrogen into the interstices of the transition metal lattice, e.g. steels, forming alloys of varying physical properties and compositions, i.e. non-stoichiometric. These interstitial alloys are of great importance in metallurgy. Substitutional alloys are those in which some of the parent atoms in the crystal lattice have been randomly replaced by other solute atoms.

1.4.2.5 Variable oxidation states

In the s-block elements the valency always equals the group number and in the p-block elements the valency equals the group number or eight minus the group number or the group number minus two (inert pair effect). Hence, limited variable valency occurs in the p-block but it always changes by two, e.g. $TlCl$ and $TlCl_3$; $SnCl_2$ and $SnCl_4$; PCl_3 and PCl_5. The term 'oxidation state' rather than valency is to be preferred in this context and it may be defined as the charge left on the central atom when all the other atoms have been removed as ions. Hence, in the above cases the oxidation states of Tl are $+1$ and $+3$, those of Sn are $+2$ and $+4$ and those of P are $+3$ and $+5$.

Many transition elements show variable oxidation states which change by units of one, e.g. Fe, $+2$ and Fe, $+3$. Variability of the oxidation states in the series Sc to Zn is closely related to the number of 3d- and 4s-electrons available. The maximum oxidation state rises from $+3$ with scandium to $+7$ with manganese and then falls to $+2$ with zinc. With the exception of scandium, all elements show an oxidation state of 2 together with varying states up to the maximum.

1.4.2.6 Complexes

The ability to form complexes is at a maximum with the d-block elements due to the small size, comparatively high charge of the ions and vacant d-orbitals of the correct energy to accept lone pairs of electrons from donor groups (ligands). In general, the most stable complexes are those where the donor atom is nitrogen or oxygen, as in amines or hydrates respectively.

1.4.2.7 The first transition series (Sc to Zn)

It is of some interest to the metallurgist and materials engineer to examine the elements and compounds of the first transition series in more detail since this group encompasses such important elements which form the more commonly used metals and alloys, e.g. steels, nickel, copper and zinc alloys. For each element discussed, the outer electron configurations are given.

(i) Titanium: $3d^2 4s^2$

The +4 compounds are the most stable; the +2 and +3 compounds are less stable and are reducing agents.

Titanium (IV) oxide (TiO_2) occurs in nature as rutile. It is a white, stable, insoluble solid and is amphoteric and as such imparts a certain corrosion resistance to titanium in oxidising conditions. On heating with carbon in the presence of chlorine, titanium (IV) chloride is formed (see Section 7.2.2). The chloride is a colourless covalent liquid which is readily hydrolysed producing TiO_2 and which can be easily reduced with a group I or II metal (Na, Mg) to extract it.

(ii) Chromium: $3d^5 4s^1$

The main oxidation states of chromium are +2 (ionic), +3 (ionic) and +6 (covalent). Oxides corresponding with these states are CrO (black), Cr_2O_3 (green) and CrO_3 (red). As the oxidation state of the metal increases, so the acid character of the oxide increases. As a result, CrO is basic (with acid gives Cr(II) salts, e.g. $CrCl_2$), Cr_2O_3 is amphoteric (with acid gives Cr(III) salts and chromate(III) with alkali) and CrO_3 is acidic. The action of alkali on CrO_3 produces chromates (VI), e.g. Na_2CrO_4 which, in acid solution, form dichromates (VI), i.e.

$$2CrO_4^{2-} + 2H^+ \rightleftharpoons Cr_2O_7^{2-} + H_2O$$

Chromate (VI) and dichromate (VI) are powerful oxidising agents. Chromium (II) chloride is a powerful reducing agent and is rapidly oxidised by air. Chromium (III) chloride is stable.

(iii) Manganese: $3d^5 4s^2$

The electronic structure of this element suggests that the element should have a maximum oxidation state of +7. Many other lower oxidation states exist, +2 and +3 being the most common.

The following oxides are known:

(a) MnO (II) which is basic and produces manganese (II) salts with acid (e.g. $MnSO_4$).
(b) Mn_2O_3 (III) which is also basic producing unstable manganese (III) salts with acids.
(c) MnO_2 (IV) is amphoteric producing unstable manganese (IV) compounds with acids and bases.
(d) The existence of MnO_3 (VI) is doubtful but corresponding manganate (VI) compounds are known, e.g. K_2MnO_4.
(e) Mn_2O_7 (VII) is strongly acidic giving rise with bases to manganate (VII) (permanganates), e.g. $KMnO_4$. These compounds are important oxidising agents.

(iv) Iron: $3d^6 4s^2$

Oxidation states of iron up to +6 are known; +2 and +3 being the most common with +3 the most stable. Oxides with +2 and +3 oxidation states are known. FeO (II) is mainly basic and from this are derived iron (II) salts, e.g. $FeSO_4$, which are reducing in nature. The corresponding hydroxide $Fe(OH)_2$ is also known. Fe_2O_3 (III) is amphoteric producing iron (III) salts with acids, e.g. $Fe_2(SO_4)_3$, and ferrates (III) with fused alkali e.g. $NaFeO_2$. The oxide Fe_3O_4 is found in nature as magnetite. It can be made by the action of steam on red hot iron. It behaves as a mixed oxide ($Fe^{II}Fe_2^{III}O_4$). Chlorides with oxidation states $+2$ and $+3$ are $FeCl_2$ (II) and $FeCl_3$ (III). The latter is covalent and is considerably associated into Fe_2Cl_6 molecules.

Iron reacts with sulphur to give iron (II) sulphide, FeS, which is non-stoichiometric (see Section 3.2.4). It readily reacts with acids forming hydrogen sulphide. Dry hydrogen sulphide reacts with Fe_2O_3 to form iron (III) sulphide, Fe_2S_3, which readily decomposes into FeS and FeS_2. The latter occurs in nature as pyrite which, on roasting in air, forms sulphur dioxide:

$$4FeS_2 + 11O_2 \longrightarrow 2Fe_2O_3 + 8SO_2$$

(v) Cobalt: $3d^7 4s^2$

The most important oxidation state of this element is $+2$ in Co^{2+}. The Co^{3+} state is unstable but in many complexes the $+3$ oxidation state is stable.

Important cobalt (II) compounds are the oxide CoO, the chloride $CoCl_2$ (blue) and $CoCl_2.6H_2O$ (red), and the sulphide CoS (precipitated in group IV analytical schemes).

(vi) Nickel: $3d^8 4s^2$

The only common stable oxidation state of nickel is $+2$. The oxide, chloride, and sulphide corresponding with this state are known.

Nickel readily forms a carbonyl when it is heated with carbon monoxide. Advantage is taken of this reaction in the purification of nickel (see Section 7.2.10.2).

(vii) Copper: $3d^{10} 4s^1$

Common oxidation states of copper are $+1$ and $+2$. Simple copper (I) covalent compounds are generally more stable than the corresponding copper (II) compound. For example, copper (II) iodide decomposes spontaneously at room temperature:

$$2CuI_2 \longrightarrow Cu_2I_2 + I_2$$

With more electronegative radicals, however, higher temperatures are required, e.g.

$$2CuCl_2 \xrightarrow[\text{heat}]{\text{red}} Cu_2Cl_2 + Cl_2$$

$$2CuS \xrightarrow[\text{heat}]{\text{white}} Cu_2S + S$$

Simple ionic compounds of copper (I), however, disproportionate in aqueous solution, e.g.

$$2Cu^+(aq) \rightleftharpoons Cu^{2+}(aq) + Cu(s)$$

Copper (I) oxide, Cu_2O, is covalent and basic, reacting with acids to give copper (I) salts which decompose forming copper (II) salts unless they are insoluble. Copper (II) oxide is also basic forming copper (II) salts with acids. It is readily reduced to the metal It melts at $1150\,^{\circ}C$ and on further heating decomposes to copper (I) oxide, i.e.

$$4CuO \longrightarrow 2Cu_2O + O_2$$

Copper (I) chloride, Cu_2Cl_2, is covalent and insoluble in water. It is soluble in concentrated hydrochloric acid and also in ammonia solution, forming the complex ions

$CuCl_2^-$ and $[Cu(NH_3)_2]^+$ respectively. These solutions will absorb carbon monoxide and are used in gas analysis.

Copper (II) chloride hydrolyses in water and decomposes on strong heating into copper (I) chloride and chlorine. Copper (II) sulphide is insoluble and is slowly oxidised in moist air to the sulphate (VI) which is used as a prior operation to leaching with H_2SO_4 in the hydrometallurgical extraction of copper. On strong heating it decomposes into copper (I) sulphide and copper.

(viii)   Zinc: $3d^{10} 4s^2$

This element has a fixed oxidation state of $+2$ since it does not use the 3d-electrons but only the two 4s-electrons. Ions of the type Zn^{2+} can be formed easily but many zinc compounds are at least partly covalent.

The oxide ZnO is amphoteric and reacts with alkali as well as acids:

$$ZnO + 2OH^-(aq) + H_2O \longrightarrow [Zn(OH)_4]^{2-}(aq)$$

$$ZnO + 2H^+(aq) \longrightarrow Zn^{2+}(aq) + H_2O$$

It is reduced by carbon monoxide to the metal. Uses of the oxide include the manufacture of paints, pharmaceuticals, and in certain catalysts.

The chloride of zinc, $ZnCl_2$, is very covalent in nature and, hence, readily hydrolyses with water:

$$ZnCl_2 + H_2O \longrightarrow Zn(OH)Cl + HCl$$

Zinc sulphide occurs naturally in the earth's crust and is used for the production of zinc (see Section 7.2.10.3). The sulphide is used for the production of fluorescent paints for coating cathode-ray tube screens.

1.4.3   The p-block elements

Although the five elements of any p-group have many common properties it should be noted that:

(a)   There is a marked difference in the properties of the first element in a group and the remaining elements. This is particularly so in the case of carbon, nitrogen and oxygen.

(b)   The middle three elements display properties that smoothly change to those of the heaviest element with the basic nature of the elements increasing and the lower oxidation state becoming more stable.

(c)   The heaviest element in the group displays stable oxidation states lower than the group state and has the most metallic character.

The chemistry of the p-block elements is varied as they include all the non-metals together with some metals. A summary of some of the important metallurgical compounds of groups III to VII is given below while their physical characteristics are given in *Table 1.12*.

(i)   Aluminium: $3s^2 3p^1$

The oxide, Al_2O_3, is an ionic, very refractory, white solid. Although amphoteric, it is essentially basic, dissolving in acids to form salts, e.g. $AlCl_3$, and in alkalis to form the aluminate (III) ion $[Al(OH)_4]^-_{(aq)}$. It is used in the manufacture of aluminium, as an abrasive and refractory material, in the manufacture of cements and as an absorbent in chromatography.

Table 1.2 PHYSICAL CHARACTERISTICS OF SELECTED p-BLOCK ELEMENTS

	Electron configuration	Metallic radius (nm)	Covalent radius (nm)	Ionic radii (nm)	Ionisation energy (kJ mol⁻¹)		Electro-negativity	Melting point (°C)	Boiling point (°C)	Density (g cm⁻³)	Principal oxidation states
					1st	2nd					
Al	$(Ne)3s^2 3p^1$	0.143	0.125	0.050(+3)	583	1826	1.5	660	2470	2.70	3
Si	$(Ne)3s^2 3p^2$		0.117	0.27(−4) 0.041(+4)	792	1586	1.8	1410	2360	2.33	4
Sn	$(Kr)4d^{10} 5s^2 5p^2$	0.162	0.140	0.112(+2) 0.071(+4)	713	1416	1.8	232	2270	7.28 (white)	2,4
Pb	$(Xe)4f^{14} 5d^{10} 6s^2 6p^2$	0.175	0.154	0.120(+2) 0.084(+4)	722	1456	1.8	327	1744	11.3	2,4
As	$(Ar)3d^{10} 4s^2 4p^3$		0.121	0.222(−3) 0.047(+5)	972	1956	2.0		613 subl.	5.72	3,5
Sb	$(Kr)4d^{10} 5s^2 5p^3$		0.141	0.245(−3) 0.062(+5)	839	1596	1.9	630	1380	6.62	3,5
Bi	$(Xe)4f^{10} 5d^{10} 6s^2 6p^3$	0.170	0.152	0.120(+3) 0.074(+5)	780	1616	1.9	271	1560	9.80	3,5

Solid aluminium chloride has a fair amount of ionic character but its vapour is covalent consisting of dimeric molecules with the structure:

It is a white solid and readily hydrolysed by water.

(ii) Silicon: $(3s)^2 (3p)^2$

Silica (SiO_2) and silicates are important compounds and are discussed in Section 1.3.2.4 (iii and iv).

(iii) Tin: $4d^{10} 5s^2 5p^2$; lead: $4f^{14} 5d^{10} 6s^2 6p^2$

These metals exhibit oxidation states of $+2$ and $+4$. The most stable state for tin being $+4$ and that for lead $+2$ (inert pair effect).

The oxides SnO_2 and PbO_2 are amphoteric and with alkali give stannates (IV), $[Sn(OH)_6]^{2-}$, and plumbates (IV), $[Pb(OH)_6]^{2-}$, respectively. The oxides are insoluble in acids except when complexing agents such as Cl^- are present when complex ions such as $[SnCl_6]^{2-}$ are formed. SnO_2 occurs in nature and is an important source of tin. PbO_2 is used in lead accumulators.

The oxides SnO and PbO are more basic and ionic than the higher oxides but also display acidic characteristics. Lead also forms Pb_3O_4 which is used in paints.

The chlorides $SnCl_4$ and $PbCl_4$ are covalent and readily hydrolysed in dilute solutions. Tin (II) chloride is also hydrolysed in dilute solution and is also oxidised as it is a good reducing agent but $PbCl_2$ is stable.

The only important sulphide of these elements is PbS which is the most important ore of lead (galena).

(iv) Arsenic: $3d^{10} 4s^2 4p^3$; antimony: $4d^{10} 5s^2 5p^3$; bismuth $4f^{14} 5d^{10} 6s^2 6p^3$

Arsenic, antimony and bismuth form a closely related series of elements. They form the oxides As_2O_5, Sb_2O_5 and Bi_2O_3 together with As_4O_6, Sb_4O_6 and Bi_2O_3. The (V) oxides of arsenic and antimony produce acid solutions with water. Bismuth (V) oxide, however, is insoluble but does show some acidic nature forming bismuthates (V), e.g. $NaBiO_3$. The (III) oxides of these elements are stable to heat and insoluble in water except for the oxide of arsenic which forms arsenic (III) acid. Alkalis dissolve the oxides of arsenic and antimony forming arsenates (III), $[AsO_2]^-$, and antimonates (III), $[SbO_2]^-$, respectively. Bismuth (III) oxide has no acidic properties thus showing metallic characteristics. The (III) oxides all dissolve in acids. Chlorides of the type XCl_3 are formed by all these elements and only antimony forms a (V) chloride. The (III) chlorides are covalent and hydrolysed by water:

$$AsCl_3 + 3H_2O \rightleftharpoons As(OH)_3 + 3HCl$$

$$SbCl_3 + H_2O \rightleftharpoons SbOCl + 2HCl$$

$$BiCl_3 + H_2O \rightleftharpoons BiOCl + 2HCl$$

Further reading

Brown, G.I. 1971 *A New Guide to Modern Valency Theory*. London: Longman.
Duffy, J.A. 1974 *General Inorganic Chemistry*. London: Longman.
Murray, P.R.S. and Dawson, B.E. 1978 *Structural and Comparative Inorganic Chemistry*. London: Heineman.
Stark, J.G. and Wallace, H.G. 1978 *Chemistry Data Book*. London: Murray.
Wilson, J.G. and Newall, A.B. 1971 *General and Inorganic Chemistry*. Cambridge: Cambridge University Press.

For self-test questions on this chapter, see Appendix 5.

2

Metallurgical Thermodynamics

Thermodynamics is the study of the energies involved in a reaction and, therefore, provides information regarding the driving force behind a reaction. By studying the energy requirements it is possible to determine the optimum conditions necessary to provide the desired reaction. There are three forms of energy which must be considered:

(i)   Heat energy, i.e. H or *enthalpy*.
(ii) Energy related to the arrangement and rearrangement of electrons, atoms, ions and molecules, i.e. S or *entropy*, often referred to as the degree of disorder.
(iii) Energy which can perform useful work, e.g. can be converted to mechanical or electrical energy, i.e. G or *free energy*.

Each of these energies must be considered when assessing a reaction and each will be explained and examined in this chapter.

2.1   Thermochemistry

Thermochemistry is the study of heat energies, H, in a reaction and as such provides important information regarding heat balances and fuel requirements in a metallurgical process. The unit of heat energy is the joule or kilojoule (J or kJ).

2.1.1   Exothermic and endothermic reactions

Chemical reactions are carried out commercially either to obtain a useful product or as a way of obtaining energy, e.g. combustion of coal:

$$C + O_2 \longrightarrow CO_2 + heat$$

This process involves conversion of chemical energy into heat energy.[*]

Every known chemical reaction involves an energy change; most reactions occur with an evolution of heat and are called *exothermic reactions*, a few reactions (which will normally only take place at high temperatures) absorb heat and are called *endothermic reactions*. If heat is lost by the system, i.e. given out to the surroundings (exothermic reaction) a negative sign $(-)$ is placed in front of the stated quantity of heat. Conversely if the system gains heat from the surroundings (endothermic reaction) a positive sign is used. Two examples are given below:

(1) $2Al + \frac{3}{2}O_2 \xrightarrow{298K} Al_2O_3$; $-1700 \text{ kJ mol}^{-1}$

---

[*]Energy changes in chemical reactions may involve other interconversions, e.g. chemical energy to electrical energy (accumulators), chemical energy to mechanical energy (explosions). Work on the development of fuel cells has (so far unsuccessfully) attempted to devise a commercial means of obtaining electrical energy directly from the combustion of coal without the intermediate costly stage of raising steam.

where the heat gained or lost is per mole of Al_2O_3 formed. For any given reaction the heat liberated or absorbed is constant under specified conditions.

The top of a mould into which metal is cast may be made of a material (e.g. powdered aluminium and an oxidising agent) which evolves heat when in contact with a source of heat such as molten metal. This allows the metal at the top of the mould to remain molten thus feeding the contraction cavity.

(2) $$ZnO + C \xrightarrow{1373 K} Zn + CO; \quad +350 \text{ kJ mol}^{-1}$$

This extraction reaction for zinc needs to overcome the difficulties of heat supply and conservation in order to be commercially operable at 1373 K, i.e. 350 kJ of heat energy must be added to the system in order that one mole of carbon will reduce one mole of zinc oxide at 1373 K. (Thermodynamic relationships refer to the absolute temperature scale, therefore K (kelvin) rather than $^\circ$C will be used extensively in this chapter.)

2.1.1.1 Standard enthalpy change for a reaction

The enthalpy, H, of a substance can be defined as its heat (energy) content. Consider the enthalpy changes for exothermic and endothermic reactions as follows:

Exothermic reaction	Endothermic reaction
Heat is evolved	Heat is absorbed
Energy content (H) is lower at the end of the reaction, i.e.	Energy content (H) is higher at the end of the reaction, i.e.
reactants $\longrightarrow$ products	reactants $\longrightarrow$ products
$H_1 \longrightarrow H_2$ (initial) (final)	$H_1 \longrightarrow H_2$ (initial) (final)
$H_2 < H_1$	$H_2 > H_1$

Now the change in enthalpy, ΔH, is given by*

$$\Delta H = H_2 - H_1$$
$$= H_{products} - H_{reactants} \qquad (2.1)$$

Equation (2.1) is written this way conventionally (instead of $\Delta H = H_{reacts} - H_{prods}$). Hence, for an exothermic reaction, as $H_2 < H_1$ then ΔH is negative, i.e. heat is *given out* during the reaction (at constant pressure). While for an endothermic reaction, as $H_2 > H_1$ then ΔH is positive and heat is *absorbed* during the reaction (at constant pressure). *Fig. 2.1* gives typical heat energy profiles for exothermic and endothermic reactions. Before a reaction can occur energy is required, called *activation energy* (E), to surmount an initial energy barrier; without this energy the reaction cannot proceed (see Section 3.5.1).

As stated earlier the enthalpy change, ΔH, for a reaction is constant for a given reaction under specified conditions ΔH is dependent upon: (a) temperature; (b) pressure; (c) physical states of reactants and products; (d) amounts of substances reacting.

The standard enthalpy change for a reaction, ΔH°, is defined as the change in enthalpy referring to the masses (in moles) of the reactants and products shown in the

---

*Δ is the symbol used to refer *generally* to the incremental difference between two states, in this case enthalpies, H_{prods} and H_{reacts}. Hence ΔH is the *change* in enthalpy.

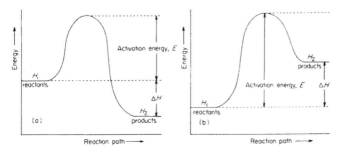

Figure 2.1 Heat energy profiles for: (a) an exothermic reaction; (b) an endothermic reaction

equation for the reaction at 101 325 Pa* (1 atm) and at a stated temperature, T (given as a subscript) with the substances in the physical states normal under these conditions. For example, ΔH°_{298} refers to the enthalpy change at standard pressure (1 atm) and a temperature of 298 K. For an arbitrary standard of pressure other than 1 atm the enthalpy change is written $\Delta H^{\ominus}_{T}$.

Data[1,2] usually refer to the *standard state* of reactants and products, i.e. the forms in which they are stable under a pressure of 1 atm at a stated temperature (usually 298 K or 25 °C). The thermodynamic standard state for gaseous reactants and products is normally taken as 1 atm partial pressure (i.e. ΔH°) while that for solids or liquids is normally taken as the pure state (i.e. $\Delta H^{\ominus}$). Since 1 atm partial pressure is also an arbitrary standard state the symbol ⊖ is also often used when considering gaseous states. For this reason *all* standard states in this text will be denoted by the symbol ⊖ from now on.

The previous example equations can now be rewritten as full thermochemical equations:

(a) $C(graphite) + O_2(g) \longrightarrow CO_2(g)$ $\Delta H^{\ominus}_{298} = -393 \text{ kJ mol}^{-1}$

(b) $2Al(s) + \frac{3}{2}O_2(g) \longrightarrow Al_2O_3(s)$ $\Delta H^{\ominus}_{298} = -1700 \text{ kJ (mol Al}_2O_3)^{-1}$

(c) $ZnO(s) + C(graphite) \longrightarrow Zn(g) + CO(g)$ $\Delta H^{\ominus}_{1373} = +350 \text{ kJ mol}^{-1}$

Note that reaction (b) could be written as:

$4Al(s) + 3O_2(g) \longrightarrow 2Al_2O_3(s)$ $\Delta H^{\ominus}_{298} = -3400 \text{ kJ}$

i.e. 3400 kJ of heat are released when 4 moles of solid Al react with 3 moles of gaseous O_2 to produce 2 moles of solid Al_2O_3. In this case the units are kJ not kJ mol^{-1} since all amounts differ from 1 mol.

There are various ways in which enthalpy changes for reactions may be stated more specifically.

(i) Standard enthalpy change of formation for a compound, $\Delta H^{\ominus}_{f}$

This is the change in enthalpy when *one mole of a compound* is formed from its elements under standard conditions. An example is given by reaction (a) above. The standard enthalpy change of formation for carbon dioxide may also be written:

$$\Delta H^{\ominus}_{f298}[CO_2(g)] = -393 \text{ kJ mol}^{-1}$$

*Pa represents the SI pressure unit called the pascal.

Standard (atmospheric) pressure = 1 atm
            = 101 325 N m^{-2}
            = 101 325 Pa (i.e. 1 N m^{-2} = 1 Pa)
            (= 760 mmHg)

Further examples include:

Reaction	$\Delta H^{\ominus}_{f298}$ (kJ mol^{-1})
$2Al(s) + \frac{3}{2}O_2(g) \longrightarrow Al_2O_3(s)$	-1700
$Ca(s) + \frac{1}{2}O_2(g) \longrightarrow CaO(s)$	-637
$Zn(s) + \frac{1}{2}O_2(g) \longrightarrow ZnO(s)$	-354
$Hg(l) + \frac{1}{2}O_2(g) \longrightarrow HgO(s)$	-91
$2Ag(s) + \frac{1}{2}O_2(g) \longrightarrow Ag_2O(s)$	-29
$H_2(g) + \frac{1}{2}O_2(g) \longrightarrow H_2O(l)$	-288
$C(s) + \frac{1}{2}O_2(g) \longrightarrow CO(g)$	-112
$\frac{1}{2}N_2(g) + \frac{1}{2}O_2(g) \longrightarrow NO(g)$	$+90$

All of the oxides formed are termed 'exothermic compounds' except for NO which is an 'endothermic compound'. The enthalpy change of formation for a compound is one of the factors which determines its stability to heat and chemical attack; in fact the more negative the value of $\Delta H^{\ominus}_f$ the greater the stability of that compound since this amount of heat must be added to decompose it.

It will be explained later that other factors (i.e. free energy and entropy) must influence the feasibility of a reaction otherwise: (a) endothermic reactions would never occur spontaneously; and (b) reversible chemical reactions would not be possible since

$$A \underset{+\Delta H^{\ominus}}{\overset{-\Delta H^{\ominus}}{\rightleftarrows}} B \qquad \text{(see Laplace's rule, Section 2.1.2.2)}$$

(ii)   Standard enthalpy change of combustion, $\Delta H^{\ominus}_c$, of a substance

This is the enthalpy change when *one mole of a substance* is completely burned in oxygen under standard conditions, e.g.

$$Mg(s) + \frac{1}{2}O_2(g) \longrightarrow MgO(s); \qquad \Delta H^{\ominus}_c = -605 \text{ kJ mol}^{-1}$$

or

$$\Delta H^{\ominus}_{c298}[Mg(s)] = -605 \text{ kJ mol}^{-1}$$

Obtaining $\Delta H^{\ominus}_c$ values for elements and compounds by use of the bomb calorimeter (i.e. burning a known mass of substance in a steel 'bomb' which will withstand high pressures and measuring the consequent temperature change) is a valuable way of accumulating thermochemical data (see *Fig. 2.3*).

(iii)   Enthalpy change of atomisation, $\Delta H^{\ominus}_A$, for an element

This is the enthalpy change occurring when *one mole of gaseous atoms* is formed from the element in the defined physical state under standard conditions, e.g.

$$\frac{1}{2}H_2(g) \longrightarrow H(g); \quad \Delta H^{\ominus}_A[H_2(g)] = +218 \text{ kJ mol}^{-1}$$

(iv)   Standard enthalpy change of transformation, $\Delta H^{\ominus}_{trans}$

This is the change in enthalpy when *one mole of a substance* undergoes a specific physical change, e.g. melting, evaporating, allotropic transformation, under standard conditions.

(a) $Zn(s) \longrightarrow Zn(l)$; $\Delta H^{\ominus}_{trans\ 693} = +7.14\ kJ\ mol^{-1}$

This is better known as the molar latent heat* of fusion of zinc.

(b) $Zn(l) \longrightarrow Zn(g)$; $\Delta H^{\ominus}_{trans\ 1180} = +114.7\ kJ\ mol^{-1}$

This is better known as the molar latent heat of evaporation of zinc.

(c) $C(diamond) \longrightarrow C(graphite)$; $\Delta H^{\ominus}_{trans\ 2273} = -2.0\ kJ\ mol^{-1}$

(d) $Ti(\alpha,\ hcp) \longrightarrow Ti(\beta,\ bcc)$; $\Delta H^{\ominus}_{trans\ 1153} = +3.36\ kJ\ mol^{-1}$

The latter two examples are allotropic transformations (hcp stands for hexagonal close packed and bcc for body-centred cubic).

(v)   Standard enthalpy change of solution, $\Delta H^{\ominus}_{sol}$

This is the change in enthalpy when *one mole of a solute* is dissolved in a given amount of solvent, under standard conditions. For example

$$NaCl(s)\ +\ 200H_2O(l)\ \rightarrow NaCl.200H_2O(aq)$$
$$\Delta H^{\ominus}_{sol} = +4.26\ kJ\ mol^{-1}$$

In general, enthalpy changes of solution are small relative to those for chemical reactions (see below for the effect of dilution on $\Delta H^{\ominus}_{sol}$ and also Section 4.3.8 for ΔH^m, enthalpy of mixing liquid metals).

(vi)   Standard enthalpy change of dilution, $\Delta H^{\ominus}_{dil}$

This is the enthalpy change when further solvent is added to a solution of specified concentration (i.e. initial and final concentrations of the solution are known), under standard conditions. In fact as solutions are further diluted the enthalpy change becomes progressively smaller. Finally at infinite dilution the ions from the original crystalline solid (solute) are so widely separated as free ions in solution that they no longer interattract.

Enthalpy changes of solution refer to the enthalpy change when *one mole of solute* is dissolved in a large enough volume of solvent to form a solution at infinite dilution. This may be called the *integral* enthalpy change of solution whilst at lower dilutions the term *partial* enthalpy change of solution is used (compare with Section 4.3.1).

The integral enthalpy change of solution for silicon in molten iron of such concentration as that used to manufacture transformer laminations (5% silicon by weight) is $-11.8\ kJ\ mol^{-1}$. This means that a temperature rise of about 55 K occurs when the silicon is added (at 1850 K) and this could lead to refractory attack.

(vii)   Standard enthalpy change of neutralisation, $\Delta H^{\ominus}_{neut}$

This is the enthalpy change when an acid and a base react together to give *one mole of water*. For a strong acid and a strong base the enthalpy change of neutralisation is effectively constant at $-56.9\ kJ\ mol^{-1}$ when considering infinite dilution. This constancy arises from the fact that strong acids, strong bases and their salts are all fully ionised in dilute solution. For instance,

$$H^+(aq) + Cl^-(aq) + Na^+(aq) + OH^-(aq) \longrightarrow Na^+(aq) + Cl^-(aq) + H_2O(l)$$

---

*Molar latent heats refer to the transformation of 1 mol of material. In physics specific latent heats are more commonly used, these refer to the transformation of 1 kg of material, the unit being $kJ\ kg^{-1}$.

In this reaction, as in any other strong acid–strong base reactions, the sole reaction is the formation of molecular water,

$$H^+(aq) + OH^-(aq) \longrightarrow H_2O(l); \qquad \Delta H^{\ominus}_{neut} = -56.9 \text{ kJ mol}^{-1}$$

from acid   from base

(or strictly:

$$H_3O(aq) + OH^-(aq) \longrightarrow 2H_2O(l); \qquad \Delta H^{\ominus}_{neut} = 2 \times (-56.9) \text{ kJ mol}^{-1}$$
$$= -113.8 \text{ kJ mol}^{-1}$$

see Section 6.1.7).

2.1.2 Calculating enthalpies and enthalpy changes

$H^{\ominus}$ and $\Delta H^{\ominus}$ values can be calculated using equation (2.1), i.e.

$$\Delta H^{\ominus} = H^{\ominus}_{prods} - H^{\ominus}_{reacts}$$

The enthalpies required for the reactants and products will be for either elements or compounds. These are determined as follows.

(i) $H^{\ominus}$ values for elements

The enthalpy of an element, by convention, is taken to be zero at the reference temperature (usually 298 K), providing the element is in its normal physical state under the conditions considered, and becomes finite at any other temperature. For example

$$H^{\ominus}_{298}[C(graphite)] = 0; \qquad H^{\ominus}_{298}[O_2(g)] = 0$$
$$H^{\ominus}_{298}[\alpha Fe(s)] = 0; \qquad H^{\ominus}_{298}[Hg(l)] = 0$$

but

$$H^{\ominus}_{298}[Hg(g)] = +60.84 \text{ kJ mol}^{-1}$$

since, under the specified conditions, mercury is a liquid rather than a gas, i.e.

$$Hg(l) \longrightarrow Hg(g); \qquad \Delta H^{\ominus}_{298} = +60.8 \text{ kJ mol}^{-1}$$

(ii) $H^{\ominus}$ values for compounds

The enthalpy of a compound is taken to be its enthalpy change of formation as previously defined, e.g.

$$C(graphite) + O_2(g) \longrightarrow CO_2(g); \qquad \Delta H^{\ominus}_{298} = -393 \text{ kJ mol}^{-1}$$

so

$$H^{\ominus}_{298}[CO_2(g)] = \Delta H^{\ominus}_{f\,298}[CO_2(g)] = -393 \text{ kJ mol}^{-1}$$

Data books[1,2] furnish values of $H^{\ominus}$ ($\Delta H^{\ominus}_f$) for compounds, usually at 298 K.

Example

Calculate the standard enthalpy change for the following reaction at 298 K:

$$2PbS(s) + 3O_2(g) \longrightarrow 2PbO(s) + 2SO_2(g)$$

Given that:

$$H^{\ominus}_{298} PbS(s) = -94.5 \text{ kJ mol}^{-1}$$
$$H^{\ominus}_{298} PbO(s) = -220.5 \text{ kJ mol}^{-1}$$
$$H^{\ominus}_{298} SO_2(g) = -298.0 \text{ kJ mol}^{-1}$$

Rewriting the equation for the reaction to show standard enthalpies we have:

$$2PbS(s) + 3O_2(g) \longrightarrow 2PbO(s) + 2SO_2(g)$$

$$2(-94.5) \qquad 3(0) \qquad 2(-220.5) \qquad 2(-298) \qquad kJ$$

Using equation (2.1) then

$$\Delta H^{\ominus} = [2(-220.5) + 2(-298)] - [2(-94.5) + 0] \text{ kJ}$$

$$= -848.0 \text{ kJ}$$

2.1.2.1 *Further calculations*

(1)   Calculate the standard enthalpy change for the following reaction at 1523 K:

$$Cu_2S(l) + 2Cu_2O(s) \longrightarrow 6Cu(l) + SO_2(g)$$

given that the values for the standard enthalpy changes of formation, $H^{\ominus}_{1523}$, in kJ mol^{-1} are:

$Cu_2S(s)$	$Cu_2O(s)$	$SO_2(g)$
−86.7	−176.4	−278.4

(2)   By reference to a suitable data book[1,2] calculate $\Delta H^{\ominus}_{298}$ for each of the following reactions:

(a) $3Zn(s) + Al_2O_3(s) \longrightarrow 2Al(s) + 3ZnO(s)$

(b) $MgCO_3(s) \longrightarrow MgO(s) + CO_2(g)$

(3)   Use the following data to calculate the standard enthalpy change for the formation of ethane:

$$C(s) + O_2(g) \longrightarrow CO_2(g); \qquad\qquad \Delta H^{\ominus}_{298} = -395 \text{ kJ mol}^{-1}$$

$$H_2(g) + \tfrac{1}{2}O_2(g) \longrightarrow H_2O(l); \qquad\qquad \Delta H^{\ominus}_{298} = -286 \text{ kJ mol}^{-1}$$

$$C_2H_6(g) + \tfrac{7}{2}O_2(g) \longrightarrow 2CO_2(g) + 3H_2O(l); \qquad \Delta H^{\ominus}_{298} = -1561 \text{ kJ mol}^{-1}$$

(4)   Given $\Delta H^{\ominus}$ for the following reactions:

$$H_2(g) + O_2(g) \longrightarrow 2H_2O(l)$$

$$Cu(s) + \tfrac{1}{2}O_2(g) \longrightarrow CuO(s)$$

what further data is needed, if any, for calculation of $\Delta H^{\ominus}$ for the following reaction?

$$CuO(s) + H_2(g) \longrightarrow Cu(s) + H_2O(g)$$

(5)   The standard enthalpy change of combustion for liquid ethanol at 25 °C is −1373 kJ mol^{-1}. Calculate its standard enthalpy change of formation at this temperature referring to a data book where necessary.

(6)   Calculate the standard enthalpy change for the following reaction:

$$2Al(s) + Cr_2O_3(s) \longrightarrow Al_2O_3(s) + 2Cr(s)$$

given that

$$H^{\ominus}_{298}[Cr_2O_3] = -1125 \text{ kJ mol}^{-1}; \quad H^{\ominus}_{298}[Al_2O_3] = -1585 \text{ kJ mol}^{-1}$$

2.1.2.2 Hess's law (of constant heat summation, 1840)

Hess's law states that the enthalpy change for a reaction is the same whether it takes place in one or several stages (i.e. ΔH for a reaction depends only on the initial and final states). The great utility of this law is in calculating enthalpy changes for reactions which cannot be carried out experimentally.

Example 1

The standard enthalpy change of formation for carbon monoxide is required. The reaction is represented as:

$$C(s) + \tfrac{1}{2}O_2(g) \longrightarrow CO(g); \quad \Delta H_I^{\ominus} = ? \tag{2.I}$$

but cannot occur stoichiometrically* because of the tendency of the carbon to form carbon dioxide giving rise to a mixture of products (CO and CO_2).

However, standard enthalpy values are experimentally obtainable for the following reactions:

$$C(s) + O_2(g) \longrightarrow CO_2(g) \qquad \Delta H_{II}^{\ominus} = -393 \text{ kJ mol}^{-1} \text{ (bomb calorimeter)} \tag{2.II}$$

$$CO(g) + \tfrac{1}{2}O_2(g) \longrightarrow CO_2(g) \qquad \Delta H_{III}^{\ominus} = -284 \text{ kJ mol}^{-1} \text{ (gas calorimeter)} \tag{2.III}$$

These reactions may be manipulated (e.g. reversed, added, subtracted, multiplied) to represent the change occurring in reaction (2.I), hence making it possible to calculate $\Delta H_I^{\ominus}$:

Restating (2.II) $C(s) + O_2(g) \rightarrow CO_2(g)$ $\Delta H_{II}^{\ominus} = -393 \text{ kJ mol}^{-1}$

Reversing† (2.III) $CO_2(g) \rightarrow CO(g) + \tfrac{1}{2}O_2(g)$ $\Delta H_{III}^{\ominus} = +284 \text{ kJ mol}^{-1}$

Adding (2.II) and (2.III) $C(s) + O_2(g) + CO_2(g) \rightarrow CO_2(g) + CO(g) + \tfrac{1}{2}O_2(g)$ $\Delta H_I^{\ominus} = \Delta H_{II}^{\ominus} + \Delta H_{III}^{\ominus}$

Cancelling CO_2 and $\tfrac{1}{2}O_2$ from each side $C(s) + \tfrac{1}{2}O_2(g) \longrightarrow CO(g)$ $\Delta H_I^{\ominus} = -109 \text{ kJ mol}^{-1}$

Hence the enthalpy of formation of carbon monoxide has been determined. This technique is the same as that used to solve simultaneous equations in algebra.

Example 2

Methane cannot be synthesised by direct combustion of carbon and hydrogen:

$$C(s) + 2H_2(g) \longrightarrow CH_4(g); \quad \Delta H_{IV}^{\ominus} = ?$$

but its enthalpy of formation, $\Delta H_f^{\ominus}$, can be calculated from the enthalpies of combustion of methane, carbon and hydrogen by application of Hess's law:

$$CH_4(g) + 2O_2(g) \longrightarrow CO_2(g) + 2H_2O(l) \qquad \Delta H_I^{\ominus} = -890 \text{ kJ mol}^{-1} \tag{I}$$

$$C(s) + O_2(g) \longrightarrow CO_2(g) \qquad \Delta H_{II}^{\ominus} = -393 \text{ kJ mol}^{-1} \tag{II}$$

$$H_2(g) + \tfrac{1}{2}O_2(g) \longrightarrow H_2O(l) \qquad \Delta H_{III}^{\ominus} = -287 \text{ kJ mol}^{-1} \tag{III}$$

---

*For an explanation of stoichiometry see Section 3.2.4.

†If it is necessary to reverse an equation, the sign of ΔH must also be reversed according to Laplace's law which states that the quantity of heat which must be supplied to decompose a substance into its elements is equal to the heat evolved when that compound is formed from its elements.

Reverse reaction (I), multiply reaction (III) by 2* and add all three together (reversing is equivalent to subtracting), i.e.

$$CO_2(g) + 2H_2O(l) \longrightarrow CH_4(g) + 2O_2(g) \qquad \Delta H_I^{\ominus} = +890 \text{ kJ mol}^{-1}$$

$$C(s) + O_2(g) \longrightarrow CO_2(g) \qquad \Delta H_{II}^{\ominus} = -393 \text{ kJ mol}^{-1}$$

$$2H_2(g) + O_2(g) \longrightarrow 2H_2O(l) \qquad 2\Delta H_{III}^{\ominus} = -574 \text{ kJ mol}^{-1}$$

$$C(s) + 2H_2(g) \longrightarrow CH_4(g) \qquad \Delta H_{IV}^{\ominus} = \Delta H_I^{\ominus} + \Delta H_{II}^{\ominus} + 2\Delta H_{III}^{\ominus}$$

$$\Delta H_{IV}^{\ominus} = -79 \text{ kJ mol}^{-1}$$

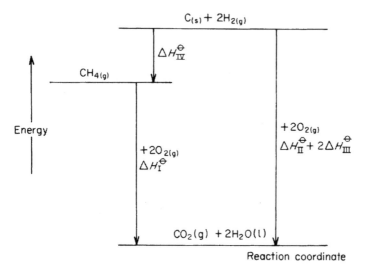

Figure 2.2 Energy diagram for calculation of the standard enthalpy change of formation of methane

Alternatively this may be represented by a simple energy diagram (*Fig. 2.2*). From *Fig. 2.2* we see that

$$\Delta H_{IV}^{\ominus} = \Delta H_{II}^{\ominus} + 2\Delta H_{III}^{\ominus} - \Delta H_I^{\ominus}$$
$$= -393 + 2(-287) - (-890) \text{ kJ mol}^{-1}$$
$$= -77 \text{ kJ mol}^{-1}$$

As the formation of methane is an exothermic process its elements $(C + 2H_2)$ are shown at a higher energy level than the resulting compound. $\Delta H_f^{\ominus}$ is thus found by considering the enthalpy changes for each intermediate step.

Calculations involving Hess's law

(1)   Calculate the standard enthalpy changes of formation at 298 K for the stated compounds from the data supplied in each case:

*If the reaction is multiplied by a factor, e.g. 2, so must its value of ΔH.

(a) For carbon(IV) sulphide(liquid), CS_2 $\Delta H^{\ominus}_{298}$ (kJ mol^{-1})

$$CS_2(l) + 3O_2(g) \longrightarrow CO_2(g) + 2SO_2(g) \qquad -1113$$

$$C(s) + O_2(g) \longrightarrow CO_2(g) \qquad -407$$

$$S(s) + O_2(g) \longrightarrow SO_2(g) \qquad -298$$

(b) For iron(II) chloride(solid), $FeCl_2$

$$Fe(s) + 2HCl(aq) \longrightarrow FeCl_2(aq) + H_2(g) \qquad -88.2$$

$$FeCl_2(s) + aq \longrightarrow FeCl_2(aq) \qquad -81.9$$

$$HCl(g) + aq \longrightarrow HCl(aq) \qquad -73.5$$

$$H_2(g) + Cl_2(g) \longrightarrow 2HCl(g) \qquad -184.8$$

(c) For lead(IV) oxide(solid), PbO_2

$$Pb(s) + \tfrac{1}{2}O_2(g) \longrightarrow PbO(s) \qquad -220.1$$

$$3PbO(s) + \tfrac{1}{2}O_2(g) \longrightarrow Pb_3O_4(s) \qquad -77.3$$

$$Pb_3O_4(s) + O_2(g) \longrightarrow 3PbO_2(s) \qquad -95.3$$

(2) The standard enthalpy changes of combustion at 298 K for $H_2(g)$, $CO(g)$ and $CH_3OH(l)$ are -286, -283 and -714 kJ mol^{-1} respectively. Make use of these data to calculate the value of $\Delta H^{\ominus}$ at the same temperature for the reaction:

$$CO(g) + 2H_2(g) \longrightarrow CH_3OH(l)$$

(3) In the iron blast furnace iron(III) oxide is reduced to iron. Coke (carbon) is charged with the ore and air (oxygen) is blown into the furnace, the overall reaction at 1600 °C possibly being:

$$Fe_2O_3 + 3C + \tfrac{3}{2}O_2 \longrightarrow 2Fe + 3CO_2$$

Calculate $\Delta H^{\ominus}$ for this reaction at 1600 °C given that the following may be feasible intermediate stages:

$$\tfrac{3}{2}C + \tfrac{3}{2}O_2 \longrightarrow CO_2 \qquad \Delta H^{\ominus} = -593.3 \text{ kJ}$$

$$\tfrac{3}{2}CO_2 + \tfrac{3}{2}C \longrightarrow 3CO \qquad \Delta H^{\ominus} = +259.6 \text{ kJ}$$

$$Fe_2O_3 + \tfrac{1}{3}CO \longrightarrow \tfrac{2}{3}Fe_3O_4 + \tfrac{1}{3}CO_2 \qquad \Delta H^{\ominus} = -17.6 \text{ kJ}$$

$$\tfrac{2}{3}Fe_3O_4 + \tfrac{8}{3}CO \longrightarrow 2Fe + \tfrac{8}{3}CO_2 \qquad \Delta H^{\ominus} = -10.5 \text{ kJ}$$

2.1.3 Measurement of enthalpy changes of reactions

In order to measure the heat gained or lost in a reaction simple calorimetry is frequently sufficient. Measured amounts of reactants are mixed and allowed to react in an insulated calorimeter and the resulting temperature change is measured. Insulation is extremely important since heat exchange with the surroundings leads inevitably to considerable loss in accuracy.

Heat loss or gain = (heat capacity of contents + calorimeter) $\times$ temperature change

where

heat capacity = mass $\times$ specific heat capacity

(see Section 2.1.4.1).

As results are stated in kJ mol^{-1} it is necessary to convert the masses of reactants used to molar quantities, e.g. 3 g of carbon are completely burned in oxygen: since the relative atomic mass of carbon is 12 then 3 g is 3/12 or 0.25 moles. Hence if the heat exchange calculation has produced a result of 95 kJ (3g)$^{-1}$ the result would be stated as 4 × 95 = 380 kJ mol^{-1} (since heat is evolved, $\Delta H = -380$, kJ mol^{-1}).

For the latter determination simple calorimetry is inadequate and use is made of the bomb calorimeter (*Fig. 2.3*). This is a small, thick-walled, steel pressure vessel ('bomb') into which oxygen can be piped to about 10^6 Pa (20 atm). The sample is ignited electrically.

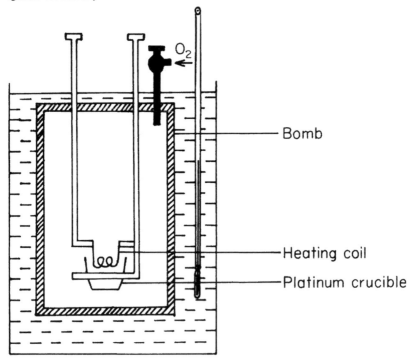

Figure 2.3 Diagrammatic representation of a bomb calorimeter

For combustion of gases a useful apparatus is the gas calorimeter. This involves passing metered amounts of gas through a tube to burn at a jet and hence heat water moving at a measured rate through a coil surrounding the flame.

2.1.4   Enthalpy changes: the effect of temperature

It is now necessary to consider how ΔH values are influenced by changes in temperature. First an understanding of heat capacities is required.

2.1.4.1   Heat capacity (C)

This is the amount of heat required to raise the temperature of a system by 1 K. Two alternative definitions are in use.

(a) Specific heat capacity

This is the amount of heat required to raise the temperature of 1 kg of a substance by 1 K. Specific heat capacities for some common substances are given in *Table 2.1*. These

Table 2.1 SOME SPECIFIC HEAT CAPACITIES (IN kJ K^{-1} kg^{-1})

Water	4.18
Copper	0.38
Steel	0.46
Sand/refractories	0.80–0.86
Iron ore sinter (typical)	0.92
Air	1.10

values can rise by 50% on increasing the temperature by 1000 °C and this must be accounted for in heat input–output calculations such as when considering heating processes or chilling effects of moulds. The relatively large specific heat capacity of air emphasises the importance of preheating air when used as a combustant as in blast furnace smelting (Section 7.2.10.5) or the large amount of fuel required to heat the air during sintering of ores (Section 7.1.4.1).

(b) Molar heat capacity

This is the amount of heat required to raise the temperature of 1 mole of a substance by 1 K. The latter definition will be used in this discussion and is expressed in J K^{-1} mol^{-1}.

Gases may have different molar heat capacities depending on whether the gas is heated at constant volume or at constant pressure, the symbols respectively being C_V and C_p. The difference between C_V and C_p is small and theoretically constant ($C_p - C_V = R$, the gas constant). Most metallurgical reactions occur at constant pressure, therefore, C_p values only will be used in this text.

2.1.4.2   Kirchoff's equation

The heat capacity of a given substance usually alters with temperature change. It is this variation in heat capacity of products and reactants which gives rise to variation of ΔH values for reactions as temperature changes, i.e. in the generalised reaction shown below there are two alternative pathways which may be considered: (i) shown by solid arrows and (ii) shown by broken arrows

$$\text{Reactants (R)} \xrightarrow[\text{(i)}]{\Delta H_1} \text{Products (P)} \qquad \text{at temperature } T_1$$

$$\text{(ii)} \downarrow C_p^R(T_2 - T_1) \qquad \text{(i)} \downarrow C_p^P(T_2 - T_1)$$

$$\text{Reactants (R)} \dashrightarrow[\text{(ii)}]{\Delta H_2} \text{Products (P)} \qquad \text{at temperature } T_2$$

where C_p^R and C_p^P are the heat capacities of the reactants (R) and products (P) respectively.

(i)   Heat required to convert R at T_1 to P at $T_1 = \Delta H_1$ and heat required to convert P at T_1 to P at $T_2 = C_p^P(T_2 - T_1)$, i.e. heat capacity × temperature rise. Thus the heat required to complete the reaction via (i) is $\Delta H_1 + C_p^P(T_2 - T_1)$.

(ii)   Following the same deductive process for pathway (ii) it may be shown that the total heat requirement to complete the reaction via (ii) is $C_p^R(T_2 - T_1) + \Delta H_2$.

As in both (i) and (ii) the initial and final states are identical then, according to Hess's law, the heat exchange for each process will be the same, i.e.

$$\Delta H_1 + C_p^P(T_2 - T_1) = \Delta H_2 + C_p^R(T_2 - T_1)$$

or

$$\Delta H_2 - \Delta H_1 = (C_p^P - C_p^R)(T_2 - T_1)$$

This is only applicable when using mean values of C_p for the temperature interval $T_1 - T_2$. This can be rewritten as:

$$\frac{d(\Delta H)}{dT} = \Delta C_p \qquad (2.2)$$

i.e. change of ΔH with T is equal to change in C_p. This is known as Kirchoff's equation.*

---

*A similar expression may be deduced for constant volume conditions, where ΔU is change in internal energy (see Section 2.2.1.1)

$$d(\Delta U)/dT = C_V$$

This equation is useful in the integrated form as in the case of high-temperature conditions which normally apply to metallurgical reactions. Integrating Kirchoff's equation between T_2 and T_1 yields:

$$\int_{\Delta H_{T_1}}^{\Delta H_{T_2}} d(\Delta H) = \Delta H_{T_2} - \Delta H_{T_1} = \int_{T_1}^{T_2} \Delta C_p\, dT$$

This may be expressed more usefully:

$$\Delta H_{T_2} = \Delta H_{T_1} + \int_{T_1}^{T_2} \Delta C_p\, dT \tag{2.3}$$

To calculate ΔH_{T_2} from this equation it is necessary to consider how C_p varies with temperature. This information has been obtained experimentally by construction of curves of the type shown in *Fig. 2.4*. This type of graph yields empirical equations giving the variation of heat capacity with temperature and they are generally of the type:

$$C_p = a + bT + cT^{-2} + \dots \tag{2.4}$$

a, b, c, etc., are constants (for a given substance) and T is the temperature (K). Data books[1, 2] furnish constants for a variety of substances (see *Table 2.2*).

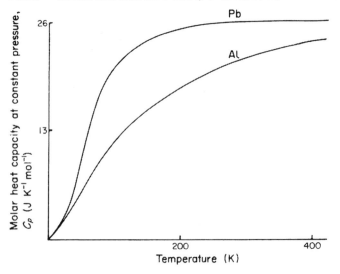

Figure 2.4 Change in heat capacity of metals (e.g. Pb and Al) with temperature

Therefore, the enthalpy change for a reaction at any temperature, ΔH_T, may be calculated if the enthalpy change for the reaction at another specified temperature, usually 298 K, is known and molar heat capacity data are available, i.e. equation (2.3) may be rewritten (considering standard conditions):

$$\Delta H_T^\ominus = \Delta H_{298\,K}^\ominus + \int_{298}^{T} \Delta C_p\, dT \pm \text{latent heats} \tag{2.5}$$

Table 2.2 TABLE OF CONSTANTS (*a*, *b*, *c*) FOR CALCULATION OF MOLAR HEAT CAPACITIES (C_p) AT A GIVEN TEMPERATURE (*T*)

Substance		$C_p = a + bT + cT^{-2}$		
		a (J K^{-1} mol^{-1})	b (10^{-3} J K^{-2} mol^{-1})	c (10^5 J K mol^{-1})
C(graphite)		17.31	4.26	−8.77
Al(s)		20.67	12.38	0
Pb(s)		22.13	11.72	0.96
Zn(s)		22.36	10.03	0
ZnO(s)		48.94	5.10	−9.11
CO(g)		28.38	4.10	−0.55
CO$_2$(g)		44.23	8.79	−8.62
H$_2$O(g)		30.54	10.29	0
Zn(s)		20.77	0	0
CH$_4$(g)		23.64	47.85	−1.92
H$_2$O(l)	(mp − bp)	75.48	0	0
Zn(l)	(mp − bp)	31.35	0	0

Molar latent heats are added if products transform; subtracted if reactants transform. In the simplest case it may be assumed that the molar heat capacities of the reactants and products do not alter appreciably with change of temperature, i.e. C_p values at 298 K from data sources may be used directly. Equation (2.5) then becomes:

$$\Delta H_T^{\ominus} = \Delta H_{298K}^{\ominus} + \Delta C_p(T_2 - T_1) \tag{2.6}$$

and ΔC_p is simply calculated as follows:

$$\Delta C_p = \sum C_p^P - \sum C_p^R \tag{2.7}$$

e.g. for the generalised reaction:

$$aA + bB \rightleftharpoons cC + dD$$
$$\Delta C_p = (cC_{p_C} + dC_{p_D}) - (aC_{p_A} + bC_{p_B})$$

For more accurate values of ΔH_T it is necessary to calculate ΔC_p using C_p values for reactants and products at the required temperature deduced from the type of empirical equation illustrated by equation (2.4) such that:

$$\Delta C_p = \Delta a + \Delta bT + \Delta cT^{-2} + \ldots \tag{2.8}$$

If the heat capacity of a substance differs in the various states considered, the calculation must be performed separately for each state, i.e.

$$\Delta H_{T_2} = \Delta H_{T_1} + \int_{T_1}^{T_f} \Delta C_p \, dT + L_f + \int_{T_f}^{T_2} \Delta C_p \, dT$$

where T_f and L_f are the melting point and latent heat of fusion of the product respectively.

Latent heat of transformation and melting and boiling points must also be considered in the above equation if the reactants or products go through a phase transformation.

The enthalpy change for a reaction may alter slightly or appreciably with change of temperature, this variation being in either the positive or negative direction.*

For example, the smelting of zinc oxide with carbon takes place according to the equation:

$$ZnO(s) + C(s) \longrightarrow Zn(g) + CO(g) \qquad \Delta H^{\ominus}_{298} = +239 \text{ kJ mol}^{-1}$$

At the working temperature of 1373 K the reaction is considerably more endothermic. Unless heat input can be maintained sufficiently the gaseous zinc leaving the plant may liquefy or solidify prematurely.

In exothermic extraction processes calculation of ΔH_T will make it possible to deduce fuel utilisation figures to allow for optimum operating temperatures.

(i) Calculations involving the use of Kirchoff's equation

The molar heat capacities (in J K^{-1} mol^{-1}) of various substances at 298 K is given in *Table 2.3*.

Table 2.3 MOLAR HEAT CAPACITIES (IN J K^{-1} mol^{-1}) OF VARIOUS SUBSTANCES AT 298 K

Substance	Al(s)	Al$_2$O$_3$(s)	CaO(s)	CaCO$_3$(s)	C(graphite)	CO(g)	CO$_2$(g)
C_p	24.3	79.0	43.8	82.0	8.6	29.1	37.0

Substance	Cu(s)	CuO(s)	H$_2$(g)	H$_2$O(g)	H$_2$O(l)	O$_2$(g)	Zn(s)	Zn(l)	Zn(g)	ZnO(s)
C_p	24.5	44.4	28.9	33.5	75.0	29.4	25.1	32.5	21.8	40.2

Example

Does the enthalpy change of formation of water vapour increase or decrease with rise in temperature?

$$H_2(g) + \tfrac{1}{2}O_2(g) \longrightarrow H_2O(g) \qquad \Delta H^{\ominus} \text{ is negative at 298 K}$$

Using Kirchoff's equation (2.6)

$$\Delta H^{\ominus}_T = \Delta H^{\ominus}_{298} + \Delta C_p(T_2 - T_1)$$
$$= (-) \quad + \Delta C_p(T_2 - T_1)$$

and from equation (2.7) and *Table 2.3*

$$\Delta C_p = \{ C_p(H_2O) - [C_p(H_2) + \tfrac{1}{2}C_p(O_2)] \}$$

i.e.,

$$\Delta C_p = [33.5 - (28.9 + 14.7)]$$
$$= 33.5 - 43.6$$
$$= -10.1 \text{ J K}^{-1} \text{ mol}^{-1}$$

Substituting this value for ΔC_p in equation (2.6) we have

$$\Delta H^{\ominus}_T = (-) + (-10.1)(T_2 - T_1)$$
$$= (-) - 10.1(T_2 - T_1)$$

*If ΔC_p is positive ($C^P_p > C^R_p$), ΔH increases with temperature.
If ΔC_p is negative ($C^P_p < C^R_p$), ΔH decreases with temperature.
If $\Delta C_p = 0$, ΔH is independent of temperature.

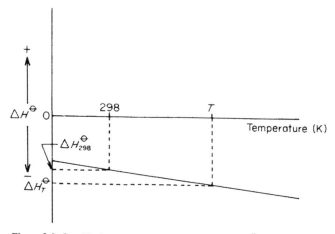

Figure 2.5 Graphical expression of the variation of $\Delta H^{\ominus}$ with temperature for the reaction $H_2(g) + \frac{1}{2}O_2(g) \rightarrow H_2O(g)$

So $\Delta H_T^{\ominus}$ is more negative than $\Delta H_{298}^{\ominus}$. The trend may be expressed graphically as in *Fig. 2.5*.

(ii) Further calculations

Assume the heat capacity values in *Table 2.3* apply at all temperatures unless otherwise stated.

(1)   For the following reaction at 298 K:

$$CaCO_3(s) \longrightarrow CaO(s) + CO_2(g) \qquad \Delta H_{298}^{\ominus} \text{ is positive}$$

(a)   Deduce how $\Delta H^{\ominus}$ alters with increase in temperature.
(b)   Indicate the trend by means of a simple sketch graph.
(c)   Calculate $\Delta H_{598}^{\ominus}$ referring to *Table 2.3* and the following data:

	$CaCO_3(s)$	$CaO(s)$	$CO_2(g)$
$H_{298}^{\ominus}$ (kJ mol^{-1})	-1207	-635	-394

(2)   Find an approximate value of $\Delta H_{1373}^{\ominus}$ for the zinc blast furnace reaction:

$$ZnO(s) + C(s) \longrightarrow Zn(g) + CO(g)$$

for which $\Delta H_{298}^{\ominus} = +238.6$ kJ mol^{-1}, assuming that the values for C_p given in *Table 2.3* are constant over the range considered, and:

Latent heat of fusion of zinc $= +7.36$ kJ mol^{-1} (mp $= 692$ K)

Latent heat of evaporation of zinc $= +114.7$ kJ mol^{-1} (bp $= 1180$ K)

(3)   For the reaction:

$$CO(g) + \frac{1}{2}O_2(g) \longrightarrow CO_2(g)$$

Calculate the standard enthalpy change at 473 K given that the standard enthalpy changes of formation (in kJ mol^{-1}) at 298 K are -110.50 for CO and -393.50 for CO_2. Molar heat capacities, C_p (in J K^{-1} mol^{-1}), are:

$CO(g)$, $C_p = 30.0 + 0.0041T$

$O_2(g)$, $C_p = 28.5 + 0.0042T$

$CO_2(g)$, $C_p = 44.2 + 0.0088T$

2.2 Thermodynamics

2.2.1 Energy: The driving force for chemical change

Observation establishes that energy in any system attempts to reduce to a minimum. Hence for any process to occur spontaneously* it must be accompanied by a decrease in the system's energy content. Thus, the change in energy during a reaction, i.e. energy of products − energy of reactants, must be negative for a spontaneous reaction. The law of conservation of energy states that the energy content of the universe is constant and:

$$Mass \rightleftharpoons Energy$$
(electrons, protons, neutrons)   (quanta)

Chemical reactions involve movements and rearrangements of electrons and quanta.

The enthalpy change, ΔH, for a given reaction might be expected to indicate the feasibility of the reaction. In fact, most reactions are exothermic but many examples of spontaneous endothermic reactions are known. This means that the spontaneity of reactions cannot be decided by consideration of enthalpy alone, a second factor at least must exist.

It is found that the tendency of bodies to move together, e.g. to form orderly crystal lattices, is countered by a fundamental tendency in the opposite direction, i.e. towards increased disorder. This tendency towards disorder (measured by the entropy, S) also involves a separate energy change. Hence the energy accompanying a chemical reaction is influenced by:

(a)   A tendency towards order, represented by the enthalpy change, ΔH, i.e. a tendency to minimise the potential energy of the system.

(b)   A tendency towards disorder, represented by the entropy change, ΔS, i.e. a tendency to maximise the kinetic energy of the system.

From a consideration of the ways in which ΔH and ΔS jointly influence the useful energy change during a chemical reaction the feasibility of the reaction occurring can be deduced. Prediction of conditions under which reactions can be made possible and estimation of equilibrium positions are achieved by application of the laws of thermodynamics.

2.2.1.1 The first law of thermodynamics

The total energy of a body, often called its internal or intrinsic energy, U, is the sum of all the differing types of energy which a body possesses:

$$U_{total} = U_{nuc} + U_{trans} + U_{rot} + U_{elec} + U_{vib}$$

U_{nuc} is the binding energy of the nucleus and is not important in chemical reactions since it remains unaltered. The other contributors to total energy, U_{total}, may alter during the progress of a chemical reaction and may be listed as:

*A *thermodynamically spontaneous* reaction is one which is energetically possible (capable of going in the direction as written of its own accord) − the phrase does not describe the rate of the reaction.

U_{trans} = energy of translation (due to the motion of atoms and molecules)

U_{rot} = energy of rotation (due to the rotation of molecules about stated axes)

U_{elec} = electronic energy (due to the arrangement of electron orbitals)

U_{vib} = vibrational energy (due to stretching and flexing of bonds)

Unlike U_{trans} and U_{rot}, which diminish to zero at 0 K, residual vibrational motion remains and at this temperature is called zero point energy.

The energies of translation, rotation and vibration can be considered as the ways an atom or molecule stores its energy. The internal energy of a system cannot be measured and for a chemical reaction the only calculable quantity is ΔU, the difference between the energy of the products (U_{prods}) and the energy of the reactants (U_{reacts}), i.e.

$$\Delta U = U_{prods} - U_{reacts} \tag{2.9}$$

(following a similar convention to that used for ΔH, see equation (2.1)).

The first law of thermodynamics is simply a statement of the law of conservation of energy which can be restated in a second way here (Hess's law being a third way) as: the total energy of an isolated system (theoretical system which can exchange neither matter nor energy with its surroundings) is constant though it can change from one form to another.

Chemical reactions may occur under conditions of constant volume or constant pressure. For reactions during which there is a change in gaseous volume, differing amounts of heat evolved (or absorbed) are found in practice depending on whether the reaction is performed under constant pressure or constant volume. This is because, under constant pressure conditions, work must be done by the gas (if expansion occurs), or on the gas (if contraction occurs); under constant volume conditions no expansion or contraction occurs. This is best illustrated by considering an endothermic reaction, e.g.

$$N_2O_4(g) \longrightarrow 2NO_2(g)$$
$$\text{1 volume} \qquad \text{2 volumes}$$

Fig. 2.6 illustrates such a reaction. Part (*a*) shows constant volume conditions in which a reaction occurs in a sealed container of gas of volume V. In this case the total amount of heat absorbed to complete the reaction is ΔU. In part (*b*) the volume of the container is allowed to vary by movement of a weightless, frictionless piston (constant pressure conditions). In this case the total energy (ΔH) absorbed to complete the

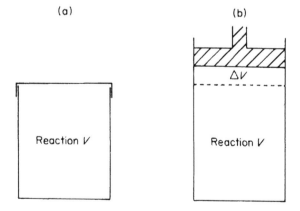

(a) (b)

Figure 2.6 Illustration of an endothermic reaction under: (a) constant volume conditions; (b) constant pressure conditions

reaction is given by ΔU with an extra amount required to do work in expanding the gas against atmospheric pressure. Thus, ΔH is larger than ΔU by an amount representing the work done (w) in expansion, i.e.

$$\Delta H - \Delta U = w$$

or
$$(2.10)$$

$$\Delta U = \Delta H - w$$

This form of the law uses the convention that work done *by* the system (e.g. in expansion against the applied pressure) is positive, but work done *on* the system (e.g. by the applied pressure during contraction) is negative.

Hence in the example used ΔU is equivalent to the heat energy absorbed from the surroundings less the energy lost to the surroundings as external work, w. ΔU is positive as energy is absorbed (heat absorbed at constant volume, sometimes designated q_V); ΔH is positive as the reaction is endothermic (heat absorbed at constant pressure, sometimes designated q_p); w is positive as work is done by the gas, i.e. $-(+w)$.

The law is equally well demonstrated by considering an exothermic reaction, in which case ΔU and ΔH are negative.

Deducing an expression for w

Fig. 2.7 represents a reaction vessel, depicted as a cylinder, in which gas volume change causes movement of a weightless, frictionless piston. We require an expression for work done, w (given by work done = force × distance). Assuming that the gas obeys the ideal gas law, i.e.

$$PV = nRT \qquad (2.11)$$

where n is the number of moles of gas, R is the gas constant and P, V, T are the pressure, volume and temperature of the gas respectively. The force on the piston is given by pressure × area, i.e. PA. If P is measured in pascals (N m^{-2}) and A in m^2 then the force is in newtons (N). When the gas expands the piston moves distance d, so increasing the volume of the gas by ΔV (assume d in metres and V in m^3). This change in volume can be expressed as

$$\Delta V = Ad$$

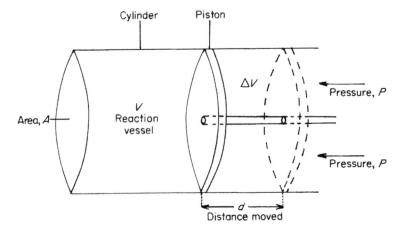

Figure 2.7 Representation of a reaction occurring at constant pressure

and so

$$d = \Delta V/A$$

Thus the work done by the system (in N m) is

$$w = \text{force on piston} \times \text{distance moved}$$
$$= PA(\Delta V/A)$$
$$= P\Delta V$$

Hence, the mathematical expression for the first law as expressed by (2.10) may be rewritten:

$$\Delta U = \Delta H - P\Delta V \tag{2.12}$$

This may be expressed in words as follows:

Increase in internal energy = heat absorbed by − work done by
the system   the system
(ΔU or q_V)   (ΔH or q_p)   ($P\Delta V$ or w)

From equation (2.11) we can write $V = nRT/P$ and since P, R and T remain constant, $\Delta V = \Delta nRT/P$. Therefore

$$\Delta U = \Delta H - \Delta nRT \tag{2.13}$$

For reactions involving only solids or liquids where the volume change is usually negligible:

$$\Delta U = \Delta H$$

This is also true for gaseous reactions involving no change in volume. However, where gaseous volumes do alter a significant difference between ΔU and ΔH exists.

Maximum work and reversibility

With regard to thermodynamics the terms 'reversible' and 'irreversible' process have a particularly specialised meaning. To illustrate the meaning of these terms consider the following simple mechanical system of a weightless rod, free to move about a frictionless fulcrum resting on a bench (*Fig. 2.8*). In each case illustrated the rod is held in a horizontal position and then released. With reference to *Fig. 2.8*:

(a)   The mass will fall to the bench losing its potential energy as heat and sound.
(b)   Some energy is lost as before but useful work is done in lifting the 0.5 kg mass.

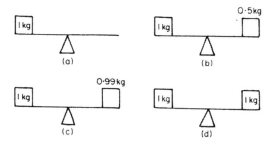

Figure 2.8 Diagrammatic representation of the fulcrum analogy for thermodynamically reversible and irreversible processes

(c)   It is of value to consider how the system can be modified to perform the maximum useful work. This can be achieved by replacing the 0.5 kg mass with one fractionally less than 1 kg.

It is apparent that in all three of the situations considered the 1 kg mass will fall in one direction only, i.e. the change is *irreversible* and spontaneous.

(d)   In this case (when the masses considered are the same) the system will be balanced. A slight displacement of the rod from the horizontal in either direction will be followed by movement in the reverse direction, i.e. the change is *reversible*. However, with identical masses the change occurs infinitely slowly (i.e. effectively no change at all).

Hence we can conclude that to obtain the *maximum work* from a system a change must occur *reversibly*, that is *infinitely slowly*. In practice this is neither possible nor desirable.

An alternative way of explaining the terms 'reversible' and 'irreversible' thermodynamic process is by considering expansion of a gas in a cylinder as in *Fig. 2.7*. If the gas is allowed to expand suddenly by moving the piston back rapidly, the temperature of the gas falls, i.e. conditions have altered and irreversible expansion has occurred. However, if the piston is moved infinitely slowly, then the pressure on both sides remains equal (no temperature change) and a reversible change has taken place.

In conclusion it may be said that a thermodynamically reversible process is one which is carried out infinitesimally slowly in order that the system remains at equilibrium with the surroundings throughout. Hence when using the first law equation (equation (2.13)) the term ΔnRT (or $P\Delta V$) refers to the maximum work done and assumes the change considered to be thermodynamically reversible. In practice the useful work done is less than ΔnRT ($P\Delta V$) and could not be utilised as maximum work (which would only be generated with infinite slowness).

Using the first law

In bomb calorimeter experiments data on ΔU are obtained since the apparatus operates under constant volume conditions. Since laboratory and industrial reactions are usually performed under constant pressure conditions a value for ΔH is required rather than one for ΔU. During most reactions no appreciable change in volume occurs so $\Delta U = \Delta H$. Constant volume (ΔU) data from the bomb calorimeter can be converted into more useful constant pressure (ΔH) data by use of the first law equation (the form in equation (2.13) being more easily applicable), where Δn is obtained from an examination of the reaction equation. For example, in the following reaction $\Delta n = +2$ since there are zero gas volumes initially and 2 volumes finally:

$$ZnO(s) \ + \ C(s) \xrightarrow{1173\,K} Zn(g) \ + \ CO(g)$$

The rest of the terms in equation (2.13), i.e. ΔU, R and T, are known and so ΔH can be calculated.

Calculations

(1)   The following compounds are subject to complete combustion in a bomb calorimeter. Assuming the mean temperature to be 298 K calculate ΔnRT and hence the difference between $\Delta H^{\ominus}$ and $\Delta U^{\ominus}$ for each reaction, given that $R = 8.314$ J K^{-1} mol^{-1}:

(a)   Oxalic acid (ethandioic acid): $(COOH)_2(s) + \frac{1}{2}O_2(g) \longrightarrow 2CO_2(g) + H_2O(l)$
(b)   Elementary calcium: $Ca(s) + \frac{1}{2}O_2(g) \longrightarrow CaO(s)$

(2)   The heat evolved when one mole of solid phenol, C_6H_5OH, is completely burned

at constant volume is 3056.7 kJ. If the mean temperature is 298 K calculate the enthalpy change for the reaction.

(3)   When tungsten carbide, WC, is burned with excess oxygen in a bomb calorimeter, the heat volved is 1195 kJ at 300 K:

$$WC(s) + \tfrac{5}{2}O_2(g) \longrightarrow WO_3(s) + CO_2(g)$$

Calculate $\Delta H^{\ominus}$ for the reaction.

2.2.1.2   Entropy: The second factor governing energy changes

Earlier it was stated that a given reaction is more likely to be spontaneous if it is accompanied by an increase in disorder (i.e. products more disordered than reactants). For such an increase in the disorder of a system, there is said to be an increase in entropy, S, i.e. ΔS is positive (by convention).

For example, during the change of state of material from solid through liquid to the gaseous state the entropy of the material increases by virtue of the increasingly random movement of the particles involved. As the converse is true, then theoretically the entropy of a perfect crystal at 0 K will be zero (see equation (2.15)). This tendency towards increasing disorder is readily observable in everyday processes: the combustion of solid and liquid fuels giving rise to disordered gaseous products; solution of ordered crystalline minerals in river water to produce relatively disordered solutions; evaporation of water masses. This general shift from a situation of order to one of disorder can also occur in processes which do not apparently involve energy changes as, for example, in the spontaneous mixing of non-reacting gases by diffusion.

The concept of entropy leads to the *second law of thermodynamics*. As with the first law there are several ways of stating the second law, the most straightforward probably being: all spontaneous processes are accompanied by an increase in the entropy of the system and its surroundings.

This does not mean that decreases in entropy cannot occur but that the *total* entropy of the system and its surroundings will increase. To illustrate the idea consider the following equation representing the oxidation of iron:

$$Fe(s) + \tfrac{1}{2}O_2(g) \longrightarrow FeO(s) \qquad \Delta H^{\ominus}_{298} = -266 \text{ kJ mol}^{-1}$$

The process involves:

(1)   conversion of disordered gas, O_2, into an ordered solid form, FeO, i.e. the entropy of the system, S_{sys}, decreases and the consequent entropy change, ΔS_1, is negative.

(2)   evolution of heat to the surroundings, the reaction being exothermic, i.e. the entropy of the surroundings, S_{surr}, increases and the consequent entropy change, ΔS_2, is positive.

For the process to be spontaneous the magnitude of $-\Delta S_1$ must be smaller than that of $+\Delta S_2$ i.e. $[-\Delta S_1] + [\Delta S_2]$ must be positive.

Applying these ideas to a grander scale the second law can also be stated in another well-known form, i.e. the entropy of the universe tends to a maximum. Again, although certain processes leading to order occur naturally, the overall tendency as exemplified by mountain erosion and combustion of fuels is inevitably towards increased disorder or chaos. The laws of thermodynamics cannot be applied to single or few molecules but only to a relatively large number in which case the study is called statistical thermodynamics.

The reason why changes occur in the direction which produces a state of disorder rather than one of order is most simply answered in that any system moves towards a state of maximum probability; disorder being statistically more probable than order.

This may be seen, in a more easily realisable way, in the result of shuffling an ordered pack of cards or vigorously shaking a box of jigsaw pieces where the more cards or pieces used the greater are the possible number of combinations of arrangements. Hence, it is apparent that the entropy, S, of the system is proportional in some way to the number of available combinations (i.e. probabilities, W) in which the system may exist. On the atomic or molecular scale it may be said that each particle in, say, a mole of substance $(6.02 \times 10^{23}$ particles)* has its own energetic orientation (see Section 2.2.1.1) or *microstate* at a given instant. The entropy of a system is determined by both the arrangement of the energy quanta within the particles and by the arrangement of the particles in the system.

The thermodynamic probability, W, of a system is equal to the number of microstates, w, which contribute to it. Since: (i) entropy is an extensive property (i.e. dependent on mass) the total entropy, S, of systems S_1 and S_2 is given by $S = S_1 + S_2$, i.e. they can be added; (ii) probability is multiplicative, the total probability, W, of systems W_1 and W_2 is given by $W = W_1 W_2$. Thus, the basic equation of statistical thermodynamics becomes:

$$S \propto \ln W$$

i.e. (2.14)

$$S = K \ln W$$

where K is the Boltzmann constant $(= R/N = 8.314/(6.023 \times 10^{23})$ J K^{-1}).

If substances are considered to be perfectly crystalline at absolute zero then the entropy of the solid will be zero, since if there is unit probability:

$$S_0 = K \ln 1 = 0 \qquad (2.15)$$

This is the third law of thermodynamics, i.e. entropy is zero at absolute zero.

Processes involving entropy changes

(1) Heating (or cooling) of a material

Expansion is caused by the increased range of movement of the particles in a material upon acquisition of increased kinetic energy.

(2) Change of state

Melting of a solid is accompanied by a moderate entropy increase while vaporisation involves a relatively large entropy increase. Solidification and condensation are accompanied by corresponding entropy decreases, e.g.

$$Zn(s) \xrightarrow{\Delta S = +10.5 \text{ J K}^{-1} \text{ mol}^{-1}} Zn(l) \xrightarrow{\Delta S = +98.3 \text{ J K}^{-1} \text{ mol}^{-1}} Zn(g)$$

$$I_2(s) \xrightarrow{\Delta S = +144 \text{ J K}^{-1} \text{ mol}^{-1}} I_2(g)$$

Hence changes in state occurring during chemical reactions, e.g. if gaseous products arise from solid or liquid reactants, effect a consequent change in entropy for the overall reaction. Three examples are shown below:

$Zn(s) + H_2SO_4(aq) \longrightarrow ZnSO_4(aq) + H_2(g)$ $\qquad \Delta S$ positive $(0 \rightarrow 1$ gas volumes)

$2C(s) + O_2(g) \longrightarrow 2CO(g)$ $\qquad \Delta S$ positive $(1 \rightarrow 2$ gas volumes)

$Zn(s) + \frac{1}{2}O_2(g) \longrightarrow ZnO(s)$ $\qquad \Delta S$ negative $(\frac{1}{2} \rightarrow 0$ volumes)

---

*Avogadro's number (N) is the number of particles (atoms, molecules, ions) contained in 1 mole of any substance, and is constant at 6.02×10^{23} particles.

(3) Mixing processes: e.g. formation of solutions (see Section 4.3.8); mixing of gases by diffusion.

(4) Reactions in which complex species are converted into simpler ones and vice versa: e.g.

Dissociation:	$Fe_2Cl_6(g) \longrightarrow 2FeCl_3(g)$	ΔS positive as simpler molecules are produced
Association (e.g. polymerisation):	$nC_2H_4(g) \xrightarrow[\text{pressure}]{200°C} (C_2H_4)_n(s)$ (ethane) (polythene)	ΔS negative as more complex, ordered molecules are produced (also n moles of gas are consumed)

As with enthalpies, $S_T^{\ominus}$ or S_T° represents standard entropy and $\Delta S_T^{\ominus}$ or ΔS_T° represents standard entropy change, i.e. under 1 atm pressure and at a stated temperature.

A quantitative expression for entropy

When an amount of heat, q, is supplied to a body at temperature T, its disorder or entropy increases. The magnitude of the entropy change can be related to these two factors as follows:

(i) $\Delta S \propto q$, i.e. the entropy change is directly proportional to the amount of heat supplied;
(ii) $\Delta S \propto 1/T$.

If the third law of thermodynamics holds, the entropy of the body will be zero at absolute zero (0 K). Hence at 1 K, by virtue of the heat supplied to cause this temperature rise, its new entropy (S_1) is infinitely larger than its original value (S_0). Obviously, if a similar amount of heat were added to the body at a higher temperature, e.g. 298 K, the *relative* entropy change would be far smaller. For this reason entropy change (ΔS) and absolute temperature, T, must vary in some form of inverse relationship as expressed above, i.e. the magnitude of the change in entropy decreases as temperature increases. Combining both variations of ΔS we have:

$$\Delta S \propto q/T$$

This relationship only holds for reversible thermodynamic processes and may be rewritten:

$$\Delta S = q_{rev}/T$$

By use of the correct units and referring to 1 mole this becomes

$$\Delta S^{\ominus} = q_{rev}/T \tag{2.16}$$

In fact processes involving changes of state, e.g. fusion and evaporation (vaporisation), are reversible and during these changes q_{rev} is equal to the appropriate molar latent heats of fusion or evaporation for the substance.

Equation (2.16) was first developed by a French engineer, Sadi Carnot, in 1824 while studying the efficiency of heat engines. He considered mathematically the performance of an ideal heat engine in which all changes were carried out reversibly. The fairly complex derivation is not considered necessary in the development of thermodynamics with regard to its metallurgical applications.

For a thermodynamic irreversible process:

$$\Delta S^{\ominus} > q/T$$

In reality most changes are of this type except for state-change processes as mentioned above.

Under constant pressure conditions, $q = \Delta H$ and $q_{rev} = \Delta H_{rev}$ (similarly under constant volume conditions, $q = \Delta U$ and $q_{rev} = \Delta U_{rev}$). If we use equation (2.16) and consider 1 mole of substance, assuming q_{rev} is expressed in joules and T in degrees kelvin, the unit of entropy becomes $J\ K^{-1}\ mol^{-1}$.

Calculating entropy changes

(1) Entropy changes during change of state: These changes occur at constant temperature and are reversible. Using equation (2.16)

$$\Delta S^{\ominus} = \begin{cases} L_f/T_m & \text{for fusion} & (2.16a) \\ L_e/T_b & \text{for vaporisation} & (2.16b) \end{cases}$$

where L_f and L_e are molar latent heats of fusion and evaporation respectively, and T_m and T_b are the melting and boiling point temperatures respectively.

(2) Entropy changes during change of temperature: For a finite change in entropy it is necessary to integrate equation (2.16) between the limits of temperature, T_1 and T_2, thus,

$$\Delta S^{\ominus} = S_2^{\ominus} - S_1^{\ominus} = \int_{T_1}^{T_2} \frac{dq_{rev}}{T} \qquad (2.17)$$

but referring to equation (2.2) and assuming that heat absorbed, q_{rev}, can be substituted by ΔH (since at constant pressure, $q_{rev} = \Delta H_{rev}$) then

$$\frac{dq_{rev}}{dT} = \Delta C_p$$

whence,

$$dq_{rev} = \Delta C_p dT$$

so that equation (2.17) becomes:

$$\Delta S^{\ominus} = S_2^{\ominus} - S_1^{\ominus} = \int_{T_1}^{T_2} \frac{\Delta C_p}{T} dT \qquad (2.18)$$

Integration of equation (2.18) yields,

$$\Delta S^{\ominus} = S_2^{\ominus} - S_1^{\ominus} = C_p \ln(T_2/T_1) \qquad (2.19)$$

Thus, given a substance's molar heat capacity, C_p, at constant pressure and the initial and final temperatures T_1 and T_2, the entropy change can be calculated. The absolute entropy can also be found by considering entropies at an arbitrary temperature T and at absolute zero. In this case

$$\Delta S^{\ominus} = S_T^{\ominus} - S_0^{\ominus}$$

but $S_0^{\ominus} = 0$ by the third law of thermodynamics and so

$$\Delta S^{\ominus} = S_T^{\ominus}$$

Data books[1,2] furnish values of $S_{298}^{\ominus}$, the standard entropies for substances at 298 K. Hence for high temperature conditions it is convenient to use equation (2.19) in the form:

$$\Delta S_T^{\ominus} = S_{298}^{\ominus} + C_p \ln(T/298)$$

(3) Entropy changes during chemical reactions: (i) at 298 K the entropy change, $\Delta S^{\ominus}_{298}$, may be calculated from:

$$\Delta S^{\ominus}_{298} = S^{\ominus}_{\text{prods}, 298} - S^{\ominus}_{\text{reacts}, 298} \qquad (2.20)$$

Values for $S^{\ominus}_{\text{prods}, 298}$ and $S^{\ominus}_{\text{reacts}, 298}$ are available from data books[1,2].

(ii) at any temperature T

$$\Delta S^{\ominus}_{T} = S^{\ominus}_{\text{prods}, T} - S^{\ominus}_{\text{reacts}, T} \qquad (2.21)$$

In this case $S^{\ominus}_{T}$ must be calculated for each individual reactant and product using equations (2.16a), (2.16b) and (2.19) for changes in state and changes in temperature respectively (from 298 K to T K usually) prior to calculating $\Delta S^{\ominus}_{T}$.

Table 2.4 STANDARD ENTROPIES OF VARIOUS SOLIDS, LIQUIDS AND GASES (IN J K^{-1} mol^{-1})

Solids	$S^{\ominus}_{298}$	Liquids	$S^{\ominus}_{298}$	Gases	$S^{\ominus}_{298}$
Al(s)	22	Br$_2$(l)	152	Ar(g)	155
C(diamond)	5.6	CCl$_4$(l)	214	CH$_4$(g)	186
C(graphite)	6.2	H$_2$O(l)	70	C$_2$H$_4$(g)	229
Cu(s)	33	H$_2$O$_2$(l)	102	CO(g)	198
CuO(s)	44	Hg(l)	77	CO$_2$(g)	214
Fe(s)	27	TiCl$_4$(l)	253	Cl$_2$(g)	223
FeO(s)	54			H$_2$(g)	131
Fe$_2$O$_3$(s)	90			H$_2$O(g)	189
Fe$_3$O$_4$(s)	146			O$_2$(g)	205
FeS(s)	67			SO$_2$(g)	249
Ni(s)	30				
NiCl$_2$(s)	97				
SiO$_2$(quartz)	42				
Sn(grey)	45				
Sn(white)	52				
TiO$_2$(s)	50				
Zn(s)	42				

Table 2.4 lists entropy values for selected substances; $S^{\ominus}$ denotes the standard entropy of a substance at 298 K and 101 325 Pa (1 atm) pressure and is stated in J K^{-1} mol^{-1}.* This table illustrates the low entropy of very ordered crystalline solids such as diamond and close packed metal crystal structures. It also shows the anticipated higher entropies of less ordered liquids and highly disordered gaseous substances.

Example

Calculate the entropy of liquid iron at its melting point, 1808 K, given that, for iron,

$$L_f = 15.4 \text{ kJ mol}^{-1}$$
$$S^{\ominus}_{298} = 27.2 \text{ J K}^{-1} \text{ mol}^{-1}$$
$$C_p = 25.2 \text{ J K}^{-1} \text{ mol}^{-1}$$

Considering the initial temperature of the iron to be 298 K then the following consecutive changes involving the supply of heat must occur: the temperature of the iron is

*These are sometimes referred to as entropy units (eu).

raised from 298 K to 1808 K; the solid iron is converted into liquid iron, i.e. from equations (2.16a) and (2.19)

$$\Delta S^{\ominus} = C_p \ln\left(\frac{T_2}{T_1}\right) + \frac{L_f}{T_m}$$

or

$$S^{\ominus}_{1808} = S^{\ominus}_{298} + C_p \ln\left(\frac{T_2}{T_1}\right) + \frac{L_f}{T_m}$$

Substituting the given values as follows:

$$S^{\ominus}_{1808} = 27.2 + 25.2 \ln\left(\frac{1808}{298}\right) + \frac{15400}{1808} \ \ \mathrm{J\,K^{-1}\,mol^{-1}}$$

$$= 27.2 + 45.4 + 8.5 \ \mathrm{J\,K^{-1}\,mol^{-1}}$$

$$= 81.1 \ \mathrm{J\,K^{-1}\,mol^{-1}}$$

As explained earlier, $C_{p,T}$ should be used in the calculation (equation (2.14)) rather than $C_{p,298}$ for completely accurate values of $\Delta S^{\ominus}$ or $S^{\ominus}$.

Calculations

In each case values of $S^{\ominus}_{298}$ for substances can be obtained by reference to *Table 2.4*.

(1)   If 30.8 kJ of heat are supplied to a mixture of benzene and its vapour at 353 K, one mole of benzene is vaporised. Calculate the change in entropy, $\Delta S^{\ominus}$, for this process.

(2)   Calculate $\Delta S^{\ominus}_{298}$ for the reactions:

(a) $2H_2(g) + O_2(g) \longrightarrow 2H_2O(l)$

(b) $Ni(s) + Cl_2(g) \longrightarrow NiCl_2(s)$

(c) $3Fe(s) + 4H_2O(g) \longrightarrow Fe_3O_4(s) + 4H_2(g)$

In each case state how the sign of $\Delta S^{\ominus}_{298}$ can be predicted by reference to the chemical equation for the reaction.

(3)   Calculate the standard entropy of copper at 1800 K, given that, for copper, $T_m = 1356$ K, $C_p = 24.5$ J K^{-1} mol^{-1} (assumed constant in the region 298–1800 K), $L_f = 13.0$ kJ mol^{-1} and $\Delta S^{\ominus}_{298}[Cu(s)] = 33.3$ J K^{-1} mol^{-1}

(4)   For the reaction

$$TiO_2(s) + 2Cl_2(g) + C(s) \xrightarrow{\ 298\ K\ } TiCl_4(l) + CO_2(g)$$

Calculate: (a) $\Delta S^{\ominus}_{298}$ and (b) $\Delta S^{\ominus}_{798}$, the temperature at which the commercial reaction is carried out. Between 298 K and 798 K only TiCl$_4$ goes through a transformation and the molar heat capacities (assumed constant over the temperature range) required are as follows:

	$TiO_2(s)$	$Cl_2(g)$	$C(s)$	$TiCl_4(g)$	$CO_2(g)$
C_p (J K^{-1} mol^{-1})	55	40	8.6	157	37

$$T_b = 413 \ \mathrm{K} \qquad L_e = 138.6 \ \mathrm{kJ\,mol^{-1}}$$

2.2.2   Free energy: The driving force of a chemical reaction

A chemical reaction will only occur if there is an overall decrease in the energy of the system and its surroundings during the reaction; a prerequisite for calculation of this energy change is a knowledge of the various factors contributing to the change. The existence of spontaneous endothermic reactions implies that enthalpy change is not the sole arbiter of the actual energy change nor hence the direction of the chemical change. The entropy change, ΔS, for the reaction must also be considered. For example,

(a) $\qquad\qquad ZnO(s) + C(s) \longrightarrow Zn(g) + CO(g)$

Spontaneous at 1373 K when:

$$\Delta S^{\ominus} = +285 \text{ J K}^{-1} \text{ mol}^{-1}$$

$$\Delta H^{\ominus} = +349.9 \text{ kJ mol}^{-1}$$

(b) $\qquad\qquad Fe(s) + \frac{1}{2}O_2(g) \longrightarrow FeO(s)$

Spontaneous at 298 K when:

$$\Delta S^{\ominus} = -71 \text{ J K}^{-1} \text{ mol}^{-1}$$

$$\Delta H^{\ominus} = -265.5 \text{ kJ mol}^{-1}$$

In each example $\Delta H^{\ominus}$ and $\Delta S^{\ominus}$ apparently work in opposition to each other in terms of the direction of the energy change, but both reactions can proceed in the direction indicated.

It is necessary to know how to predict which of these two contributing factors, ΔH and ΔS, outweighs the other and thus decides the direction of a spontaneous reaction when they are opposed to each other. To do this both factors must be incorporated in such a way as to introduce a single function expressing the combined effect of a change in both ΔH and ΔS.

A system, including chemical reactions at constant pressure, which turns heat energy into mechanical energy is a form of heat engine though chemical reactions are rarely harnessed for this purpose. It is never practically possible to obtain maximum work, i.e. 100% efficiency, from such a device. Hence, of the total enthalpy (heat content) of the system only a part can be converted into useful work, this fraction of the total enthalpy being called the free energy, G. As with H, absolute values of G cannot be measured, only changes in free energy, ΔG. This new function ΔG, can be defined in terms of ΔH, ΔS and T, the temperature at which the reaction occurs and is a measure of the feasibility* of a reaction.

2.2.2.1   The Gibbs–Helmholtz (second law) equation

The relationship between ΔH, ΔS and ΔG may be deduced as follows. The difference between ΔH and ΔG is the unavailable energy used to cause disorder. This difference is related to ΔS and may be derived from the expression for entropy change occurring during change of state. In this event all of the heat supplied to the system, i.e. q_{rev}, is used to create disorder since the temperature remains unchanged. From equation (2.16), $q_{rev} = T\Delta S^{\ominus}$, i.e. the energy used to create disorder. So now the difference between ΔH and ΔG may be stated:

$$\underset{\substack{\text{enthalpy} \\ \text{change}}}{\Delta H^{\ominus}} \quad - \quad \underset{\substack{\text{energy available} \\ \text{to do work}}}{\Delta G^{\ominus}} \quad = \quad \underset{\substack{\text{unavailable energy} \\ \text{used to cause disorder}}}{T\Delta S^{\ominus}}$$

*A reaction is said to be feasible if ΔG is negative (or zero) or if the equilibrium constant, K (Section 2.2.3.6) equals or exceeds 1.

This may be rearranged:

$$\Delta G^{\ominus} = \Delta H^{\ominus} - T\Delta S^{\ominus} \qquad (2.22)$$

and

$$\Delta G = \Delta H - T\Delta S$$

if one or more reactant or product are not under standard thermodynamic conditions.

This important relationship is sometimes referred to as the Gibbs–Helmholtz equation (or the second law equation) and ΔG is a measure of the work obtainable from a reversible, isothermal process occurring at constant pressure* and gives a direct indication of the possibility of chemical reaction. Hence, we have defined the driving force for chemical change as ΔG, the Gibbs free energy change, so named after J. Willard Gibbs in honour of his work in this field.

Reactions tend to go in the direction which results in a decrease in free energy which means that the feasibility of reaction may now be defined as follows:

(i)   if ΔG is negative a reaction is feasible;
(ii)   if ΔG is positive a reaction is not feasible;
(iii)   if ΔG is zero an equilibrium mixture is obtained, i.e. equal feasibility for both the forward and reverse reactions.

Even if ΔG is negative other factors such as position of equilibrium and rate of attainment of this equilibrium influence the readiness of formation of products and hence the utility of the reaction. It will be useful to consider how calculations of ΔG can help in evaluation of commercial processes, e.g. the feasibility of reduction of metal oxides with hydrogen.

Example

Calculate $\Delta G^{\ominus}$ for the following reductions at 500 K by using the data provided and equation (2.22):

(i) $\qquad CuO(s) + H_2(g) \longrightarrow Cu(s) + H_2O(g) \qquad$ at 500 K

$\qquad \Delta H^{\ominus}_{500} = -87 \text{ kJ mol}^{-1}$

$\qquad \Delta S^{\ominus}_{500} = +47 \text{ J K}^{-1} \text{ mol}^{-1}$

$\qquad \Delta G^{\ominus} = \Delta H^{\ominus} - T\Delta S^{\ominus}$

that is,

$$\Delta G^{\ominus} = -87\,000 - (500 \times 47) \text{ J mol}^{-1}$$
$$= -110.5 \text{ kJ mol}^{-1}$$

Since $\Delta G^{\ominus}$ is negative (and, incidentally, fairly large) the reaction is feasible at 500 K.

(ii) $\qquad ZnO(s) + H_2(g) \longrightarrow Zn(s) + H_2O(g) \qquad$ at 500 K

$\qquad \Delta H^{\ominus}_{500} = +104 \text{ kJ mol}^{-1}$

$\qquad \Delta S^{\ominus}_{500} = +60 \text{ J K}^{-1} \text{ mol}^{-1}$

$\qquad \Delta G^{\ominus} = \Delta H^{\ominus} - T\Delta S^{\ominus}$

*Under constant volume isothermal conditions it may be stated that: $\Delta F = \Delta U - T\Delta S$, where ΔF (ΔA in USA) is called the maximum work function or Helmholtz free energy.

that is,

$$\Delta G^{\ominus} = +104\,000 - (500 \times 60) \text{ J mol}^{-1}$$

$$= +74 \text{ kJ mol}^{-1}$$

Since $\Delta G^{\ominus}$ is positive the reaction cannot occur at 500 K.

2.2.2.2   The effect of temperature on the feasibility of a reaction

The value of ΔG depends on whether ΔS is positive or negative; a low temperature makes $T\Delta S$ small and vice versa.

Consider the following reaction at say 1180 K, the boiling point of zinc:

$$ZnO(s) + C(s) \longrightarrow Zn(g) + CO(g)$$

Since for this reaction ΔH and ΔS are both positive the sign of ΔG depends on the temperature of the reaction, i.e. low temperature makes $T\Delta S$ small and negative, but high temperature makes $T\Delta S$ more negative, possibly sufficiently to outweigh the positive $\Delta H^{\ominus}$ and make $\Delta G^{\ominus}$ negative.

Substituting approximate, averaged values for $\Delta H^{\ominus}$ and $\Delta S^{\ominus}$, i.e.

$$\Delta H_T^{\ominus} = +295 \text{ kJ mol}^{-1}$$

$$\Delta S_T^{\ominus} = +249 \text{ J K}^{-1} \text{ mol}^{-1}$$

and assuming that the temperature of reversal, T, occurs when $\Delta G^{\ominus} = 0$, this reversal temperature may be calculated. Using equation (2.22) we can write

$$0 = +295\,000 - 249T$$

and so

$$T = 1184 \text{ K}$$

Thus, within the accuracies used: if $T < 1184$ K, $\Delta G^{\ominus}$ is positive, and so the reaction cannot occur; but if $T > 1184$ K, $\Delta G^{\ominus}$ is negative, and the reaction can proceed.

Summary of the effect of the sign and magnitude of ΔH, ΔS and T on ΔG:

(i) $\Delta H = 0$; ΔS positive:

$$\Delta G = 0 - T\Delta S$$

$$\Delta G = -T\Delta S$$

i.e. ΔG is negative (feasible), e.g. the mixing of non-reacting gases.

(ii) ΔH negative or positive; $\Delta S = 0$:

$$\Delta G = \Delta H \quad \text{(feasible if } -\Delta H)$$

i.e. reactions occurring at 0 K or when entropy remains constant, e.g.

$$C(s) + O_2(g) \longrightarrow CO_2(g)$$

(iii) ΔH negative; ΔS positive:

$$\Delta G = \Delta H - T\Delta S$$

Both ΔH and $-T\Delta S$ terms are negative giving a negative ΔG at all temperatures (feasible), i.e. exothermic reactions involving entropy increases, e.g.

$$C(s) + \tfrac{1}{2}O_2(g) \longrightarrow CO(g)$$

(iv) ΔH negative; ΔS negative:

$$\Delta G = \Delta H - T\Delta S$$

ΔH is negative but $-T\Delta S$ is positive as $-T(-\Delta S) = T\Delta S$. Thus at low temperatures $-\Delta H$ exceeds $+T\Delta S$ in magnitude but at higher temperatures the increasingly larger $T\Delta S$ will eventually outweigh $-\Delta H$. Hence, the reaction is feasible at low temperatures but not at higher temperatures, e.g.

$$2Ag(s) + \tfrac{1}{2}O_2(g) \longrightarrow Ag_2O(s)$$

(v) ΔH positive; ΔS positive:

$$\Delta G = \Delta H - T\Delta S$$

This is a reverse analogy of (iv) since at high temperatures the increasing magnitude of $-T\Delta S$ will begin to outweigh $+\Delta H$, i.e. the reaction is not feasible at low temperature but becomes so at higher temperatures, e.g.

$$ZnO(s) + C(s) \longrightarrow Zn(g) + CO(g)$$

(vi) ΔH positive; ΔS negative:

$$\Delta G = \Delta H - T\Delta S$$

If both ΔH and $-T\Delta S$ are positive ΔG will be positive at all temperatures and the reaction is never feasible, e.g.

$$2Au(s) + \tfrac{3}{2}O_2(g) \longrightarrow Au_2O_3(s)$$

2.2.2.3   Calculating $\Delta G^\ominus$ for reactions

(1)   At 298 K

(a) $\Delta G^\ominus_{298} = \Delta H^\ominus_{298} - T\Delta S^\ominus_{298}$

Values of $\Delta H^\ominus_{298}$ and $\Delta S^\ominus_{298}$ are available from data books[1,2].

(b) $\Delta G^\ominus = \Delta G^\ominus_{prods} - G^\ominus_{reacts}$ (2.23)

$G^\ominus$, the standard free energy for elements, is taken, by convention, to be zero (provided the element is in its standard state under the specified conditions). $G^\ominus$, the standard free energy for compounds, is taken to be equal to the standard free energy change of formation for the compound, $\Delta G^\ominus_f$, i.e. the free energy change when one mole of a compound is formed from its elements at a stated temperature e.g. 298 K and one atmosphere pressure (see *Table 2.5*).

The units of free energy are kJ mol^{-1}. As with $\Delta H^\ominus_{298}$ and $\Delta S^\ominus_{298}$, values of $\Delta G^\ominus_{298}$ are readily available from standard data books[1,2].

(2)   At any temperature, T

Strictly:

$$\Delta G^\ominus_T = \Delta H^\ominus_T - T\Delta S^\ominus_T$$

$\Delta H^\ominus_T$ can be calculated by use of Kirchoff's law (see Section 2.1.4.2) and, being measured in kJ mol^{-1}, will usually appreciably alter the value of $\Delta G^\ominus$. $\Delta S^\ominus_T$ can be calculated by considering the various phase changes and heating processes involved from 298 to T K (see Section 2.2.1.2) but, being measured in J K^{-1} mol^{-1}, may be neglected unless precise accuracy is required, i.e. it is often convenient to use:

Table 2.5 TABLE OF STANDARD FREE ENERGIES, $\Delta G_{298}^{\ominus}$ (OR $\Delta G_{f,298}^{\ominus}$) IN kJ mol^{-1}

Substance	$\Delta G_{298}^{\ominus}$ ($\Delta G_{f,298}^{\ominus}$)	Substance	$\Delta G_{298}^{\ominus}$ ($\Delta G_{f,298}^{\ominus}$)
$Al_2O_3(s)$	-1576	$PbO(s)$	-188
$CaO(s)$	-604	$PbS(s)$	-93
$CaCO_3(s)$	-1207	$MgO(s)$	-570
$CO(g)$	-137	$HgO(s)$	-59
$CO_2(g)$	-395	$NiO(s)$	-216
$HCl(g)$	-95	$SiO_2(s)$	-805
$Cr_2O_3(s)$	-1047	$SO_2(g)$	-300
$CuO(s)$	-127	$TiO_2(s)$	-853
$Cu_2O(s)$	-146	$TiCl_4(l)$	-674
$Cu_2S(s)$	-86	$ZnO(s)$	-318
$H_2O(l)$	-237		
$H_2O(g)$	-229		
$FeO(s)$	-244		
$Fe_2O_3(s)$	-741		
$Fe_3O_4(s)$	-1015		
$FeS(s)$	-98		

$$\Delta G_T^{\ominus} = \Delta H_T^{\ominus} - T\Delta S_{298}^{\ominus} \qquad (2.24)$$

The variation of ΔG with temperature may be expressed:

(a)   Mathematically, by the Gibbs–Helmholtz equation, which may be restated, without lengthy derivation, by saying that at constant pressure it can be shown that the rate of change of ΔG with temperature is related to ΔS by the equation

$$\frac{d(\Delta G)}{dT} = -\Delta S$$

Then, substituting for ΔS from equation (2.24), we can write

$$\frac{d(\Delta G)}{dT} = \frac{\Delta G - \Delta H}{T} \qquad (2.25)$$

For a full derivation see Mackowiak[3], p. 90 or Ives[4], p. 122.

(b)   Graphically, by *standard* free energy–temperature diagrams as shown in *Fig. 2.9*. Since $\Delta G^{\ominus} = \Delta H^{\ominus} - T\Delta S^{\ominus}$ is of the form $y = mx + c$ (i.e. $\Delta G^{\ominus} = -\Delta S^{\ominus}T + \Delta H^{\ominus}$), a plot of $\Delta G^{\ominus}$ versus T will give a straight line graph with:

(i)   an intercept, c, on the y-axis of $\Delta H_0^{\ominus}$, where $\Delta H_0^{\ominus}$ is the standard enthalpy at 0 K; thus $\Delta G_0^{\ominus} = \Delta H_0^{\ominus}$.

(ii)   a slope, m, of $-\Delta S^{\ominus}$. Hence when entropy is increasing ($+\Delta S^{\ominus}$) the slope of the graph will be negative since $-(+\Delta S^{\ominus}) = -\Delta S^{\ominus}$ (negative slope). This is illustrated by lines B and D in *Fig. 2.9*. Line B shows an endothermic reaction, with entropy increasing (reaction becomes feasible where the line crosses the T-axis); line D shows an exothermic reaction, with entropy increasing (reaction increasingly feasible with increase in temperature).

Similarly when entropy is decreasing ($-\Delta S^{\ominus}$) the slope will be positive since $-(-\Delta S^{\ominus}) = +\Delta S^{\ominus}$ (positive slope). This is illustrated by lines A, C and E in *Fig. 2.9*. Line A shows an endothermic reaction, with entropy decreasing (reaction never feasible); line C shows an exothermic reaction, with entropy decreasing (reaction ceases to be feasible at intercept on the T-axis).

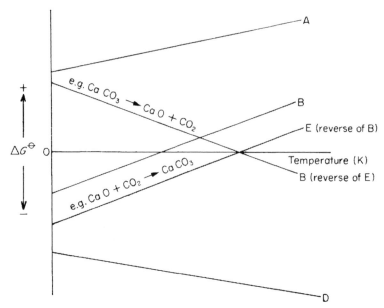

Figure 2.9 Effect of the sign of $\Delta H^{\ominus}$ and $\Delta S^{\ominus}$ on $\Delta G^{\ominus}$-T diagrams

If the reverses of these reactions were plotted on the diagram they would have the same values of ΔS, ΔH and ΔG but of opposite sign and cross the T-axis at the same point, thus line E is the reverse reaction of line B, e.g. $CaO + CO_2 \longrightarrow CaCO_3$ (line E) and $CaCO_3 \longrightarrow CaO + CO_2$ (line B).

Diagrams of this type (*Fig. 2.9*) are important in their use for considering the feasibility of extraction reactions for metals from oxides, sulphides, etc. (see $\Delta G^{\ominus}$-T diagrams in Section 2.2.3.10).

Calculation of $\Delta G^{\ominus}$ by the method of Hess's law

Calculations of $\Delta G^{\ominus}$ can be made for required reactions given $\Delta G^{\ominus}$ values for other relevant reactions, as shown in the thermochemistry calculations on Hess's law (see Section 2.1.2.2). As an example we calculate $\Delta G^{\ominus}$ for the reaction

$$Fe(l) + CO_2(g) \longrightarrow CO(g) + FeO(l)$$

given the following data at 1600 °C (1873 K):

$$2CO(g) + O_2(g) \longrightarrow 2CO_2(g) \qquad \Delta G^{\ominus} = -242 \text{ kJ } (\text{mol } O_2)^{-1}$$

$$2Fe(l) + O_2(g) \longrightarrow 2FeO(l) \qquad \Delta G^{\ominus} = -292 \text{ kJ } (\text{mol } O_2)^{-1}$$

Reversing the first reaction and rewriting the second:

$$2CO_2(g) \longrightarrow 2CO(g) + O_2(g) \qquad \Delta G^{\ominus} = +242 \text{ kJ } (\text{mol } O_2)^{-1}$$

$$2Fe(l) + O_2(g) \longrightarrow 2FeO(l) \qquad \Delta G^{\ominus} = -292 \text{ kJ } (\text{mol } O_2)^{-1}$$

Adding these we get

$$2Fe(l) + 2CO_2(g) \longrightarrow 2CO(g) + 2FeO(l) \qquad \Delta G^{\ominus} = -50 \text{ kJ } (\text{mol } O_2)^{-1}$$

Since these units refer to one mole of oxygen this result can be written

$$Fe + CO_2 \longrightarrow CO + FeO \qquad \Delta G^{\ominus} = -25 \text{ kJ}$$

As $\Delta G^{\ominus}$ is calculated it is assumed that the reaction takes place in equilibrium with 1 atm of O_2. Hence the forward reaction is feasible at this temperature (1600 °C). Calculations of this type may be easily performed for a variety of reactions by extracting $\Delta G^{\ominus}$ values at the required temperatures from the comprehensive free energy–temperature diagrams illustrated and discussed later in Section 2.2.3.10.

2.2.2.4 Calculations of free energy

(1) Calculate $\Delta G^{\ominus}$ for the following reactions at the stated temperatures:

(a) $\qquad\qquad ZnO(s) + C(s) \longrightarrow Zn(s) + CO(g)$

at 298 K, when $\Delta H^{\ominus}_{298} = +238.6 \text{ kJ mol}^{-1}$; $\Delta S^{\ominus}_{298} = +213 \text{ J K}^{-1} \text{ mol}^{-1}$.

(b) $\qquad\qquad ZnO(s) + C(s) \longrightarrow Zn(g) + CO(g)$

at 1373 K, when $\Delta H^{\ominus}_{1373} = +349.9 \text{ kJ mol}^{-1}$; $\Delta S^{\ominus}_{1373} = +285 \text{ J K}^{-1} \text{ mol}^{-1}$.

(c) $\qquad\qquad Fe(s) + \frac{1}{2}O_2(g) \longrightarrow FeO(s)$

at 298 K, when $\Delta H^{\ominus}_{298} = -265.4 \text{ kJ mol}^{-1}$; $\Delta S^{\ominus}_{298} = -71 \text{ J K}^{-1} \text{ mol}^{-1}$.

(2) Using the values for standard free energies given in *Table 2.5*, calculate $\Delta G^{\ominus}_{298}$ for each of the following reactions:

(a) $\qquad\qquad CaCO_3(s) \longrightarrow CaO(s) + CO_2(g)$

(b) $\qquad\qquad H_2(g) + CO_2(g) \longrightarrow H_2O(l) + CO(g)$

(c) $\qquad\qquad Fe_2O_3(s) + 2Al(s) \longrightarrow Al_2O_3(s) + 2Fe(s)$

(d) $\qquad\qquad 2Cu_2O(s) + Cu_2S(s) \longrightarrow 6Cu(s) + SO_2(g)$

(3) For the following reactions at 500 K:

(a) $\qquad\qquad PbS(s) + O_2(g) \longrightarrow Pb(s) + SO_2(g)$

(b) $\qquad\qquad PbS(s) + 2PbO(s) \longrightarrow 3Pb(s) + SO_2(g)$

calculate: (i) $\Delta H^{\ominus}_{500}$, (ii) $\Delta S^{\ominus}_{500}$, (iii) $\Delta G^{\ominus}_{500}$, given the following information:

	PbS(s)	$O_2(g)$	Pb(s)	$SO_2(g)$	PbO(s)
$\Delta H^{\ominus}_{500}$ (kJ mol^{-1})	−94	0	0	−296	−219
$\Delta S^{\ominus}_{500}$ (J K^{-1} mol^{-1})	91	205	65	249	68

In each case discuss briefly the feasibility of the reaction under the stated conditions and how the feasibility is influenced by temperature change. Calculate the reversal temperature where relevant assuming that $\Delta H^{\ominus}$ and $\Delta S^{\ominus}$ remain constant over the temperature range considered.

(4)   From the data given below calculate $\Delta G^{\ominus}_{1200}$ and hence comment on the feasibility of use of hydrogen as a reductant for copper(I) oxide:

$$Cu_2O(s) + H_2(g) \longrightarrow 2Cu(s) + H_2O(g)$$

Standard enthalpy changes of formation:

$$\Delta H^{\ominus}_{298}[Cu_2O(s)] = -164 \text{ kJ mol}^{-1}$$

$$\Delta H^{\ominus}_{298}[H_2O(g)] = -242 \text{ kJ mol}^{-1}$$

Heat capacities, C_p in J K^{-1} mol^{-1} (assumed to remain constant in the range 298–1200 K):

Cu$_2$O(s)	H$_2$(g)	Cu(s)	H$_2$O(g)
70.0	29.0	24.2	33.5

Standard entropies, $\Delta S^{\ominus}_{298}$ in J K^{-1} mol^{-1} (assume that $\Delta S^{\ominus}_{298}$ remains constant in the range 298–1200 K):

Cu$_2$O(s)	H$_2$(g)	Cu(s)	H$_2$O(g)
94	131	33	188

2.2.3   Chemical equilibrium

We have now seen how thermodynamic quantities can be calculated for chemical reactions, and consequently the feasibility of any given reaction at any temperature can be deduced using equation (2.22). It is now appropriate to consider the extent or degree of completion of chemical reactions.

The simple chemical equation for a reaction gives an incomplete and, up to a point, false statement of the prevailing situation. The true picture is discovered only from a study of thermodynamics (feasibility), kinetics (rate of reaction–see Chapter 3) and reaction mechanisms (bond reforming steps).

Thermodynamics allows prediction of the direction in which a reaction will proceed under given conditions; in fact many reactions do not go to completion and then cease since the products of the reaction react together to reform the original reactants. For example, consider heating calcium carbonate in a sealed vessel:

$$CaCO_3(s) \longrightarrow CaO(s) + CO_2(g)$$

Ultimately the sealed vessel contains all three substances and at any stated temperature the composition of this mixture will be the same irrespective of whether the starting point was on the left (CaCO$_3$) or the right (CaO + CO$_2$) of the equation–assuming equimolar quantities in each case. The reaction is said to be (chemically) reversible and is represented by the symbol $\rightleftharpoons$. This is not to be confused with thermodynamic reversibility as discussed in Section 2.2.1.1. Other reversible reactions include:

$$ZnO(s) + CO(g) \rightleftharpoons Zn(s) + CO_2(g)$$

$$N_2(g) + 3H_2(g) \rightleftharpoons 2NH_3(g)$$

$$3Fe(s) + 4H_2O(g) \rightleftharpoons Fe_3O_4(s) + 4H_2(g)$$

In each of these reactions occurring in closed vessels, at a given temperature, chemical equilibrium is established after a certain time (the reactants and products existing together in equilibrium), the mixture at this point being called the equilibrium mixture. When a state of chemical equilibrium has been established there is no bulk change in reactants or products. The equilibrium existing is dynamic (rather than static), that is on attainment of equilibrium:

$$\text{rate of forward reaction, } r_F = \text{rate of reverse reaction, } r_R$$

2.2.3.1 Law of mass action

This is concerned with the effect of the relative concentrations of reactants and products on the equilibrium position finally obtained. The law is based on the experimental results observed for many reactions and states that the rate of a chemical reaction is proportional to the product of the molar concentrations* (*active masses* was the term formerly used) of the reacting substances.

For the general reaction:

$$A + B \rightleftharpoons C + D$$

By the law of mass action:

$$\text{rate of forward reaction, } r_F \propto [A][B]$$

$$\text{rate of reverse reaction, } r_R \propto [C][D]$$

where [ ] represents molar concentration in dilute solution (for the law to hold). Hence

$$r_F = k_1[A][B] \quad \text{and} \quad r_R = k_2[C][D]$$

where k_1, k_2 are constants of proportionality or rate constants (see Section 3.2.2). But at equilibrium, $r_F = r_R$, thus,

$$k_1[A][B] = k_2[C][D]$$

and this may be rearranged such that, by convention, product concentrations appear as the numerator:

$$\frac{[C][D]}{[A][B]} = \frac{k_1}{k_2} = K_c \tag{2.26}$$

where K_c is called the equilibrium constant in terms of concentration. This may be more completely expressed considering the reaction,

$$pA + qB \rightleftharpoons rC + sD$$

where p,q,r,s are numbers of moles of A,B,C,D respectively, in which case K_c may be restated:

$$K_c = \frac{[C]^r[D]^s}{[A]^p[B]^q} \tag{2.27}$$

e.g. for

$$3Mg + Al_2O_3 \rightleftharpoons 3MgO + 2Al$$

$$K_c = \frac{[MgO]^3[Al]^2}{[Mg]^3[Al_2O_3]}$$

The concentrations shown represent the *equilibrium concentrations* of the reactants and products at a given temperature. In instances which refer to pure solids or liquids

---

*Molar concentrations of solutions are often expressed in mol dm^{-3} but the strict SI unit is mol m^{-3}. For an explanation of the term active mass or activity see Sections 2.2.3.3, 4.3.2.

then by convention the active mass is equal to 1 (substances in standard states). The importance of equation (2.27) is in its indication of the position of equilibrium such that:

(i)   If K_c is large, i.e. greater than 1, equilibrium lies well to the right since products predominate in the equilibrium mixture (numerator > denominator).
(ii)   If K_c is small, i.e. less than 1, equilibrium lies to the left since reactants predominate, (numerator < denominator).

The active masses (concentrations) of the reactants and products which are gases may be more usefully interpreted in terms of their partial pressures, p.* Considering a reaction in which all reactants and products are gases:

$$A(g) + B(g) \rightleftharpoons C(g) + D(g)$$

$$K_p = \frac{p_C p_D}{p_A p_B} \tag{2.28}$$

where K_p is the equilibrium constant in terms of partial pressures. For example, for

$$2CO(g) + O_2(g) \longrightarrow 2CO_2(g)$$

$$K_p = \frac{p_{CO_2}^2}{p_{CO}^2 p_{O_2}}$$

If different phases are involved in a reaction, e.g.

$$Cu(s) + \tfrac{1}{2}O_2(g) \longrightarrow CuO(s)$$

then a 'compromise' between K_c and K_p may be written as follows:

$$K = \frac{[CuO]}{[Cu] p_{O_2}^{1/2}}$$

If Cu and CuO are pure solids their concentrations remain constant allowing their active masses to be taken as 1, thus simplifying the expression to:

$$K = \frac{1}{p_{O_2}^{1/2}}$$

Similarly for $CaCO_3(s) \rightleftharpoons CaO(s) + CO_2(g)$, assuming the solids to be pure then:
$K_p = p_{CO_2}$, e.g. at 1073 K, $p_{CO_2} = 22\,265$ Pa $= K_p$.

As a broad guide the figures given in *Table 2.6* apply at 298 K and 1 atm pressure with respect to the magnitude of the equilibrium constant. Hence to summarise

---

*Partial pressure of a gas: the total pressure, P, of a gas is the sum of the partial pressures of the constituent gases (Dalton's law of partial pressures), i.e.

$$P = p_A + p_B + p_C + p_D$$

where p_A, p_B, p_C, p_D are partial pressures of A, B, C, D respectively. Considering A, for example,

$$p_A = P n_A / N$$

where

$$\frac{n_A}{N} = \frac{\text{number of molecules of } A}{\text{total number of molecules}} = \text{mole fraction of } A$$

Table 2.6

$\Delta G^{\ominus}$ (kJ mol^{-1})	K	*Equilibrium position*
< -60	10^{10} and greater	effective completion of reaction
-60 to 0	10^{10} to 1	products predominate
0 to $+60$	1 to 10^{-10}	reactants predominate
$> +60$	10^{-10} and less	effectively no reaction

conveniently: a reaction is feasible if $\Delta G^{\ominus} \leqslant 0; K \geqslant 1$. This explains the dual feasibility of reversible reactions in which both forward and reverse reactions occur simultaneously, i.e.

$$A \underset{\Delta G^{\ominus} = 0}{\overset{\Delta G^{\ominus} = 0}{\rightleftharpoons}} B$$

2.2.3.2 Reactions in progress

The value of ΔG is an indication of the energy available to do useful work where $\Delta G^{\ominus}$ gives the maximum work obtainable from the system when the change is performed thermodynamically reversibly (i.e. infinitely slowly and hence not, in general, in a practically utilisable way).

For a reversible chemical reaction the position of minimum free energy (towards which the system tends) occurs when:

$$G_{\text{prods}} = G_{\text{reacts}} \quad \text{i.e.} \quad \Delta G = 0$$

This condition exists when the equilibrium position for a reaction (at constant temperature) is established. It should be noted that $\Delta G = 0$, rather than $\Delta G^{\ominus} = 0$, since once the reaction has commenced all of the products and reactants cannot, at the same time, be in their standard states (one mole of substance at one atmosphere pressure) except in rare cases. Hence calculation of $\Delta G^{\ominus}$ gives valid information regarding the initial (and final) point of a reaction (for pure reactants) and prediction of feasibility and equilibrium position is possible. However, as soon as the reaction is in progress ΔG applies rather than $\Delta G^{\ominus}$.

2.2.3.3 Activities: the effective concentrations of solutions

So far we have referred to the concentrations of substances, and the partial pressures of gases, throughout implying that equilibrium constants (K) are independent of concentration. However, the departure of the behaviour of solutions from ideality leads to the introduction of a more precise concentration term called activity since in actuality it can only be said that equilibrium constants are 'almost independent' of concentration. In practice the concentration of a given component (1) in a solution may vary from its apparent concentration as a result of the somewhat complex way in which charged particles interact with one another (that is solute–solute and solute–solvent interactions). This means that the concentration of the component present in a solution may not indicate the amount which is available for reaction, i.e. its *effective* concentration, or activity, a. Thus to obtain the activity (or effective active mass) of the component the molecular or ionic concentration must be multiplied by an activity coefficient (γ), i.e.

activity = concentration $\times$ activity coefficient

or

$$\gamma_1 = \frac{\text{effective concentration of 1}}{\text{true concentration of 1}}$$

In many cases the differences are slight and may be neglected. Thus $\gamma_1 = 1$ in the case of ideal solutions and for pure substances.

With regard to gases, the 'effective pressure' is normally called the fugacity, f, of the gas. This deals with the non-ideality of gases and the fugacity coefficient is analogous to the activity coefficient. It approaches unity as p approaches zero and, since most metallurgical processes are carried out at relatively low pressures, i.e. atmospheric pressure or less, then fugacity and true pressure approximate fairly closely.

For a more detailed treatment of activities and activity coefficient, see Sections 4.3.2, 4.3.4 and 4.3.6.

2.2.3.4 Factors affecting the position of equilibrium

(1) Concentration

Reconsidering the general reaction:

$$A + B \rightleftharpoons C + D$$

at equilibrium, for which

$$K_c = \frac{[C][D]}{[A][B]}$$

if the concentration of any one of the reactants or products is altered the concentration of the others must alter to keep K_c constant. If for example $[D]$ is increased by adding more D to the system then $[C]$ will decrease (by reaction with D to form A and B) thus preserving the value of K_c. Addition or removal of any given reactant or product may be similarly considered.

These tendencies illustrate Le Chatelier's principle, i.e. when a system in equilibrium is subjected to a constraint, the system will alter in such a way as to oppose that constraint. Hence in the above generalised case addition of a reactant to the system in equilibrium restores the equilibrium by driving the reaction to the right. The converse is true if a product is added to the system. It is apparent that the displacement of the position of equilibrium may be used to advantage in some commercial processes since continuous removal of a product will drive the reaction to completion. This is most easily achieved if one product is a gas: e.g.

$$CaCO_3(s) \rightleftharpoons CaO(s) + CO_2(g)$$

In the lime kiln, removal of carbon dioxide in an air current drives the reaction to the right. Also in the production of magnesium

$$2MgO(s) + Si(s) \longrightarrow Mg(g) + SiO_2(s)$$

at the temperature employed magnesium gas can be evacuated (see Section 7.2.10.4).

(2) Pressure

Changes in pressure have no effect on the position of equilibrium when only solids or liquids are involved or in gaseous reactions incurring no volume change. For instance, pressure change does not alter the equilibrium position for:

$$Fe(s) + CO_2(g) \rightleftharpoons FeO(s) + CO(g)$$

or

$$3Fe(s) + 4H_2O(g) \rightleftharpoons Fe_3O_4(s) + 4H_2(g)$$

since equal volumes of gas appear on both sides of the equations. However, the equilibrium position for a reaction in which changes in gaseous volume occur may be displaced by pressure changes. Again recourse to Le Chatelier's principle may be made, using the following reaction as an example:

$$N_2(g) + 3H_2(g) \rightleftharpoons 2NH_3(g)$$

$$1 \text{ volume} + 3 \text{ volumes} \rightleftharpoons 2 \text{ volumes}$$

Here, an overall decrease in volume occurs ($4 \rightarrow 2$ volumes). If the system in equilibrium is subjected to a pressure increase the system will move in such a direction as to lessen this increase in pressure in order to maintain the value of K_p. By moving to the right the volume diminution results in a consequent reduction in pressure, i.e. increasing pressure drives the reaction to the right increasing the product yield and vice versa. In the commercial production of ammonia by this method (Haber process) a pressure of about 350 atm is usual. It may be seen in, for example, the 'water gas' reaction:

$$H_2O(g) + C(s) \rightleftharpoons CO(g) + H_2(g)$$

that here the forward reaction is favoured by low pressures.

(3) Temperature

Although unaffected by concentration or pressure changes, K is temperature dependent. This is because a temperature change influences both forward and reverse reactions but rarely to the same extent. This causes a shift in equilibrium and hence a change in the value of K.

Reactions may be exothermic or endothermic and again, according to Le Chatelier's principle:

(i) Exothermic reaction: $-\Delta H^{\ominus}$. For example

$$N_2 + 3H_2 \rightleftharpoons 2NH_3 \qquad \Delta H^{\ominus} = -92.37 \text{ kJ}$$

If the temperature is increased the system reacts in such a way as to oppose this constraint by removing heat, which it can do by shifting to the left, $\Delta H^{\ominus}$ in this direction being $+92.37$ kJ.

Hence, exothermic reactions are favoured by low temperatures. Low temperatures however affect rates (see Section 3.5) and in commercial processes a compromise must be struck which provides a reasonable yield in a reasonable time.

(ii) Endothermic reaction: $+\Delta H^{\ominus}$. For example

$$ZnO + C \rightleftharpoons Zn + CO \qquad \Delta H^{\ominus}_{1373} = +349 \text{ kJ mol}^{-1}$$

Increase in temperature will again shift the equilibrium in the direction which absorbs heat, i.e. to the right in this case. Endothermic reactions are therefore favoured by high temperatures.

Summary of factors affecting equilibrium constant

	Composition of equilibrium mixture	*Equilibrium constant*
Concentration of reactant increased or product decreased	shift to right	no effect
Concentration of reactant decreased or product increased	shift to left	no effect
Pressure increased or decreased	no effect unless reaction involves a volume change	no effect
Temperature increased or decreased	Change depending on sign of ΔH	K alters
Addition of catalyst*	no effect	no effect

*Catalysts alter the rates of forward and reverse reactions to the same extent so reaction yield is unaltered but is obtained more rapidly (see Section 3.5.5).

2.2.3.5   Questions on equilibrium

(1)   Calculate the equilibrium constant for the following reaction:

$$FeO(s) + CO(g) \rightleftharpoons Fe(s) + CO_2(g)$$

if at 298 K the equilibrium amounts present are 2.5 mol FeO, 0.2 mol Fe, 3.0 mol CO_2 and 4.0 mol CO.

(2)   For the reaction:

$$C(s) + CO_2(g) \rightleftharpoons 2CO(g)$$

at 1120 K and 1 atm, the equilibrium mixture contains 92.63% CO and 7.37% CO_2 by volume. Calculate K_p for the reaction. Calculate the partial pressure of CO_2 if p_{CO} is changed to 10^{-4} atm.

(3)   State the effect of: (i) increasing pressure, (ii) increasing temperature on the equilibrium yield of products for reactions:

(a) $CO(g) + H_2O(g) \rightleftharpoons CO_2(g) + H_2(g)$ $\Delta H^{\ominus} = -40$ kJ   at 770 K

(b) $N_2(g) + O_2(g) \rightleftharpoons 2NO(g)$ $\Delta H^{\ominus} = +190$ kJ   at 2300 K

(c) $4Ag(s) + O_2(g) \rightleftharpoons 2Ag_2O(s)$ $\Delta H^{\ominus} = -62$ kJ   at 298 K

(d) $ZnO(s) + C(s) \rightleftharpoons Zn(g) + CO(g)$ $\Delta H^{\ominus} = +349$ kJ   at 1373 K

2.2.3.6   Relationship between free energy and equilibrium constant

It was indicated earlier (Section 2.2.3.2) that once a reaction is in progress ΔG should be used rather than $\Delta G^{\ominus}$. The contribution made by one mole of any constituent, A, to the total free energy, G, of the mixture (often called its chemical potential, μ) is $G_A(\mu_A)$ which may be represented:

$$G_A = G_A^{\ominus} + RT \ln a_A$$

where a_A is the activity of the substance, T the absolute temperature and R the gas constant.

An expression for G may be developed as follows. From the Gibbs–Helmholtz equation (2.22)

$$G = H - TS$$

assuming constant pressure, taking $G_A = G$ for convenience. Substitution for H from the first law equation (2.12) gives

$$G = U + PV - TS$$

which on full differentiation becomes:

$$dG = dU + PdV + VdP - TdS - SdT$$

Equation (2.12) may be restated in the form:

$$dU = q_{rev} - PdV$$

and q_{rev}, from equation (2.16), may be expressed as

$$q_{rev} = TdS$$

Hence substitution of $(TdS - PdV)$ for dU now yields,

$$dG = TdS - PdV + PdV + VdP - TdS - SdT$$

i.e.

$$dG = VdP - SdT$$

At constant pressure this becomes:

$$\frac{dG}{dT} = -S$$

or at constant temperature,

$$dG = VdP$$

However the ideal gas equation (2.11) for 1 mole of reaction states that

$$PV = RT$$

so

$$V = \frac{RT}{P} \quad \text{and} \quad dG = \frac{RT}{P} dP$$

Integration between the limits G'_A, G_A and P'_A, P_A yields

$$\int_{G'_A}^{G_A} dG = \int_{P'_A}^{P_A} \frac{RT}{P} dP$$

i.e.

$$G_A - G'_A = RT \ln (P_A/P'_A)$$

and if G'_A is taken to refer to standard conditions it becomes $G_A^{\ominus}$ and P'_A becomes 1 atm, hence

$$G_A - G_A^{\ominus} = RT \ln P_A$$

or

$$G_A = G_A^{\ominus} + RT \ln P_A$$

If activities are concerned instead of pressures, then:

$$G_A = G_A^{\ominus} + RT \ln a_A$$

Applying equation (2.23) to the following generalised reaction:

$$aA + bB \rightleftharpoons cC + dD$$

gives:

$$\Delta G = (cG_C + dG_D) - (aG_A + bG_B)$$

and using the expression above for G_A, G_B, G_C and G_D

$$\Delta G = [c(G_C^{\ominus} + RT \ln a_C) + d(G_D^{\ominus} + RT \ln a_D)]$$

$$- [a(G_A^{\ominus} + RT \ln a_A) + b(G_B^{\ominus} + RT \ln a_B)]$$

$$= (cG_C^{\ominus} + dG_D^{\ominus}) - (aG_A^{\ominus} + bG_B^{\ominus}) + (cRT \ln a_C + dRT \ln a_D)$$

$$- (aRT \ln a_A + bRT \ln a_B)$$

Hence, for 1 mole of forward reaction,

$$\Delta G = \Delta G^{\ominus} + RT \ln \left(\frac{a_C^c \, a_D^d}{a_A^a \, a_B^b} \right) \tag{2.29}$$

As, at equilibrium, $\Delta G = 0$ and $(a_C^c a_D^d / a_A^a a_B^b) = K$ (the equilibrium constant) then

$$0 = \Delta G^{\ominus} + RT \ln K$$

i.e.

$$\Delta G^{\ominus} = -RT \ln K \qquad (2.30)$$

Equation (2.30) holds true only under equilibrium conditions whereas equation (2.29) must be applied when the reactants and products are under non-equilibrium conditions. Similarly, it may be shown that when gaseous reactants and products are involved equation (2.29) may be rewritten,

$$\Delta G = \Delta G^{\ominus} + RT \ln \left(\frac{p_C p_D}{p_A p_B} \right) \qquad (2.31)$$

where p_C, p_D, p_A and p_B are the *actual* partial pressures and not those representing the equilibrium ratio.

These equations linking ΔG and K are all forms of the *van't Hoff isotherm* (also known as the reaction isotherm).

ΔG at a given temperature is made up of a *fixed term* ($\Delta G^{\ominus}$ or $-RT \ln K$) and a *variable term*, i.e.

$$RT \ln \left(\frac{a_C^c a_D^d}{a_A^a a_B^b} \right) \qquad \text{or} \qquad RT \ln \left(\frac{p_C p_D}{p_A p_B} \right)$$

The van't Hoff isotherm is extensively applied in the study of extraction processes and impurity removal from melts in Chapter 7. Here it will suffice simply to consider how equilibrium constants may be calculated from free energy data and vice versa.

Example

For the equilibrium

$$ZnO(s) + CO(g) \rightleftharpoons Zn(s/g) + CO_2(g)$$

The standard enthalpy and entropy changes at 300 K (zinc solid) and 1200 K (zinc gas) are as follows:

$$\Delta H_{300}^{\ominus} = +65.0 \text{ kJ mol}^{-1} \qquad \Delta H_{1200}^{\ominus} = +180.9 \text{ kJ mol}^{-1}$$
$$\Delta S_{300}^{\ominus} = +13.7 \text{ J K}^{-1} \text{ mol}^{-1} \qquad \Delta S_{1200}^{\ominus} = +288.6 \text{ J K}^{-1} \text{ mol}^{-1}$$

By calculation deduce the direction in which the reaction will be feasible at: (i) 300 K and (ii) 1200 K, when all reactants and products are in their standard states. Deduce K_p for the reaction at each temperature ($R = 8.314$ J K^{-1} mol^{-1}; $\ln x = 2.303 \log_{10} x$).

Using the relationship $\Delta G^{\ominus} = \Delta H^{\ominus} - T\Delta S^{\ominus}$, for the considered reaction:

(i)   at 300 K,

$$\Delta G_{300}^{\ominus} = 65\,000 - (300 \times 13.7)$$
$$= 60\,890 \text{ J mol}^{-1}$$

(ii)   at 1200 K,

$$\Delta G_{1200}^{\ominus} = 180\,900 - (1200 \times 288.6)$$
$$= -165\,420 \text{ J mol}^{-1}$$

Hence, the forward reaction is feasible at 1200 K ($\Delta G^{\ominus}$ negative) but not at 300 K ($\Delta G^{\ominus}$ positive).

K_p may be calculated from equation (2.30), i.e.

$$\ln K_p = -\Delta G^{\ominus}/RT$$

At 300 K,

$$\log_{10} K_p = \frac{-(60\ 890)}{2.303 \times 8.314 \times 300} = 10.60$$

Hence,

$$K_p = 2.512 \times 10^{-11}$$

At 1200 K,

$$\log_{10} K_p = \frac{-(-165\ 420)}{2.303 \times 8.314 \times 1200} = 7.197$$

Hence,

$$K_p = 1.574 \times 10^7$$

The relative magnitudes of the values of K_p at the different temperatures show that the reaction proceeds well to the right at 1200 K but negligibly at 300 K.

Calculations

In the following calculations assume $R = 8.314$ J K^{-1} mol^{-1} and $\ln x = 2.303 \log_{10} x$.

(1) Calculate K for the reaction:

$$Cu_2O(s) + H_2(g) \rightleftharpoons 2Cu(s) + H_2O(g)$$

at 1200 K given the following data:

$$\Delta H^{\ominus}_{1200} = -93.42 \text{ kJ} \qquad \Delta S^{\ominus}_{1200} = +29 \text{ J}$$

(2) If for the following reaction

$$3Fe(s) + 4H_2O(g) \rightleftharpoons Fe_3O_4(s) + 4H_2(g)$$

at 1173 K, $\Delta G^{\ominus} = -14.76$ kJ, calculate:

(a) the equilibrium constant at 1173 K,
(b) the equilibrium pressure of hydrogen if that of the water vapour is 6.58 $\times$ 10^2 Pa (0.0065 atm) at 1173 K.

NB: Use 0.0065 atm rather than 6.58 $\times$ 10^2 Pa since thermodynamic data and constants have been calculated with respect to the unit of atmospheres rather than Pa.

(3) Calculate K for the following reaction at 298 K and 1.01325 $\times$ 10^5 Pa (1 atm):

$$SnO(s) + CO_2(g) \longrightarrow SnO_2(s) + CO(g)$$

given the following data:

$$2Sn(s) + O_2(g) \longrightarrow 2SnO(s) \qquad \Delta G^{\ominus}_{298} = -514 \text{ kJ}$$
$$Sn(s) + O_2(g) \longrightarrow SnO_2(s) \qquad \Delta G^{\ominus}_{298} = -520 \text{ kJ}$$
$$C(s) + O_2(g) \longrightarrow CO_2(g) \qquad \Delta G^{\ominus}_{298} = -395 \text{ kJ}$$
$$2C(s) + O_2(g) \longrightarrow 2CO(g) \qquad \Delta G^{\ominus}_{298} = -274 \text{ kJ}$$

(4)   Deduce, by calculation of ΔG, whether the smelting of aluminium oxide, according to the reaction:

$$Al_2O_3(s) + 3C(s) \longrightarrow 2Al(l) + 3CO(g) \quad \Delta G^{\ominus}_{1773} = +150 \text{ kJ}$$

is feasible at 1500 $^{\circ}$C if a vacuum unit provides a p_{CO} value of 1×10^{-3} atm.

2.2.3.7 Equilibria in metallurgical processes

CO/CO₂ ratios

In many smelting or heat treatment processes the primary constituents of the gas mixture are CO and CO_2 as shown by the generalised equation:

$$M(s) + CO_2(g) \rightleftharpoons MO(s) + CO(g)$$

Since the object is isolation of the pure metal in the case of extraction or avoidance of the corrosion product (oxide), e.g. in bright annealing, it is desirable that the reverse reaction should be favoured. For the following reaction:

$$Zn(s) + CO_2(g) \rightleftharpoons ZnO(s) + CO(g)$$

in which zinc and zinc oxide are pure solids, and hence (by convention) possess unit activities, the equilibrium constant may be written:

$$K = \frac{a_{ZnO} p_{CO}}{a_{Zn} p_{CO_2}}$$

which becomes, $K = p_{CO}/p_{CO_2}$. If the actual p_{CO}/p_{CO_2} ratio exceeds the equilibrium ratio the reverse reaction (reduction of ZnO) will take place. Therefore, to avoid the oxidation of zinc the *actual ratio* of p_{CO}/p_{CO_2}, at a given temperature, must exceed the *equilibrium ratio* under the working conditions. For example, at 1373 K (1100 $^{\circ}$C), $\Delta G^{\ominus}$ for this reaction is -50 kJ.

Hence, applying equation (2.30) for equilibrium conditions:

$$\log_{10} K = \frac{-(50\ 000)}{8.314 \times 1373 \times 2.303} = 1.902$$

$$\therefore K = 17.98$$

Using a p_{CO}/p_{CO_2} ratio which exceeds 17.98 at 1373 K the oxidation product (ZnO) will not be formed.

Gaseous mixes of CO and CO_2 in the appropriate proportions are used to prevent oxidation of iron to iron (II) oxide in bright annealing. The same principle applies in smelting processes in which the temperature of the ore-reduction zone must be known for calculation of p_{CO}/p_{CO_2}.

Dissociation pressures or equilibrium partial pressures

Considering metal oxide formation of a divalent metal M:

$$2M + O_2 \longrightarrow 2MO$$

and again taking the pure metal and oxide to have unit activities: then at equilibrium

$$K = \frac{1}{p_{O_2}}$$

and by substitution in equation (2.30)

$$\Delta G^{\ominus}_T = -RT \ln \frac{1}{p_{O_2}}$$

which becomes

$$\Delta G_T^{\ominus} = RT \ln p_{O_2}$$

allowing the dissociation or equilibrium p_{O_2} values to be calculated:

Reactions (all at 1073 K)	$K_{1073}(1/p_{O_2})$	$p_{O_2}{}^*$ (atm) (since $K = 1/p_{O_2}$)	$\Delta G_{f,1073}^{\ominus}$ (kJ mol O_2)$^{-1}$
$2Pd(s) + O_2(g) \rightarrow 2PdO(s)$	10	10^{-1}	−16
$2Pb(l) + O_2(g) \rightarrow 2PbO(s)$	10^{12}	10^{-12}	−221
$\frac{4}{3}Cr(s) + O_2(g) \rightarrow \frac{2}{3}Cr_2O_3(s)$	10^{28}	10^{-28}	−556
$2Mg(s) + O_2(g) \rightarrow 2MgO(s)$	10^{40}	10^{-40}	−982

*These values are obtainable from the p_{O_2} grid on the $\Delta G^{\ominus} - T$ diagrams, see page 85

From this list the following conclusions may be formed:

(i) the higher the value of K, the more stable the oxide (as reflected in the corresponding value of $\Delta G_f^{\ominus}$).

(ii) the values of p_{O_2}, which are the reciprocal of K represent the so called dissociation pressure or equilibrium p_{O_2} of the metal oxide: that is the pressure of oxygen below which the oxide will decompose and above which the metal will oxidise.

For example, PdO decomposes at pressures below 10^{-1} atm and thus is a relatively unstable oxide easily decomposed to its elements. On the other hand, MgO is stable down to the effectively unattainable pressure of 10^{-40} atm. These considerations are of importance in studies of conditions required for the vacuum processing of metals. Obviously similar deductions may be made with respect to the stabilities of other compounds, e.g. sulphides, halides, carbonates.

Calculations

(1) Calculate the dissociation pressure of: (a) $Al_2O_3(s)$, (b) $FeO(s)$, at 1873 K (1600 °C) given that for

$$\frac{4}{3}Al(l) + O_2(g) \longrightarrow \frac{2}{3}Al_2O_3(s) \qquad \Delta G_{1873}^{\ominus} = -718 \text{ kJ}$$

$$2Fe(l) + O_2(g) \longrightarrow 2FeO(l) \qquad \Delta G_{1873}^{\ominus} = -287 \text{ kJ}$$

(2) Calculate $\Delta G^{\ominus}$ for the formation of TiO_2 at 1373 K (1100 °C) given that its dissociation pressure at this temperature is 10^{-26} atm.

2.2.3.8 The effect of temperature on equilibrium

The variation of equilibrium constant with temperature is described by an equation known as the *van't Hoff isochore* (1884). In fact the term 'isochore' which refers to equal volumes, is a misnomer arising because van't Hoff originally considered constant volume conditions. The relationship is hence often called the *van't Hoff equation* although the original term persists through familiar use. The isochore is derived by combining the van't Hoff isotherm (equation (2.30)) and the Gibbs–Helmholtz equation (equation (2.25)). Differentiating equation (2.30) with respect to temperature yields,

$$\frac{d(\Delta G^{\ominus})}{dT} = -R \ln K_p - RT \frac{d(\ln K_p)}{dT}$$

Multiplying both sides by T gives

$$T \frac{d(\Delta G^{\ominus})}{dT} = -RT \ln K_p - RT^2 \frac{d(\ln K_p)}{dT}$$

and substitution of

$$T \frac{d(\Delta G^{\ominus})}{dT} = \Delta G^{\ominus} - \Delta H^{\ominus} \qquad \text{and} \qquad -RT \ln K_p = \Delta G^{\ominus}$$

in the above equation produces:

$$\Delta G^{\ominus} - \Delta H^{\ominus} = \Delta G^{\ominus} - RT^2 \frac{d(\ln K_p)}{dT}$$

and hence rearranging, gives the van't Hoff isochore

$$\frac{d(\ln K_p)}{dT} = \frac{\Delta H^{\ominus}}{RT^2} \qquad (2.32)$$

The isochore is generally more useful in the integrated form, i.e.

$$\ln K_p = -\frac{\Delta H^{\ominus}}{RT} + \text{constant} \qquad (2.33)$$

The following points are noteworthy concerning the use of the isochore:

(1) Equation (2.33) is of straight line form ($y = mx + c$) and if $\ln K_p = y$ and $1/T = x$, a plot of $\ln K_p$ versus $1/T$ should be a straight line of slope $-\Delta H^{\ominus}/R$ and the y-axis intercept will be the constant of integration. The integrated isochore is also a quantitative justification of Le Chatelier's principle since it is apparent that: (a) if the reaction is endothermic ($\Delta H^{\ominus}$ positive) K_p increases with increasing temperature, i.e. forward reaction favoured; (b) if the reaction is exothermic ($\Delta H^{\ominus}$ negative) K_p decreases with increasing temperature, i.e. reverse reaction favoured (see Section 2.2.3).

  The integrated isochore provides another method for determination of $\Delta H^{\ominus}$ values. Deviations from linearity of $\ln K_p$ versus $1/T$ arise from the variation of $\Delta H^{\ominus}$ with temperature. Over narrow temperature ranges $\Delta H^{\ominus}$ may be taken to be approximately constant for the range considered. If departure from constancy is large or accuracy is required Kirchoff's equation (2.5) and the empirical heat capacity equations stated earlier (equation (2.4)) can be used to deduce the temperature dependence of $\Delta H^{\ominus}$.

(2) A generally more useful form of the integrated isochore is obtained when integration is between limits of temperature T_1 and T_2. The isochore now becomes:

$$\ln \left(\frac{(K_p)_2}{(K_p)_1} \right) = -\frac{\Delta H^{\ominus}}{R} \left(\frac{1}{T_2} - \frac{1}{T_1} \right)$$

or   (2.34)

$$\ln \left(\frac{(K_p)_2}{(K_p)_1} \right) = \frac{\Delta H^{\ominus}}{R} \left(\frac{1}{T_1} - \frac{1}{T_2} \right)$$

where $(K_p)_1$ is the value of the equilibrium constant at temperature T_1, and $(K_p)_2$ is the value at temperature T_2.[*]

---

[*]$\ln [(K_p)_2/(K_p)_1]$ may be written as $\ln (K_p)_2 - \ln(K_p)_1$, e.g. $\ln(K_p)_T - \ln(K_p)_{298}$.

Hence, given the value of the equilibrium constant, K_p, at T_1 — say 298 K (since at this temperature $(K_p)_1$ is readily calculated from data book information) — then $(K_p)_2$ is easily deduced using equation (2.34).

Example

Calculate the equilibrium constant, K, at 773 K (500 °C), for the reaction:

$$NiO(s) + H_2(g) \longrightarrow Ni(s) + H_2O(g)$$

given the following information:

	NiO(s)	H$_2$(g)	Ni(s)	H$_2$O(l)	H$_2$O(g)
$\Delta G^{\ominus}_{f,298}$ (kJ mol^{-1})	−216	0	0	−237	−
$\Delta H^{\ominus}_{f,298}$ (kJ mol^{-1})	−241	0	0	−286	−
$(C^{\ominus}_p)_{298 - 773}$ (J K^{-1} mol^{-1})	44.4	28.8	26.0	−	33.6

K_{773} may be calculated by use of equation (2.34) in the form:

$$\log_{10} K_{773} - \log_{10} K_{298} = -\frac{\Delta H^{\ominus}_{773}}{R}\left(\frac{1}{T_2} - \frac{1}{T_1}\right)$$

Before substitution in this equation it is necessary to calculate: (i) K_{298} (or $\ln K_{298}$); (ii) $\Delta H^{\ominus}_{298}$; (iii) $\Delta H^{\ominus}_{773}$ (via ΔC_p); then (iv) K_{773} can be calculated.

(i)

$$NiO(s) + H_2(g) \longrightarrow Ni(s) + H_2O(l)$$

$\Delta G^{\ominus}_{f,298}$ (kJ)   −216   0   0   −237

From equation (2.23)

$$\Delta G^{\ominus}_{298} = G^{\ominus}_{\text{prods, 298}} - G^{\ominus}_{\text{reacts, 298}}$$

i.e.

$$\Delta G^{\ominus}_{298} = (-237) - (-216)$$
$$= -21 \text{ kJ}$$

From equation (2.30), i.e. $\Delta G^{\ominus} = RT \ln K$

$$\ln K_{298} = \frac{\Delta G^{\ominus}_{298}}{RT} = \frac{-(-21000)}{8.314 \times 298} = 8.476$$

(ii)

$$NiO(s) + H_2(g) \longrightarrow Ni(s) + H_2O(l)$$

$\Delta H^{\ominus}_{f,298}$ (kJ mol^{-1})   −241   0   0   −286

From equation (2.1)

$$\Delta H^{\ominus}_{298} = H^{\ominus}_{\text{prods, 298}} - H^{\ominus}_{\text{reacts, 298}}$$

i.e.,

$$\Delta H^{\ominus}_{298} = -286 - (-241)$$
$$= -45 \text{ kJ}$$

(iii)

$$NiO(s) + H_2(g) \longrightarrow Ni(s) + H_2O(g)$$

$(C^{\ominus}_p)_{298-773} \ (\text{J K}^{-1} \text{ mol}^{-1}) \quad 44.4 \quad 28.8 \quad 26.0 \quad 33.6$

From equation (2.7)

$$\Delta C^{\ominus}_p = \Sigma C^{\ominus}_{p, \text{ prods}} - \Sigma C^{\ominus}_{p, \text{ reacts}}$$
$$= (33.6 + 26.0) - (44.4 + 28.8)$$
$$= -13.6 \text{ J}$$

This value of ΔC_p can now be substituted into the simplified Kirchoff equation (2.6):

$$\Delta H^{\ominus}_T = \Delta H^{\ominus}_{298} + \Delta C_p(T_2 - T_1) \pm \text{ latent heats}$$

and so

$$\Delta H^{\ominus}_{773} = -45000 + (-13.6)(773-298) + 41100$$
$$= -10360 \text{ J}$$

where the latent heat of evaporation of water is 41 100 J mol^{-1}.

(iv)   From equation (2.34)

$$\ln K_{773} = \ln K_{298} - \frac{\Delta H^{\ominus}_{773}}{R} \left(\frac{1}{T_2} - \frac{1}{T_1} \right)$$

and substitution from (i) and (iii) gives:

$$\ln K_{773} = 8.476 - \left(\frac{-10360}{8.314} \right) \left(\frac{1}{773} - \frac{1}{298} \right)$$
$$= 8.476 + 1246(-0.00206)$$
$$= 5.909$$

so that

$$K_{773} = 367 = 3.67 \times 10^2$$

2.2.3.9   Variation of vapour pressure with temperature

The metallurgical importance of variation of vapour pressure of a substance with temperature concerns the loss of constituents from melts, e.g. manganese from steel during vacuum melting, and in distillation metallurgy. The isochore (equation 2.32) may be developed to express the temperature dependence of vapour pressure. Consider the evaporation of zinc, such that the liquid has attained equilibrium with its vapour at a given temperature,

$$Zn(l) \rightleftharpoons Zn(g)$$

For this reaction it is apparent that:

(a) $\Delta H^{\ominus}$ for the reaction = latent heat of evaporation for zinc, L_e;
(b)   the equilibrium constant for the reaction may be expressed:

$$K = \frac{p_{Zn(g)}}{a_{Zn(l)}}$$

and since $a_{Zn(l)} = 1$ if the zinc is pure,

$$K_p = \overset{\circ}{p}_{Zn}$$

or, for any liquid, $K_p = \overset{\circ}{p}$, where $\overset{\circ}{p}$ is the vapour pressure of the pure liquid metal.
Substituting $\overset{\circ}{p}$ for K_p, and L_e for $\Delta H^{\ominus}$ in equation (2.32)

$$\frac{d(\ln \overset{\circ}{p})}{dT} = \frac{L_e}{RT^2} \tag{2.35}$$

This is known as the *Clausius–Clapeyron* equation. On integration this becomes:

$$\ln \overset{\circ}{p} = -\frac{L_e}{RT} + \text{constant} \tag{2.36}$$

The Clausius–Clapeyron equation may be utilised as follows:

(i) A plot of $\ln \overset{\circ}{p}$ versus $1/T$ yields a straight line graph of slope $-L_e/R$. This allows determination of molar latent heat of evaporation, subject to its constancy with temperature (as indicated for $\Delta H^{\ominus}$ in the isochore).

(ii) If integration of the Clausius–Clapeyron equation is between the limits T_1 and T_2 then:

$$\ln \left(\frac{\overset{\circ}{p}_2}{\overset{\circ}{p}_1} \right) = -\frac{L_e}{R} \left(\frac{1}{T_2} - \frac{1}{T_1} \right) \tag{2.37*}$$

$$= \frac{L_e}{R} \left(\frac{1}{T_1} - \frac{1}{T_2} \right)$$

where $\overset{\circ}{p}_1$ is the vapour pressure of the pure liquid at temperature T_1 and $\overset{\circ}{p}_2$ is its vapour pressure at temperature T_2. If temperature T_1 is taken as the boiling point of the liquid then $\overset{\circ}{p}_1 = 1$ atm, since the vapour pressure of any liquid at its boiling point equals one atmosphere (see *Fig. 7.9*). Thus equation (2.37) becomes, since $\ln \overset{\circ}{p}_1 = 0$,

$$\ln \overset{\circ}{p}_2 = -\frac{L_e}{R} \left(\frac{1}{T_2} - \frac{1}{T_b} \right)$$

where T_b is the boiling point of the pure liquid. Hence the vapour pressure of the liquid, $\overset{\circ}{p}_2$, at any other temperature can be conveniently calculated as can the variation of boiling point with change in vapour pressure.

The Clausius–Clapeyron equation can also be applied to other phase change processes such as melting, sublimation and precipitation in each case suitably modifying $\overset{\circ}{p}$ and L_e to suit the considered transformation. The equation can also be used to deduce boiling point elevation when a solute is dissolved in a liquid.

Example

Calculate the vapour pressure of pure lead at 1273 K (1000 °C) given that:

Boiling point of lead = 2017 K

Latent heat of evaporation for lead = 177 kJ mol^{-1}

(assumed constant in the range 1273–2017 K). From the integrated Clausius–Clapeyron equation (2.37)

* $\ln [(\overset{\circ}{p}_2)/(\overset{\circ}{p}_1)]$ may be written as $\ln \overset{\circ}{p}_2 - \ln \overset{\circ}{p}_1$, e.g. $\ln \overset{\circ}{p}_T - \ln \overset{\circ}{p}_{\text{boiling point}}$.

$$\ln\left(\frac{\overset{o}{p}_{1273}}{\overset{o}{p}_{2017}}\right) = -\frac{177\ 000}{8.314}\left(\frac{1}{1273} - \frac{1}{2017}\right) \tag{2.38}$$

but $\overset{o}{p}_{2017} = 1$ atm so

$$\ln \overset{o}{p}_{1273} = -21289 \times 0.00029$$

Hence
$$= 6.1738$$

$$\overset{o}{p}_{Pb,1273} = 0.00208 \text{ atm}$$

Notes

(i)   The result indicates that there would be a considerable loss of lead due to volatilisation on vacuum processing liquid lead since vacuums of 10^{-4} to 10^{-6} atm
are attainable in practice. In fact impure lead may be refined by vacuum
distillation techniques (see Section 7.2.7).

(ii)   For accurate determinations L_e should be calculated for the temperature
considered.

(iii) If the metal is a constituent of an alloy calculations of the actual pressure are
determined using equation (4.8).

Calculations

In the following calculations assume $R = 8.314$ J K^{-1} mol^{-1} and $\ln x = 2.303 \log_{10} x$.

(1)   Calculate the equilibrium constant at 600 K for the reaction:

$$\text{Hg(g)} + \tfrac{1}{2}O_2(g) \longrightarrow \text{HgO(s)}$$

given that $K_{298} = 1.72 \times 10^{10}$ and the standard enthalpy change for the reaction,
$\Delta H^{\ominus}_{600}$, is 48.6 kJ.

(2)   The following table gives the equilibrium constant, K_p, at different temperatures
for the oxidation of metal oxide, MO, according to:

$$2MO(s) + O_2(g) \longrightarrow 2MO_2(s)$$

T (K)	600	700	800	900	1000
K_p (atm^{-1})	140	5.14	0.437	0.0625	0.0131

Plot $10^3/T$ against $\ln$(or $\log_{10}$) K_p and obtain a value for $\Delta H^{\ominus}$ from the slope of the
graph.

(3)   The boiling point of manganese under normal conditions is 2097 °C. Calculate
the vapour pressure of manganese at 1600 °C given that its latent heat of evaporation
is 225 kJ mol^{-1}.

2.2.3.10 Standard free energy–temperature diagrams: application to metal extraction

Metal ores include oxides, sulphides, carbonates and halides; extraction is primarily
from oxides to which sulphides and carbonates are easily converted by roasting or calcination (Sections 7.2.2 and 7.2.1). Reduction of the oxide may be achieved by chemical
or electrolytic reduction techniques as discussed in Sections 7.2.5 and 7.4.
    The feasibility of these or, indeed, any reactions, can be gauged by determining $\Delta G^{\ominus}$
for the reaction (which must be negative) and K (which must be reasonably large).
This is best achieved by a study of standard free energy–temperature diagrams (or

Ellingham diagrams, after H.J.T. Ellingham who constructed the original diagrams) which are plots of the standard free energy of formation of oxides (or sulphides, etc.) against temperature and hence conveniently allow $\Delta G^{\ominus}$ to be read off directly at any required temperature. The diagrams have the great advantage of conveying visually information which is more difficult to interpret from a series of separate calculations.

Construction of $\Delta G^{\ominus}-T$ diagrams

A fairly comprehensive $\Delta G^{\ominus}-T$ diagram for oxide formation is shown in *Fig. 2.11* but explanation is facilitated by first considering a simplified diagram (*Fig. 2.10*) involving only four lines representing oxide formation, namely for CO, CO_2, Al_2O_3 and ZnO.

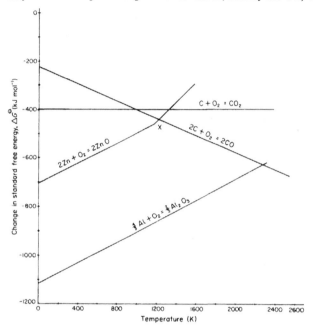

Figure 2.10 $\Delta G^{\ominus}-T$ diagram for formation of various oxides

It was indicated in Section 2.2.2.2 that equation (2.22) is of straight line form ($y = mx + c$) such that when $\Delta G^{\ominus}$ is plotted against T the slope is $-\Delta S^{\ominus}$, and since $\Delta S^{\ominus}$ for a reaction can be positive, negative or zero the corresponding slopes will be negative, positive or zero respectively. The intercept on the y-axis is $\Delta H_0^{\ominus}$.

Since oxygen is the common reactant and, by definition, $\Delta G^{\ominus}$ applies to one mole of reactant (or product) and a condition of one atmosphere pressure, then the diagrams are standardised on reaction of the element with *one mole of oxygen* gas. This simplifies the situation for comparative purposes.

The reactions between solid metals and oxygen are normally accompanied by negative $\Delta H^{\ominus}$ and negative $\Delta S^{\ominus}$ values since most oxidation processes are exothermic with one mole of oxygen being consumed. It is perhaps surprising that the $\Delta G^{\ominus}-T$ lines are straight since neither $\Delta H^{\ominus}$ nor $\Delta S^{\ominus}$ are independent of temperature. However, it is reasonable to base plots on approximate constancy with temperature since for 100 K rise in temperature $\Delta H^{\ominus}$ varies roughly by 1% and $\Delta S^{\ominus}$ by 5% and in such a way that the variations tend to cancel rather than reinforce each other. This means that within the accuracies required the $\Delta G^{\ominus}-T$ curves are virtually straight lines, their slopes remaining constant except when transformation occurs (see below). It is now useful to consider the reason for the varying slopes of the lines representing oxide formation in *Fig. 2.10.*

(i) Al₂O₃ formation:

$$\tfrac{4}{3}Al(s) \ + \ O_2(g) \longrightarrow \tfrac{2}{3}Al_2O_3(s)$$

For this reaction, as for all others involving metal oxide formation, the standard entropy change, $\Delta S^{\ominus}$, is negative, since during the reaction one mole of oxygen gas is used up (i.e. gaseous volume changes: $1 \to 0$ volumes). Hence:

$$\text{Entropy change} \ = \ -\Delta S^{\ominus}$$

$$\text{Slope of line} \ = \ -(-\Delta S^{\ominus}) \ = \ +\Delta S^{\ominus}$$

i.e. positive. It can thus be seen that in plots of the standard free energy change of formation for metal oxides against temperature the invariable $-\Delta S^{\ominus}$ value for the reaction gives rise to a positive slope for the resulting graph.

(ii) ZnO formation: As indicated in (i) the slope of this line is positive but there is a sudden and appreciable change in slope at the point (X) which corresponds, on the T-axis, to the boiling point of zinc (1180 K). At higher temperatures the standard entropy change for the reaction becomes even more negative, this being apparent from the equations:

below 1180 K $\quad 2Zn(s \text{ or } l) + O_2(g) \longrightarrow 2ZnO(s) \quad (1 \to 0 \text{ volumes of gas})$

above 1180 K $\quad 2Zn(g) + O_2(g) \longrightarrow 2ZnO(s) \quad (3 \to 0 \text{ volumes of gas})$

A more positive slope naturally follows this more markedly negative value of $\Delta S^{\ominus}$ for the reaction at temperatures in excess of 1180 K. Similar slope variations are evident for other metals at their appropriate boiling points, these points being designated on the graph according to the particular key used for each $\Delta G^{\ominus}-T$ diagram. Although corresponding entropy change effects occur at the melting points of the metals, their magnitude is less significant producing less marked alterations in slope.

Gradient variations will also take place as a consequence of the following transformations (see *Fig. 2.11*):

(a) allotropic modifications of metal, e.g. $Ti_{(\alpha)} \longrightarrow Ti_{(\beta)}$ (880 °C);
(b) melting point of oxide, e.g. PbO (890 °C);
(c) sublimation point of element or oxide, e.g. S (540 °C); Li_2O (1300 °C);
(d) boiling point of oxide, e.g. PbO (1470 °C);
(e) allotropic modification of oxide, e.g. SiO_2 (870 °C; 1470 °C).

By considering the volume change for each transition it is straightforward to deduce the effect of such a change on $\Delta S^{\ominus}$ for the reaction and hence, in the manner shown above, on the geometry of the line.

(iii) CO₂ formation:

$$C(s) \ + \ O_2(g) \longrightarrow CO_2(g)$$

During this reaction the gaseous volume remains unchanged at one volume throughout. As a result $\Delta S^{\ominus}$ for the reaction is zero or negligibly small in value and the line representing carbon dioxide formation is approximately parallel to the temperature axis, i.e. $\Delta G_f^{\ominus}$ for $CO_2(g)$ does not vary with change in temperature.

(iv) CO formation: The slope of the line representing carbon monoxide formation is

exceptional since the gaseous volume increases ($1 \rightarrow 2$ volumes) during the reaction as shown:

$$2C(s) + O_2(g) \longrightarrow 2CO(g)$$

This means that the standard entropy change for the reaction is positive: so

$$\text{Slope} = -(+\Delta S^\ominus) = -\Delta S^\ominus$$

i.e. negative. Thus the slope of this reaction line is negative and illustrates the increasing stability of carbon monoxide with increasing temperature. Since $\Delta G_f^\ominus$ for carbon monoxide becomes more negative with temperature rise, carbon has an increasing affinity for oxygen as temperature increases indicating the increasing effectiveness of carbon as a reducing agent for oxides at high temperatures.

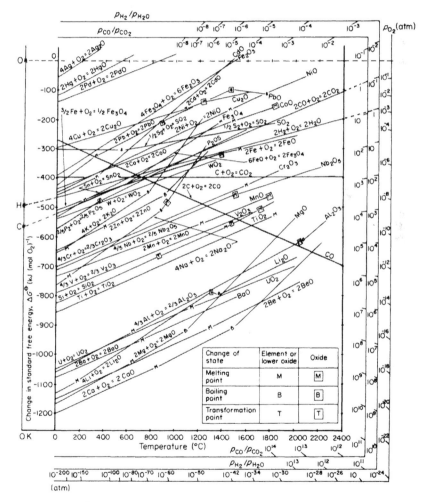

Figure 2.11 $\Delta G^\ominus - T$ diagram for the formation of oxides. For equilibrium p_{O_2}, p_{CO}/p_{CO_2} and p_{H_2}/p_{H_2O} values use points O, C and H respectively on the $\Delta G^\ominus$ axis at 0 K

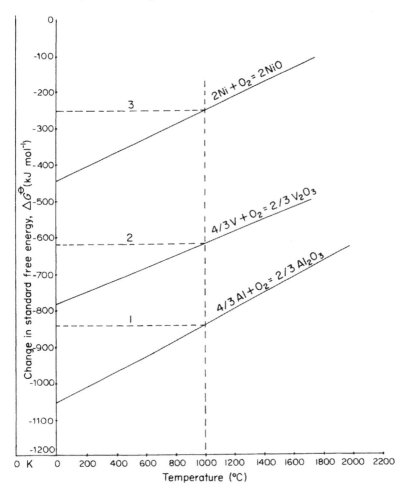

Figure 2.12 $\Delta G^{\ominus}-T$ diagram for three selected reactions

Use of $\Delta G^{\ominus}-T$ diagrams

The following discussion refers to $\Delta G^{\ominus}-T$ oxide formation diagrams shown in *Figs 2.11* and *2.12*.

(i) Within the temperature range considered (see *Fig 2.12*), i.e. 0 to 2200 °C (273–2473 K), $\Delta G_f^{\ominus}$ for oxides is invariably negative, so that oxide formation is feasible at all temperatures and, conversely, decomposition of the oxide is not feasible. However, if the reaction temperature is sufficient for the oxide formation curve to cross the horizontal $\Delta G^{\ominus} = 0$ line the oxide will spontaneously decompose. At $\Delta G^{\ominus} = 0$ there is an equal tendency for the metal to oxidise or the oxide to decompose. The following reaction lines (see *Fig. 2.11*) intersect the $\Delta G^{\ominus} = 0$ line at the temperatures stated:

$$4Ag(s) + O_2(g) \longrightarrow 2Ag_2O(s) \qquad \text{at } 200\ ^\circ C$$

$$2Pd(s) + O_2(g) \longrightarrow 2PdO(s) \qquad \text{at } 900\ ^\circ C$$

$$2Ni(l) + O_2(g) \longrightarrow 2NiO(l) \qquad \text{at } 2400\ ^\circ C$$

Above each of these temperatures the overall tendency is in favour of the reverse reaction, i.e. reduction (decomposition) since the oxide is less stable than the metal (positive value of $\Delta G^\ominus$ for the oxide formation reaction), while oxidation of the metal will occur at any temperature below it. The temperature at which the oxide formation line crosses the $\Delta G^\ominus = 0$ line is called the *decomposition temperature, T_D*.

(ii) An element can reduce the oxide of any element with a $\Delta G_f^\ominus$ line appearing above it at a given temperature. The distance between the two lines considered for the redox reaction at the given temperature represents the standard free energy change for the reaction.

Referring to a simplified diagram (*Fig. 2.12*) showing only three oxide formation lines:

(a) Since the line representing Al_2O_3 formation lies below that for V_2O_3 formation, elementary aluminium can reduce V_2O_3, i.e.

$$\tfrac{4}{3}Al + \tfrac{2}{3}V_2O_3 \longrightarrow \tfrac{4}{3}V + \tfrac{2}{3}Al_2O_3$$

$\Delta G^\ominus$ for the reaction is obtained directly in kilojoules as the vertical distance between the two lines at the temperature required, e.g. at 1000 $^\circ C$

$$\Delta G_{1273}^\ominus = (1) - (2)$$

(where (1) and (2) are the $\Delta G_f^\ominus$ values for Al_2O_3 and V_2O_3 respectively at 1000 $^\circ C$), i.e.

$$\Delta G_f^\ominus = (-846) - (-616) = -230 \text{ kJ}$$

i.e. $\Delta G^\ominus$ for the reduction of vanadium (III) oxide by aluminium at 1000 $^\circ C$. The reasoning behind this may be more fully understood by considering the individual oxide formation reactions contributing to the overall reaction:

$$\tfrac{2}{3}V_2O_3 \longrightarrow \tfrac{4}{3}V + O_2 \qquad \Delta G_{1273}^\ominus = +616 \text{ kJ}$$

$$\tfrac{4}{3}Al + O_2 \longrightarrow \tfrac{2}{3}Al_2O_3 \qquad \Delta G_{1273}^\ominus = -846 \text{ kJ}$$

$$\overline{\tfrac{4}{3}Al + \tfrac{2}{3}V_2O_3 \longrightarrow \tfrac{4}{3}V + \tfrac{2}{3}Al_2O_3 \quad \Delta G_{1273}^\ominus = -230 \text{ kJ}}$$

All $\Delta G^\ominus$ values are in kilojoules per mole of oxygen in accordance with the construction of the diagram. Thus the above reaction's $\Delta G^\ominus$ value assumes the presence of one mole of oxygen.

(b) Similarly reduction of NiO by vanadium will be accompanied by a standard free energy change at 1000 $^\circ C$ deduced as follows:

$$\Delta G_{1273}^\ominus = (2) - (3)$$

$$2NiO \longrightarrow 2Ni + O_2 \qquad \Delta G_{1273}^\ominus = +260 \text{ kJ}$$

$$\tfrac{4}{3}V + O_2 \longrightarrow \tfrac{2}{3}V_2O_3 \qquad \Delta G_{1273}^\ominus = -616 \text{ kJ}$$

$$\overline{\tfrac{4}{3}V + 2NiO \longrightarrow 2Ni + \tfrac{2}{3}V_2O_3 \quad \Delta G_{1273}^\ominus = -356 \text{ kJ}}$$

(c)   Logically it follows that the reduction of NiO by Al at 1000 °C is accompanied by a standard free energy change equal to the sum of the $\Delta G^{\ominus}$ values obtained in (a) and (b), i.e.

$$\Delta G^{\ominus}_{1273} = (-230) + (-356) = -586 \text{ kJ (mol } O_2)^{-1}$$

(iii)   The top third of the full diagram shown in *Fig. 2.11* comprises the lines of the less reactive metals bounded by the H_2/H_2O line indicating that the former are reducible by hydrogen. The middle third contains less easily reducible elements such as Si, Mn, Cr; the higher relative stabilities of these oxides making these elements important deoxidisers in steel melts. The lower third of the diagram is occupied by elements forming very stable oxides, e.g. Al, Ca, Mg, which are difficult to reduce and find application as refractories.

(iv)   Where oxide formation lines intersect at a particular temperature it follows that the relative reducing power of the elements concerned reverse with respect to one another. The following examples illustrate the point and refer to the detailed $\Delta G^{\ominus}-T$ diagram (*Fig. 2.11*).

(a)   Magnesium will reduce aluminium oxide at all temperatures below the point of intersection of the MgO line with the Al_2O_3 line, i.e. 1550 °C. At temperatures above this aluminium will reduce magnesium oxide; $\Delta G^{\ominus}$ for the reaction becoming increasingly more negative with rise in temperature.

(b)   The products of reduction processes using carbon vary with increase in temperature, i.e.

$$C(s) + O_2(g) \longrightarrow CO_2(g) \tag{I}$$

$$2C(s) + O_2(g) \longrightarrow 2CO(g) \tag{II}$$

Examination of *Fig. 2.10* shows that the lines for reactions (I) and (II) intersect at 973 K, at lower temperatures the line for (I) being below that for (II) so that CO_2 will be more stable and thus predominates in the reduction reaction products. For example, at 773 K (*Fig. 2.11*)

$$2NiO + C \longrightarrow 2Ni + CO_2 \qquad \Delta G^{\ominus}_{773} = -38 \text{ kJ}$$

$$NiO + C \longrightarrow Ni + CO \qquad \Delta G^{\ominus}_{773} = -6 \text{ kJ}$$

but at 1473 K the line (II) is lower such that:

$$2NiO + C \longrightarrow 2Ni + CO_2 \qquad \Delta G^{\ominus}_{1473} = -180 \text{ kJ}$$

$$NiO + C \longrightarrow Ni + CO \qquad \Delta G^{\ominus}_{1473} = -263 \text{ kJ}$$

showing that in this case CO is more stable than CO_2 and is hence more likely to form at this temperature. (For 'direct' and 'indirect' carbon reduction, see Section 7.2.5).

The increasing stability of carbon monoxide (more negative $\Delta G^{\ominus}_f$) with increasing temperature is such that it will reduce all oxides appearing on the diagram, even such refractory oxides as MgO, CaO and Al_2O_3 providing the temperature is sufficiently high (1850 °C, 2150 °C and 2000 °C respectively).

The extraction of these reactive elements by smelting methods is, however, precluded by the cost of running plant at such elevated temperatures and difficulty of preventing impurity pick-up (e.g. carbon, oxygen) by the melts. The extraction methods employed in practice are considered in detail in Chapter 7.

(v)   Standard free energy—temperature diagrams may be equally well constructed though rather less extensively used for any other type of compound formation reaction; compounds of interest in this context include sulphides, chlorides and carbonates. A common reactant is used for each reaction, e.g. S_2, Cl_2, CO_2 and one mole of this

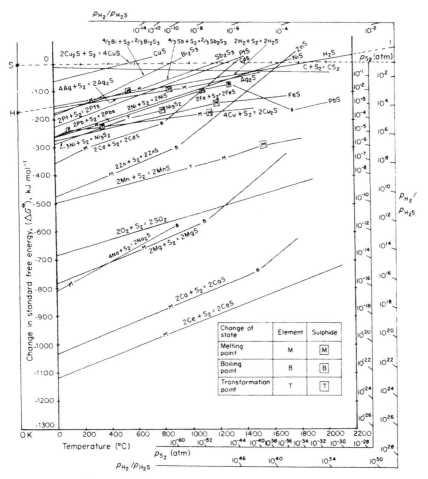

Figure 2.13 $\Delta G^{\ominus} - T$ diagram for sulphide formation. For calculation of equilibrium p_{S_2} and equilibrium p_{H_2}/p_{H_2S} use points S and H respectively on the $\Delta G^{\ominus}$ axis at 0 K

reactant is always considered. An example of such a diagram is *Fig. 2.13* representing sulphide formation (see also *Figs 7.6* and *7.10*).

Nomographic scales on $\Delta G^{\ominus}$-T diagrams

The diagrams already described can be made more useful by superimposing grids or nomographic scales around them.

(a) The p_{O_2} grid: Considering the reaction involving the oxidation of a pure solid metal to produce a pure solid oxide as:

$$2M(s) + O_2(g) \longrightarrow 2MO(s)$$

equilibrium will be established between MO, M and O_2 at a certain value (the equilibrium value) of p_{O_2}. This can be calculated from the van't Hoff isotherm (equation (2.30)). Since $a_M = 1 = a_{MO}$ (see Section 2.2.3.7), thus

$$\Delta G^{\ominus} = -RT \log_{10} \left(\frac{1}{p_{O_2}} \right)$$

$$= RT \log_{10} p_{O_2}$$

$$= 2.303\, RT \log_{10} p_{O_2} \tag{2.39}$$

Hence knowing $\Delta G^{\ominus}$ and T, the corresponding equilibrium p_{O_2} can be calculated for any such reaction.

A nomographic scale can be constructed around the $\Delta G^{\ominus} - T$ diagram from which the value of $\log_{10} p_{O_2}$ may be deduced. As seen from equation (2.39), when $\Delta G^{\ominus} = 0$, $p_{O_2} = 1$ atm and the equilibrium p_{O_2} values radiate from the point 'O' on the $\Delta G^{\ominus}$ axis ($\Delta G^{\ominus} = 0$ at 0 K). Therefore, to find an equilibrium p_{O_2} value for a metal–metal oxide reaction at a certain temperature a straight line is drawn from the point 'O' to the appropriate reaction line at the required temperature, T, giving the $\log_{10} p_{O_2}$ value where the projected line cuts the nomograph (see *Fig. 2.14*).

Since the p_{O_2} scale is a logarithmic scale the equilibrium p_{O_2} value is obtained from the antilogarithm of the scale reading. This method of obtaining equilibrium p_{O_2} values is more rapid but usually less accurate than performing the corresponding numerical calculation. If the *ambient* p_{O_2} value is greater than the equilibrium p_{O_2} value at a

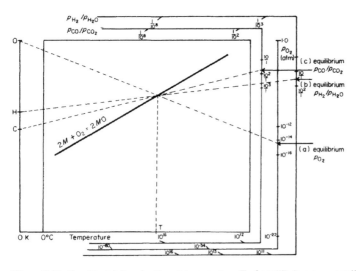

Figure 2.14 Graphical determination at temperature T of equilibrium p_{O_2}, equilibrium p_{H_2}/p_{H_2O} ratio, equilibrium p_{CO}/p_{CO_2} ratio for the reactions: (a) $2M + O_2 \rightarrow 2MO$; (b) $MO + H_2 \rightarrow M + H_2O$ and (c) $MO + CO \rightarrow M + CO_2$ respectively, using the nomographic scales around the $\Delta G^{\ominus} - T$ diagrams. Thus, the equilibrium p_{O_2} for reaction (a) = antilog of scale reading at (a) = antilog of 0.7×10^{-15} atm (i.e. antilog $\overline{15}.7$) = 5.0×10^{-15} atm; the equilibrium p_{H_2}/p_{H_2O} ratio for reaction (b) = antilog of scale reading at (b) = antilog of 0.4×10^1 (i.e. antilog 1.4) = 25.1; the equilibrium p_{CO}/p_{CO_2} ratio for reaction (c) = antilog of $0.7 \times 10/1$ (i.e. antilog of 1.7) = 50.1.

Notes
(1) In each determination the reaction line $2M + O_2 \rightarrow 2MO$ is used irrespective of the fact that in reactions (b) and (c) the *metal oxide* is being reduced with H_2, and CO respectively.
(2) The nomographic scales are read from *low to high values*, i.e. p_{O_2} (reaction (a)) is antilog of 0.7×10^{-15} atm *not* 0.3×10^{-14} atm.
(3) Determinations of equilibrium values from nomographic scales will vary in accuracy depending on the accuracy of the reaction lines and each individual's interpretation of the scale reading.
(4) Equilibrium p_{S_2} and p_{H_2}/p_{H_2S} can be obtained in a similar way from *Fig. 2.13*.

given temperature oxidation of M occurs. Conversely, if the ambient p_{O_2} value is less than the equilibrium p_{O_2} value, dissociation of MO takes place, so that, logically, the equilibrium p_{O_2} value is sometimes called the *dissociation pressure* of the metal oxide (see Section 2.2.3.7). In conclusion: low equilibrium p_{O_2} values favour more ready formation of the oxide whilst high equilibrium p_{O_2} values lead to greater ease of dissociation of MO.

(b) The CO/CO₂ grid: In a similar fashion the equilibrium p_{CO}/p_{CO_2} ratio can be deduced for the reduction of metal oxide, MO, with carbon monoxide according to:

$$MO(s) + CO(g) \rightleftharpoons M(s) + CO_2(g)$$

Explanation is made easier by considering this as a two stage reaction as follows:

$$2M + O_2 \longrightarrow 2MO \qquad \Delta G_I^{\ominus} \qquad \text{(I)}$$

$$2CO + O_2 \longrightarrow 2CO_2 \qquad \Delta G_{II}^{\ominus} \qquad \text{(II)}$$

Subtracting (I) from (II) gives

$$2MO + 2CO \longrightarrow 2M + 2CO_2 \qquad \Delta G_{II}^{\ominus} - \Delta G_I^{\ominus} \text{ kJ (mol } O_2)^{-1} \qquad \text{(III)}$$

or $\qquad MO + CO \longrightarrow M + CO_2$

If MO and M are pure solids and $p_{O_2} = 1$ atm the equilibrium CO/CO₂ ratio for reaction (III) above will be the same as that for reaction (II); both reactions being in equilibrium in the presence of one atmosphere of oxygen.

Thus from reaction (III):

$$\Delta G_{III}^{\ominus} = -RT \ln \left(\frac{p_{CO_2} a_M}{p_{CO} a_{MO}} \right) = \Delta G_{II}^{\ominus} - \Delta G_I^{\ominus}$$

since $a_M = 1 = a_{MO}$

$$\Delta G_{III}^{\ominus} = 2.303 \, RT \log_{10} (p_{CO}/p_{CO_2})$$

When $\Delta G_{II}^{\ominus} = 0$, CO/CO₂ = 1 and the line $2CO + O_2 \longrightarrow 2CO_2$ can be extrapolated back to point 'C' on the $\Delta G^{\ominus}$ axis at 0 K. Hence a second nomographic scale can be constructed around the $\Delta G^{\ominus}-T$ diagram from which the equilibrium CO/CO₂ ratio can be deduced for any reaction in which a pure metal oxide is reduced by carbon monoxide. To find the equilibrium CO/CO₂ ratio for such a reaction a line is drawn from point 'C' to the appropriate $2M + O_2 \rightarrow 2MO$ reaction line at the required temperature T and then continued to cut the nomographic scale thus giving the equilibrium $\log_{10}$(CO/CO₂) ratio (see *Fig. 2.14*). As with the p_{O_2} scale the CO/CO₂ ratio is obtained from the antilogarithm of the scale reading.

If the ambient CO/CO₂ ratio is greater than the equilibrium ratio, since it contains excess CO, the forward reaction of:

$$MO + CO \rightleftharpoons M + CO_2$$

will be favoured as the system tends towards the equilibrium ratio. Conversely if excess CO₂ is present metal oxide formation is favoured.

(c) *Other grids:* A scale for equilibrium p_{H_2}/p_{H_2O} ratios may be constructed as previously described using a set of lines radiating from point 'H' on the $\Delta G^{\ominus}$ ordinate at 0 K, i.e. $\Delta G^{\ominus} = -496$ kJ.

It is possible, where appropriate, to design similar nomographic scales for determination of, for example, equilibrium p_{S_2}, p_{Cl_2}, p_{H_2}/p_{H_2S} ratios and so on (see *Figs 2.11, 2.13*).

Example of use of nomographic scales

By reference to *Fig. 2.11* determine:

(a)   The equilibrium oxygen pressure at 1025 $^\circ$C for the reaction:

$$2Ni + O_2 \rightleftharpoons 2NiO$$

Construct a straight line from point 'O' through the $2Ni + O_2 \rightarrow 2NiO$ line at the point corresponding to 1025 $^\circ$C and continue until it cuts the p_{O_2} scale. The scale shows a value of 10^{-10} atm for p_{O_2} at the stated temperature.

(b)   The equilibrium CO/CO_2 ratio at 800 $^\circ$C for the reaction

$$TiO_2 + 2CO \rightleftharpoons Ti + 2CO_2$$

Construct a straight line from point 'C' through the $Ti + O_2 \rightarrow TiO_2$ reaction line at 800 $^\circ$C when it will be seen that the line cuts the CO/CO_2 ratio scale between $10^8/1$ and $10^9/1$. The point of intersection of the line drawn is in fact about 0.15 of the distance between the $10^8/1$ and $10^9/1$ divisions and since the scale is logarithmic the antilogarithm of 0.15, i.e. 1.413 ($\sim$ 1.4), must be taken to find the estimated distance between the points such that the CO/CO_2 ratio now becomes 1.4×10^8. The nomographic scale must be read from a low to a high value, i.e. scale reading is 0.15×10^8 *not* 0.85×10^9.

Disadvantages of $\Delta G^\ominus - T$ diagrams

Despite the useful information which these diagrams yield their application is limited in several ways.

(a) As standard free energy changes are considered, the diagrams refer to substances in their standard thermodynamic states only but in practical situations concentrations may continuously alter and activities deviate from unity. This means that processes apparently impossible under standard conditions may be satisfactorily utilised in extraction processes operated under non-standard conditions, e.g. extraction of magnesium with silicon (see Pidgeon process, Section 7.2.10.4 as explained by application of equation (2.29) rather than (2.30)).

(b) Compounds whose formation lines are represented in the diagrams are assumed to be stoichiometric which often is not true.

(c) Although the feasibility of a reaction may be deduced from the diagram no information concerning the rate of the reaction is obtained. A feasible but very slow reaction may not be viable commercially.

(d) The diagrams do not show the conditions under which the reactions tend to occur, e.g. several fairly reactive metals oxidise only superficially; powdered elements may ignite spontaneously.

(e) Where oxide formation lines in the diagrams are very close together accurate measurement and subsequent calculation is difficult (accuracy ranges from ±4 kJ at best to errors of 40 kJ).

(f) The possibility of formation of intermetallic compounds between reactants and products is ignored.

Calculations

Questions (1) to (12) refer to the $\Delta G^\ominus - T$ oxide formation diagram (*Fig. 2.11*).

(1)   At what temperatures will carbon reduce: (a) $SnO_2(s)$, (b) $Cr_2O_3(s)$, (c) $SiO_2(s)$?

(2) Steam blown through hot coke gives rise to the fuel gas mixture called 'water gas' ($CO + H_2$):

$$C(s) + H_2O(g) \longrightarrow CO(g) + H_2(g)$$

Calculate the temperature at which the coke must be maintained for the reaction to be feasible.

(3) At what temperature does Ag_2O just begin to decompose at one atmosphere oxygen pressure?

(4) In what temperature range can hydrogen be used to reduce SnO_2 to Sn?

(5) Deduce the standard free energy change for the reduction of Al_2O_3 by Mg at $1000\,°C$.

(6) Explain the reason for the changes in slope of the following reaction lines:

(a) $2Mg + O_2 \longrightarrow 2MgO$ at $1100\,°C$

(b) $2Pb + O_2 \longrightarrow 2PbO$ at $1470\,°C$

(c) $4Li + O_2 \longrightarrow 2Li_2O$ at $1300\,°C$

(7) Estimate the standard free energy changes for the following reactions at 1200 K:

(i)   reduction of copper(I) oxide with hydrogen;
(ii) reduction of copper(I) oxide with carbon.

Hence, compare the reducing powers of hydrogen and carbon monoxide in this instance.

(8) Write a brief account of the use of carbon as a reductant of iron oxides.

(9) At what temperature is the reaction

$$\tfrac{4}{3}Cr + O_2 \longrightarrow \tfrac{2}{3}Cr_2O_3$$

at equilibrium when p_{O_2} 10^{-14} atm?

(10) What is the equilibrium CO/CO_2 ratio at $1100\,°C$ for the following reaction?

$$MnO + CO \longrightarrow Mn + CO_2$$

(11) At what temperature is the reaction

$$PbO + H_2 \longrightarrow Pb + H_2O$$

at equilibrium when the H_2/H_2O ratio is $1/10^4$?

(12) Sketch the $\Delta G_f^{\ominus}-T$ plots (using arbitrary scales and the same axes) for the formation of TiO_2 from Ti at p_{O_2} values of: (i) 0.1 atm; (ii) 1 atm; (iii) 10 atm. Hence, describe the effect of increasing external oxygen pressure on the system.

Further reading

Glasstone, S. and Lewis, D. 1963 *Elements of Physical Chemistry*. London: Macmillan.
Moore, W.J. 1972 *Physical Chemistry*. London: Longman.
The Open University 1974 *Principles of Chemical Processes* (Thermodynamics, Parts 2, 3 ST294 Blocks 2, 3). Milton Keynes: Open University Press.
1 Kubaschewski, O. and Evans, U. 1958 *Metallurgical Thermochemistry*. Oxford: Pergamon Press.

2 Stark, J.G. and Wallace, H.G. 1973 *Chemistry Data Book.* London: Murray.
   Parker, R.H. 1967 *An Introduction to Chemical Metallurgy.* Oxford: Pergamon Press.
   Gilchrist, J.D. 1967 *Extraction Metallurgy.* Oxford: Pergamon Press.
3 Mackowiak, J. 1966 *Physical Chemistry for Metallurgists.* London: George Allen & Unwin.
4 Ives, D.J.G. 1971 *Chemical Thermodynamics.* London: Macdonald.
   Smith, E. Brian 1973 *Basic Chemical Thermodynamics.* Oxford: Clarendon Press.
   Gaskell, D. 1973 *Introduction to Metallurgical Thermodynamics.* New York: McGraw-Hill.
   Bodsworth, C. and Appleton, C. 1965 *Problems in Applied Thermodynamics.* London: Longman.
   Ives, D.J.G. 1972 *Principles of the Extraction of Metals.* London: Chemical Society.
   Johnstone, A.H. and Webb, G. 1977 *Energy Chaos and Chemical Change.* London: Heinemann.
   Thompson, J.J. 1969 *An Introduction to Chemical Energetics.* London: Longman.

For self-test questions on this chapter, see Appendix 5.

3

Reaction Kinetics

A study of thermodynamics provides information regarding the chemical equilibrium of a metallurgical reaction and indicates the direction in which a reaction will proceed under the stated conditions. However, thermodynamics does not provide any information on the rate of a reaction.

For example, there is a large negative free energy change when aluminium is exposed to air at room temperature on oxidising to Al_2O_3. The initial formation of the tenacious film of Al_2O_3 acts as a barrier to further contact between the atmosphere and the aluminium substrate resulting in an extremely low overall rate of reaction. In fact, aluminium is used as an oxidation resistant material under these conditions (Section 9.2).

Similarly, the reaction between carbon and oxygen in liquid steel has a large thermodynamic driving force but requires the bubbling of oxygen, an inert gas, e.g. argon, or some other form of turbulence in the liquid steel to initiate the carbon 'boil' reaction:

$$C_{(Fe)} + O_{(Fe)} \longrightarrow CO$$

In both the above examples there is a kinetic factor which restricts the rates at which the reactions proceed. Therefore, consideration of both thermodynamics and reaction kinetics is required to provide a full understanding of any metallurgical process.

Homogeneous reactions are those which take place within one single phase, such as a gas or a liquid solution, while heterogeneous reactions are those between different phases such as a gas–solid reaction. Most metallurgical reactions are heterogeneous reactions since a large proportion of apparently homogeneous reactions occurring in metallurgical processes require a different phase on which the products may nucleate. Catalytic reactions provide an example of this latter point. The main types of reactions that take place in process metallurgy and corrosion are:

(i)   gas–solid, e.g. reduction of iron ore with CO in the blast furnace, or oxidation of a metal;
(ii) gas–liquid, e.g. oxygen lancing in steelmaking or gaseous reduction in hydrometallurgy;
(iii) liquid–liquid, e.g. metal–slag refining reactions, or solvent extraction;
(iv) liquid–solid, e.g. leaching and corrosion reactions;
(v)   solid–solid, e.g. reduction of zinc oxide with coke;

The overall rates of these reactions will be dependent on three main process steps which are present in all reactions:

(i)   rate of transport of reactants to the reaction site;
(ii) chemical reaction rate at the reaction site;
(iii) rate of transport of products away from the reaction site.

Reaction kinetics is a study of the rate of reactions by consideration of these individual reaction steps, the mechanisms by which they proceed and the effect of certain process variables such as concentration of reactants and reaction temperature on the reaction rate.

The present chapter endeavours to discuss these factors and the methods by which the reaction rate may be experimentally determined.

The considerable experimental difficulties in obtaining reaction kinetic data for metallurgical processes at high temperatures account for the restricted range of experimental data quoted in the following discussions.

3.1 Rate of reaction

The rate of reaction may be defined as the rate at which the concentration of a reactant decreases or the rate at which the concentration of a product increases. Thus for a reaction of the type,

$$A + B \longrightarrow C$$

$$\text{rate} = -\frac{d[A]}{dt} = -\frac{d[B]}{dt} = \frac{d[C]}{dt}$$

where [ ] refers to concentration so that $d[\ ]/dt$ refers to the rate of change of concentration with time. Note that the rate is a positive quantity but as $[A]$ is decreasing $d[A]/dt$ is negative, and hence the need for the negative sign.

The rate of a reaction can also be expressed in terms of the rate of change of a volume of a gas produced, of a pressure of a gas produced, of a weight of a sample and in other ways, but their use is more limited.

3.1.1 Effect of conditions on rate of reaction

The relationships between rate and concentration and temperature are discussed in some detail in this chapter. Some of the other factors are worthy of a brief mention here.

3.1.1.1 Surface area

Owing to the fact that heterogeneous reactions occur at an interface, the area of the interface will affect the rate of reaction. For solid–liquid or solid–gas reactions, the state of division of the solid affects the rate of reaction. In the zinc blast furnace, zinc oxide is introduced not in the form of powder which would be blown out by escaping flue gases but in another form (sinter) with a high surface area.

Sinters are relatively heavy and robust and are produced (Section 7.1.4) to provide a highly permeable charge for smelting in a blast furnace. In some furnaces the reduction of a solid metal oxide may be with solid carbon:

$$MO(s) + C(s) \longrightarrow M(l) + CO(g) \tag{3.I}$$

Alternatively, reduction can be with carbon monoxide:

$$MO(s) + CO(g) \longrightarrow M(l) + CO_2(g) \tag{3.II}$$

Reaction (3.I) will be slow because of the low area of contact between two solids. Reaction (3.II) is likely to be more rapid because it involves a gas, that is a gas–solid reaction which benefits from the high diffusivity of a gas allowing contact with the solid reactant over its whole surface.

This is an important consideration because reactions (3.I) and (3.II) require quite different conditions in order to provide a high thermodynamic driving force (large negative ΔG) for each reaction: the first simply a low partial pressure of carbon monoxide, the second a high CO/CO_2 ratio (see Section 2.2.3.4).

In the carbon 'boil' reaction during steelmaking the area of the metal–slag interface is increased by the turbulence caused by bubbles of carbon monoxide rising through the liquid metal and slag, thereby creating a 'boiling' effect.

3.1.1.2   Catalysts

A catalyst is a substance which can alter the rate of a reaction but remains unchanged in amount at the end. The more familiar catalysts such as manganese (IV) oxide which catalyses the decomposition of hydrogen peroxide:

$$2H_2O_2 \xrightarrow{MnO_2} 2H_2O + O_2$$

and vanadium (V) oxide which catalyses the oxidation of sulphur dioxide:

$$2SO_2 + O_2 \xrightarrow{V_2O_5} 2SO_3$$

are positive catalysts in that they increase the rate of reaction. Negative catalysts (inhibitors) are also known. For example, glycerol decreases the rate of the reaction:

$$2H_2O_2 \longrightarrow 2H_2O + O_2$$

All catalysts take part in the reaction: the heterogeneous (surface) catalysts by adsorption of reactants and the homogeneous (chemical) catalysts by reaction to form a reactive intermediate (see Section 3.5.5)[1,2,3].

3.1.2   Concentration–time graphs

A useful way of observing the effect of various conditions on reaction rate is to draw graphs of concentration (weight, volume or other suitable parameter) against time, as shown in *Figs 3.1* and *3.2*. Curve A in *Fig. 3.1* shows how the volume of carbon dioxide varies with the time for the reaction:

$$CaCO_3 + 2HCl \longrightarrow CaCl_2 + H_2O + CO_2$$

where the acid is in excess; curve B is for the same reaction at a higher temperature, with a greater acid concentration, more finely divided calcium carbonate or on using a catalyst. *Fig. 3.2* shows how the concentration of hydrochloric acid varies for the same reaction as given in *Fig. 3.1*, curve A.

3.1.3   Kinetics and mechanism

The simple picture of a chemical reaction is one in which reactant molecules moving at random collide by chance and interact. Consider the reaction:

$$A + B \longrightarrow products$$

The probability of an A–B collision, that is the collision rate, will be proportional to $[A]$ and to $[B]$. Hence we can write:

$$collision\ rate \propto [A][B]$$

If it is assumed that reaction rate is proportional to collision rate:

$$rate \propto [A][B]$$

Expressions of this type are in fact obtained by experiment in order to deduce the collisions which are taking place and hence obtain some idea of the mechanism.

If, for example, the rate is indeed found to be proportional to the product $[A][B]$ the conclusion is that one molecule of A collides with one molecule of B.

The use of kinetics in determining mechanism is discussed in detail in Section 3.6.

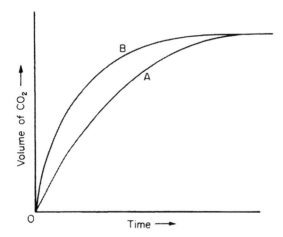

Figure 3.1 Volume–time graph for the reaction $CaCO_3 + 2HCl \rightarrow CaCl_2 + H_2O + CO_2$

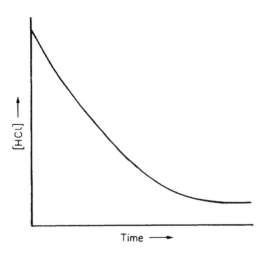

Figure 3.2 Concentration–time graph for the reaction $CaCO_3 + 2HCl \rightarrow CaCl_2 + H_2O + CO_2$

3.2   Experimental rate laws

For many reactions, it is found that there is a fairly simple relationship between rate and concentration of reactants in that the reaction rate is proportional to simple powers of concentration. Such a relationship, found by experiment, is known as the *experimental rate law*, for example, for the reaction:

$$2H_2O_2 \longrightarrow 2H_2O + O_2$$

it is found that the rate is proportional to the concentration of H_2O_2, i.e.

$$rate \propto [H_2O_2]$$

In many cases however, the experimental rate law is more complex, e.g. for the reaction:

$$H_2(g) + Br_2(g) \longrightarrow 2HBr(g)$$

$$\text{rate} = \frac{k[H_2][Br_2]^{\frac{1}{2}}}{1 + k'[HBr]/[Br_2]}$$

where k and k' are constants. In such cases a complex or multistep reaction is suggested. In other cases an improved relationship is obtained if activities* are used instead of concentration. It should also be understood that any simple relationship while holding true over a reasonable range of conditions may not be valid at extremely high or low concentrations, pressures or temperatures.

3.2.1   Order of reaction

The order of a reaction is the total power to which the concentration is raised in the experimental rate law. Thus, if

$$\text{rate} \propto [A]^2 \qquad\qquad (3.1)$$

for the reaction:

$$2A \longrightarrow \text{products}$$

the order is *two*: it is a second-order reaction. The same is true if rate $\propto [A][B]$ since the sum of the powers of $[A]^1$ and $[B]^1$ is two. This could be for a reaction of the type:

$$A + B \longrightarrow \text{products}$$

In this case it is said that the reaction is first order with respect to A and first order with respect to B. In general if:

$$\text{rate} \propto [\text{concentration}]^n$$

the reaction is nth order. Similarly if,

$$\text{rate} \propto [A]^x[B]^y$$

the reaction is pth order where $p = x + y$. Orders of reaction are usually 1, 2 or 3 but other orders (including fractional ones) are possible. For example the rate of oxidation of Cu to Cu_2O is proportional to the eighth root of the partial pressure of oxygen[4]:

$$4Cu + O_2 = 2Cu_2O \quad \text{rate} \propto [p_{O_2}]^{1/8}$$

3.2.2   The rate constant

In the above relationships, there must be a constant of proportionality so that equation (3.1) becomes:

$$\text{rate} = k[A]^2$$

where the constant k is called the (specific) rate constant. The units of concentration are mol dm^{-3} (for some metallurgical reactions per cent by weight may be more convenient), those of rate are mol dm^{-3} s^{-1} and the rate constant has units which depend on the order of the reaction. The rate constant units can of course be found by dimensional analysis. Since $k = \text{rate}/[\text{reactants}]^n$, where n is the order of reaction, the unit of rate constant, using the units given above, is given by:

$$\frac{\text{rate}}{[\text{reactants}]^2} = \frac{\text{mol dm}^{-3}\text{ s}^{-1}}{(\text{mol dm}^{-3})^2} = \frac{\text{mol dm}^{-3}\text{ s}^{-1}}{\text{mol}^2\text{ dm}^{-6}} = \text{mol}^{-1}\text{ dm}^3\text{ s}^{-1}$$

for a second-order reaction.

---

*For simplicity concentration terms will be used in this chapter, though strictly speaking, activities should be used. Activities are discussed in Sections 2.2.3.3 and 4.3.2.

3.2.3 Molecularity

When kinetic data have been analysed the number of molecules of each reacting species colliding together can be proposed. For example, studies of the reaction:

$$NO + O_3 \longrightarrow NO_2 + O_2$$

show that one molecule of nitrogen monoxide and one molecule of ozone collide: two molecules are involved and the molecularity is two–the reaction is bimolecular. The term *molecularity* applies to a single step in a *theoretical* reaction mechanism and unlike order, which is an experimental result, it must be a whole number.

3.2.4 Stoichiometry

The stoichiometric equation for a chemical reaction gives the simplest ratio of the number of molecules of reactants and products. In no way does it represent a mechanism for the reaction: neither molecularity nor order can be deduced from the stoichiometry. For example, the reaction:

$$2H_2O_2 \longrightarrow 2H_2O + O_2$$

is first order despite $2H_2O_2$ appearing in the equation. Nevertheless the stoichiometry is an essential prerequisite for determining a reaction mechanism.

3.2.5 Integrated equations

Owing to the fact that the concentration of reactants changes as the reaction proceeds, a more useful form of the rate equation gives concentration as a function of time. This enables the extent of the reaction to be determined at a given time and vice versa. It also provides an equation which can be used for determining the rate constant k.

3.2.5.1 Zero order

For the zero order reaction:

$$A \longrightarrow products$$

the rate is constant, i.e.

$$d[A]/dt = -k \tag{3.2}$$

This can occur in some heterogeneous reactions or in surface catalysed reactions, e.g. the reaction,

$$N_2 + 3H_2 \longrightarrow 2NH_3$$

occurs in the presence of a solid tungsten catalyst[5]. Under a range of gas pressures, the catalyst surface is completely covered with reactant molecules. Since almost all the reaction occurs at the surface, the rate depends on the number of reactant molecules adsorbed which cannot increase. Thus the reaction rate is constant (except at low pressures) for a given sample of catalyst and is independent of the concentrations of the reactants. Another example[6] is the reaction which takes place when solid titanium is immersed in liquid aluminium at 1073 K:

$$Ti(s) + 3Al(l) \longrightarrow TiAl_3(s)$$

The solid $TiAl_3$ forms a crust on the titanium surface. The reaction can occur only at the surface of the Ti and therefore takes several hours; the surface area remains virtually constant. Since [Ti] and [Al] are also constant, the rate is constant. Strictly speaking the reaction is pseudo-zero order, that is to say it is not a true zero-order process but over the range of time when measurements are taken, the rate appears constant. Eventually, the titanium will react extensively and its surface area decreases so that the rate must fall. Thus, the initial rate of the reaction only is being examined (see Section 3.3.2). Also, the concentrations of Ti and Al are fixed for each experiment but samples have been used having different titanium concentrations. It is then found that rate is a function of [Ti]. Integrating equation (3.2) between the limits $t = t$ and $t = 0$ when the concentrations of A are [A] and [A]$_0$ respectively allows the change in concentration of A with time to be calculated or assessed graphically when t is plotted against concentration:

$$\int_{[A]}^{[A]_0} d[A] = -k \int_t^0 dt$$

hence,

$$kt = [A]_0 - [A] \tag{3.3}$$

A typical concentration–time graph for a zero-order reaction is found in *Fig. 3.3*.

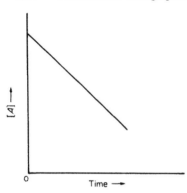

Figure 3.3 Concentration–time graph for the zero-order reaction $A \longrightarrow$ products

3.2.5.2 First order

For the first-order reaction:

$$A \longrightarrow \text{products}$$

$$d[A]/dt = -k[A] \tag{3.4}$$

There are many physical and chemical examples of first-order processes. The rate of charging a capacitor, for example, is given by

$$dQ/dt \propto Q$$

where Q is the charge at time t. The rate of nuclear decay is given by:

$$dN/dt \propto N$$

where N is the number of atoms of a radioactive isotope at time t. An example of a chemical first-order reaction is:

$$SO_2Cl_2 \longrightarrow SO_2 + Cl_2$$

for which,

$$\frac{d[SO_2Cl_2]}{dt} = -k[SO_2Cl_2]$$

The transfer of sulphur from iron to slag[7] and decarburising[8] have been shown to be first-order processes.

Integrating equation (3.4) between $[A]$ and $[A]_0$ in the same way as for equation (3.2):

$$\int_{[A]}^{[A]_0} \frac{d[A]}{[A]} = -k \int_t^0 dt$$

whence

$$kt = \ln\left(\frac{[A]_0}{[A]}\right) \tag{3.5}$$

Rearranging, this becomes:

$$[A] = [A]_0\, e^{-kt} \tag{3.6}$$

A typical concentration–time graph for a first-order reaction is shown in *Fig. 3.4*. Equation (3.5) can be used to find the rate constant from experimental data. Thus, rearranging equation (3.5):

$$\ln[A] = \ln[A]_0 - kt$$

Hence a plot of $\ln[A]$ versus t gives a straight line of slope $-k$ and intercept $\ln[A]_0$. If logarithms to the base 10 are used, the slope becomes $-k/2.303$.

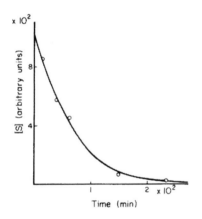

Figure 3.4 Concentration–time graph for sulphur transfer from iron to slag at 1775 K (after Ward and Salmon[7])

Note that in this case the units of k are time^{-1} and $[A]$ may be in any units; therefore the units and powers of ten of concentration of A may be ignored for the purpose of finding the slope in a first-order plot.

Example

The data shown in *Tables 3.1* and *3.2* were obtained by Ward and Salmon[7] from studies of sulphur transfer from iron to slag at 1775 K. To calculate the first-order rate constant, the data in *Table 3.1* are retabulated in *Table 3.2* using $\ln[S]$. The results are plotted in *Fig. 3.5*. From *Fig. 3.5*:

$$\text{slope} = -k = -1.58 \times 10^{-2}\ \text{min}^{-1} = -2.63 \times 10^{-4}\ \text{s}^{-1}$$

Table 3.1 EXPERIMENTAL DATA FOR TRANSFER OF SULPHUR FROM IRON TO SLAG AT 1775 K

Time/min	[S] (arbitrary units)
0	1000
4	860
15	825
33	568
63	448
91	208
151	72
234	27

For [S], 1000 ≡ 0.16 wt%

Table 3.2 RETABULATION OF DATA GIVEN IN *TABLE 3.1*

Time/min	ln[S]
0	6.91
4	6.76
15	6.72
33	6.34
63	6.10
91	5.34
151	4.28
234	3.30

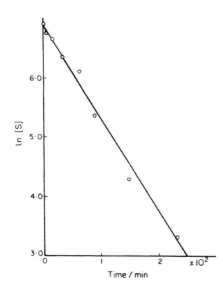

Figure 3.5 First-order plot for the transfer of sulphur from iron to slag at 1775 K (after Ward and Salmon[7])

Therefore the actual rate of transfer of sulphur from iron to slag for the same sample at 1775 K can be calculated at any stage in the process by:

$$\text{rate} = 2.63 \times 10^{-4} \times [S]$$

3.2.5.3 Second order

For the second-order reaction:

$$2A \longrightarrow \text{products} \quad d[A]/dt = -k[A]^2 \tag{3.7}$$

This equation is also valid for the second-order reaction:

$$A + B \longrightarrow \text{products}$$

where $[A] = [B]$. For example the reaction,

$$CH_3COOC_2H_5 + OH^- \longrightarrow CH_3COO^- + C_2H_5OH$$

is second order and the rate constant can be found easily in an experiment where reactant concentrations are equal[9]. The denitriding of iron* is also second order[10]:

$$2N_{(in\ Fe)} \longrightarrow N_2 \quad \text{and} \quad d[N]/dt = -k[N]^2$$

Again, integrating equation (3.7) between $[A]$ and $[A]_0$ as in equation (3.2):

$$\int_{[A]}^{[A]_0} \frac{d[A]}{[A]^2} = -k \int_t^0 dt$$

then:

$$\frac{1}{[A]} - \frac{1}{[A]_0} = kt$$

A typical second-order plot is shown in *Fig. 3.6*. The intercept is $1/[A]_0$ and the slope is k. Note that the units of $[A]$ and of t are both important since the units of k in this case are concentration $\times$ time^{-1}.

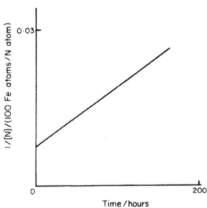

Figure 3.6 Second-order plot for the denitriding of iron at 623 K (after Goodeve and Jack[10])

For a reaction of the type:

$$A + B \longrightarrow \text{products}$$

in which $[A] \neq [B]$ integration yields[1]

$$\frac{1}{[A]_0 - [B]_0} \ln \left(\frac{[B]_0[A]}{[A]_0[B]} \right) = kt \tag{3.8}$$

where $[B]$ is the concentration of B at time t and $[B]_0$ is the initial concentration of B.

3.2.5.4 *Reversible reactions*

Reversible reactions† are common in pure chemistry and in metallurgy. The gas carburising of thin foils of iron is one example. Carbon atoms enter the sample as a result of gas phase reactions. At the same time carbon atoms leave the surface of the sample. The

---

*More strictly, ϵ-iron nitride.

†In a sense all reactions are reversible but here the term is used for reactions in which there are substantial amounts of reactants and products at equilibrium.

result is that the carbon concentration in the sample reaches a maximum when equilibrium is established.

Here only the simplest case will be considered, that is reversible first-order reactions:

$$A \underset{k_{-1}}{\overset{k_1}{\rightleftharpoons}} B$$

where k_1 is the rate constant for the forward reaction (forward rate constant) and k_{-1} is the rate constant for the back reaction (back rate constant). The integrated rate equation is:

$$\ln \left(\frac{[B]_e}{[B]_e - [B]} \right) = \frac{k_1 [A]_0}{[B]_e} \, t \tag{3.9}$$

where $[B]_e$ is the equilibrium concentration of B, $[B]$ is the concentration of B at time t and $[A]_0$ is the initial concentration of A. Since $[A]_0$ and $[B]_e$ are constants, equation (3.9) is essentially a first-order equation in which the function $([B]_e - [B])$, not $[B]$ or $[A]$, is changing exponentially with time. Thus, there is an exponential

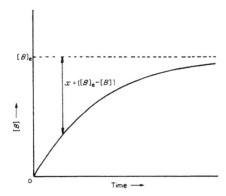

Figure 3.7 Concentration–time plot for the reversible reaction $A \rightleftharpoons B$

approach to the equilibrium position. This is made clearer in *Fig. 3.7*, where $x = [B]_e - [B]$. A plot of $\ln x$ versus time will be a straight line with slope $-k_1 [A]_0/[B]_e$. Rearranging equation (3.9):

$$\ln[B]_e - \ln([B]_e - [B]) = \frac{k_1 [A]_0}{[B]_e} \, t$$

i.e., $\tag{3.9a}$

$$\ln[B]_e - \frac{k_1 [A]_0}{[B]_e} \, t = \ln x$$

Example

The following data (*Table 3.3*) were adapted from results obtained by Collin *et al.*[8] for the carburising of thin foils of steel at 925 °C and can be used to calculate the first-order rate constant. For the carburising of thin foils it can be shown that:

$$\frac{2kt}{\delta} = \ln \left(\frac{C_g - C_i}{C_g - C_f} \right) \tag{3.9b}$$

where C_g is the equilibrium concentration of carbon in steel, C_f is the concentration of

Table 3.3 CARBURISING DATA AFTER
COLLIN *et al.* [8]

Time/min	Carbon
0	0.1
15	0.39
30	0.65
60	1.08
∞	$6.01 = C_g$

Table 3.4 RETABULATION OF DATA IN
TABLE 3.3

Time/min	$C_g - C_f$/(wt %)	$\ln(C_g - C_f)$
0	5.91	1.777
15	5.62	1.726
30	5.36	1.679
60	4.93	1.595

carbon at time t, C_i is the initial concentration of carbon (that is, before carburising), k is the rate constant and δ is the foil thickness.

Taking logarithms provides the data in *Table 3.4* which are used in *Fig. 3.8* to plot $\ln(C_g - C_f)$ versus time. The slope of the straight line graph, $-2k/\delta$, is -3.14 min^{-1}.

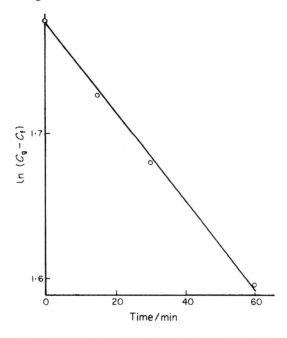

Figure 3.8 First-order plot for the carburising of a thin foil at 1198 K (after Collin *et al.* [8])

If the foil thickness is 5×10^{-5} m, k is 1.31×10^{-9} ms^{-1}. Substituting this value into equation (3.9b) enables the extent of carburisation to be determined for any foil thickness.

3.2.5.5   Consecutive reactions

Most processes in chemical metallurgy consist of several steps and even when there is only one chemical reaction, there may be physical steps such as diffusion of reactants to the site of reaction. For example in the denitriding of iron, the chemical reaction is $2N \longrightarrow N_2$, which seems to occur at the surface of the iron. Obviously, in a previous step, nitrogen atoms must diffuse through the sample to the surface and after the

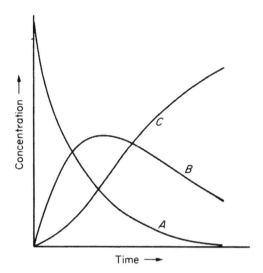

Figure 3.9 Concentration–time graph for the reaction $A \rightarrow B \rightarrow C$

chemical reaction producing N_2 molecules these must be transported away from the surface reaction site in order that no N_2 build up is allowed to occur which would reduce the overall reaction rate. If one step in a multistep reaction is very much slower* than the others, the problem simplifies to a consideration of the kinetics of that one step (see rate determining step, Section 3.6.1). The overall kinetics for consecutive reactions are usually complex and only the simplest case will be considered here, that is two first-order reactions:

$$A \xrightarrow{k} B \xrightarrow{k'} C$$

It can be shown, too, that the concentrations of A, B and C are expressed in terms of exponential functions and a typical concentration–time graph for k and k' of the same order of magnitude is shown in *Fig. 3.9*.

3.3 Determination of order of reaction

It is clear from the foregoing discussion that if the reaction mechanism is to be elucidated it is essential to obtain the order of the reaction. Three methods for obtaining the order of reaction from experimental data will be described.

3.3.1 Integral method

This method could be called the method of trial and error. After examining the data it may be possible to guess the order correctly. For example, use can be made of the half-life method (Section 3.3.3) which measures the time required to halve the concentration of the reactant during the reaction. If the guess was correct, a plot of the appropriate integrated rate equation will give a straight line. If not, the result, in theory at least, is a curve. For example, consider a second-order reaction. The first guess is, let us say, first order where $\ln[A]$ is plotted against time. Appropriate functions assuming third order, then second order, are also plotted against time. Typical results are shown in *Fig. 3.10*. Only the second-order plot gives a straight line. Difficulties can arise as it is sometimes

*See footnote to Section 3.6.1

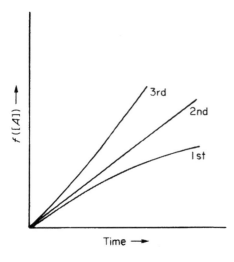

Figure 3.10 First-, second- and third-order plots for second-order data

very difficult to distinguish which is the straight line when the data are poor and all the plots seem to be gentle curves. This method is not satisfactory when the order of reaction is a fraction.

3.3.2 Differential method

If there is one reactant, the rate law may be written:

$$d[A]/dt = k[A]^n \tag{3.10}$$

where n is the order of reaction. Taking logarithms to the base ten:

$$\log \frac{d[A]}{dt} = \log k + n \log[A] \tag{3.10a}$$

Then, plotting $\log(d[A]/dt)$ versus $\log[A]$ gives a straight line of slope n. The problem is to obtain values of $d[A]/dt$ from experimental data since $[A]$ varies with time. To do

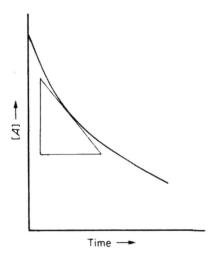

Figure 3.11 A concentration–time curve for the nth-order reaction $A \rightarrow$ products

this, tangents are drawn on concentration–time graphs as shown in *Fig. 3.11*. This procedure is a major source of inaccuracy though computer calculations of $d[A]/dt$ can be more satisfactory. An alternative is to use the initial rate method.

3.3.2.1 Initial rate method

During the first 5% or so of reaction, so little reactant has been used up that its concentration is virtually constant. Thus the rate is then constant (see *Fig. 3.12*) and is known as the *initial rate*. For a single reactant,

$$\text{initial rate} = \left(\frac{d[A]}{dt}\right)_0 = k[A]_0^n \tag{3.11}$$

where the zero indicates initial conditions. Taking logarithms to the base ten:

$$\log(\text{initial rate}) = \log k + n \log[A]_0 \tag{3.11a}$$

A plot of log (initial rate) versus $\log[A]_0$ gives a straight line of slope n and intercept log k. To find n and k from one set of data the initial rate is measured for a series of solutions

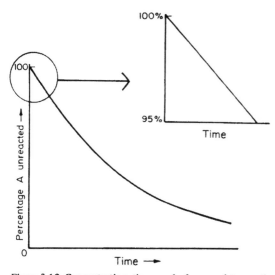

Figure 3.12 Concentration–time graphs for complete reaction and first 5% of reaction

or samples of different initial concentrations of A. The advantage of the initial rate method is that a linear rate is much easier to determine than a continuously changing one as described above. Also the concentrations of $[A]_0$ can be chosen at will, but using a concentration–time graph as in *Fig. 3.11* the values of $[A]$ where tangents can be drawn may be restricted. An example where the initial rate method has been successfully applied is the reaction:

$$2H_2O_2 \longrightarrow 2H_2O + O_2$$

Volumes of oxygen are measured at intervals of time during the initial rate period for each solution and plotted against time. The slope is the initial rate. Indeed, the rate can be found for each solution from a single measurement in the initial rate period: the time taken for a given extent of reaction, for example 50 cm^3 of oxygen. If the same extent of reaction is timed for all solutions,

$$\text{rate} \propto \frac{1}{\text{time}}$$

Therefore using equation (3.11): initial rate $= k[A]_0^n$, which is proportional to time^{-1}. Then equation (3.11a) simplifies to:

$$\log k + n \log[A]_0 \propto \log(\text{time}^{-1})$$

or

$$\log(\text{time}) = \text{constant} - n \log[A]_0 \tag{3.11b}$$

This is a suitable equation if only the value of n is required (though k can be calculated from the intercept).

The same argument applies to the units of $[A]_0$ which can be arbitrary. If there are two reactants (e.g. A and B) the initial rate is given by:

$$\text{initial rate} = k[A]_0^p [B]_0^q \tag{3.12}$$

where p and q are the orders with respect to A and B respectively. To find p and q, $\log$ (initial rate) is plotted versus $\log[A]_0$ for a series of solutions containing different concentrations of A but the same concentrations of B. In this case equation (3.12) becomes:

$$\log(\text{initial rate}) = \underbrace{\log k + q \log[B]_0}_{\text{constant}} + p \log[A]_0$$

where p is the slope. From a similar set of data, q can also be found.

3.3.3   Half-life method

Consider equation (3.5). In the case where $t = t_{1/2}$, the half-life, i.e. the time required to decrease the concentration of A to half its initial value:

$$[A] = [A]_0/2$$

and so

$$kt_{1/2} = \ln\left(\frac{[A]_0}{[A]_0/2}\right) = \ln 2$$

Thus, for first-order reactions $t_{1/2}$ is a constant for a given reaction, and it is independent of $[A]$. For example, consider the data tabulated in *Table 3.5* for an imaginary first-order reaction. It can be seen that $[A]$ is halved every 20 s regardless of value taken for

Table 3.5   DATA FOR IMAGINARY FIRST-ORDER REACTION

$[A]$/(arbitrary units)	100	84	70.7	50	35.4	25	17.7	13
Time/s	0	5	10	20	30	40	50	59

$[A]_0$: 100, 70.7, 50, 35.4, or 25 s. This behaviour is found only in first-order processes. If the reaction had been second order, the half-life would have depended on the value of $[A]_0$ used. For an nth order reaction:

$$t_{1/2} \propto 1/[A]_0^{n-1} \tag{3.13}$$

from which n can be obtained using appropriate data[1]. Other fractional lives can also be used, for example for first order:

$$t_{1/2}/t_{1/4} = \frac{1}{2}$$

but for second order:

$$t_{1/2}/t_{1/4} = \frac{1}{3}$$

3.4 Experimental techniques

In this section some of the methods used for obtaining kinetic data are outlined.
The methods cited serve as examples to give an idea of how to approach the problem.
The list of methods is not intended to be exclusive nor are all the methods necessarily
suitable for metallurgical processes but they do serve to show the techniques used.

3.4.1 General considerations

(a) To follow a reaction it must be slow enough to allow: (i) mixing of reactants;
(ii) readings to be taken. For conventional* methods, the half-life should be about
30 seconds.

(b) The concentration of reactant or product or some physical property related to the
concentration of either (e.g. optical density) is measured at intervals.

(c) Initial rate method should be used if possible.

(d) Since the rate constant is a function of temperature, the reaction mixture should
be thermostatically controlled at a constant temperature.

(e) Experiments should be carried out under a variety of conditions, e.g. for studies on
transfer of sulphur from iron to slag, experiments should be carried out using different
slags.

3.4.2 Techniques—in general

(a) If a reaction is sufficiently slow it may be possible to remove portions of the
reaction mixture from time to time for direct analysis. For example the reaction:

$$CH_3COOC_2H_5 + OH^- \longrightarrow CH_3COO^- + C_2H_5OH$$

can be studied by removing samples, quenching them in ice-cold water to slow further
reaction and then titrating with acid to find $[OH^-]$.

(b) Equivalent samples can each be allowed to react (e.g. in a furnace), then quenched
after different lengths of time before analysis. Carburising can be studied in this way.
Samples removed from the furnace after different lengths of time are subjected to photo-
micrography to obtain the hardness profiles.

(c) If there is a change in a physical property as reaction proceeds, this can be monitored
directly or recorded continuously on a chart recorder. For example there is an increase
in pressure as the following reaction occurs:

$$C_2H_6 \longrightarrow C_2H_4 + H_2$$

this can be recorded using a Pirani gauge which produces an electrical signal. It can be
shown that the partial pressure of C_2H_6 at time t is given by $(2P_0 - P)$ where P_0 is the
initial pressure and P is total pressure at time t.

3.4.3 Techniques—in detail

3.4.3.1 Weight

A sample which reacts to evolve or absorb a gas on heating can, after quenching, be
weighed after a given length of time. This is also used in corrosion rate determinations
where weight loss or weight gain is plotted against time.

*Special techniques are available[11] for following reactions with half-lives in the millisecond $(10^{-3}$ s)
microsecond $(10^{-6}$ s) or even nanosecond $(10^{-9}$ s) ranges.

When weight changes are small, the method is likely to be inaccurate and results should be checked using another method such as those described below.

3.4.3.2   Volume of gas

If a gas is evolved from a reaction, this can be collected and the volume measured. An arrangement such as the one described by Avery[1] can be used. This method can be applied to such reactions as:

$$2H_2O_2 \longrightarrow 2H_2O + O_2$$

or

$$C(steel) + O_2 \longrightarrow CO_2$$

In the latter case, the CO_2 can be absorbed by potassium hydroxide solution and the loss of volumes estimated (e.g. Ströhlein apparatus).

3.4.3.3   Pressure

Sometimes it is more convenient to ..easure the pressure of a gas produced (at a constant volume). In principle, gas pressure can always be used to monitor a reaction where a change in pressure occurs.

3.4.3.4   Analysis

Samples of reaction mixture can be removed from time to time for direct analysis as was described in Section 3.4.2. For metal samples, after quenching or removal from the site of reaction, suitable analysis such as microhardness testing can be used (hardness is related to carbon and nitrogen content).

Alternatively, a gas produced by a chemical reaction can be estimated directly as in the case when iron is denitrided in hydrogen, ammonia is produced. This can be absorbed in water and titrated. Also, gas samples can be analysed by gas chromatography[12], infra-red, ultra-violet or mass spectrometry[13].

3.4.3.5   Radioactivity

For heterogeneous systems where one species transfers from one phase to another, radioactive 'labelling' can be used. Changes in the radioactive levels in one phase will then be a measure of changes in concentration in that species. Ward and Salmon[7] used this method to investigate sulphur transfer from iron to slag. They used iron samples containing FeS labelled with ^{35}S. During the reaction iron samples were removed and the β-radiation measured.

3.4.3.6   Dilatometric method

A dilatometer[14] is used for measuring small changes in the volume of liquids. A change in the volume as reaction proceeds results in a change in height of a liquid in the capillary tube and this is easily measured. A reaction which can be investigated by this method is the hydrolysis of acetal*:

$$CH_3CH(OC_2H_5)_2 + H_2O \xrightarrow{\text{H}^+ \text{ catalyst}} CH_3CHO + 2C_2H_5OH$$

The volume increases during reaction. This technique is also commonly used for solids whose length and volume changes are accurately recorded against time.

---

*1,1-diethoxyethane.

3.4.3.7 Absorbance

If a reactant or product absorbs radiation in the ultra-violet/visible region of the spectrum the absorbance can be measured at intervals of time (or continuously) and if Beer's law* is obeyed over the range of the concentration involved, the results represent directly the change in concentration. Consider the reaction:

$$6H^+(aq) + 5Br^-(aq) + BrO_3^-(aq) \longrightarrow 3Br_2(aq) + 3H_2O$$

The absorbance of the solution at 267 nm is measured and gives a measure of the Br_2 concentration.

3.5 Kinetics and temperature

It is a common misapprehension that reaction rates increase with temperature only because molecules collide more frequently. Collision frequency (Z) increases with temperature according to the expression:

$$Z \propto T^{\frac{1}{2}}$$

where T is the temperature in kelvin (see Section 3.7). Thus an increase in temperature from 320 K to 330 K, increases Z by $(330/320)^{\frac{1}{2}} \approx 1.015$, that is an increase in collision frequency of $(1.015/1.000) \times 100\%$ or 1.5%. For many reactions, however, the rate increases by 200% or more for this temperature rise. For example, the reaction rate for the first stage in the tempering of martensite in a 1% carbon steel increases by about 290% between 320 K and 330 K. Clearly there is another factor to consider apart from collision frequency. This factor is the proportion of collisions which actually lead to a reaction ('fruitful' collisions). The proportion of 'fruitful' collisions rises steeply with temperature for many reactions. This is discussed again in Sections 3.5.3, 3.5.4 and 3.7.1. In 1889, Arrhenius obtained a relationship between rate and temperature from thermodynamic functions rather than by considering collisions (see Section 3.5.1). It will be shown, however, that both approaches produce equations of the same format.

3.5.1 The Arrhenius equation

Consider the reaction:

$$A + B \underset{k_{-1}}{\overset{k_1}{\rightleftharpoons}} C + D$$

which is first order in A, B, C and D. Using equation (2.26) at equilibrium

$$K_c = \frac{k_1}{k_{-1}} = \frac{[C][D]}{[A][B]} \tag{3.14}$$

where K_c is the equilibrium constant and k_1 and k_{-1} are the rate constants for the forward and reverse reactions. Now, the dependence of K_c on temperature by an equation analogous to equation (2.32):

$$\frac{d(\ln K_c)}{dT} = \frac{\Delta U}{RT^2}$$

Substituting for K_c from equation (3.14) gives:

$$\frac{d(\ln k_1)}{dT} - \frac{d(\ln k_{-1})}{dT} = \frac{\Delta U}{RT^2} \tag{3.15}$$

---

*optical density $\propto$ concentration.

If E_1 is an energy term associated with the forward reaction, and E_{-1} an energy term associated with the reverse reaction such that $\Delta H = E_1 - E_{-1}$, equation (3.15) can be written as two separate equations:

$$\frac{d(\ln k_1)}{dT} = \frac{E_1}{RT^2} + I \tag{3.16a}$$

and

$$\frac{d(\ln k_{-1})}{dT} = \frac{E_{-1}}{RT^2} + I \tag{3.16b}$$

such that the equation (3.15) is given by equation (3.16a) minus equation (3.16b). The term I is an integration constant which Arrhenius found to be zero for many reactions. In general then, for one reaction,

$$\frac{d(\ln k)}{dT} = \frac{E_A}{RT^2} \tag{3.17}$$

where E_A is an energy term called the *activation energy*. Integration of equation (3.17) yields the Arrhenius equation:

$$\ln k = -\frac{E_A}{RT} + \text{constant} \quad \text{or} \quad k = A e^{-E_A/RT} \tag{3.18}$$

where the term A, the constant of integration, is known as the pre-exponential, or frequency, factor.

3.5.2 Determination of activation energy

The activation energy is a useful parameter in determining reaction mechanisms and in predicting reaction rates at given temperatures. To obtain the activation energy, the values of k (the rate constant) at two different temperatures at least are required, though preferably more data should be available so that calculation can be done graphically. From equation (3.18)

$$\ln k = \ln A - \frac{E_A}{RT} \tag{3.19}$$

Plotting $\ln k$ versus $1/T$ (the Arrhenius plot) a straight line of slope $-E_A/R$ is produced. In general the range of temperature for which k has been obtained is not practical for obtaining an accurate value of A. This should be calculated using equation (3.18). The data shown in *Table 3.6* were obtained by Goodeve and Jack[10] for the denitriding of iron *in vacuo* at various temperatures. The data are retabulated in *Table 3.7* and the results are plotted in *Fig. 3.13*, where the slope, $-E_A-R$, is $-22\,000 \text{ K}^{-1}$. Thus, E_A is 183 kJ mol^{-1}.

Table 3.6 DATA FOR DENITRIDING OF IRON *IN VACUO* AFTER GOODEVE AND JACK[10]

Temperature /°C	Rate constant/(N atoms per 100 Fe atoms)$^{-1}$ min^{-1}
350	6.46×10^{-7}
400	1.03×10^{-5}
450	8.52×10^{-5}
500	5.64×10^{-4}

Table 3.7 RETABULATION OF DATA GIVEN IN *TABLE 3.6*

Temperature /K	$1/T/\text{K}^{-1}$	$\ln (k \times 10^7)$
623	1.605×10^{-3}	1.87
673	1.486×10^{-3}	4.63
723	1.383×10^{-3}	6.75
773	1.294×10^{-3}	8.64

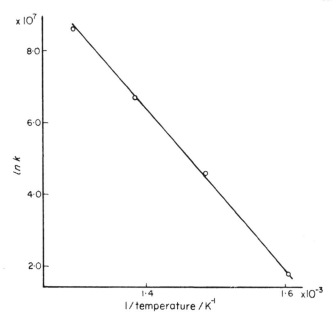

Figure 3.13 Arrhenius plot for denitriding of iron (after Goodeve and Jack[10])

3.5.3 Potential energy profiles

To obtain some idea of the physical significance of the activation energy, consider the collision and interaction of the molecules A and B in the reaction:

$$A + B \rightleftharpoons C + D$$

As the molecules approach, the forces of repulsion (due to their associated electron clouds) increase and the kinetic energies of A and B are converted to potential energy. If the average kinetic energy is sufficiently high A and B will approach closely enough to allow interaction producing an overlap of orbitals in such a way that 'new' bonds can form while 'old' ones break. If the kinetic energy is insufficient, the collision is 'unfruitful' and no reaction occurs. The minimum average kinetic energy which is sufficient to produce a reaction is the activation energy.

At the point of closest approach, corresponding to the maximum potential energy, (see *Fig. 3.14*) the molecules are said to have formed an *activated complex* or *transition state*. This is extremely short-lived: it breaks up to give $C + D$ or $A + B$ and the molecules fly apart again with increasing velocity as potential energy is reconverted to kinetic energy. The reaction $C + D \longrightarrow A + B$ also occurs via the same transition state but in reverse (principle of microscopic reversibility). The variation of potential energy for one pair of molecules is shown in *Fig. 3.14*. The x-axis, sometimes called the reaction coordinate, allows the plots of potential energy of reactants, products and transition state to be spaced horizontally and no great significance can be attached to the shape of the curve. Note that at constant pressure the enthalpy of reaction, ΔH, is given by:

$$\Delta H = E_A - E_B \tag{3.20}$$

where E_B is the activation energy of the reverse reaction. To illustrate the nature of the transition state, consider the reaction:

$$H_2 + I_2 \longrightarrow 2HI$$

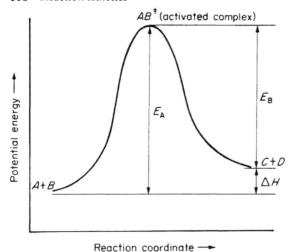

Figure 3.14 Potential energy profile for the reaction $A + B \rightleftharpoons C + D$

The proposed transition state is:

```
H-----I
 |    |
 |    |
 |    |
 |    |
H-----I
```

where the broken lines represent bonds which are in the process of being made (reactants forming activated complex) or broken (activated complex breaking down to form the products). In a diffusion process (e.g. carbon in iron) the atom moves

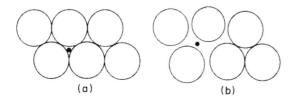

Figure 3.15 Carbon atom in an iron lattice: (a) equilibrium position; (b) transition state

from one equilibrium position where repulsions are at a minimum to the next via a potential energy barrier. This barrier (the transition state) occurs at the closest approach of the atom (see *Fig. 3.15*).

3.5.4   Effect of temperature

On the basis of the above discussion, it is clear that for many reactions only a small proportion of molecules have sufficient kinetic energy for 'fruitful' collisions. For example, in the reaction:

$$2HI \longrightarrow H_2 + I_2$$

at 700 K and 1 atm pressure only about 1 collision in 10^{14} leads to a reaction. (Note that when these molecules have reacted, new molecules will obtain the minimum required kinetic energy through intermolecular collisions.)

As the temperature is increased, the average kinetic energy increases and the proportion of molecules with kinetic energy greater than or equal to E_A also increases. This is clearly shown by comparing the Maxwell–Boltzmann distributions at two temperatures (*Fig. 3.16*). The Maxwell–Boltzmann distribution shows the number of molecules of different kinetic energies plotted against the average kinetic energy. The numbers of molecules having kinetic energy greater than or equal to E_A are shown by the hatched areas and are obviously greater at the higher temperature. It will be shown in Section 3.7.1 that the proportion of molecules having kinetic energy greater than or equal to E_A is given by $e^{-E_A/RT}$.

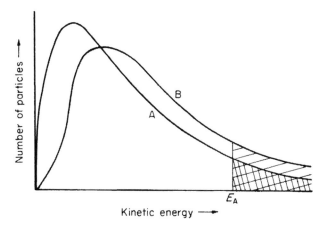

Figure 3.16 Maxwell–Boltzmann distribution at: Curve A, low temperature; curve B, high temperature

To illustrate the foregoing, consider again the first stage of the tempering of martensite of a 1% carbon steel for which the activation energy[15] is 119.5 kJ mol^{-1}. Thus, for a temperature rise from 320 K to 330 K, since rate is proportional to rate constant:

$$\frac{\text{rate at 330 K}}{\text{rate at 320 K}} = \frac{\exp(-E_A/330R)}{\exp(-E_A/320R)} = \frac{\exp(-119\,500/330R)}{\exp(-119\,500/320R)} \simeq 3.9$$

i.e. an increase of 290% as stated at the beginning of this section. The pre-exponential factor, A, is considered to be the same at both temperatures and therefore cancels out.

3.5.5 Catalysis

It is interesting to compare the potential energy profiles for a catalysed and uncatalysed reaction. Consider for example the reaction

$$HCOOH \longrightarrow H_2O + CO$$

for which the catalyst is H^+. The reaction in the presence of the catalyst occurs by a different path and at each stage the activation energy is lower than for the uncatalysed reaction (see *Fig. 3.17*). This means that for the catalysed reaction, a larger proportion of molecules have sufficient kinetic energy to react thereby increasing the reaction rate.

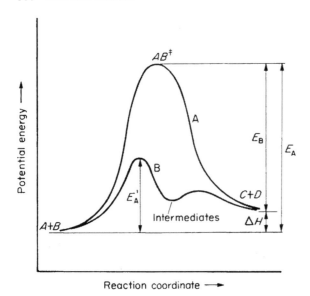

Figure 3.17 Potential energy profiles for the reaction $A + B \rightleftharpoons C + D$. Curve A, without catalyst; curve B, with catalyst in two stages. The activation energy for the first stage is E'_A

3.6 Mechanism

Understanding the mechanisms by which reactions proceed allows fuller control of the metallurgical process to be achieved. One of the most important applications of reaction kinetics is in the elucidation of mechanisms. The relationship between the order of reaction and the collisions taking place has already been referred to and there are a number of simple reactions for which a mechanism can be easily deduced from a limited amount of kinetic data. For example, the reaction

$$R\,Br + OH^- \longrightarrow ROH + Br^-$$

where R stands for $CH_3.CH_2.CH_2.CH_2$, is found to be first order in both $R\,Br$ and OH^- with the logical conclusion that the reaction occurs in one simple step: one molecule of $R\,Br$ and one OH^- ion collide to form a transition state:

$$Br^- - - - - R - - - - - OH^-$$

resulting in the products. This is the so-called concerted or $S_N 2$ mechanism[16]. Similarly, for the reaction:

$$2NO + O_2 \longrightarrow 2NO_2$$

the rate is given by:

$$\text{rate} \propto [NO]^2 [O_2]$$

and the conclusion is that collisions occur between two NO molecules and one O_2 molecule*.

---

*If collisions and order are related as indicated, first-order reactions cannot exist since collisions occur between at least two molecules. Unimolecular gas reactions can be explained in that collisions between reactant molecules result in activated molecules which break down to form the products. This forms the basis of Lindermann's theory[1,17].

3.6.1 Rate determining step

Now consider the reaction,

$$R'Br + OH^- \rightarrow R'OH + Br^-$$

where $R' = (CH_3)_3C$. The rate is given by

$$\text{Rate} \propto [R'Br]$$

The OH^- concentration does not appear in the experimental rate law although OH^- takes part in the reaction. The mechanism is clearly more complicated than for the hydrolysis of RBr and is explained in terms of a *rate determining step* involving $R'Br$. The rate determining step (RDS) is the step in a multistep reaction which is considerably slower* than any of the others: the bottleneck. The overall rate is thus controlled by the rate of this step and the overall rate law simplifies to a rate law for the RDS.

In the above example, the RDS is

$$R'Br \longrightarrow R'^+ + Br^- \qquad \text{Step 1}$$

and this is followed by

$$R'^+ + OH^- \longrightarrow R'OH \qquad \text{Step 2}$$

This mechanism is consistent with the evidence: one slow step involving one molecule of $R'Br$ only and a fast step involving the remaining reactant (the OH^- ion). It is expected that R'^+ should form as it is a cation easily stabilised by the three CH_3 groups and it is reasonable that step 2 should be fast as it is purely ionic.

The foregoing shows how the experimental rate laws are used to formulate a mechanism and the importance of other chemical knowledge with which the postulated mechanism must of course be consistent. Only a few simple ideas are introduced here. For a fuller discussion consult Edwards *et al.*[18].

3.6.1.1 Diffusion control

For some reactions the RDS is the diffusion of reactant molecules through a solvent. Since this is a fairly rapid process it implies that the reaction itself must be very fast. Such is the case with ionic reactions, e.g.

$$Ag^+(aq) + Cl^-(aq) \longrightarrow AgCl(s)$$

for which there is a negligible activation energy. The rate is controlled by the preceding step: the migration of ions through the solution. The activation energy for diffusion in non-metal systems is of the order[11] of $5-12$ kJ mol^{-1} while for metal systems values[19] may be as high as 200 kJ mol^{-1}. Most metal–slag refining reactions during steelmaking have relatively low activation energies, e.g. about 105 kJ mol^{-1}. Exceptions to this are the transfer of sulphur from iron to slag in steelmaking and the reduction of SiO_2 with coke in the blast furnace (approximately 500 kJ mol^{-1}). In the latter two cases the rate controlling step is likely to be the chemical reaction itself.

However, in process metallurgy where high temperatures in excess of the melting points of metals are used the chemical reaction rate is not normally the RDS. The transport of reactants to and/or products away from the reaction site tend to be the rate controlling steps. To this end increased turbulence and large slag–metal interfacial areas in liquid metal refining processes, increased fineness of ore particles providing

---

*That is *intrinsically* slower, having a lower rate constant than for any other step. It is assumed in fact that all the steps will occur at the same rate – those with high rate constants having correspondingly low reactant concentrations – and the concentrations of all intermediates are constant. This is the basis of the *steady state approximation*.[1]

large surface areas in roasting reactions and the use of gaseous reductants with their improved diffusivity compared with solid reductants in metal oxide reduction reactions are used to improve the reaction kinetics in process metallurgy.

Mass transfer coefficient

When the rate controlling step is the transport of a substance to or away from the reaction interface the study of mass transfer theory is useful.

Mass transfer in the bulk of liquids and gases is mainly by convection or turbulence (forced convection), but within a small distance, δ, of the interface (the effective boundary layer) molecular diffusion is the main mode of transport. The rate of molecular diffusion is given by Fick's laws. His first law of diffusion states that

$$J = -D \, dc/dt$$

where J (the flux) is the mass (m) of substance transported in unit time (t) through unit area (A) of a plane at right angles to the direction of diffusion and can be represented as $(dm/dt)A^{-1}$. The term D is the diffusion coefficient and dc/dx the concentration (c) gradient within a distance x from the interface (*Fig. 3.18*).

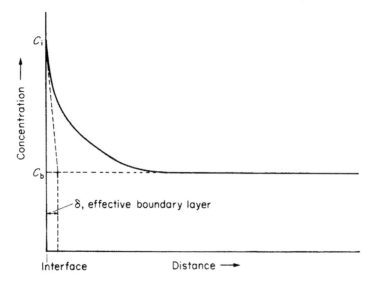

Figure 3.18 Concentration gradient of a solute adjacent to an interface and graphical construction of the associated effective boundary layer, δ

Hence within the effective boundary layer, δ,

$$\frac{dm/dt}{A} = \frac{D}{\delta} \cdot \frac{dc}{dx} \tag{3.21}$$

If the limits of concentration of the diffusing substance at the interface and at the extent of the effective boundary layer (bulk concentration) are c_i and c_b respectively, equation (3.21) can be rewritten:

$$\frac{dm}{dt} = \frac{DA}{\delta} (c_i - c_b) \tag{3.22}$$

From equation (3.22) it can be seen that the rate of mass transfer, dm/dt, within the effective boundary layer is dependent on D, A and δ.

To this end decreasing δ by increasing turbulence, e.g. inert gas, air or oxygen blowing, and increasing the interfacial area will improve the rate of mass transfer and therefore the reaction kinetics of the process.

DA/δ is called the mass transfer coefficient, k_m, i.e.

$$\frac{dm}{dt} = k_m(c_i - c_b) \tag{3.23}$$

k_m is easier to determine experimentally than δ which tends to be small, e.g. of the order of 10^{-2} mm. Thus

$$k_m = \frac{dm}{dt}\left(\frac{1}{c_i - c_b}\right) = \frac{dm}{dt}(c_b - c_i)$$

This effective boundary layer theory applies to the transfer of substances to or away from interfaces where the viscosity difference between the two phases at the interface is sufficiently large, e.g. $\sim 10^2$. In this respect liquid metal–liquid slag and liquid metal–gas interfaces, as well as liquid–solid and aqueous liquid–gas interfaces, in extraction and refining processes have sufficiently different values for the above theory to apply.

3.6.2 Elucidation of mechanism

Most metallurgical reactions are too complex for the kind of rapid elucidation that has been illustrated above and usually more data are required. Even then there is no guarantee that a completely satisfactory mechanism can be postulated. Below is a list of items which a kineticist would automatically attempt to measure in order to throw some light on the mechanism of a reaction:

(i) *Stoichiometry* (see Section 3.2.4).
(ii) *Order of reaction* (see Section 3.2).
(iii) *Rate constant*. Measured at different temperatures, this yields activation energy (see item (v) below).
(iv) *Activation energy*. Can sometimes indicate the RDS, for example if the value is high, it may be that the breaking of a strong bond is a controlling factor.
(v) *Correlation of data*. The absolute values of rate constant (k), activation energy (E_A) and pre-exponential factor (A) will not in themselves suggest a mechanism but can be compared with the parameters for analogous reactions for which there are established mechanisms. An example of this is a reaction which is thought to be diffusion-controlled. The measured values of E_A and k should then agree with the values found for the diffusion process alone, determined from separate diffusion experiments. Of course it is not always as straightforward as this but the method is still valuable. In the case of sulphur transfer from iron to slag[7], the activation energies for sulphur transfer to the slag and silicon transfer from the slag to the metal agree quite well and it may be concluded that there is a complex mechanism in which silicon transfer is the RDS or that both processes have a common controlling step. This could easily be the breaking of the Si–O bonds, an idea which is consistent with the value of the measured E_A.
(vi) *Other information*. Apart from obvious facts about chemical reactions (e.g. ionic reactions are fast), other data may be useful. For example, in the denitriding of iron[10], it is supposed that nitrogen atoms, N, diffuse to the surface of the sample where recombination to molecular nitrogen (N_2) occurs and that diffusion is too fast for this to be the RDS. This is confirmed by X-ray photography which shows that there is apparently no nitrogen concentration gradient in the sample, indicating that diffusion can easily maintain loss of nitrogen from the surface.

3.7   Theories of reaction rates

3.7.1   Collision theory

According to the collision theory, the rate is given by the frequency of the 'fruitful' collisions, i.e.

$$\text{rate} = \text{collision frequency} \times \text{fraction of 'fruitful' collisions} \qquad (3.24)$$

An attempt will be made here to quantify the factors in equation (3.24).

3.7.1.1   Collision frequency (Z)

The collision frequency for bimolecular gas reactions is obtained from first principles from the kinetic theory of gases and it can be shown[2] that for collisions between molecules A and B the collision frequency per unit volume (Z_{AB}) is given by:

$$Z_{AB} = n_A n_B \sigma_{AB}^2 \left(\frac{8\pi RT}{\mu} \right)^{1/2} \qquad (3.25)$$

where n_A and n_B are the number of molecules per unit volume of A and B respectively, σ_{AB} ($= (\sigma_A + \sigma_B)/2$) is the mean molecular diameter, and μ is the reduced mass, given by

$$\mu = \frac{M_A M_B}{M_B + M_A}$$

where M_A and M_B are the molar masses (molecular weights) of A and B respectively. Thus, collision frequency can be found from fundamental physical properties of the gases, measured in separate experiments. Molecular diameters for example may be found from viscosity measurements[2].

Maxwell–Boltzmann theory

The fraction of molecules with kinetic energy greater than or equal to E_A is found from the Maxwell–Boltzmann theory[2] which gives the fraction of molecules, dn_V, within the velocity range V to $V + dV$ in terms of velocity:

$$dn_V = 4\pi \left(\frac{M}{2\pi RT} \right)^{3/2} \exp(-MV^2/RT)V^2 \, dV \qquad (3.26)$$

A more convenient form* gives the fraction of molecules dn_E in the kinetic energy range E to $E + dE$:

$$dn_E = \frac{1}{RT} e^{-E/RT} dE \qquad (3.27)$$

Hence, the fraction of molecules between $E = E_A$ and $E = \infty$ can be found by integrating equation (3.27) between these limits:

$$\int_{n_{E_A}}^{n_\infty} dn_E = \int_{E_A}^{\infty} \frac{\exp(-E/RT)}{RT} \, dE$$

$$= \exp(-E_A/RT) \qquad (3.28)$$

Now, combining equations (3.28) and (3.25) an expression is obtained for the rate of reaction:

---

*This assumes two degrees of freedom, i.e. velocity in two dimensions (the head-on collision), and while this yields a more convenient result it is not easily justified[2].

$$\text{rate} = n_A n_B \sigma_{AB}^2 \left(\frac{8\pi RT}{\mu}\right)^{1/2} \exp(-E_A/RT)$$

Since the rate of reaction is proportional to the number of reacting molecules, i.e. rate $= k n_A n_B$, where k is in $m^3 \, s^{-1}$. Thus

$$k = \sigma_{AB}^2 \left(\frac{8\pi RT}{\mu}\right)^{1/2} \exp(-E_A/RT) \tag{3.30}$$

Comparing equation (3.27) with the Arrhenius equation (3.18) it can be seen that

$$A = \sigma_{AB}^2 \left(\frac{8\pi RT}{\mu}\right)^{1/2} \tag{3.31}$$

It is now possible to compare theoretical and experimental values of A. For some reactions there is good agreement, for example the reaction[1]:

$$2HI \longrightarrow H_2 + I_2$$

Goodeve and Jack[10] applied the collision theory to the denitriding of iron to good effect. They calculated the rate constant using a form of equation (3.30), the value of E_A being taken from the experimental data and assuming that σ_N (the nitrogen atomic diameter) is 1.5×10^{-10} m, quite a reasonable value. Their experimental rate constant and the calculated value agreed extremely well. However, for many reactions the experimental rate is a good deal less than that calculated on the basis of simple collision theory. Sometimes discrepancies are explained in terms of steric* effects[2], but this cannot be quantified on a theoretical basis. Clearly, a more sophisticated theory is required: the transition state theory.

3.7.2 Potential energy surfaces

Attempts have been made to calculate a theoretical value of activation energy. Consider a reaction of the type:

$$A + BC \longrightarrow AB + C$$

Detailed calculations must be carried out to determine the potential energy as a function of interatomic distances between A and B and B and C. This was first done by Eyring and Polanyi in 1931 for the reaction:

$$H + H_2 \longrightarrow H_2 + H$$

Their work and the work of others in the field are fully described by Laidler[17] and Moore[2]. Here, only a general outline of the results will be mentioned.

The variation of potential energy can be shown in the form of a three-dimensional model[2] or can be plotted as contours (*Fig. 3.19*). It can be seen that there are two 'valleys' at right angles meeting at a 'col' or 'saddle point', X. This is the lowest point at which it is possible to cross from one 'valley' to the next. The easiest pathway QXR is along the 'valley' floor and over point X (the transition state). The potential energy at point X is the activation energy. Away from the 'valley' floor, the potential energy is greater and a cross section of the 'valley' (*Fig. 3.20*) shows how potential energy varies with interatomic distance σ_{BC} or σ_{AB} as BC or AB vibrates. At Z the molecules are completely dissociated and path QZR represents the most difficult path:

$$A + BC \longrightarrow A + B + C \longrightarrow AB + C$$
$$\text{transition}$$
$$\text{state (Z)}$$

---

*Steric effects are related to unusual shapes and sizes of molecules which physically hinder the approach of reacting molecules.

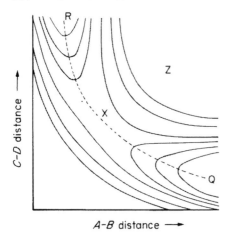

Figure 3.19 Potential energy contours for the reaction $A + B \rightarrow C + D$. The easiest reaction path is QXR

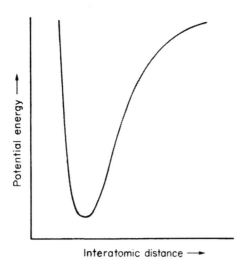

Figure 3.20 Variation of potential energy with interatomic distance

3.7.3   Transition state theory

Transition state theory still assumes that molecular collisions take place but rejects rate as a function of collision frequency. According to the theory, the rate of reaction is given by the rate of decomposition of the transition state. Consider now the reaction:

$$A + BC \longrightarrow ABC^{\ddagger} \longrightarrow AB + C$$
$$\text{transition} \atop \text{state}$$

where $\ddagger$ indicates the transition state. It is assumed that the reactants and the transition state are in equilibrium, i.e.

$$A + BC \rightleftharpoons ABC^{\ddagger}$$

for which the equilibrium constant, $K^{\ddagger}$, is given by:

$$K^{\ddagger} = \frac{[ABC^{\ddagger}]}{[A][BC]}$$

Thus

$$[ABC^{\ddagger}] = K^{\ddagger}[A][BC] \tag{3.32}$$

Now, rate is given by:

$$\text{rate} = f[ABC^{\ddagger}] \tag{3.33}$$

where f is the frequency of break-up of $ABC^{\ddagger}$, and if $ABC^{\ddagger}$ lasts for only one molecular vibration, f is the vibrational frequency of $ABC^{\ddagger}$.

Therefore, combining equations (3.32) and (3.33):

$$\text{rate} = fK^{\ddagger}[A][BC] \tag{3.34}$$

But rate $= k_1[A][BC]$ where k_1 is the rate constant for the forward reaction, and so

$$k_1 = fK^{\ddagger} \tag{3.35}$$

From statistical thermodynamics[2], it can be shown that $K^{\ddagger}$ is given by:

$$K^{\ddagger} = \frac{Q^{\ddagger}}{Q_A Q_{BC}} \cdot \frac{kT}{hf} \exp(-E_0/RT) \tag{3.36}$$

where Q is used to represent functions known as *partition functions* and these are obtained, in terms of physical constants and physical properties of the molecules, from statistical thermodynamics. $Q^{\ddagger}$, Q_A, and Q_{BC} are the partition functions of the transition state, A and BC respectively. The terms k and h which appear on the right-hand side of equation (3.36) are the Boltzmann and Planck constants respectively. E_0 is the energy difference between the lowest energy for $ABC^{\ddagger}$ and the sum of the lowest energy levels of A and BC.

Substituting equation (3.36) in equation (3.35) gives:

$$k_1 = \frac{kT}{h} \cdot \frac{Q^{\ddagger}}{Q_A Q_{BC}} \exp(-E_0/RT) \tag{3.37}$$

which is the Eyring equation. The term

$$\frac{kT}{h} \cdot \frac{Q^{\ddagger}}{Q_A Q_{BC}}$$

corresponds to A, the pre-exponential factor. It is now possible to calculate A using known molecular parameters and spectroscopic data.

Estimates must be made for parameters for the transition state using data on known molecules with similar structures. For many reactions a good agreement is obtained between theoretical and actual values of the pre-exponential factor. For example[2], for the reaction:

$$NO + Cl_2 \longrightarrow NOCl + Cl$$

$\log A$ (observed) $= 12.6$ cm^3 mol^{-1} s^{-1}

$\log A$ (collision theory) $= 14.0$ cm^3 mol^{-1} s^{-1}

$\log A$ (transition state theory) $= 12.1$ cm^3 mol^{-1} s^{-1}

There are other approaches to transition state theory and one of these gives rate in terms of the standard entropy of activation, $\Delta S^{\ominus\ddagger}$ (the standard entropy difference between

$A + BC$ and $ABC^{\ddagger}$). Experimental values of $\Delta S^{\ominus \ddagger}$ give a good indication of the nature of the transition state. If for example $\Delta S^{\ominus \ddagger}$ is large and negative, it indicates a tightly bound transition state, that is the transition state is more ordered than the reactant from which it is formed.

REFERENCES

1. Avery, H.E. 1974 *Basic Reaction Kinetics and Mechanisms*. London: MacMillan.
2. Moore, W.J. 1972 *Physical Chemistry*. London: Longman.
3. Heys, H.L. 1985 *Physical Chemistry*, 5th edn. London: Harrap.
4. Kubaschewski, O. and Hopkins, B.E. 1967 *Oxidation of Metals and Alloys*. London: Butterworths.
5. Barrow, G.M. 1979 *Physical Chemistry*, 4th edn. New York: McGraw-Hill.
6. Mackowiak, J. 1966 *Physical Chemistry for Metallurgists*. London: George Allen & Unwin.
7. Ward, R.G. and Salmon, K.A. 1960 *J. Iron Steel Inst.*, **196**, 393.
8. Collin, R., Brachaczek, M. and Thulin, D. 1969 *J. Iron Steel Inst.*, **207**, 1122.
9. Atherton, M.A. and Lawrence, J.K. 1970 *An Experimental Introduction to Reaction Kinetics*. London: Longman.
10. Goodeve, C. and Jack, K.H. 1948 *Dis. Faraday Soc.*, **4**, 82.
11. Caldin, E.F. 1964 *Fast Reactions in Solution*. Oxford: Blackwell Scientific Publications.
12. Browning, D.R. 1973 *Chromatography*. London: Harrap.
13. Whitfield, R.C. 1970 *Spectroscopy in Chemistry*. London: Longman.
14. *Findlay's Practical Physical Chemistry*, 1973, revised by B.P. Levitt. London: Longman.
15. Kurdjumov, G.V. 1960 *J. Iron Steel Inst.*, **195**, 41.
16. Murray, P.R.S. 1977 *Principles of Organic Chemistry*, 2nd edn. London: Heinemann Educational.
17. Laidler, K.J. 1973 *Chemical Kinetics*, 2nd edn. New York: McGraw-Hill.
18. Edwards, J.O., Greene, E.F. and Ross, J. 1968 *J. Chem. Educ.*, **45**, 381.
19. Smithells, C.J. 1983 *Metals Reference Book*, 6th edn. London: Butterworths.

For self-test questions on this chapter, see Appendix 5.

4

Liquid Metal Solutions

4.1 Solution and composition

A solution may be defined as a homogeneous mixture of two or more substances; it is a single phase. The substance present in the larger amount is the solvent and the substance added to the solvent is the solute. In contrast, a compound is homogeneous but whereas a solution can accommodate changes in concentration a compound cannot accommodate such changes without precipitating a new phase. The composition of the solution may be measured in:

(i)   weight per cent (wt %); i.e.

$$\frac{\text{wt of solute}}{\text{wt of solute} + \text{wt of solvent}} \times 100\%$$

(ii)   molal composition (M) is

$$\frac{\text{no. of moles* of solute}}{\text{total no. of moles (solvent + solute)}}$$

This is usually expressed as a percentage. *Molality* is the number of moles of solute per 1000 g of solvent and *molarity* is the number of moles of solute per litre (dm^{-3}) of solution. These terms are generally used in analysis work.

(iii)   Mole fraction (N) of a component in a solution is the number of gram molecules (n) of that component expressed as a fraction of the total number of gram molecules; i.e.

$$N = \frac{\text{no. of gram molecules of solute } (n)}{\text{total no. of gram molecules (solvent + solute)}}$$

The number of gram molecules of a substance (n) is calculated by dividing the weight of the substance in grams by its molecular weight.

The mole fraction is used most extensively in theoretical work because many physical properties of solutions are expressed most simply in these terms. Thus the mole fraction of copper in a liquid metal solution containing 6 kg of Cu and 1 kg of Zn may be determined by:

$$N_{Cu} = \frac{6000/63.55}{(6000/63.55) + (1000/65.37)} = 0.86$$

Similarly,

$$N_{Zn} = \frac{1000/65.37}{(6000/63.55) + (1000/65.37)} = 0.14$$

*1 mole = 1 gram molecule or formula weight.

or, since $N_{Cu} + N_{Zn} = 1$,

$$N_{Zn} = 1 - N_{Cu}$$
$$= 1 - 0.86$$
$$= 0.14$$

A system (solution) composed of two chemical substances is a binary solution, and a ternary solution is one composed of three substances while a solution containing many substances may be termed a multicomponent system.

4.2 Surface and interfacial energy

The surface particles (atoms, molecules, ions), of a liquid or solid will have a higher free energy than those in the bulk of the substance. This can be explained diagrammatically in *Fig. 4.1* by considering the simplest case of a particle, A, with four direction bonds in a simple cubic arrangement. It is clear that all four bonds have been satisfied while for particle B the surface bonds are not satisfied.

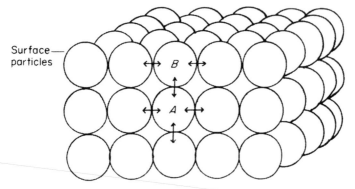

Surface—
particles

B

A

Figure 4.1 Diagrammatic representation of bonding between particles in the bulk (A) and in the surface (B) of a solid or liquid. The unsatisfied bonds of particles in the surface give rise to a surface free energy, γ, which is larger than the free energy of particles in the bulk of the solid or liquid

Bonding between particles will only take place with a lowering of free energy. Therefore, A-particles will have a lower free energy than B-particles since each particle in the surface has one unsatisfied bond and will therefore have a higher or more positive free energy value. This gives rise to a resultant *surface free energy* (SFE), the magnitude of which will be dependent on the total surface area. Thus, SFE may be defined as the energy required to create unit area of new surface. Similarly, if two surfaces meet adsorption of one component on another will only take place if there is a lowering of the surface free energies. If there is no lowering of SFE adsorption will not take place. Each liquid will try to become more stable by reducing its surface area in order to lower its surface free energy. Hence, there will be a contracting force which produces a resultant tension in the surface. This is called *surface tension* and is the force acting at right angles to a line drawn of unit length in the surface, and as such will have the same value as surface free energy. By the same considerations liquids will tend to form drops providing there are no constrictions at the surface, since a sphere has the smallest surface area to volume ratio of any shape. Surface tension is normally used when referring to liquids, while surface free energy and *interfacial energy* (the resultant SFE between two phases at an interface) are used when referring to liquids and solids. They are each given the same symbol (γ) and units (joules per unit area) and as such provide

Table 4.1 SURFACE FREE ENERGIES OF VARIOUS MATERIALS

Liquid	Temperature (°C)	Surface free energies $(\times 10^{-9} \text{ J mm}^{-2})$
Water	50	73
Water	100	59
Sodium chloride	910	106
Steel making slag	1600	400
Mercury	20	480
Mercury	39	467
Zinc	419	754
Zinc	640	761
Tin	232	531
Tin	1000	497
Lead	232	444
Lead	1000	401
Cadmium	320	630
Antimony	1000	355
Copper	1131	1103
Copper	1150	1034
Silver	995	1128
Silver	1120	923
Iron	1600	1360
Many organic liquids		20–40
Fused salts		100
Liquid slags		300–500*
Solid oxides		100–1000
Solid metals		1000–2000

*Decreases with increasing SiO_2 and additions of alkali oxides, boric acid, calcium sulphide, phosphoric oxide, and increases with additions of lime, iron oxide, alumina

a constant property of the liquid or solid. Metals exhibit strong bonding between molecules in highly coordinated structures. A comparison of surface energies for various metals and compounds is given in *Table 4.1*.

Some metallurgical processes in which interfacial energy becomes important include: (a) froth flotation; (b) removal of deoxidation products and solid impurities from a liquid melt; (c) nucleation of gas within a liquid melt, e.g. CO in steelmaking; (d) boundary lubrication in metal processing; (e) precipitate shape in solid–solid transformations; (f) metal–mould penetration. In examples (a), (c), (d), (e) a low interfacial energy between the two components concerned will facilitate the process, while a high interfacial energy will aid (b) and prevent (f).

4.3 Thermodynamics of solutions

The properties of liquid metal solutions become important when considering pyrometallurgical extraction and refining processes, melting operations and corrosion phenomena. The transfer of components between a liquid metal and a liquid slag will largely depend on their thermodynamic properties. The main thermodynamic functions to be considered are free energy (G), enthalpy (H), entropy (S), and volume (V). The variation of G, H and S on melting a pure metal is represented in *Fig. 4.2*.

During melting, a metal's long-range atomic order, e.g. FCC, BCC, HCP, is broken down to provide a much smaller range of order. This can be shown by comparing the coordination number, that is the number of atoms in near proximity to each other. The coordination number in most solid metals is 12 while that for most liquid metals is 11. Therefore, there has been a decrease in the amount of order due to melting. This indicates that the entropy of a liquid metal is greater than that for a solid metal and

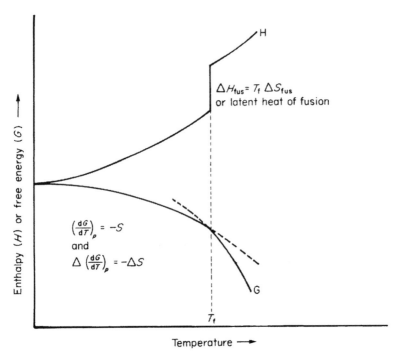

Figure 4.2 Enthalpy and free energy relationships around the melting point, T_f, of a metal. (From *Principles of Extractive Metallurgy* by T. Rosenqvist, © 1974 McGraw-Hill Inc. Used with permission of McGraw-Hill Book Co.)

the entropy change (ΔS_{fus}) on fusion or melting is positive. The more complex solid structures, such as silicon and bismuth, exhibit larger positive values of ΔS_{fus}. On the other hand, strongly disordered solids such as Ag_2S provide only a small positive ΔS_{fus} value.

Most substances melt with a volume expansion, again emphasising an increase in the amount of disorder and therefore in entropy. The relationship between transformation temperature (T) and pressure (p) and volume (V) is given in one form of the Clausius Clapeyron equation:

$$\frac{dp}{dT} = \frac{\Delta H}{T \Delta V}$$

since for an ideal gas, $\Delta V = RT/p$ (see equation (2.35)). Thus, increasing the pressure on melting requires a corresponding increase in dT or melting point.

4.3.1   Partial and integral quantities

Apart from the changes in G, H and S on melting the liquid metal's thermodynamic functions will be dependent upon the components present in solution. In this respect there are two main thermodynamic quantities which need to be known and explained, i.e. partial molar and integral molar quantities.

The partial molar quantities are those which relate to each individual component or part of the solution. For example, in a liquid metal solution containing zinc and cadmium, zinc will have its own thermodynamic functions denoted by $\bar{H}_{Zn}$, $\bar{S}_{Zn}$, $\bar{G}_{Zn}$, $\bar{V}_{Zn}$ and copper will have its own thermodynamic functions $\bar{H}_{Cu}$, $\bar{S}_{Cu}$, $\bar{G}_{Cu}$, $\bar{V}_{Cu}$. A partial molar quantity has been defined as any thermodynamic function which is characteristic of the state of the system and which depends on the amount of that substance present in the system. At the same time the liquid metal solution will have its total or integral thermodynamic functions of H, S, G and V, which are dependent upon the partial thermodynamic functions. The relationships between integral and partial functions for a solution containing components 1, 2, 3 are:

$$H = n_1 \bar{H}_1 + n_2 \bar{H}_2 + n_3 \bar{H}_3$$
$$S = n_1 \bar{S}_1 + n_2 \bar{S}_2 + n_3 \bar{S}_3$$
$$G = n_1 \bar{G}_1 + n_2 \bar{G}_2 + n_3 \bar{G}_3$$

where

$$n_1 = \text{no. of moles of 1}$$
$$n_2 = \text{no. of moles of 2}$$
$$n_3 = \text{no. of moles of 3}$$

The integral *molar* function is determined by dividing by the total number of moles, $n_{tot} (= n_1 + n_2 + n_3)$,

$$\frac{H}{n_{tot}} = \frac{n_1}{n_{tot}} \bar{H}_1 + \frac{n_2}{n_{tot}} \bar{H}_2 + \frac{n_3}{n_{tot}} \bar{H}_3$$

where H/n_{tot} is given the symbol H_m (integral molar enthalpy), i.e.

$$H_m = N_1 \bar{H}_1 + N_2 \bar{H}_2 + N_3 \bar{H}_3$$

where

$$N_1 = \frac{n_1}{n_{tot}}, \quad N_2 = \frac{n_2}{n_{tot}}, \quad N_3 = \frac{n_3}{n_{tot}}$$

The *relative* integral molar function, i.e. change in integral molar function is therefore given by:

$$\Delta H_m = N_1 \Delta \bar{H}_1 + N_2 \Delta \bar{H}_2 + N_3 \Delta \bar{H}_3 \qquad (4.1)$$

The same relationship may be used to relate ΔG_m to $\Delta \bar{G}$, ΔS_m to $\Delta \bar{S}$, and ΔV_m to $\Delta \bar{V}$:

$$\Delta G_m = N_1 \Delta \bar{G}_1 + N_2 \Delta \bar{G}_2 + N_3 \Delta \bar{G}_3 \qquad (4.2)$$

$$\Delta S_m = N_1 \Delta \bar{S}_1 + N_2 \Delta \bar{S}_2 + N_3 \Delta \bar{S}_3 \qquad (4.3)$$

$$\Delta V_m = N_1 \Delta \bar{V}_1 + N_2 \Delta \bar{V}_2 + N_3 \Delta \bar{V}_3 \qquad (4.4)$$

Hence, it is possible to determine the integral molar functions of a metal solution knowing all the partial molar functions and compositions (mole fractions). In some circumstances $\bar{G}$, the partial molar free energy, is denoted by μ, the chemical potential of a liquid (see Sections 6.5.1 and 6.7.1).

4.3.2 Ideal solutions and activity

An ideal solution is one in which its component parts act individually of each other and do not interact. Therefore, an ideal solution may be adequately described by considering the individual concentrations of its components. However, when these components interact simple consideration of their concentrations will not provide a sufficiently accurate expression of the solution.

The active mass or *activity* of the components must be used in place of concentration (see Section 2.2.3.3) for most solutions since the majority of liquid metal systems are non-ideal (i.e. real). It is possible to explain the concept of activity, a, of an element or component as being the amount of that element or component which is *free* to react with any other element or component added to the solution; the amount being expressed as a *fraction* of a *chosen standard state* or condition. Normally, the standard state is considered to be an activity of one such that this will apply to a pure element or compound. Therefore, in an ideal solution where there is no interaction between the components, the activity of a component will be the same value as its mole fraction in that solution: both describing the amount as a fraction which is free to react. Thus for component 1 in solution:

$$a_1 = N_1 \qquad (4.5)$$

However, most solutions will contain constituents which react with each other. Indeed, reactions between different solutions and their components is desirable in most metal-slag refining reactions. A simple example may be used to explain the difference between activity and concentration. Since an acid will react with a base, a liquid slag composed entirely of an acidic component, SiO_2, and a basic component, CaO, will result in a reaction between these two producing calcium silicate, $CaSiO_3$:

$$CaO + SiO_2 \longrightarrow CaSiO_3$$

The slag now no longer consists of 'free' components of CaO and SiO_2 acting individually in the solution but rather as $CaSiO_3$. If the slag is made up of equal molecules of CaO (molecular weight 56.1) and SiO_2 (molecular weight 60.1), the weight per cent of CaO is

$$\frac{56.1}{56.1 + 60.1} \times 100 = 48.3\%$$

and of SiO_2 is

$$\frac{60.1}{56.1 + 60.1} \times 100 = 51.7\%$$

However, the amount of CaO which is *free* to react with any SiO_2 which is added is negligible since all the lime has reacted with silica to form $CaSiO_3$. Similarly, the amount of *free* SiO_2 present is negligible and hence the activities of CaO and SiO_2 are extremely small. In this case activity and concentration show a marked contrast to each other. Adding further lime or silica will eventually result in an excess of that compound and hence increase its activity in the solution. Therefore, consideration of the activity of a component in a solution is more informative than its concentration or mole fraction.

It may be more meaningful to change the standard state if conditions are far remote from the normal standard state of unit activity for a pure component. For example, it is usual to adopt a standard state of 1 Wt% when solute elements are present to less than 1 Wt% (see Section 4.3.6).

4.3.3 Raoult's law

Ideal solutions obey Raoult's law which states that the relative lowering of the vapour pressure, p, of the solvent due to the addition of a solute is equal to the mole fraction of the solute in solution. Thus in a solution containing components 1 and 2:

$$\frac{\overset{\circ}{p}_1 - p_1}{\overset{\circ}{p}_1} = N_2$$

where $\overset{\circ}{p}_1$ is the vapour pressure of pure component 1, p_1 is the vapour pressure of component 1 (solvent) in solution and N_2 is the mole fraction of component 2 (solute) in solution. Similarly,

$$\frac{\overset{o}{p}_2 - p_2}{\overset{o}{p}_2} = N_1$$

where $\overset{o}{p}_2$ is the vapour pressure of pure component 2, p_2 is the vapour pressure of component 2 in solution and N_1 is the mole fraction of component 1 in solution. Since $N_1 + N_2 = 1$, Raoult's law may be rewritten:

$$\frac{\overset{o}{p}_1 - p_1}{\overset{o}{p}_1} = 1 - N_1$$

or

$$\overset{o}{p}_1 - p_1 = \overset{o}{p}_1 - \overset{o}{p}_1 N_1$$

and so

$$-p_1 = -\overset{o}{p}_1 N_1$$

or

$$p_1 = \overset{o}{p}_1 N_1 \tag{4.6}$$

Similarly,

$$p_2 = \overset{o}{p}_2 N_2 \text{ , or } \frac{p_2}{\overset{o}{p}_2} = N_2$$

Equation (4.6) is more usually applied since it directly relates $\overset{o}{p}_1$, p_1 and N_1.

An ideal gas obeys Raoult's law which may be represented graphically by plotting vapour pressure against mole fraction. This gives a straight line with $\overset{o}{p}_1$ and $\overset{o}{p}_2$ being

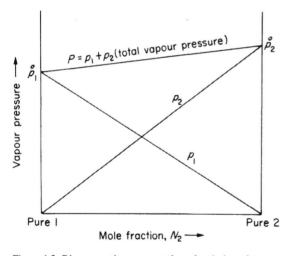

Figure 4.3 Diagrammatic representation of variation of vapour pressure of a liquid metal mixture (components 1 and 2) with mole fraction

the intersection of the line with the vapour pressure axes as shown in *Figure 4.3*. Since the solution is ideal the gas pressures p_1 and p_2 do not interact, they obey the ideal gas equation ($PV = RT$) and their physical properties will be additive, i.e.

$$\text{total vapour pressure} = P = p_1 + p_2$$

4.3.4   Non-ideal or real solutions

Deviations from Raoult's law occur when the attractive forces between the molecules of components 1 and 2 of the solution are stronger or weaker than those that exist between 1 and 1 or 2 and 2 in their pure states. For example if there were a stronger attractive force between components 1 and 2 in solution than the mutual attraction between molecules of 1 or molecules of 2 there would be less tendency for these components to leave the solution as a vapour. In this case the vapour pressure would be less than that predicted by Raoult's law and the vapour pressure–mole fraction graph would be as shown in *Fig. 4.4*. This is called *negative deviation* from Raoult's law. Intermetallic compounds and solid solutions exhibit negative deviation since the attractive force between the components is high.

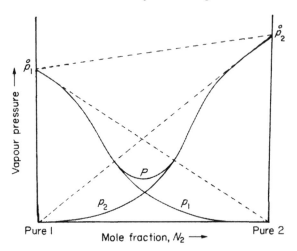

Figure 4.4 Diagrammatic representation of negative deviation from Raoult's law. Broken lines represent ideal behaviour

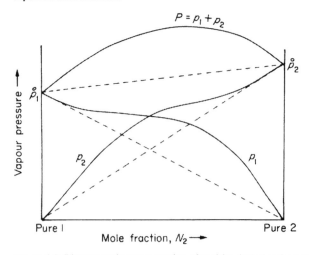

Figure 4.5 Diagrammatic representation of positive deviation from Raoult's law. Broken lines represent ideal behaviour

Using a similar argument, if the attractive force between 1 and 2 was less than that between 1 and 1 or 2 and 2 in the their pure states there would be a greater tendency for these components to leave the solution as a gas, thereby increasing the vapour pressure above the liquid. This is represented graphically in *Fig. 4.5*, and is called *positive deviation* from Raoult's law. Immiscible liquids exhibit positive deviation since the attractive force between the components in the liquid is low.

Not only will the respective attractive or bonding forces between molecules lead to deviation from Raoult's law but so also will differences in size of molecules since these will result in differences in distances between centres of molecules leading to a change in the bonding forces. Similarly, interaction between vapour molecules will result in changes in vapour pressure and a deviation from Raoult's law. For a solution which deviates from Raoult's law, the activity of a component in the solution will not be equal to its mole fraction. In this case an *activity coefficient*, γ, must be used to relate the two and which will be greater than unity for solutions producing positive deviation and less than unity for solutions producing negative deviation, i.e.

$$a_1 = \gamma_1 N_1 \tag{4.7}$$

However, for non-ideal or *real* solutions the relationship between the vapour pressure of a component in the solution and its concentration or mole fraction (equation (4.6)), must be rewritten replacing mole fraction with activity.

$$p_1 = \overset{o}{p}_1 a_1 \tag{4.8}$$

and hence substituting equation (4.7) into equation (4.8):

$$p_1 = \overset{o}{p}_1 \gamma_1 N_1 \tag{4.9}$$

It is therefore possible to determine the activity and activity coefficient of a component in solution by measuring its vapour pressure (p_1), knowing its mole fraction (N_1), and its vapour pressure in the pure state ($\overset{o}{p}_1$). Plotting a graph of activity of a component (1) in a solution against its mole fraction (*Fig. 4.6*) describes the effect of deviation from Raoult's law.

From the foregoing requirements of an ideal solution, *viz.* component parts have similar sized molecules and no interaction between the component molecules, it

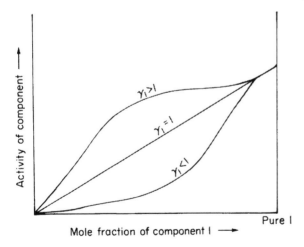

Figure 4.6 Diagrammatic representation of negative ($\gamma_1 < 1$) and positive ($\gamma_1 > 1$) deviation from Raoult's law ($\gamma_1 = 1$). (The curves would take the same form if vapour pressure of component 1 was plotted against mole fraction—see *Fig. 4.10*.)

seems unlikely that ideal solutions would be anything but rare. This is indeed the case. An example of a liquid metal mixture which produces near ideal solutions is bismuth–tin.

The term real solutions is often applied to non-ideal solutions. Nevertheless, ideal behaviour may be assumed as a first approximation for solutions of chemically similar components or elements, while Raoult's law ($p_1 = \overset{o}{p}_1 N_1$) may be obeyed for a certain concentration range or at a certain temperature even though the solution on the whole is not ideal.

4.3.5   The Gibbs–Duhem equation

Of some interest to the more advanced student will be the Gibbs–Duhem equation which allows calculation of the activity or activity coefficient of one component knowing that of the other component in a binary solution.

Subjecting equation (4.2) to total differentiation for the binary solution consisting of components 1 and 2 gives:

$$d(\Delta G_m) = N_1 d(\Delta \bar{G}_1) + \Delta \bar{G}_1 dN_1 + N_2 d(\Delta \bar{G}_2) + \Delta \bar{G}_2 dN_2$$

If the composition of the solution is maintained constant dN_1, dN_2, $d(\Delta G_m)$ are each zero, and hence:

$$N_1 d(\Delta \bar{G}_1) + N_2 d(\Delta \bar{G}_2) = 0 \qquad (4.10)$$

On forming a binary liquid metal solution the activity of component 1 will change from its pure or standard value (unity) to a_1. Therefore the molar free energy of component 1 changes from its standard molar free energy, $G_1^{\ominus}$, to its partial molar free energy, $\bar{G}_1$, which is given in Section 2.2.3.6 as $\bar{G}_1 = G_1^{\ominus} + RT \ln a_1$. Hence the change in partial molar free energy

$$\Delta \bar{G}_1 = G_1^{\ominus} + RT \ln a_1 - G_1^{\ominus}$$
$$= RT \ln a_1$$

and, similarly for component 2

$$\Delta \bar{G}_2 = RT \ln a_2 \qquad (4.10)$$

Substituting these values for $\Delta \bar{G}_1$ and $\Delta \bar{G}_2$ into equation (4.10),

$$N_1 d(RT \ln a_1) + N_2 d(RT \ln a_2) = 0$$

Dividing by RT gives

$$N_1 d(\ln a_1) + N_2 d(\ln a_2) = 0 \qquad (4.11)$$

or

$$d(\ln a_2) = -\frac{N_1}{N_2} d(\ln a_1)$$

Integrating between $N_2 = 1$ and $N_2 = X$ (X is the mole fraction at which a_2, the activity of component 2, is needed to be calculated):

$$\ln a_2 = -\int_1^X \frac{N_1}{N_2} d(\ln a_1) \qquad (4.12)$$

Thus by plotting N_1/N_2 against $-\ln a_1$ (*Fig. 4.7*) and measuring the area under the curve (hatched) it is possible to determine the value $\ln a_2$ between mole fractions $N_2 = X$ and $N_2 = 1$. The usefulness of this integration is restricted by the fact that when N_1/N_2 is zero (i.e. pure component 2) $\ln a_1$ will be minus infinity and the measurement of the area under the curve becomes difficult: when the solution is composed totally of pure component 1, N_1/N_2 will be plus infinity. This will not create a problem in measuring the area under the curve since interest lies only in determining a_2, i.e. those solutions

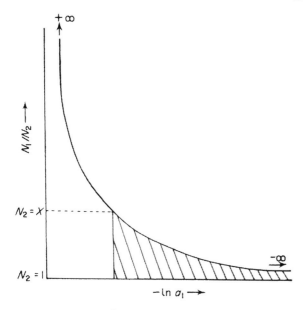

Figure 4.7 Plot of N_1/N_2 against $-\ln a_1$ for a binary liquid metal solution

containing component 2, although some difficulty may be experienced when N_2 is extremely small. Replacing activity with activity coefficient in equation (4.12) surmounts the problem mentioned above (i.e. $N_1/N_2 = 0$, $\ln a_1 = -\infty$) since

$$\ln \gamma_2 = -\int_1^X \frac{N_1}{N_2} \, d(\ln \gamma_1) \tag{4.13}$$

in which case the lower integration limit is set by the pure component 2, i.e. $N_1/N_2 = 0$ and $\ln \gamma_2 = 0$ (*Fig. 4.8*). *Fig. 4.8* represents a solution which exhibits negative deviation from Raoult's law. For a solution which exhibits positive deviation

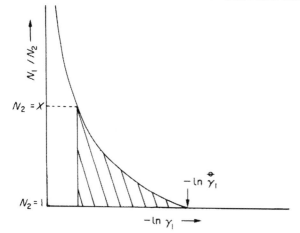

Figure 4.8 Plot of N_1/N_2 against $-\ln \gamma_1$ for a binary solution which exhibits negative deviation

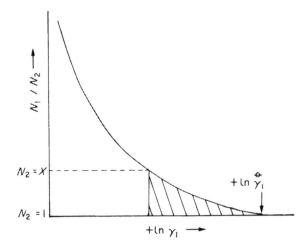

Figure 4.9 Plot of N_1/N_2 against $+\ln \gamma_1$ for a binary solution which exhibits positive deviation

the value of $\ln \gamma_1$ will be positive and can be determined by plotting N_1/N_2 versus $+\ln \gamma_1$ (*Fig. 4.9*).

Determining the unknown activity coefficient of the component 2 allows its activity to be calculated using $a_2 = \gamma_2 N_2$.

Application of this technique is particularly useful in determining the activity of a non-volatile component mixed with a volatile component. The activity of the volatile component is measured from its vapour pressure ($p_1 = \overset{o}{p}_1 a_1$), converting a_1 to γ_1 using $a_1 = \gamma_1 N_1$ and subsequently determining γ_2 and hence a_2 using the graphical plot of the integrated Gibbs–Duhem equation as described above.

It is also possible to apply the integrated form of the Gibbs–Duhem equation to ternary or more complicated solutions[1]. From *Figs 4.8* and *4.9*, γ_1 is shown to reach a finite value when N_1/N_2 equals zero (pure component 2). This finite value is used as a standard state ($\overset{\leftrightarrow}{\gamma}_1$), with which other activity coefficients are compared.

4.3.6 Henry's law and dilute solutions

Whereas Raoult's law accounts for the lowering of the vapour pressure of the *solvent* by addition of the solute to a solution, Henry's law is concerned with the vapour pressure of the solute. Henry's law states that the mass of a gas, i.e. *solute*, dissolved in a given volume of solvent at constant temperature is proportional to the pressure of the gas in equilibrium with the solution. Hence, comparing Raoult's law and Henry's law:

$$p_1 \propto N_1 \text{ for the solute} \qquad \text{Henry's law}$$
$$p_2 = \overset{o}{p}_2 N_2 \text{ for the solvent} \qquad \text{Raoult's law}$$

(4.14)

Henry's law does not hold for components which undergo dissociation or association in solution. When Henry's law is valid for the solute, Raoult's law is valid for the solvent.

As mentioned earlier, with increasing dilution the activity coefficient of the solute is found to approach a finite value, $\overset{\leftrightarrow}{\gamma}$. At these dilutions the mole fraction of the solute is often less than 0.01 and therefore to compare properties at this dilution with those of a standard state of unit Raoultian activity would be somewhat misleading. Hence, for dilute metal solutions a new standard state of 1 wt% solute* is used with which

*In aqueous metal systems a molar solution (molecular weight in grams of component in 1 litre of water) is used as the standard state. In some texts Henrian activity is given the symbol, h, but in this book the symbol, a, is used for both Henrian and Raoultian activity.

comparison of various properties may be made more meaningful. Since a new standard state is used when considering Henry's law and dilute solutions, the values of activity and activity coefficient must refer to this standard state and as such the Henrian activity coefficient, f, replaces the Raoultian activity coefficient, γ, i.e.

$$a_1 = f_1 \text{ wt\% } 1$$

By examination of *Fig. 4.10* it is seen that Henry's law is obeyed for dilute solution ($p_1 \propto N_1$) and Raoult's law is obeyed when the mole fraction approaches 1 ($p_1 = \overset{\circ}{p}_1 N_1$).

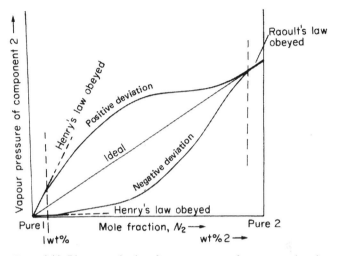

Figure 4.10 Diagrammatic plot of vapour pressure of component 1 against mole fraction for solutions which exhibit positive deviation, negative deviation or ideal behaviour, showing that each type of solution obeys Henry's law at infinite dilution while Raoult's law is obeyed at high mole fractions

From *Fig. 4.10* it is shown that solutions which exhibit either positive or negative deviation from Raoult's law may obey Henry's law at concentrations of less than 1 wt%. As was seen in Raoult's law, the activity coefficient (f in this case) is unity for ideal solutions, is less than one for solutions which exhibit negative deviation and greater than one for solutions which exhibit positive deviation.

4.3.7 Multicomponent solutions and interaction coefficients

It has been assumed up to this point that when there is more than one solute present there is no interaction between them. *Interaction coefficients* can be introduced to indicate the effect of each solute on that under consideration but the assumption is made that interaction between the remaining solutes is negligible. In this respect for a solution containing solutes 1, 2, 3 and 4 interaction coefficients may be used to show the effect of 2 on 1, 3 on 1, and 4 on 1, but no interaction is assumed between 2, 3 and 4. This hypothetical situation may be used to a first approximation to obtain a more accurate determination of the activity of a solute in a multicomponent system. It is often the case that a liquid metal contains several solutes in dilute solution, i.e. less than 1 wt%. The relationship between the Henrian activity coefficient of solute 1 due to interaction with solute 2 (f_1^2) can be represented by

$$\log f_1^2 = e_1^2 \text{ wt\% } 2 \tag{4.15}$$

where e_1^2 is the Henrian interaction coefficient of solute 2 on solute 1. Similarly the effect of solute 3 on solute 1 will be:

$$\log f_1^3 = e_1^3 \text{ wt\% 3}$$

and of solute 4 on solute 1:

$$\log f_1^4 = e_1^4 \text{ wt\% 4}$$

where e_1^3 and e_1^4 are the appropriate interaction coefficients of solute 3 on solute 1 and solute 4 on solute 1 respectively.

If the activity coefficient of solute 1 alone in solution is f_1^1, the Henrian activity of solute 1 in the multicomponent system will be:

$$a_1 = f_1^1 f_1^2 f_1^3 f_1^4 \text{ wt\% 1}$$

or

$$\log a_1 = \log f_1^1 + \log f_1^2 + \log f_1^3 + \log f_1^4 + \log \text{ wt\% 1}$$

Hence,

$$\log a_1 = \log f_1^1 + e_1^2 \text{ wt\% 2} + e_1^3 \text{ wt\% 3} + e_1^4 \text{ wt\% 4} + \log \text{ wt\% 1} \qquad (4.16)$$

If solute 1 reacts more strongly with solute 2 than with the solvent, e_1^2 will be negative since the amount of 'free' solute 1 will be reduced thereby decreasing the activity of solute 1. On the other hand, if solute 1 reacts more strongly with the solvent than solute 2, e_1^2 will be positive resulting in an increase in activity of solute 1 on addition of solute 2.

This theoretical approach has been supported by empirical data. In the same way Raoultian interaction coefficients, ϵ, can be applied to determine the activity of the solvent in a solution.

$$a_1 = \gamma_1^1 \gamma_1^2 \gamma_1^3 \gamma_1^4 N_1$$

where

$$\log \gamma_1^2 = \epsilon_1^2 N_2; \quad \log \gamma_1^3 = \epsilon_1^3 N_3; \quad \log \gamma_1^4 = \epsilon_1^4 N_4$$

Thus,

$$\log a_1 = \log \gamma_1^1 + \epsilon_1^2 N_2 + \epsilon_1^3 N_3 + \epsilon_1^4 N_4 + \log N_1 \qquad (4.17)$$

For a deeper understanding of interaction coefficients the student should consult the references[2,3] at the end of this chapter. The point to be made here, however, is that the concept of activity becomes increasingly complex as the number of components in a solution increases. In practice, of course, these solutions are far more common than simple binary solutions.

4.3.8 Thermodynamics of mixing of solutions

When two or more liquid metals are mixed together there is a change in both the integral and partial thermodynamic functions of the solution. This is due to the contribution of the partial thermodynamic functions of the components to the new solution and to the change in composition.

If there is to be a complete mixing of the two solutions the change in free energy must be negative otherwise the solutions would be immiscible. In other words, the free energy of the mixed solution must be less than the sum of the free energies of its components. Mixing may occur with either evolution (exothermic) or absorption (endothermic) of heat while the entropy change on mixing may also be positive or negative depending on the arrangement and order of the individual components of the solution.

An ideal solution is produced on mixing two liquid components providing:

(i)   there is no change in internal energy, i.e. $\Delta U = 0$ on mixing;
(ii) there is no change in enthalpy, i.e. $\Delta H = 0$ on mixing;
(iii) the physical properties of the components are additive.

The relationship between the free energy change on mixing (ΔG^m) and the mole fractions of the solution's components (1, 2 and 3) for an ideal solution can be deduced from equation (4.2) in which $\Delta \bar{G}_1 = RT \ln N_1$, i.e.

$$\Delta G^m = RT(N_1 \ln N_1 + N_2 \ln N_2 + N_3 \ln N_3) \qquad (4.18)$$

The entropy change on mixing of an ideal solution is given by equation (4.3) in which $\Delta \bar{S} = -R \ln N^*$, i.e.

$$\Delta S^m = -R(N_1 \ln N_1 + N_2 \ln N_2 + N_3 \ln N_3) \qquad (4.19)$$

It has already been stated that for an ideal solution $\Delta H^m = 0$.

As mentioned earlier most real solutions are not ideal. However, there are a number of solutions (*regular* solutions), which vary only slightly from ideal behaviour in that the value of ΔS^m is the same as for an ideal solution while ΔG^m may be obtained by replacing mole fraction by activity in equation (4.18), i.e.

$$\Delta G^m = RT(N_1 \ln a_1 + N_2 \ln a_2 + N_3 \ln a_3) \qquad (4.20)$$

Since $\qquad \Delta G^m = \Delta H^m - T\Delta S^m$ for a real solution,

$$\Delta H^m = \Delta G^m + T\Delta S^m$$

and so

$$\Delta H^m = RT(N_1 \ln a_1 + N_2 \ln a_2 + N_3 \ln a_3) - RT(N_1 \ln N_1 + N_2 \ln N_2 + N_3 \ln N_3)$$

$$= RT\left[N_1 \ln \left(\frac{a_1}{N_1}\right) + N_2 \ln \left(\frac{a_2}{N_2}\right) + N_3 \ln \left(\frac{a_3}{N_3}\right)\right]$$

Since $\qquad a = \gamma N$, or $a/N = \gamma$, hence

$$\Delta H^m = RT(N_1 \ln \gamma_1 + N_2 \ln \gamma_2 + N_3 \ln \gamma_3) \qquad (4.21)$$

Therefore, if the component in solution exhibits negative deviation, γ will be less than one and ΔH^m will be negative, resulting in evolution of heat (exothermic) on mixing. Positive deviation will result in heat absorption (endothermic) on mixing. This provides a means of measuring the activity coefficient and therefore the activity of components on mixing but will only apply to regular solutions. Examples of regular or near regular solutions include: Ag–Cu; Al–Zn; Cd–Zn; Fe–Mn; Fe–Ni; Hg–Sn. Although the above relationships have been used for real solutions to a first approximation, an accurate determination of the thermodynamic functions of real solutions may only be found by empirical assessment of activities and activity coefficients.

4.3.9   Excess thermodynamic quantities

For real solutions the activity coefficient, γ, indicates the extent of departure of a solution from ideal. Scatchard[4] introduced the idea of excess partial molar quantities which

*

$$\Delta \bar{S}_1 = \frac{\Delta \bar{H}_1 - \Delta \bar{G}_1}{T} \qquad .$$

(from equation 2.22), but $\Delta \bar{H}_1 = 0$ for an ideal solution thus,

$$\Delta \bar{S}_1 = \frac{-\Delta \bar{G}_1}{T} = -R \ln N_1$$

may be defined as the difference between the partial molar quantity of a component in a real solution and its partial molar quantity assuming ideal mixing. Thus, the excess partial molar free energy of component 1 in solution is $\bar{G}_1^E$ and is equated to the partial molar free energy, $\bar{G}_1$, and the ideal partial molar free energy, $\bar{G}_1^I$, for the component in solution by:

$$\bar{G}_1^E = \bar{G}_1 - \bar{G}_1^I \tag{4.22}$$

Since

$$\bar{G}_1 = RT \ln a_1 \quad \text{and} \quad \bar{G}_1^I = RT \ln N_1$$

thus,

$$\bar{G}_1^E = RT \ln a_1 - RT \ln N_1$$
$$= RT \ln \left(\frac{a_1}{N_1}\right)$$
$$= RT \ln \gamma_1 \tag{4.23}$$

Therefore, the excess partial molar free energy of a component in a real solution represents an alternative way of expressing its activity coefficient. A solution which exhibits positive deviation from Raoult's law will have a positive value of $\bar{G}_1^E$.

Since the excess partial molar quantities are thermodynamic quantities they will obey all the thermodynamic relationships, e.g.

$$\bar{G}_1^E = \bar{H}_1^E - T\bar{S}_1^E$$

if the solution is regular:

$$\bar{S}_1 = \bar{S}_1^I$$

thus,

$$\bar{S}_1^E = 0$$

and

$$\bar{G}_1^E = \bar{H}_1^E = \bar{H}_1 - \bar{H}_1^I$$

But

$$\bar{H}_1^I = 0$$

thus,

$$\bar{G}_1^E = \bar{H}_1 = RT \ln \gamma_1 \tag{4.24}$$

(compare with equation (4.21)).

For a real solution:

$$\bar{S}_1^E = \bar{S}_1 - \bar{S}_1^I$$

where

$$\bar{S}_1^I = -R \ln N_1, \text{ thus}$$
$$\bar{S}_1^E = \bar{S}_1 - R \ln N_1 \tag{4.25}$$

Excess functions can also be used to express integral thermodynamic quantities for a solution containing components 1 and 2, i.e.

$$G^E = G - G^I$$

whence,

$$G^E = G - RT(N_1 \ln N_1 + N_2 \ln N_2) \tag{4.26}$$

$$S^E = S - R(N_1 \ln N_1 + N_2 \ln N_2) \tag{4.27}$$

4.3.10   Construction of equilibrium phase diagrams from integral free energy–composition curves

The relationship between partial and integral molar quantities may be determined graphically as described below.

On mixing two components to produce a solution the partial molar quantity of a component in the solution may not be the same as that predicted by the pure state of the components. For instance, a certain molar volume of pure component 1 may occupy a different partial molar volume after mixing with component 2. The partial molar volume of a component may be defined as the volume it occupies in the solution. Thus, the partial molar volume of component 1 in a solution containing components 1 and 2 may be defined as the change of volume (dV) of solution with respect to the change in the number of moles of component 1 (dn_1), i.e.

$$\overline{V}_1 \; = \; dV/dn_1 \tag{4.28}$$

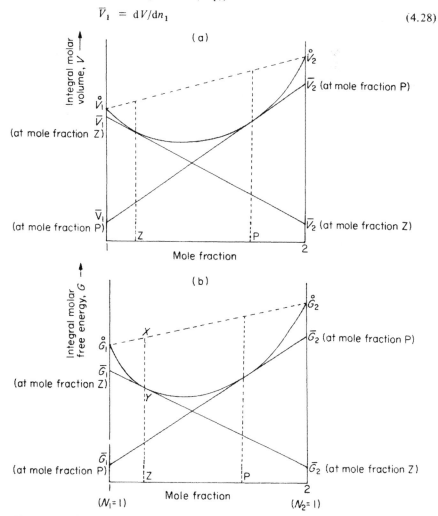

Figure 4.11 Plot of: (a) integral molar volume, and (b) integral molar free energy against mole fraction for a binary metal solution

Similarly, the partial molar free energy of component 1 in a solution of components 1 and 2 may be defined as the change in integral free energy of the solution with respect to the change in the number of moles of component 1:

$$\bar{G}_1 = dG/dn_1 \tag{4.29}$$

Therefore, as shown in *Figs 4.11(a)* and *(b)*, plotting integral molar volumes and free energies against concentration (mole fraction) allows the partial molar function to be determined from the tangent to the curve (dV/dn or dG/dn) at the appropriate mole fraction required. The point at which the tangent cuts the integral volume or integral free energy axis provides the partial molar quantity for each component at that mole fraction.

On mixing components 1 and 2 to provide a composition described by the line XYZ in *Fig. 4.11(b)*, X represents the free energy of mixing if both components behaved as pure substances while point Y represents the free energy of mixing for the real solution of 1 and 2. Hence, the change in free energy on mixing (ΔG^m) is the difference between these two points, i.e. XY. From the lever rule the amount of component 1 to component 2 at this composition would be:

$$\frac{\text{amount of 1}}{\text{amount of 2}} = \frac{Z2}{1Z}$$

The above two curves show complete solution of components 1 and 2 throughout the whole of their composition. If there is a limited range of solution between the two components, two solution curves will result; one being component 1-rich representing limited solution of 2 in 1 (α-phase) the other being component 2-rich representing limited solution of 1 in 2 (β-phase). This type of solution is shown in *Fig. 4.12* by a plot of integral molar free energy against mole fraction.

Drawing a common tangent to the two curves represents the limiting ranges within which the two phases, α and β, may coexist. In this respect the composition range is GJ and the range of partial molar free energies is from $\bar{G}_1$ to $\bar{G}_2$. The point X represents the free energy of a mixture of pure 1 and pure 2 at composition XYZ. The value of ΔG^m is XY while the free energy of solution α is A and the free energy of solution β is

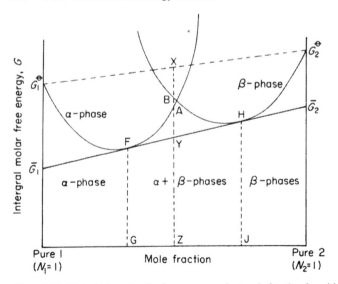

Figure 4.12 Plot of integral molar free energy against mole fraction for a binary metal solution which exhibits a limited range of solid solution

B. Again using the lever rule the amount of α to β at composition Z is ZJ/GZ consisting of α of composition FG and β of composition HJ. Any composition between pure 1 and G will be α and any composition between J and pure 2 will be β.

The free energy–composition curves discussed above assume equilibrium, constant temperature conditions and provide a means of determining the equilibrium phases for different compositions within a binary alloy system. Both solid and liquid metal solutions may be plotted on integral free energy–mole fraction graphs.

It now becomes possible by constructing integral molar free energy–composition curves for an alloy system at different temperatures to construct the equilibrium phase diagrams which are so important to the processing of metals.

Fig. 4.13 represents diagrammatically integral free energy–composition curves at different temperatures for a simple eutectic system.

At T_1 : α and β solid solutions are less stable (more positive free energy values) than the liquid solution therefore only liquid metal is present at this temperature.

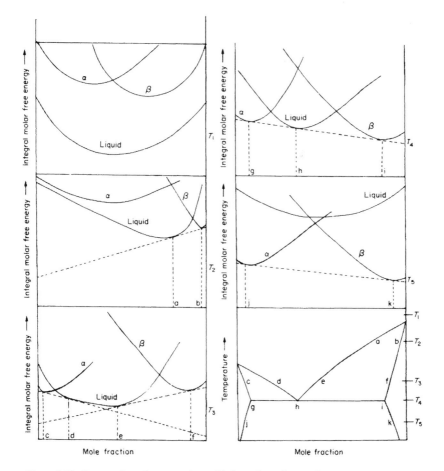

Figure 4.13 Construction of an eutectic equilibrium phase diagram from integral molar free energy–mole fraction curves

At T_2: β solid solution becomes more stable than the liquid at high mole fractions. Drawing a common tangent to β solid solution and liquid curves determines the range of β and β + liquid and liquid phase fields at this temperature.

At T_3: α solid solution becomes more stable than the liquid at low mole fractions. Common tangents can now be drawn for the α solid solution and liquid curves and the β solid solution and liquid curves which determine the α, α + liquid, liquid, liquid + β, and β phase fields at this temperature.

At T_4 (eutectic temperature): A common tangent can be drawn between the α, liquid and β curves indicating at this temperature all three phases are in equilibrium. Hence the α, α + eutectic, eutectic + β, and β phase fields are determined at this temperature.

At T_5: The liquid curve is now less stable than the α and β solid solution curves and hence the liquid phase will not be present at this temperature. The common tangent to the α and β curves determines the α, α + β (eutectic) and β phase fields at this temperature.

This technique may be extended to more complex systems and also used in conjunction with other thermochemical and physical measurement techniques. For further information on this matter the reader should consult references[5,6,7] at the end of this chapter.

4.3.11   Free energy of nucleation

The free energy of nucleation of a solid sphere of radius r on solidification from a liquid metal involves two factors;

(i)   the change in free energy associated with creating unit volume, ΔG_{vol}, of new phase, and

(ii)   the energy required to create an interface with interfacial energy, γ, of the new phase in the liquid metal.

If nucleation is to take place ΔG_{vol} must always be negative* and will be dependent on the actual volume of the sphere, $\frac{4}{3}\pi r^3$. At the same time, the interfacial energy will always have a positive value dependent upon the surface area of the sphere $4\pi r^2$.

Hence, the free energy of nucleation may now be written as:

$$\Delta G_{nucleation} = \frac{4}{3}\pi r^3 \Delta G_{vol} + 4\pi r^2 \gamma$$

But $\Delta G_{vol} = \Delta H_{vol} - T\Delta S_{vol}$ (from equation (2.22)) where, at the melting point, T_f, of the liquid, $\Delta H_{vol} = -L_{f,vol}$, the latent heat of fusion per unit volume and $\Delta S_{vol} = -L_{f,vol}/T_f$. Thus, cooling the liquid metal below T_f to temperature T:

$$\Delta G_{nucleation} = \frac{4}{3}\pi r^3 \left[-L_{f,vol} - T\left(\frac{-L_{f,vol}}{T_f}\right)\right] + 4\pi r^2$$

$$= \frac{4}{3}\pi r^3 \frac{L_{f,vol}(T - T_f)}{T_f} + 4\pi r^2 \gamma$$

But $T - T_f = -\Delta T$, ΔT being the degree of undercooling, thus,

$$\Delta G_{nucleation} = -\frac{4}{3}\pi r^3 \frac{L_{f,vol}\Delta T}{T_f} + 4\pi r^2 \gamma$$

---

*This does not mean that there must be a decrease in volume but rather a decrease in free energy associated with creating unit volume of solid.

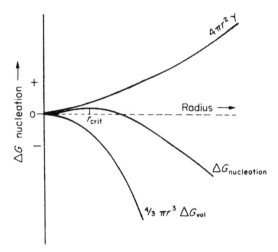

Figure 4.14 Diagrammatic representation of free energy of nucleation

The above relationship is plotted in *Fig. 4.14* where it is seen that there is a critical radius, r_{crit}, for the nucleating sphere below which nucleation is thermodynamically impossible (positive value of $\Delta G_{nucleation}$) and above which the nucleating sphere becomes thermodynamically stable. At the critical radius the gradient of the curve for $\Delta G_{nucleation}$, $d\Delta G_{nucleation}/dr$, is zero. Thus differentiating $\Delta G_{nucleation}$ with respect to r gives:

$$\frac{d\Delta G_{nucleation}}{dr} = -4\pi r_{crit}^2 \frac{L_{f,vol}\Delta T}{T_f} + 8\pi r_{crit}\gamma = 0$$

and solving this quadratic equation for r_{crit} gives a value which is also valid for nucleating from gaseous and solid phases, i.e. $r_{crit} = 2\gamma T_f/\Delta T L_{f,vol}$.

Therefore, increasing the degree of undercooling, ΔT, and decreasing the interfacial energy, γ, between the liquid metal and the nucleating phase will favour nucleation by reducing the critical radius size. The value of γ can be lowered by effecting *heterogeneous nucleation* on a third phase which lowers the interfacial energy between the nucleating and parent phases. The third phase must have similar crystallographic dimensions to the nucleating phase such that nucleation may readily take place. Examples of heterogeneous nucleation include the seeding of nickel shot in the decomposition of nickel carbonyl (Section 7.2.10.2) and the nucleation of carbon monoxide in refractory crevices or vortices in the liquid steel created by turbulence during the carbon boil reaction in steel refining (Section 7.2.10.6). Homogeneous nucleation, that is nucleation within the parent phase, generally has a high value of γ and therefore an increased critical radius size.

The rate of nucleation can be given from equation (3.20) as

$$\text{rate} \propto \exp\left(\frac{-\Delta G_{nucleation}^*}{RT}\right)$$

i.e.

$$\log(\text{rate}) \propto -\frac{\Delta G_{nucleation}^*}{RT}$$

where $\Delta G_{nucleation}^*$ is the free energy of activation (activation energy E_A) for nucleation and T the temperature to which the melt is cooled. Decreasing T, the melt temperature (increasing ΔT), decreases r_{crit} and $\Delta G^*_{nucleation}$ and therefore increases the rate of nucleation. Once nucleation has started the rate controlling steps tend to be those involving the transport of reactants to and products away from the reaction site.

4.4 Gases in metals

Inadequate removal of gases from liquid metals during solidification results in porosity and a consequent reduction in mechanical properties.

There are two main types of gases in liquid metals: (i) simple gases, i.e. the single gases O_2, N_2, H_2 which dissolve in the liquid metal in their atomic form, O, N and H; (ii) complex gases, such as CO_2, CO, H_2O, SO_2, which are formed by the reaction between elements present in the melt (e.g. dissolved oxygen, carbon and sulphur) or moisture present in the furnace atmosphere or refractories.

Complex gas formation often produces higher gas porosity levels than if the component gaseous elements were separately present in the melt. Reducing the level of oxygen in the liquid metal provides the best means of control of complex gas solution.

The presence of gas porosity in the solidified metal is caused by the large change in gas solubility between the liquid and solid metal (*Fig. 4.15*) and the difficulty of removing this relatively large volume of gas at the solidification front. A rapid rate of solidification will increase the likelihood of gas entrapment in the solidified metal.

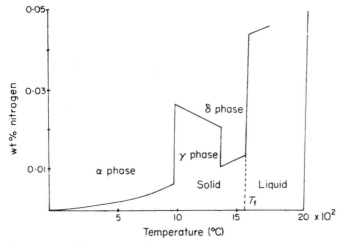

Figure 4.15 Variation of solubility of nitrogen in iron under a partial pressure of 1 atm of nitrogen

Simple gases dissolve in most metals by an endothermic reaction so that as the liquid metal temperature decreases, gas solubility decreases. This increases the possibility of gas entrapment during solidification due to the large difference in gas solubility between the solid and liquid metal.

However, gas solubility in a few metals occurs with an exothermic reaction—such as H_2 in Ti, Zr, Ta, V, Nb—in the formation of hydrides or nitrides as in the case of nitrogen solubility. Thus, gas solubility in these metals increases on decreasing the melt temperature and creates fewer problems of gas removal during solidification.

A few parts per million of H_2 in solid steels will result in hairline cracks due to the pressure created by the entrapped gas trying to diffuse out of the solid metal lattice. Gas solubility in certain phases can be less than in other phases of the same alloy components, e.g. N_2 in δ iron compared with that in γ iron (see *Fig. 4.15*). This is due to the difference in the interatomic spaces within the crystal structure of the solid metal.

Since simple gases dissolve in liquid metals in their atomic form, the dissolution reaction may be written:

$$H_{2\,(\text{molecular gas})} = 2H_{(\text{in liquid metal})}$$

The corresponding equilibrium constant is seen to be:

$$K = \frac{a_H^2}{p_{H_2}}$$

(Henrian activity is assumed since gases are normally present in liquid metals to less than 1 wt%.) This may be rewritten:

$$a_H^2 = K p_{H_2} \quad \text{or} \quad a_H = (K p_{H_2})^{\frac{1}{2}}$$

Since $a_H = f_H$ wt% H, the above equation may be rewritten:

$$\text{wt\% H} = \frac{(K p_{H_2})^{\frac{1}{2}}}{f_H}$$

or

$$\text{wt\% H} = k \sqrt{p_{H_2}} \tag{4.31}$$

where k is a constant at constant temperature related to the equilibrium constant and activity coefficient of hydrogen in the liquid metal. The value of f_H will depend on the metal's composition.

The above relationship is commonly known as Sievert's law, which states that the amount of gas dissolved in a liquid metal is proportional to the square root of the partial pressure of that gas above the melt. Sievert's law holds true for O_2, N_2, H_2 gases dissolved in liquid metals providing oxide-, nitride- or hydride-forming elements are absent, and atomic gas concentrations are less than 1 wt%.

Since k in equation (4.31) is dependent upon the activity coefficient, $\sqrt{K}/f$, solute elements which affect f will also control the solubility of a gas in the liquid melt. In this respect carbon, silicon, phosphorus and sulphur increase the activity coefficient of nitrogen (f_N) in liquid iron and therefore reduce the solubility of nitrogen in this metal while manganese has the opposite effect. This accounts for the initial decrease in N_2 solubility in Bessemer steelmaking followed by an increase after C, Si, P and S have been largely removed (see *Table 7.3(a)*). In general, elements which form stable compounds with the gas will decrease f_N and increase the gas content of the metal and vice versa. Entrapment of N_2 can cause pinhole porosity in iron alloys while the formation of nitrides in solidified metals can substantially alter the mechanical properties of a metal.

From the application of Sievert's law it is clear that placing the liquid metal in a vacuum will result in lower gas solubility and thus reduce the problem of gas porosity on subsequent solidification.

This technique relies on a reasonable diffusion rate of the atomic gas in the liquid. The diffusion rates of hydrogen and nitrogen are normally of a sufficient value in liquid metals to make their removal by vacuum processing viable. However, the low diffusivity of oxygen enables little of this element to be removed under vacuum. Vacuum equipment is expensive so that it is only used when no other technique provides efficient degassing.

Also, bubbling an inert gas, e.g. Ar, through the melt allows the dissolved impurity gas to diffuse from the liquid metal into the inert gas bubble since the partial pressure of the impurity gas in the inert gas bubble will be zero. The inert gas containing the impurity gas rises to the surface of the liquid metal thereby effecting gas removal and reducing the problem of gas porosity. The formation of the complex gas, CO, during the carbon 'boil' reaction in steelmaking and in nickel melting aids the removal of H_2 and N_2 since these gases diffuse into the CO gas bubble according to Sievert's law and rise out of the melt. Inert gas bubbling is often used during vacuum degassing to improve gas removal from the melt and create turbulence which in turn increases the removal rates and also provides a mechanical flushing action.

Oxygen is mainly present as oxide inclusions which can also have an important effect upon the metal's properties.

Deoxidation is achieved by effecting a chemical reaction between the oxygen present in the liquid metal and a deoxidant which is added to the melt and which forms an oxide in preference to the melt. Hence the deoxidant is an element or combination of elements which has a more negative value of free energy of oxidation than the melt.

The choice of deoxidant can be made from examination of the $\Delta G^{\ominus}-T$ diagram for oxide formation (*Figure 2.11*). Suitable deoxidants for steel include aluminium, silicon, manganese (added as ferromanganese and ferrosilicon), a combination of silicon and manganese, and, to a lesser extent due to expense, titanium, zirconium and rare earths. Phosphorus in the form of a P–Cu (15% P) alloy, lithium, and carbon are used as deoxidants for copper.

A second important aspect when selecting a deoxidant is the rate of removal from the melt of the deoxidation product. To some extent the rate may be predicted from Stoke's law in which the velocity (v) of removal of the deoxidation product from the liquid metal will be:

$$ v = \frac{2}{9}\frac{gr^2(\rho_{\text{liq}} - \rho_{\text{dp}})}{\eta} $$

where g stands for acceleration due to gravity, r is the radius of the deoxidation product, ρ_{liq} and ρ_{dp} are the densities of the liquid metal and deoxidation product respectively, and η is the viscosity of the liquid metal.

In most cases the density of the liquid metal is greater than that of the deoxidation product resulting in a positive velocity and the deoxidation product rising into the slag.

Bodsworth[9] has compared the rate of rise of different sized inclusions from liquid steel in both a 40 cm steelmaking bath and a 400 cm transfer ladle, assuming the density of the liquid steel to be 7.0 g cm^{-3}, and a viscosity of 0.1 g cm^{-1} s^{-1} and the density of the inclusion of 3.0 g cm^{-3}. These figures are reproduced in *Table 4.3*.

Table 4.3 RATE OF RISE OF DIFFERENT SIZED PARTICLES IN LIQUID STEEL (Bodsworth[9])

Particle radius ($\times 10^{-3}$ cm)	Rate of rise (cm s^{-1})	Time to rise through	
		40 cm s^{-1}	400 cm s^{-1}
0.1	0.00087	46000	460000
1.0	0.087	460	4600
10.0	8.72	4.6	46

It is clear from the above figures that the r^2 term has a significant effect on the time required to remove these deoxidation products. At the same time it would be impractical to hold the ladle of liquid steel for long times in order to remove the smaller inclusions. However, Stoke's law assumes the inclusions to be spherical particles which is not always the case—Al_2O_3 in steel can be present as flakes or plates. The interfacial energy

Table 4.4 SOME ESTIMATED VALUES OF INTERFACIAL ENERGIES IN LIQUID IRON
(after Turpin *et al.*[10])

System	Interfacial energy ($\times 10^{-4}$ J cm^{-2})
solid Al_2O_3 –liquid Fe (pure)	2.3
solid Al_2O_3 –liquid Fe + 3% C	1.4
liquid FeO–liquid Fe (pure)	0.25–0.5
solid $FeO.Al_2O_3$ –liquid Fe (pure)	1.0–1.5
solid SiO_2 –liquid Fe (pure)	1.0–1.5

between the deoxidation product and liquid melt will also have a significant effect on removal rates. If the interfacial energy is low the two will wet each other and the corresponding attractive forces will provide a dragging effect of the liquid metal on the particle being removed. This will retard the removal rate. On the other hand if there is a high interfacial energy between the particle and the liquid the dragging effect will be minimal and the rate of removal will be high. Some estimated values of interfacial energies[10] of certain deoxidation products and liquid iron at 1536 °C are given in *Table 4.4*.

REFERENCES

1 Bodsworth, C. and Appleton, A.S. 1965 *Problems in Applied Thermodynamics*, chap. 8. London: Longman.
2 Bodsworth, C. and Bell, H.B. 1972 *Physical Chemistry of Iron and Steel Manufacture*, pp. 59–63 and 486–505, 2nd edn. London: Longman.
3 Gaskell, D. 1973 *Introduction to Metallurgical Thermodynamics*. New York: McGraw-Hill.
4 Scatchard, G. 1949 *Chem. Rev.,* **447**, 17.
5 Mackowiak, J. 1966 *Physical Chemistry for Metallurgists*, chap. X. London: George Allen and Unwin, for the Institution of Metallurgists.
6 Darken, L.S. and Gurry, R.W. 1953 *Physical Chemistry of Metals*, chap. 13, New York: McGraw-Hill.
7 Kubaschewski, O. and Evans, E.Ll. 1958 *Metallurgical Thermochemistry*, pp. 32–72, 399–403, 3rd edn. London: Pergamon Press.
8 Gilchrist, J.D. 1969 *Extraction Metallurgy*, p. 189. London: Pergamon Press.
9 Bodsworth, C. and Bell, H.B. 1972 *Physical Chemistry of Iron and Steel Manufacture*, p. 424, 2nd edn. London: Longman.
10 Turpin, M.L. and Elliott, J.F. 1966 *J. Iron & Steel Industry,* **204**, 217–225.

5

Slag Chemistry

5.1 Functions and properties of a slag

The majority of refining reactions required in liquid melts occur around the liquid slag–liquid metal interface and are commonly called *slag–metal reactions*. This being the case it is of some importance to examine the functions, properties and structure of liquid slags in order to be able to select suitable metal–slag conditions for each refining operation. The functions of a slag used in liquid metal processing may be summarised as:

  (i)   protection of the melt from contamination from the furnace atmosphere and combustion products of the fuel;
  (ii) insulation of the melt;
  (iii) accepting unwanted liquid and solid components;
  (iv) controlling the supply of refining media to the melt through additions to the slag.

To be able to apply these functions slags must have certain fundamental properties which may be summarised as:

  (i)   Lower melting point than the melt in order to maintain a liquid slag of high fluidity which covers the whole of the liquid metal surface, and allows good contact with the liquid metal. Fluxes such as lime, fluorspar, quartz or iron oxide may be added which lower the melting point of the slag and thereby increase its fluidity.

  (ii) Lower specific gravity than and immiscible in the melt, so that it rests above the melt as a separate liquid layer and can therefore accept impurities which are generally lighter than the liquid metal.

  (iii) Correct composition so that it can accept and react with impurities from and be immiscible in the melt.

Most slags are formed by the solution of mixed oxides and silicates, sometimes with aluminates, phosphates and borates, in an endothermic (heat absorbing) reaction. Therefore, the greater the slag volume the greater the heat input required, and the larger the fuel costs. Increasing slag volume also increases the danger of metal entrapment, reducing the metal yield and the rate of heat transfer from the furnace into the melt. The area of the slag–metal interface will also be important with respect to the rate of reaction, while the rate controlling steps are often found to be the diffusion of the reactants within the slag to the slag–metal interface and the transfer of reactants and products across the slag–metal interface. Hence the viscosity (or fluidity) of the slag and wettability (or interfacial energy) between the slag and the melt are important considerations.

Liquid slags are ionic in structure, thus the molten oxides forming the slag are present as cations and anions. These oxides are present as two main types:

(i) Basic oxides, which donate O^{2-} ions to the slag, e.g. CaO, FeO, MgO, Cu_2O, Na_2O, K_2O, and which ionise as follows:

$$CaO = Ca^{2+} + O^{2-}$$

(ii) Acidic oxides which accept O^{2-} ions, e.g. SiO_2, P_2O_5, Al_2O_3, and which produce complex cations by reaction with the oxygen anions, i.e.

$$SiO_2 + 2O^{2-} = SiO_4^{4-} \quad \text{(silicate or orthosilicate ion)}$$

$$P_2O_5 + 3O^{2-} = 2PO_4^{3-} \quad \text{(phosphate or orthophosphate ion)}$$

$$Al_2O_3 + 3O^{2-} = 2AlO_3^{3-} \quad \text{(aluminate or orthoaluminate)}$$

Hence, one molecule of SiO_2 is neutralised by two molecules of basic oxide (two O^{2-} ions), while one molecule of Al_2O_3 and P_2O_5 are each neutralised by three molecules of basic oxide. When the slag is composed of these ratios of basic and acidic oxides it has neither basic nor acidic properties and is said to be a *neutral slag*, i.e.

$$SiO_2 + 2CaO = Ca_2SiO_4 \text{ or } 2CaO.SiO_2 \text{ or } 2Ca^{2+} + SiO_4^{4-}$$

(calcium orthosilicate)

$$P_2O_5 + 3CaO = Ca_3(PO_4)_2 \text{ or } 3CaO.P_2O_5 \text{ or } 3Ca^{2+} + 2PO_4^{3-}$$

(calcium orthophosphate)

$$Al_2O_3 + 3CaO = Ca_3(AlO_3)_2 \text{ or } 3CaO.Al_2O_3 \text{ or } 3Ca^{2+} + 2AlO_3^{3-}$$

(calcium orthoaluminate)

A *basic slag* is one which contains more basic oxide than the neutral ortho-compound composition and in this state the slag will have an excess of O^{2-} ions. An *acidic slag* is one which contains more acidic oxide than the neutral ortho-compound composition and will therefore have an excess of SiO_2, P_2O_5 or Al_2O_3.

A typical basic slag composition would be three moles of CaO and one mole of SiO_2 while an acidic slag composition would contain one mole of CaO to one volume of SiO_2.

Various means are available for measuring a slag's basicity, the most accurate being given by the basicity ratio B, where:

$$B = \frac{\text{no. moles of basic oxide} - 3 \times \text{no moles} (Al_2O_3 + P_2O_5)}{2 \times \text{no. moles of } SiO_2} \quad (4.30)$$

This B ratio is so arranged that when it is unity a neutral slag composition is achieved, less than one an acidic slag and greater than one a basic slag.

A metal melter or refiner who is dealing with similar charge materials and producing a limited alloy range may not need to analyse the slag so thoroughly in order to determine slag basicity. In this case the ratio of wt% basic oxides to wt% acidic oxides or even more simply the wt% CaO to wt% SiO_2 ratio may be sufficient. In general, an acid slag is more viscous than a basic slag and sometimes the CaO/SiO_2 ratio is referred to as a V (viscosity) ratio.

5.2 Slag metal reactions

Slag metal reactions can be largely controlled by consideration of slag basicity and hearth temperature. Examination of how temperature affects the free energy and equilibrium constant of the reaction can be useful, as can the effect of basicity of the slag on the unwanted species being removed by it. If the unwanted species, say an oxide MO, is a basic oxide, employing an acidic slag will reduce the activity of MO in that slag, thereby favouring the retention of the oxide in the slag. Partitioning of the element between the slag and the metal in this way can therefore be controlled to some extent by consideration of slag basicity and hearth temperature, as discussed in the following example for manganese in liquid iron. The equilibrium can be represented by the following reaction:

$$MnO(\text{slag}) + C(\text{Fe}) = Mn(\text{Fe}) + CO(\text{gas}); \Delta G^{\ominus} = 290\,300 - 173.22T \text{ J}$$

$$\log K = \frac{a_{Mn(\text{Fe})} \times p_{CO}}{a_{MnO\,(\text{slag})} \times a_{C(\text{Fe})}} = \frac{-15\,180}{T} + 9.05$$

Examination of both $\Delta G^{\ominus}/T$ and K/T relationships indicate that increasing the temperature will make $\Delta G^{\ominus}$ more negative and K larger. MnO is a basic oxide, so employing a basic slag will increase the activity of MnO, a_{MnO}, in the slag. Both these trends, i.e. increased temperature and slag basicity, result in favouring the forward reaction and reducing MnO in the slag and increasing Mn in the iron. Hence manganese recovery from the slag to the iron is increased by increasing the temperature of the slag and slag basicity.

Fig. 5.1 reflects these trends in that increasing the CaO/SiO_2 ratio and the temperature increases the activity coefficient of MnO, γ_{MnO}, in the slag. It is also noted from this figure that increasing the Al_2O_3 content of this acid blast furnace slag increases γ_{MnO} in the slag. Under these acid slag conditions the amphoteric Al_2O_3 is acting as a basic oxide producing an increase in activity of MnO.

5.3 Structure of liquid slags

Physical property measurements, e.g. surface tension, viscosity, electrical conductivity, transport numbers, indicate that slags are mixtures of anions and cations. The small changes in entropy and viscosity on fusion of an oxide indicate minimal structural changes from the solid to the liquid. Therefore, an indication of slag structures can be gained from the structure of the solid components.

Silica is a constituent of most slags. It has a very weak tendency to crystallise and rapid cooling produces a glass with no long-range ionic ordering.

5.3.1 The cation-O^{2-} bond

Basic oxides such as CaO, MgO, MnO, FeO exhibit a close packed arrangement of the O^{2-} ions with cations in the interstices. The attraction of the cation to the O^{2-} ion can be represented by the bonding force, F, i.e.

$$F \text{ (Coulombic force)} = \frac{Z^+ Z^- e^2}{(R_a + R_c)^2} \tag{5.1}$$

where R_c and R_a are the radii of the cation and anion respectively, Z^+ and Z^- are the valencies (oxidation states) of the cation and anion respectively and e is the electron charge $(0.1602 \times 10^{-18} \text{ C})$.

Most basic oxides exhibit a valency of two, e.g. Fe^{2+}, Ca^{2+}, Mg^{2+}, Mn^{2+}, whereas acidic oxides exhibit higher charges, e.g. Al^{3+}, Si^{4+}, P^{5+}. However, oxides exhibit a combination of both ionic and covalent bonding between the cation and O^{2-} ion.

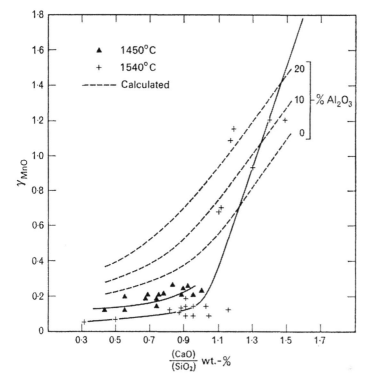

Figure 5.1 Variation of the activity coefficient of manganese oxide with CaO/SiO₂ ratio, temperature and the addition of Al_2O_3 for blast-furnace slags (from Bodsworth and Bell[13], p.168)

This will depend on the electronic structure of the elements which form the oxide and the size of the ions. Therefore, each oxide will have an ionic bond fraction and a covalent bond fraction. Those oxides which have a high ionic bond fraction will tend to be basic since ionic bonding is more conducive to ionisation and, therefore, more likely to donate O^{2-} ions to the liquid slag, whereas covalent bonding provides a strong bond between the cation and the O^{2-} ion due to shared electrons in the outer shells of the elements.

Examination of the periodic table of the elements (*Table 1.3*) indicates the tendency to form basic or acidic oxides. Cations from groups I and II combining with anions from groups VI or VII are all likely to form compounds with a high ionic bond fraction due to the ease of transfer of one or two electrons in the outer shell of the cation to the anion. An example here can be made with CaO which has an electronic structure of $1s^2$; $2s^2$, $2p^6$; $3s^2$, $3p^6$; $4s^2$; and oxygen which has an electronic structure of $1s^2$; $2s^2$, $2p^4$. Therefore, the two outer electrons in the 4s orbital of calcium can transfer to the 2p orbital of oxygen to complete the 2p shell, leaving both ions in a stable configuration of electrons. The case for similar sharing of electrons between silicon, with two outer electrons in the 3p shell, and oxygen, with four outer electrons in the 2p shell, is less favourable and, therefore, covalent bonding will predominate due to the sharing of these outer electrons to provide a stable configuration of electrons in each element.

Examination of the groups in the periodic table show that the atomic radius of the elements increases from top to bottom of each group since the number of electron shells increases in this direction. The atomic radius decreases on moving from left to right in each group since the electrons are attracted by an increased number of nucleus protons. Therefore, the value of F in equation (5.1) for oxides produced from cations in groups I and II is low and these elements are more likely to form strong basic oxides which readily donate O^{2-} ions to the slag. Elements from intermediate groups provide only a small fraction of their bond as ionic and that which is ionic has a high value of F. Therefore, oxides of these elements are likely to be strong acidic oxides.

Since electrons have been removed from atoms to produce cations and added to produce anions the radii of cations, R_c, are generally larger than those of anions, R_a. The values of R_c and R_a of selected ions are given in *Table 5.1*.

Table 5.1 RADII OF CATIONS, R_c, AND ANIONS, R_a

Cations	K^+	Ca^{2+}	Na^{2+}	Mn^{2+}	Fe^{2+}	An^{2+}	Mg^{2+}	Fe^{3+}	Cr^{3+}	Al^{3+}	Si^{4+}	P^{5+}
R_c (nm)	0.133	0.099	0.097	0.08	0.074	0.074	0.066	0.064	0.063	0.051	0.042	0.035

Anions	I^-	S^{2-}	Cl^-	O^{2-}	F^-
R_a (nm)	0.220	0.184	0.181	0.140	0.133

The binding force, F, decreases as the coordination number with O^{2-} ions increases. The coordination number is controlled by the relative sizes of O^{2-} ions and cations. Basic oxides generally have a coordination number of 6 to 8, with some being as high as 12. A typical coordination number for acidic oxides is 4.

Electronegativity is the measure by which an electron can be removed from an atom to produce an ion. Therefore basic oxides exhibit high electronegativities. Both the bond enthalpy and the electronegativity increase as the ionic bond fraction increases. As the electronegativity difference between elements increases so will the ionic bond fraction between those elements increase. Pauling stated that elements that have an electronegativity difference of greater than 1.7 tend to be predominantly ionically bonded, and those which have a difference of less than 1.7 are predominantly covalently bonded. Selected electronegativity values are given in *Table 1.6*.

The relative sizes of the cation and anion and the coordination number will control the structural arrangement of ions. Using simple geometrical considerations it is, therefore, possible to determine the ionic structure of the oxide. *Table 5.2* indicates the structure dependency on coordination number and R_c/R_a ratio.

It is seen from *Table 5.2* that SiO_2 has a tetrahedral structure in which four O^{2-} ions provide the frame of the tetrahedron and the smaller Si^{4+} ion is situated within

Table 5.2 EFFECT OF R_c/R_a RATIOS AND COORDINATION NUMBER ON THE STRUCTURE OF SOLID OXIDES

Structure	Coordination	R_c/R_a	Examples
Cubic	8	1–0.732	
Octahedral	6	0.732–0.414	CaO, MgO, MnO, FeO
Tetrahedral	4	0.414–0.225	SiO_2, P_2O_5
Triangular	3	0.255–0.155	

the tetrahedron as indicated in *Fig. 5.2*. Pauling's second law states that the interval between two neighbouring cations, e.g. Si^{4+}, is a maximum. Hence, the neighbouring cations are mutually repellent, and therefore the joining of two tetrahedra must occur at each vertex, producing a three-dimensional hexagonal structure as shown in *Fig. 5.3*. In this structure the Si^{4+} ion is bonded to four O^{2-}

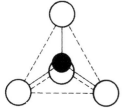

Figure 5.2 Silica tetrahedron of four O^{2-} ions (large open circles) and one Si^{4+} ion (smaller, black circle)

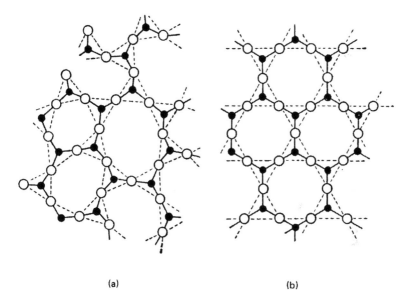

(a)   (b)

Figure 5.3 Schematic representation of the tetrahedral network structure of pure silica: (a) solid crystalline silica; (b) silica in the liquid state (from Richardson[14])

ions and each O^{2-} ion is bonded to one Si^{4+} ion[1]. In the solid state this provides a formula structure of $(SiO_2)_n$. The largely covalent bonding between silicon and oxygen is very strong which leads to the application of silica as a refractory material.

It is apparent from the hexagonal structure of silica that two tetrahedra are joined by one O^{2-} ion. On increasing the temperature, these bonds are broken one by one by thermal agitation, and the viscosity of silica decreases. Initially, with only a few bonds broken, large anions, e.g. $(Si_9O_{21})^{6-}$, are formed together with Si^{4+} cations. As more bonds are broken smaller anions are produced, e.g. $(Si_3O_9)^{8-}$, with a further increase in Si^{4-} cations. Eventually, all the bonds are broken, resulting in an equal number of anions, SiO_4^{4-}, and cations, Si^{4+}. For this reason, silica does not exhibit a fixed melting point but rather its viscosity decreases as the temperature is raised due to the increasing number of broken bonds. *Table 5.3* gives the ionic bonding force, $F = Z^+ Z^- e^2 / (R_c + R_a)^2$, the ionic bond fraction, the coordination number and the character of the oxide for certain selected oxides.

Table 5.3 TYPES OF BONDING AND ATTRACTION FORCES BETWEEN CATIONS AND O^{2-}

Oxide	$z/(R_c + R_a)$	Ionic fraction of bond	Coordination number solid–liquid		Character of the oxide
Na_2O	0.18	0.65	6	6 to 8	
BaO	0.27	0.65	8	8 to 12	
SrO	0.32	0.61	8		
CaO	0.35	0.61	6		
MnO	0.42	0.47	6	6 to 8	Network breakers or basic oxide
FeO	0.44	0.38	6	6	
ZnO	0.44	0.44	6		
MgO	0.48	0.54	6		
BeO	0.69	0.44	4		
Cr_2O_3	0.72	0.41	4		
Fe_2O_3	0.75	0.36	4		Amphoteric oxides
Al_2O_3	0.83	0.44	6	4 to 6	
TiO_2	0.93	0.41	4		
SiO_2	1.22	0.36	4	4	Network formers or acid oxides
P_2O_5	1.66	0.28	4	4	

The addition of each mole of basic oxide to silica provides a donation of one O^{2-} ion, e.g.

$$CaO = Ca^{2+} + O^{2-}, Na_2O = 2Na^+ + O^{2-}$$

Each O^{2-} ion is able to break one of the four common O^{2-} bonds between two tetrahedra by positioning itself at the join, thereby creating a breakage in the structure at this position. This is diagrammatically represented in *Fig. 5.4*.

Basic oxides are therefore termed network breakers since they donate O^{2-} ions to break up the stable three-dimensional hexagonal network of SiO_2, i.e. $SiO_2 + 2O^{2-} \rightarrow SiO_4^{4-}$. Other strong network formers include P_2O_5 and, although amphoteric, Al_2O_3 in basic slags. Elements which exhibit a variable valency, e.g. Fe^{2+} and Fe^{3+}, can possess both basic and acidic characteristics. Since the cation (R_c) of the lower valency state will be larger, the metal-O^{2-} ion bond is more electrovalent (ionic) and therefore more basic. In this respect, Fe_2O_3 is acidic and FeO is basic and, in an aqueous medium, Cr^{3+} is amphoteric, Cr^{6+} is acidic.

An acidic oxide, e.g. SiO_2, can be looked upon as an ionising solvent which increases the ionised fraction of the oxide in solution by the addition of a basic oxide. The dissociation of a basic oxide is not complete but is more complete than acidic oxides and depends largely on the ionising power of the acid solvent. Increasing the basic oxide addition to the liquid silica slag increases the number of common O^{2-} bonds broken, as indicated in *Table 5.4*. Therefore, in contrast to silica, melting of silicates generally results in a well defined fusion point with no high-viscosity range. The weak electrovalent bonds between the simple cations, e.g. Fe^{2+}, and complex anions, e.g. SiO_4^{4-}, are broken first. As the temperature increases, the solid silicate structures completely disappear. The bond numbers and bond strengths are important with respect to intersolubility of oxides. Basic cations can more readily substitute for each other than can the cation elements in SiO_4^{4-}, PO_4^{3-} and AlO_3^{3-}. In general a mixture of two basic oxides is capable of providing extensive solid solutions, as shown in *Fig. 5.5* for the FeO-MgO and FeO-MnO systems.

Therefore, addition of a basic oxide (network breakers) to an acidic oxide (ionising solvent) will result in a lower melting point mixture, as will the addition of an acidic oxide to a basic oxide. This is demonstrated in the CaO-SiO2 phase diagram, *Fig. 5.6*.

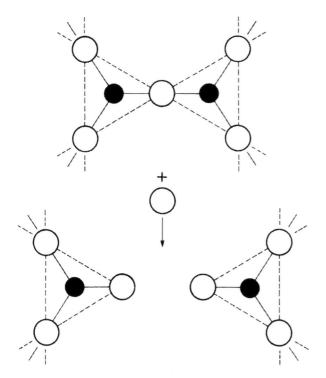

Figure 5.4 Schematic representation of the breaking of common oxygen ion bond in the silica tetrahedra by the addition of an oxygen ion donated from a basic oxide. The open circles are oxygen ions and the smaller black circles represent silicon ions

Table 5.4 EFFECT OF THE ADDITION OF A BASIC OXIDE ON THE STRUCTURE OF SILICATES

Ratio O/Si	Molecular formula	Structure
2 : 1	SiO_2	Each corner of tetrahedra joined by one O^{2-} ion bond to produce 3D hexagonal network.
5 : 2	$(MO.2SiO_2)$	One broken O^{2-} bond per tetrahedron.
3 : 1	$(MO.SiO_2)$	Two broken O^{2-} bonds per tetrahedron to produce 2D, lamellar structure.
7 : 2	$(3MO.2SiO_2)$	Three broken O^{2-} bonds per tetrahedron to produce a fibrous form.
4 : 1	$(2MO.SiO_2)$	All O^{2-} bonds broken. Discrete SiO_4^{4-} tetrahedra.

In the liquid state the regular arrangement of ions is destroyed and each cation remains coordinated with approximately the same number of anions as in the solid state, with the cations rapidly interchanging, as are the anions. The extent of mobility of the cations is dependent on the valency or state of oxidation and the coordination number. In this respect basic oxides tend to be more mobile than the

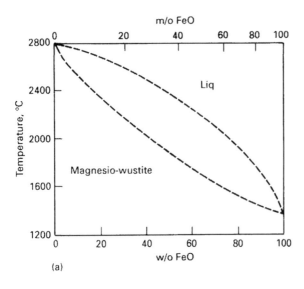

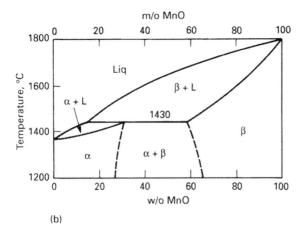

Figure 5.5 Phase diagrams for: (a) FeO–MgO; and (b) FeO–MnO systems

oxyacid groups of acidic oxides which exhibit higher valency charges and low coordination numbers.

5.3.2 Aluminate and phosphate structures

Al, P and B can each provide similar oxygen anion structures to SiO_2 in the solid state and can substitute for Si^{4+} in the joined tetrahedra structures provided that the anion/cation ratio is adjusted to preserve electroneutrality. Hence, replacing Si^{4+} by P^{5+} increases the electron field strength and the attraction of the O^{2-} ion to the P^{5+} ion. Replacement of Si^{4+} with Al^{3+} reduces the attraction between the cation and anion. Hence, the activity of the basic oxide decreases in the order of equal additions of Al_2O_3, SiO_2, P_2O_5 as outlined in *Table 5.5*.

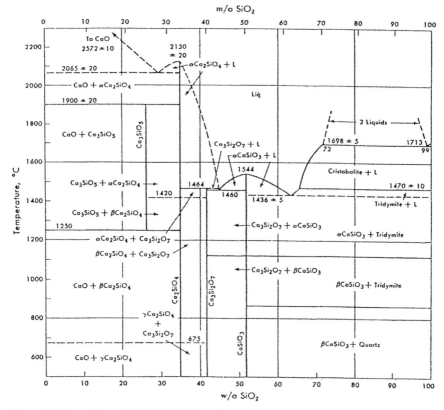

Figure 5.6 Phase diagram for the CaO–SiO$_2$ system

Table 5.5 THE ATTRACTION OF THE METAL–O^{2-} BOND FOR THREE ACIDIC OXIDES

	Charge	O/M	Ionic bond fraction	$Z/(R_c + R_a)$
Al$_2$O$_3$	Al^{3-}	3/2	0.44	0.83
SiO$_2$	Si^{4+}	2/1	0.36	1.22
P$_2$O$_5$	P^{5+}	5/2	0.28	1.66

5.4 Liquid immiscibility

5.4.1 Binary silicate melts

Unlike SiO$_2$, binary silicates exhibit a well defined melting point. The weaker electrovalent bonds between the cation and the silicate ion are broken first and any remaining hexagonal SiO$_2$ type structure disappears on increasing the temperature. In general, the addition of a divalent basic oxide to the complex acid ion structures can result in immiscibility, whereas the addition of a monovalent basic oxide to silica is more likely to provide an extensive range of miscibility. The immiscibility can be explained by examining the difficulty in retaining normal coordination of divalent cations with the O^{2-} ion of the basic oxide, compared with that which it must adopt if it enters a hole in the SiO_2 lattice. Although the coordination number decreases as the network is broken it falls to only six at the orthosilicate neutral slag composition. Increasing the addition of the divalent oxide, the silica increases the cation-O^{2-} bond length which in turn increases the energy of the system. This leads to an instability which, on reaching a critical level, will result in a second liquid phase, i.e. immiscibility. Monovalent oxides exhibit a lower value of F and a higher ionic bond fraction compared with divalent oxides and are more readily ionised than divalent oxides. This aids in extending the miscibility with silica. As the attractive force F increases, immiscibility limits may produce specific silicate complexes. Since a large cation size results in a low value of F, oxides with large cations provide an extended range of miscibility with silica. This trend is observed on examining *Fig. 5.7*, where the larger monovalent cations, e.g. K^+ and Na^+, provide a full range of miscibility with silica whereas the smaller divalent cations such as Zn^{2+}, Mg^{2+}, Fe^{2+} provide a two-liquid immiscibility region.

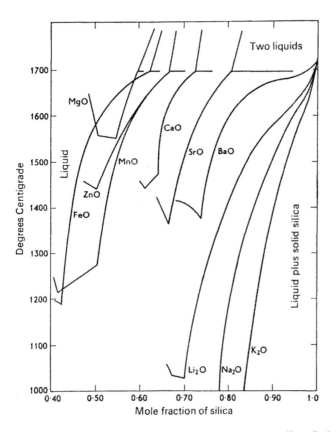

Figure 5.7 Miscibility limits for various binary silicate systems (from Bodsworth and Bell[13], p. 75)

Table 5.6 gives the values of F/e^2 for various binary silicates and their associated liquid miscibilities. Oxides produced with those cations at the top of the table provide a generally weak cation-O^{2-} ion bond and are said to be weakly polarised, whereas oxides formed from cations of Mg and Fe provide a higher cation-O^{2-} bond

Table 5.6   EFFECT OF ION O^{2-} BOND ON LIQUID IMMISCIBILITY OF CERTAIN SILICATES

Cation	Cation radius (nm)	F/e^2	Limit of liquid immiscibility (mole fraction of silica)	O/Si ratio	Mole fraction of silica at this ratio
K^+	0.133	0.24	1.0		
Na^+	0.097	0.36	1.0		
Li^+	0.060	0.50	1.0		
Ba^{2+}	0.135	0.53	1.0		
Sr^{2+}	0.113	0.63	0.81	9/4	0.80
Ca^{2+}	0.099	0.70	0.72	7/3	0.75
Mn^{2+}	0.080	0.83	0.66	5/2	0.67
Zn^{2+}	0.074	0.91	0.66	5/2	0.67
Fe^{2+}	0.074	0.87	0.62	8/3	0.60
Mg^{2+}	0.066	0.95	0.60	8/3	0.60
Al^{3+}	0.051	1.31			
Si^{4+}	0.042	2.70			
P^{5+}	0.035	4.74			

and are said to be highly polarised. Decreasing the cation size increases the F/e^2 value and results in less ionisation of the oxide and decreased miscibility with silica. It is apparent from these data that the alkali metal oxides provide a low value of F which is insufficient to detach the oxygen ion from the silicate complexes. This in turn results in extended miscibility.

Oxides that exhibit a strong cation–O^{2-} ion bond, e.g. the divalent oxides, result in immiscibility at a certain mole fraction of silica which corresponds to a critical oxygen/silicon ratio (*Table 5.6*). Once immiscibility occurs at this critical mole fraction it is energetically more favourable for a second liquid to form. Increasing the silica concentration beyond the critical point results in two liquids, one of which will correspond to the composition at the critical oxygen/silicon ratio and the other liquid will be composed of almost pure silica.

5.4.2   Ternary silicate

Mixing silica with two different basic oxides, each of which provide immiscibility, will result in an immiscibility region which joins the immiscible stoichiometry of each of the binary melts. This is shown in *Fig. 5.8* for the SiO_2–CaO–FeO ternary system, in which the immiscibility point for the binary CaO–SiO_2 is at approximately 27wt% CaO and that for the binary FeO–SiO_2 is at approximately 42wt% FeO. The discontinuous line linking these two points on the ternary diagram represents the immiscibility limit of the ternary system. Compositions richer in silica will result in a two-liquid region. The straight line linking these two points indicates that the change in the coordination number of cations is approximately a linear function of the ratio of the concentration of the two oxides, i.e. [CaO]/[FeO], in the liquid. Deviation from this relationship is indicated in some systems as indicated in *Fig. 5.9* for the FeO–MnO–SiO_2 system, where the discontinuous line is slightly curved. If only one of the oxide–silica binary systems exhibits immiscibility, ternary immiscibility is generally restricted to a small region on that binary side of the ternary diagram. An example of this is given in *Fig. 5.10* for the SiO_2–CaO–Al_2O_3 system in which CaO–SiO_2 exhibits immiscibility whereas Al_2O_3–SiO_2 exhibits total miscibility.

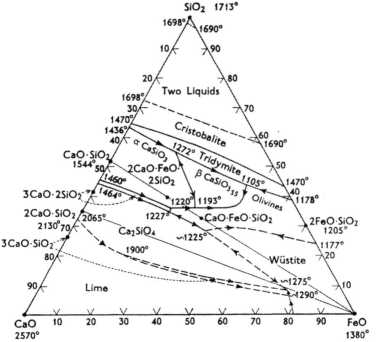

Figure 5.8 Liquidus surfaces and immiscibility region in the ternary system CaO–FeO–SiO₂ (from Bowen *et al.*[15])

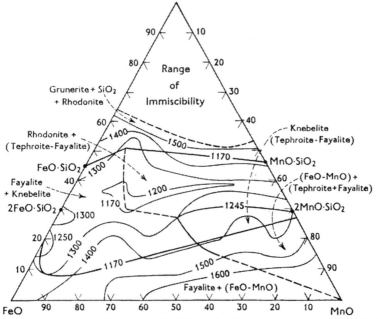

Figure 5.9 Liquidus surfaces and immiscibility region in the ternary system FeO–MnO–SiO₂ (from Maddocks[16])

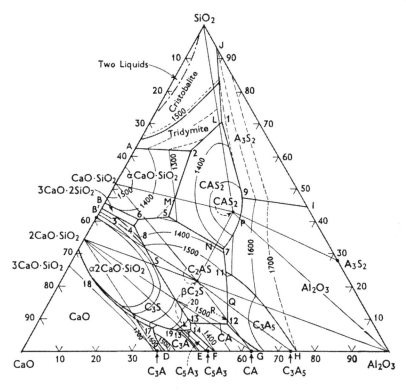

Figure 5.10 Liquidus surfaces and immiscibility region in the ternary system CaO–Al$_2$O$_3$–SiO$_2$ (from Grey[17] and Rankin and Wright[18])

5.4.3 Phosphate melts

For liquid slag basicities greater than or equal to the orthophosphate composition, i.e. $3MO.P_2O_5$, PO_4^{3-} ions exist as separate units in the melt, whereas for lower basicities a hexagonal network of anion complexes similar to that in silicate melts exists. However, immiscibility is normally produced at a lower basic oxide concentration in phosphate melts than in the silicate melt since the tetrahedral $P^{5+}-O^{2+}$ bonding is stronger than that of the $Si^{4-}-O^{2-}$ bond. Examination of the P_2O_5–CaO–FeO ternary diagram, *Fig. 5.11*, shows the effect of silica on liquid immiscibility of the system. Liquid compositions which lie on the join for the orthophosphate composition for each of the binary systems, i.e. $3FeO.P_2O_5$ and $3CaO.P_2O_5$, provide complete miscibility. Increasing the FeO content of a slag mixture of $3CaO.P_2O_5$ will result in immiscibility, as indicated in the lenticular area on the ternary diagram in *Fig. 5.11*. Addition of FeO or MnO to the calcium orthophosphate results in immiscibility in the generation of two liquids, one being of the calcium orthophosphate composition, i.e. Ca^{2+} and PO_4^{3-} ions, and the other liquid being largely FeO (Fe^{2+} and O^{2-} ions).

Increasing the O^{2-} ion concentration over and above that needed to satisfy all the P^{5+} ions will result in the formation of the basic oxide which has the highest F value. Since CaO is more easily ionised than FeO, FeO would form the main constituent of the second liquid phase.

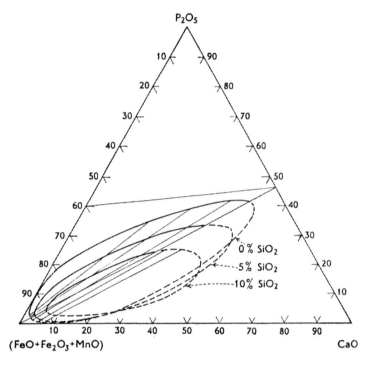

Figure 5.11 The region of liquid immiscibility in the system CaO–(FeO + Fe$_2$O$_3$ + MnO)–P$_2$O$_5$ showing the effect of various concentrations of silica (from Oelson and Maetz[19])

A schematic summary of the previous discussion on the addition of basic oxides to silica and vice versa is given in *Fig. 5.12*.

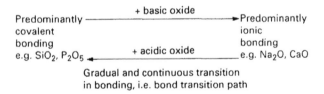

Figure 5.12 Schematic representation of gradual transition from the predominantly covalent bonding of strong acidic oxides to predominantly ionic bonding of basic oxides by the addition of basic oxides to acidic oxides and vice versa for the addition of acidic oxides to basic oxides

5.5   The transition from ionic to covalent bonding and polarisation

It has been shown that solid oxides exhibit mixed ionic and covalent bonding and the ionic bond fraction decreases as the cation size and coordination number decrease, and the oxidation state of the ion increases. Deviation from purely ionic bonding can be explained in terms of polarisation that occurs between the cation and the O^{2-} ion.

Polarisation is due to the distortion of the electron field of the ion by adjacent ions. This results in a decrease in bond length between the cation and O^{2-} ion and a consequent increase in bond strength. An increase in polarisation occurs until the energy level is sufficient to provide electron transfer between the ions, resulting in covalent bonding. In this way there is a continuous transition in bonding from ionic to covalent and vice versa. This is also referred to as the bond transition pattern, schematically represented in *Fig. 5.12*.

Therefore, the polarising power of the cation is dependent on the electron field strength and increases as the cation radius decreases and the valency increases. It is also affected by any electron shell which has an energy overlap with the valence or outer shell of the cation, e.g. transition elements. The extent of polarisation increases, therefore, as the coordination number of the cation decreases and the electron field becomes more asymmetrical.

The polarising force of ZnO, MnO, FeO in practice is greater than that predicted from consideration of cation size and valency alone. An explanation of this anomaly will be given later. An example of the effect of cation size and asymmetric electron field is given by comparing CaO and SiO_2 at room tmperature. The $Ca^{2+}-O^{2-}$ bond length is approximately as predicted by consideration of the Ca^{2+} and O^{2-} radii (*Table 5.1*) whereas the $SiO_4{}^{4+}-O^{2-}$ bond length is very much less than that predicted by consideration of the Si^{4+} ion and O^{2-} ion radii alone.

In the liquid state basic oxides exhibit high mobility and interchange of cations. This indicates a reduction in F for liquid basic oxides compared with the solid state. Therefore basic oxides can be said to be more ionic in the liquid state than in the solid state. On the contrary, liquid silicates retain their complex non-ionic three-dimensional structures for longer times. Therefore, the ionic characteristics of liquid silicates do not increase in the liquid state.

Considering a liquid comprising basic oxide, MO, and SiO_2, the linkage between $SiO_4{}^{4-}-O^{2-}-M^{2+}$ is very much stronger than that between M^{2+} and O^{2-}.

This results in a decrease in the bond strength between M^{2+} and O^{2-} remaining as MO and a decrease in polarisation of the $M^{2+}-O^{2-}$ bond.

5.5.1 Thermodynamic properties — effect of non-ionic bonding

The heat of formation, ΔH_f, of binary silicates becomes more negative for increased miscibility (*Table 5.7*). Decreasing the cation size results in an increase in bond strength but also provides increased difficulty in accommodating the smaller cation

Table 5.7 HEATS OF FORMATION OF BINARY SILICATE COMPOUNDS AT 25°C

Basic oxide		BaO	SrO	CaO	MnO	ZnO	FeO	MgO
Cation radius		1.35	1.15	0.99	0.80	0.74	0.75	0.65
$-\Delta H$ kJ/mol^{-1}	(orthosilicate)	270	209	126	49.4	29.3	43.1	63.2
$-\Delta H$ kJ/mol^{-1}	(metasilicate)	159	131	90	24.7	–	–	35.6

in the interstices of the SiO_2 tetrahedra. This results in reduced stability of the total structure, and therefore a less negative (i.e. less exothermic) heat of formation. If this trend continues a positive ΔH_f and immiscibility occur eventually. Increasing the addition of basic oxide to SiO_2 results in the more stable orthosilicate, $2MO.SiO_2$, rather than the metasilicate, $MO.SiO_2$. This is reflected in a more negative ΔH_f for both ortho- and metasilicates formed from strong basic oxides, e.g. BaO, compared with weaker basic oxides, the orthosilicate providing a more negative ΔH_f than the corresponding metasilicate composition. The anomalous behaviour of ZnO, MnO and FeO can be explained in an increased polarisation of these three oxides, which is not predicted by consideration of the cation size alone.

5.5.1.1 Thermodynamics and mixing of silicates

The enthalpy of mixing, ΔH^m, exhibits the same trends as ΔH_f for binary silicates. ΔH^m becomes less negative on decreasing the cation size, coordination number and ionic bond fraction and on increasing the values of F, e and Z. The immiscible region will be represented by a positive value of ΔH^m and positive deviation from Raoult's law, i.e. $\gamma > 1$ (Chapter 4).

Fig. 5.13 presents the work by Richardson, who compared the entropy of mixing, ΔS^m, of various CaO–B_2O_3 melts with calculated values, assuming ideal mixing. It is apparent that the ΔS^m for the calcium borate melts is not ideal. This can be explained in that (i) Ca^{2+} ions cannot regularly interchange with B^{3+} and (ii) ΔS^m is increased as the silica tetrahedra network is broken down on the addition of CaO. These general trends are common for most binary silicate melts.

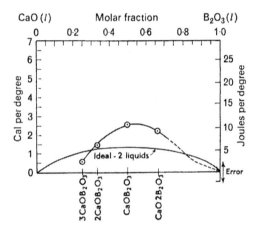

Figure 5.13 Comparison of ideal and calculated entropies of mixing for calcium borate melts (from Richardson[1], p. 75)

At a constant mole fraction of SiO_2, the increase in ΔS^m is dependent on the valency of the cation of the basic oxide. Divalent basic oxides, e.g. CaO, ionise to provide one divalent cation and an O^{2-} ion. On the other hand, monovalent basic oxides, e.g. Na_2O, are ionised to produce two Na^+ ions and one O^{2-} ion. Therefore, for one mole of the basic oxide, two monovalent ions need to be accommodated in the SiO_2 tetrahedra structure whereas only one divalent cation needs to be accommodated in the SiO_2 structure. This results in an increased ΔS^m for the monovalent basic oxide compared with the divalent basic oxide. In general, there is a decrease in the ΔS^m up to 12 mol% addition of the divalent basic oxide, at which point a limit of stability of the three-dimensional SiO_2 network is reached. Thereafter any further addition produces an increase in ΔS^m due to the breakdown of the network.

The previous discussion has taken into account only the positional effects of entropy, i.e. S_p. However, the entropy of mixing is also composed of vibrational, S_v, and magnetic, S_{mag}, effects, i.e. $S^m = S_p + S_v + S_{mag}$. S_v and S_{mag} are affected when changes in bonding occur. Therefore, a decrease in non-ionic bonding, i.e. polarisation, is likely to effect increases in S^m. The anomalous behaviour of ZnO, FeO and MnO is probably due to the S_v and S_{mag} terms rather than S_p. Table 5.8 compares the calculated ΔS^m values assuming ideal mixing for various oxides in the form of their

Table 5.8 ENTROPIES OF FORMATION OF SILICATE COMPOUNDS AT 298 K
($J K^{-1} mol^{-1}$)

Compound	Entropy sum of constituent oxides	Measured entropy	Difference (ΔS)
$2BaO.SiO_2$	182	183	+1
$2CaO.SiO_2$	127	122	−5
$2FeO.SiO_2$	145	159	+14
$2ZnO.SiO_2$	131	140	+9
$3MgO.SiO_2$	95.0	97.5	+2.5
$BaO.SiO_2$	112	113	+1
$CaO.SiO_2$	82.0	82.4	+0.4
$MnO.SiO_2$	89	102	+14
$MgO.SiO_2$	67.2	69.9	+2.7

binary silicates with those measured experimentally. The major disparities are those of iron, zinc and manganese silicates.

5.6 Physical properties of slags

5.6.1 Electrical and thermal conductivities

Molten silica is a weak electrical conductor with a conductivity dependent on the extent of ionisation achieved by adding basic oxides as fluxes. In this respect CaO and especially FeO or MnO provide increased electrical conductivity of the slag. The extent of ionisation of the slag can therefore be determined by conductivity measurements. Electrical conductivity of slags depends on (i) the number of ions present and (ii) ionic mobility, which is a function of the size of the ions and the viscosity of the liquid slag in which they exist.

Although the thermal conductivity of liquid slags is generally very low, heat losses are greater than expected due to convection.

5.6.2 Viscosity of slags

Increasing slag viscosity will reduce mass and thermal transfer characteristics of the slag, leading to a reduction in the overall slag–metal reaction rate. Increased viscosity may also result in entrapment of gas bubbles in the slag, producing 'foaming slags'. Such slags lead to a high slag volume which requires large vessel volumes, resulting in low heat transfer characteristics.

The viscosity of a slag depends largely on two factors: (i) composition and (ii) temperature. Slag composition will also determine the size of the ionic or molecular species and the structural complexity. An increase in either of these will result in increased viscosity. Viscosity can be represented by an Arrhenius equation:

$$\eta = A \exp (E_\eta/RT)$$

i.e.
$$\log \eta = \log A + E_\eta/RT$$

where E_η is the activation energy for viscous flow, and A is a constant.

The decrease in viscosity with increasing temperature for molten silica is small due to the large value of E_η. However, E_η rapidly decreases on the addition of a basic oxide flux to molten silica, the initial flux addition having a much greater effect than subsequent additions. Basic oxides or halides with a high ionic bond fraction are more effective in reducing viscosity than those with a lower ionic bond fraction. *Fig. 5.14* shows these trends for the addition of CaO, Na$_2$O and CaF$_2$ on silica. Na$_2$O and CaF$_2$ have a higher ionic bond fraction than CaO.

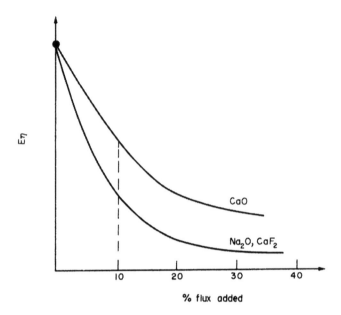

Figure 5.14 Influence of the addition of a flux (i.e. basic oxide or halide) on the activation energy of a silica-rich slag (from Coudurier *et al.*[20], p. 191)

The effect of the CaO/SiO$_2$ ratio on viscosity is shown in *Fig. 5.15*, where it is observed that the lowest viscosity is associated with a CaO/SiO$_2$ ratio of approximately 1.35. This is due to the eutectic composition at this ratio (see *Fig. 5.6*). It is also seen that viscosity decreases drastically with temperature for both basic and acidic slags, but basic slags with generally higher melting points are more sensitive to temperature. The E_η for basic slags is generally much smaller than for acid slags.

The addition of CaF$_2$ is more effective in reducing the viscosity of basic slags than acidic slags (*Fig. 5.16*). This is due to the ability of F^- ions to break the SiO$_2$ network and the low melting point of undissociated CaF$_2$.

Al$_2$O$_3$ is an amphoteric oxide and therefore behaves as a network breaker in acid slags, i.e. Al$_2$O$_3$ = 2Al^{3+} + 3O^{2-} and a network former in basic slags, i.e. Al$_2$O$_3$ + 3O^{2-} = 2AlO$_3$. In the latter case, Al$_2$O$_3$ tetrahedra replace those of SiO$_2$. As shown in *Fig. 5.17*, the addition of Al$_2$O$_3$ to a basic slag increases the viscosity since it is acting as an acidic oxide and therefore network former. The addition of Al$_2$O$_3$ to an acid slag decreases the viscosity since it is now acting as a basic oxide and therefore a network breaker.

The stronger the electrostatic force between the cation and anion complexes, the lower the sensitivity of the slag to conditions which reduce its viscosity, e.g. temperature and addition of basic oxides to acid slags. Additions to slags can also affect its surface properties, e.g. surface tension. In this respect, addition of SiO$_2$ or P$_2$O$_5$ to a basic slag will reduce the surface tension of that slag due to the absorption of a thin layer of anions, e.g. SiO$_4$$^{4-}$ or PO$_4$$^{3-}$, on the surface.

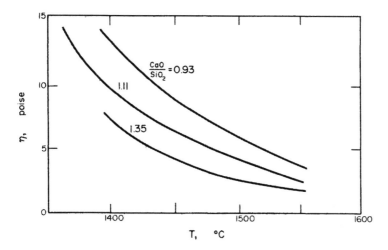

(a)

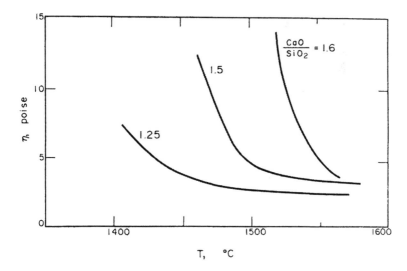

(b)

Figure 5.15 The influence of the CaO/SiO$_2$ ratio on: (a) acid; and (b) basic blast furnace slags (from Coudurier *et al.*[20], p. 192)

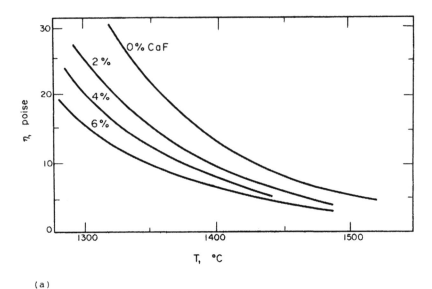

(a)

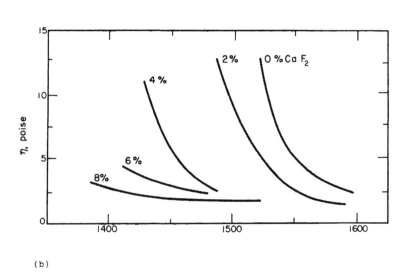

(b)

Figure 5.16 Influence of the addition of CaF$_2$ on the viscosity of: (a) acid slag: 44% SiO$_2$, 12% Al$_2$O$_3$, 41% CaO, 3% MgO; (b) basic slag: 32% SiO$_2$, 13% Al$_2$O$_3$, 51% CaO, 3% MgO (from Coudurier et al.[20], p. 193)

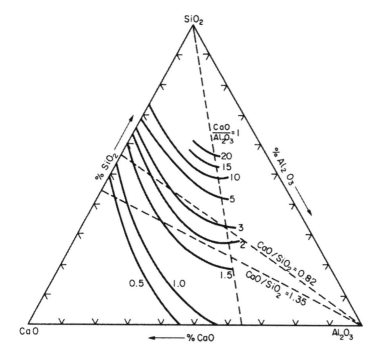

Figure 5.17 Isoviscosity lines for $CaO-SiO_2-Al_2O_3$ system at $1600\,°C$. Viscosity in poises and composition in wt% (from Coudrier et al.[20], p. 194)

5.7 Slag theories

In order to characterise a slag, the activities of the components of the slag need to be determined. Several theories have been presented which allow calculation of the activities of the slag components from calculation of their thermodynamic functions, e.g. free energy and entropy. However, certain assumptions are made which can limit the application of these theories.

The most important slag theories can be classified into molecular or ionic theories. However, none is truly accurate and they are often difficult to apply. Therefore, there is a need to determine activities experimentally, such as determined from equilibrium considerations between metal–slag and atmosphere. The ionic theory of slags mainly applies to basic slags since, in acid slags, complex anion structures provide major complications.

5.7.1 Ionic slag theories

5.7.1.1 Temkin's theory

The more simple ionic theories were developed assuming completely random mixing of cations and O^{2-} ions[2,3]. This is not true since cations will fill only cation sites and anions will fill only anion sites of similar charge and size. This latter provision was made subsequently by Temkin[4], whose ionic theory assumed that there is mixing only between cations on cation sites and anions on anion sites. Other assumptions made in these earlier simple ionic theories of slags include:

(i)   The slag is completely dissociated into ions.
(ii) There is no interaction between similarly charged ions.
(iii) There is complete randomness of ions.
(iv) The slag is composed of two ideal solutions of cations and anions.
(v)   All ion species in the slag are known.

Since slags are assumed to be composed of ions, ionic fractions need to be used rather than mole fractions, e.g. if A^{2+}, B^{2+}, C^{2-}, D^{2-} ions are present in a liquid slag, the corresponding ion fractions of A^{2+} ions, $N_{A^{2+}}$, and C^{2-} ions, $N_{C^{2-}}$, are given by:

$$N_{A^{2+}} = \frac{n_{A^{2+}}}{n_{A^{2+}} + n_{B^{2+}}} \quad \text{and} \quad N_{C^{2-}} = \frac{n_{C^{2-}}}{n_{C^{2-}} + n_{D^{2-}}}$$

where $n_{A^{2+}}$ is the number of A^{2+} ions in solution.
  Considering the reaction:

$$A^{2+}C^{2-} + B^{2+}D^{2-} = A^{2+}D^{2-} + B^{2+}C^{2-}$$

$$n_{A^{2+}} = n_{A^{2+}C^{2-}} + n_{A^{2+}D^{2-}}$$

$$n_{B^{2+}} = n_{B^{2+}D^{2-}} + n_{B^{2+}C^{2-}}$$

$$n_{C^{2-}} = n_{A^{2+}C^{2-}} + n_{B^{2+}C^{2-}}$$

$$n_{D^{2-}} = n_{B^{2+}D^{2-}} + n_{A^{2+}D^{2-}}$$

where $n_{A^{2+}}$, $n_{B^{2+}}$, $n_{C^{2-}}$, $n_{D^{2-}}$ are measured from chemical analysis.
  ΔS^m for ideal mixing of cations (ΔS_+^m) is calculated using equation (4.19), i.e.

$$\Delta S_+^m = -R(N_{A^{2+}} \ln N_{A^{2+}} + N_{B^{2+}} \ln N_{B^{2+}})$$

$$\therefore \quad \Delta S_+^m = \frac{-R}{n_{A^{2+}} + n_{B^{2+}}} \left[n_{A^{2+}} \ln \left(\frac{n_{A^{2+}}}{n_{A^{2+}} + n_{B^{2+}}} \right) + n_{B^{2+}} \ln \left(\frac{n_{B^{2+}}}{n_{A^{2+}} + n_{B^{2+}}} \right) \right]$$

$$= -k_1 \left[n_{A^{2+}} \ln \left(\frac{n_{A^{2+}}}{n_{A^{2+}} + n_{B^{2+}}} \right) + n_{B^{2+}} \ln \left(\frac{n_{B^{2+}}}{n_{A^{2+}} + n_{B^{2+}}} \right) \right]$$

Similarly, for anions ΔS_-^m can be calculated:

$$\Delta S_-^m = \frac{-R}{n_{C^{2-}} + n_{D^{2-}}} \left[n_{C^{2-}} \ln \left(\frac{n_{C^{2-}}}{n_{C^{2-}} + n_{D^{2-}}} \right) + n_{D^{2-}} \ln \left(\frac{n_{D^{2-}}}{n_{C^{2-}} + n_{D^{2-}}} \right) \right]$$

$$= -k_2 \left[n_{C^{2-}} \ln \left(\frac{n_{C^{2-}}}{n_{C^{2-}} + n_{D^{2-}}} \right) + n_{D^{2-}} \ln \left(\frac{n_{D^{2-}}}{n_{C^{2-}} + n_{D^{2-}}} \right) \right]$$

Since $\Delta S^m = \Delta S_+^m + \Delta S_-^m$ and, for an ideal solution ($\Delta H^m = 0$):

$$\Delta G^m = -T \Delta S^m$$

applying this to sulphur removal in steel where mixing involves only anions:

$$FeS_{(m)} + CaO_{(s)} = CaS_{(s)} + FeO_{(m)}$$

where (m) and (s) refer to species in the metal and slag respectively, or the ionic equation: $S_{(m)} + Ca^{2+}O^{2-}_{(s)} = Ca^{2+}S^{2-}_{(s)} + O_{(m)}$

$$\Delta \bar{S}_+^m = 0$$

$$\therefore \Delta \bar{S}_-^m = -\frac{R}{n_{O^{2-}} + n_{S^{2-}}} \left[n_{O^2} \ln\left(\frac{n_{O^{2-}}}{n_{O^{2-}} + n_{S^{2-}}}\right) + n_{S^2} \ln\left(\frac{n_{S^{2-}}}{n_{O^{2-}} + n_{S^{2-}}}\right) \right]$$

For reactions where mixing involves only cations, e.g.

$$Fe_{(m)} + MnO_{(s)} = Mn_{(m)} + FeO_{(s)}$$

or

$$Fe_{(m)} + Mn^{2+}O^{2-}{}_{(s)} = Mn_{(m)} + Fe^{2+}O^{2-}{}_{(s)}$$

since only one species of anion (O^{2-}) is used. Therefore,

$$\Delta S^m{}_{O^{2-}} = 0$$

$$\therefore \Delta S^m = \Delta S_+^m = -\frac{R}{n_{Fe^{2+}} + n_{Mn^{2+}}} \left[n_{Fe^{2+}} \ln\left(\frac{n_{Fe^{2+}}}{n_{Fe^{2+}} + n_{Mn^{2+}}}\right) \right.$$

$$\left. + n_{Mn^{2+}} \ln\left(\frac{n_{Mn^{2+}}}{n_{Fe^{2+}} + n_{Mn^{2+}}}\right) \right]$$

If ideal, $\Delta G^m = -\Delta T S^m$

$$\therefore \Delta G^m = -\frac{RT}{n_{Fe^{2+}} + n_{Mn^{2+}}} \left[n_{Fe^{2+}} \ln\left(\frac{n_{Fe^{2+}}}{n_{Fe^{2+}} + n_{Mn^{2+}}}\right) \right.$$

$$\left. + n_{Mn^{2+}} \ln\left(\frac{n_{Mn^{2+}}}{n_{Fe^{2+}} + n_{Mn^{2+}}}\right) \right]$$

But also $\Delta G^m = N_{Fe^{2+}} \Delta \bar{G}_{Fe^{2+}} + N_{Mn^{2+}} \Delta \bar{G}_{Mn^{2+}}$ (equation (4.2)) and

$$\Delta G^m{}_{Fe^{2+}} = RT \ln N_{Fe^{2+}} = RT \ln\left(\frac{n_{Fe^{2+}}}{n_{Fe^{2+}} + n_{Mn^{2+}}}\right)$$

Since $n_{Fe^{2+}} = n_{FeO}$ and $n_{Mn^{2+}} = n_{MnO}$ for this reaction:

$$\frac{n_{Fe^{2+}}}{n_{Fe^{2+}} + n_{Mn^{2+}}} = \frac{n_{FeO}}{n_{FeO} + n_{MnO}} = N_{FeO} \quad \text{and} \quad \frac{n_{Mn^{2+}}}{n_{Mn^{2+}} + n_{Fe^{2+}}} = N_{MnO}$$

$\therefore \Delta \bar{G}_{Fe^{2+}} = RT \ln N_{FeO}$. Similarly $\Delta \bar{G}_{Mn^{2+}} = RT \ln N_{Mn^{2+}} = RT \ln N_{MnO}$ and for non-ideal solutions,

$$\Delta \bar{G}_{Fe^{2+}} = RT \ln a_{Fe^{2+}} \text{ (in slag)}$$

Since $a_{Fe^{2+}} = a_{FeO}$ for this reaction:

$$\Delta \bar{G}_{Fe^{2+}} = RT \ln a_{FeO} \quad \text{and} \quad \Delta \bar{G}_{Mn^{2+}} = RT \ln a_{MnO}$$

Hence,

$$\Delta G^m = N_{FeO} RT \ln a_{FeO} + N_{MnO} RT \ln a_{MnO}$$

$$= RT (N_{FeO} \ln a_{FeO} + N_{MnO} \ln a_{MnO})$$

In cases where mixing involves both cations and anions, e.g.

$$FeS_{(s)} + MnO_{(s)} = FeO_{(s)} + MnS_{(s)}$$

or

$$Fe^{2+}S^{2-} + Mn^{2+}O^{2-} = Fe^{2+}O^{2-} + Mn^{2+}S^{2-}$$

the ideal ΔG^m for an ideal solution is obtained from the sum of the partial ionic entropies of mixing for cations and anions, i.e.

$$\Delta G^m = -T(\Delta \bar{S} \underset{+}{\,m} + \Delta \bar{S} \underset{-}{\,m}) = -T(\Delta \bar{S}^m_{Fe^{2+}} + \Delta \bar{S}^m_{Mn^{2+}} + \Delta \bar{S}^m_{O^{2-}} + \Delta \bar{S}^m_{S^{2-}})$$

where

$$n_{Fe^{2+}} = n_{FeO} + n_{FeS}; \; n_{Mn^{2+}} = n_{MnO} + n_{MnS}$$

$$n_{O^{2-}} = n_{FeO} + n_{MnO}; \; n_{S^{2-}} = n_{FeS} + n_{MnS}$$

However, considering only the Fe^{2+} and O^{2-} produced from the FeO component of the slag, i.e.

$$FeO = Fe^{2+} + O^{2-}$$

$$\Delta \bar{G}^m_{Fe^{2+}} = RT \ln N_{Fe^{2+}} = RT \ln \left(\frac{n_{Fe^{2+}}}{n_{Fe^{2+}} + n_{Mn^{2+}}} \right)$$

and

$$\Delta \bar{G}_{O^{2-}} = RT \ln N_{O^{2-}} = RT \ln \left(\frac{n_{O^{2-}}}{n_{O^{2-}} + n_{S^{2-}}} \right)$$

Since $\Delta G^m_{FeO} = \Delta \bar{G}_{Fe^{2+}} + \Delta \bar{G}_{O^{2-}}$ adding these and considering one mole of solution:

$$\Delta \bar{G}^m_{FeO} = RT \ln N_{Fe^{2+}} . N_{O^{2-}} = RT \ln N_{FeO}$$

or

$$\Delta \bar{G}^m_{FeO} = RT \ln a_{FeO} \text{ (real solution)}$$

Using the same considerations for FeS, MnS and MnO components in the slag:

$$\Delta G^m = RT(N_{FeO} \ln N_{FeO} + N_{MnO} \ln N_{MnO} + N_{FeS} \ln N_{FeS} + N_{MnS} + \ln N_{MnS})$$

and for a real solution:

$$\Delta G^m = RT (N_{FeO} \ln a_{FeO} + N_{MnO} \ln a_{MnO} + N_{FeS} \ln a_{FeS} + N_{MnS} \ln a_{MnS})$$

It is useful to provide an example here of the application of these simple ionic theories, i.e. the calculation of the activity of FeO, a_{FeO}, in a slag of the following composition:

	N		N
CaO	0.31	SiO_2	0.02
MnO	0.21	Fe_2O_3	0.04
FeO	0.42		
	0.94		0.06
$N_{total} = 1.0$			

Assuming the solution is perfectly ionised as follows:

$$CaO = Ca^{2+} + O^{2-}, SiO_2 + 2O^{2-} = SiO_4^{4-}$$

$$MnO = Mn^{2+} + O^{2-}, Fe_2O_3 + 2O^{2-} = Fe_2O_5^{4-}$$

$$FeO = Fe^{2+} + O^{2-}$$

And applying Temkin's model:

$$a_{FeO} = N_{Fe^{2+}} . N_{O^{2-}}$$

where

$$N_{Fe^{2+}} = \frac{n_{Fe^{2+}}}{n_{Ca^{2+}} + n_{Mn^{2+}} + n_{Fe^{2+}}} = \frac{0.42}{0.94} = 0.447$$

and

$$N_{O^{2-}} = \frac{n_{O^{2-}}}{n_{SiO_4^{4-}} + n_{Fe_2O_5^{4-}} + n_{O^{2-}}}$$

and

$$n_{O^{2-}} = n_{CaO} + n_{MnO} + n_{FeO} - 2n_{SiO_2} - 2n_{Fe_2O_5} = 0.82$$

It should be noted that $2O^{2-}$ ions are needed to react with both SiO_2 and Fe_2O_3 to produce SiO_4^{4-} and $Fe_2O_5^{4-}$ ions in the slag. Therefore,

$$N_{O^{2-}} = \frac{n_{O^{2-}}}{n_{SiO_4^{4-}} + n_{Fe_2O_5^{4-}} + n_{O^{2-}}} = \frac{0.82}{(0.06 + 0.82)} = \underline{0.932}$$

$$\therefore a_{FeO} = 0.447 \times 0.932 = 0.417$$

Since $a = \gamma N$:

$$\gamma_{FeO} = a_{FeO}/N_{FeO}$$
$$= \frac{0.417}{0.42} = 0.993$$

In this case γ is slightly less than unity which is typical for highly basic slags.

Temkin's theory assumes all ions species in the slag are known. In strongly basic slag, as in the example, it is well known that simple SiO_4^{4-}, PO_4^{3-} and AlO_3^{3-} ions are formed, but in acid slags $Si_2O_4^{6-}$, $Si_3O_{10}^{8-}$, etc. may exist, and application of Temkin's theory is difficult.

5.7.1.2 Electrical equivalent ion fraction

The concept of electrical equivalent ion fraction is used for slags containing ions of mixed valency, e.g. addition of CaO to a Na_2O slag. The number of cation and anion sites is assumed to be constant, and ionisation of the oxides is as follows:

$$Na_2O = 2Na^+ + O^{2-}$$

Adding one mole of CaO to the Na_2O slag provides ionisation of CaO, i.e.

$$CaO = Ca^{2+} + O^{2-}$$

However, the Ca^{2+} ions can occupy only the Na^+ sites in the slag. Therefore, in order to maintain electrical neutrality, for every one Na^+ ion replaced by a Ca^{2+} ion there must be generated a vacant cation site (n_C). The number of ions in the slag can then be represented as:

$$n_{Na^+} + n_{Ca^{2+}} + n_C$$

where n_C represents a vacant Na^+ site such that:

$$n_C = n_{Ca^{2+}}$$

Similarly, replacing one Na^+ ion by M^{3+} (trivalent cation) results in two vacant cation sites in order to maintain charge neutrality in the slag, i.e.

$$M_2O_3 = 2M^{3+} + 3O^{2-}$$

and the number of cations can be represented by:

$$Na^+ + n_{M^{3+}} + 2n_C$$

where $2n_C = n_{M^{3+}}$. From statistical thermodynamics $S_+^m = k \ln W_1$ is the number of possible combinations of arranging the cation on available cation sites, where W_1 = number of cation sites!/number of cations! Therefore, for pure Na_2O:

$$W_1 = \frac{n_{Na^+}!}{n_{Na^+}!} = 1$$

Therefore, $S_+^m = 0$ for pure Na_2O. The number of possible combinations of arranging the divalent Ca^{2+} ions on the Na_2O lattice can be represented by

$$S_+^m = k \ln W_2$$

where

$$W_2 = \frac{2n_{Ca^{2+}}!}{n_{Ca^{2+}}! \, n_C!} = \frac{2n_{Ca^{2+}}!}{n_{Ca^{2+}}! \, n_{Ca^{2+}}!}$$

since $n_C = n_{Ca^{2+}}$. Mixing these two ions on the Na_2O lattice provides $W_{mixture}$ possible combinations, where $W_{mixture} = W_1 + W_2$. But also

$$W_{mixture} = \frac{\text{total no. of cation sites!}}{\text{total no. of cations!}} = \frac{(n_{Na^+} + 2n_{Ca^{2+}})!}{n_{Na^+}! \, n_{Ca^{2+}}! \, n_{Ca^{2+}}!}$$

Hence,

$$\Delta \bar{S}_+ (\text{mixture}) = k \ln \left(\frac{(n_{Na^+} + 2n_{Ca^{2+}})!}{n_{Na^+}! \, n_{Ca^{2+}}! \, n_{Ca^{2+}}!} \right)$$

Thus, the change in entropy of mixing these two cations on the Na_2O lattice, ΔS_+^m (mixing) can be represented as:

$$\Delta \bar{S}_+^m (\text{mixing}) = \Delta \bar{S}_+ (\text{mixture}) - \underset{Na_2O + CaO}{\Delta \bar{S}_+ (\text{pure oxides})}$$

where

$$\Delta \bar{S}_+ (\text{pure oxides}) = k \left[\ln \left(\frac{n_{Na^+}!}{n_{Na^+}!} \right) + \ln \left(\frac{2n_{Ca^{2+}}!}{n_{Ca^{2+}}! \, n_{Ca^{2+}}!} \right) \right] = k \ln \left(\frac{2n_{Ca^{2+}}!}{n_{Ca^{2+}}! \, n_{Ca^{2+}}!} \right)$$

$$\therefore \Delta \bar{S}_+^m = k \ln \left(\frac{(n_{Na^+} + 2n_{Ca^{2+}})!}{n_{Na^+}! \, n_{Ca^{2+}}! \, n_{Ca^{2+}}!} \right) - k \ln \left(\frac{2n_{Ca^{2+}}!}{n_{Ca^{2+}}! \, n_{Ca^{2+}}!} \right)$$

$$= k \ln \left(\frac{(n_{Na^+} + 2n_{Ca^{2+}})!}{n_{Na^+}! \, n_{Ca^{2+}}! \, n_{Ca^{2+}}!} + \frac{n_{Ca^{2+}}! \, n_{Ca^{2+}}!}{2n_{Ca^{2+}}!} \right)$$

$$= k \ln \left(\frac{(n_{Na^+} + 2n_{Ca^{2+}})!}{n_{Na^+}! \, 2n_{Ca^{2+}}!} \right)$$

$$= k \ln (n_{Na^+} + 2n_{Ca^{2+}})! - (k \ln n_{Na^+}! + k \ln 2n_{Ca^{2+}}!)$$

Using the solution $\ln X! = X \ln X - X$

$$\Delta \bar{S}_+^m = k\,[(n_{Na^+} + 2n_{Ca^{2+}})\ln(n_{Na^+} + 2n_{Ca^{2+}}) - (n_{Na} + 2n_{Ca^{2+}})]$$

$$\qquad - k\,[(n_{Na^+}\ln n_{Na^+} - n_{Na^+}) + (2n_{Ca^{2+}}\ln 2n_{Ca^{2+}} - 2n_{Ca^{2+}})]$$

$$\qquad = k\,[n_{Na^+}\ln(n_{Na^+} + 2n_{Ca^{2+}}) + 2n_{Ca^{2+}}\ln(n_{Na^{2+}} + 2n_{Ca^{2+}}) - n_{Na^+} - 2n_{Ca^{2+}}$$

$$\qquad - n_{Na^+}\ln n_{Na^+} + n_{Na^+} - 2n_{Ca^{2+}}\ln 2n_{Ca^{2+}} + 2n_{Ca^{2+}}]$$

$$\qquad = k\left[n_{Na^+}\,\frac{\ln(n_{Na^+} + 2n_{Ca^{2+}})}{\ln n_{Na^+}} + 2n_{Ca^{2+}}\,\frac{\ln(n_{Na^+} + 2n_{Ca^{2+}})}{2n_{Ca^{2+}}}\right] \qquad (5.2)$$

$$\qquad = -k\left[n_{Na^+}\ln\left(\frac{n_{Na^+}}{n_{Na^+} + 2n_{Ca^{2+}}}\right) + 2n_{Ca^{2+}}\ln\left(\frac{2n_{Ca^{2+}}}{n_{Na^+} + 2n_{Ca^{2+}}}\right)\right]$$

$$\qquad = k\,(n_{Na^+}\ln N'_{Na^+} + 2n_{Ca^{2+}}\ln N'_{Ca^{2+}})$$

Considering 1 mole of solution, and since $R = Nk$, where R is the gas content, N is Avogadro's number and k is Boltzmann's constant,

$$\Delta \bar{S}_+^m = -R\,(N'_{Na^+}\ln N'_{Na^+} + N'_{Ca^{2+}}\ln N'_{Ca^{2+}}) \qquad (5.3)$$

where N' = electrically equivalent ion fraction. Therefore, in a mixture of A^+, B^{2+}, C^{3+}, D^{4+} ions, the electrical equivalent ion fraction of B^{2+} ions $(N'_{B^{2+}})$ is given by:

$$N'_{B^{2+}} = \frac{2n_{B^{2+}}}{n_{A^+} + 2n_{B^{2+}} + 3n_{C^{3+}} + 4n_{D^{4+}}} \qquad (5.3)$$

The advantage is that A^+, etc. are measurable species using chemical analysis, whereas AO is not so easily measured.

Thus the activities of the ions are not proportional to the mole fractions and are not a linear function of the concentration of ions. If the anions also exhibit different valencies, as in the case of mixing oxides with halides, a similar treatment can be applied to the anion fractions.

5.7.1.3 Flood's theory[5-7]

Flood's theory is based partly on Temkin's theory and on the consideration of equilibrium of elements dissolved in the metallic and slag phases, e.g. the sulphur-oxygen equilibrium in steelmaking:

$$S_{(m)} + O^{2-}_{(s)} = S^{2-}_{(s)} + O_{(m)}$$

where (m) and (s) refer to metal and slag phases respectively, and

$$K = \frac{a_{O_{(m)}}\,a_{S^{2-}_{(s)}}}{a_{S_{(m)}}\,a_{O^{2-}_{(s)}}}$$

Assuming ideal solution in the slag phase and a real solution in the metal phase, i.e.

$$a_{S^{2-}_{(s)}} = N_{S^{2-}_{(s)}}\,;\ a_{S_{(m)}} = f_{S_{(m)}}\ wt\%S$$

and

$$a_{O^{2-}_{(s)}} = N_{O^{2-}_{(s)}}\,;\ a_{O_{(m)}} = f_{O_{(m)}}\ wt\%O$$

where f = Henrian activity coefficient, then

$$K = \frac{N_{S^{2-}_{(s)}} \times \%O_{(m)} \, f_{O_{(m)}}}{N_{O^{2-}_{(s)}} \times \%S_{(m)} \, f_{S_{(m)}}}$$

$$= K'g(f) \tag{5.4}$$

where

$$K' = \frac{N_{S^{2-}_{(s)}} \times \%O_{(m)}}{N_{O^{2-}_{(s)}} \times \%S_{(m)}} \quad \text{(the apparent equilibrium constant)}$$

and

$$g(f) = \frac{f_{O_{(m)}}}{f_{S_{(m)}}}$$

Thus, $K = K'$ when $f_{O_{(m)}} = 1 = f_{S_{(m)}} = g(f)$ while Henry's law is being followed, i.e. ideal dilute solutions in the liquid metal phase.

In complex slags, sulphides and oxides of Na, Ca, Mg, Fe, Mn, etc. are present — but to simplify it is useful to consider only CaO and Na_2O are present so that the overall reaction can be represented as follows:

$$(CaO, Na_2O)_{(s)} + S_{(m)} = (CaS, Na_2S)_{(s)} + O_{(m)}$$

This reaction can be divided into the following steps:

Step I–Separation of oxides, i.e. $CaO.Na_2O = CaO + Na_2O$; $\Delta G^m_{CaO/Na_2O}$

Step II–Reactions of each oxide $\rightarrow$ sulphide ($CaS + Na_2S$); $\Delta G_{(oxide \rightarrow sulphide)}$

Step III–Mixing of products $CaS + Na_2S = \Delta G^m_{(CaS/Na_2S)}$

Therefore, the ΔG for the overall reaction, $\Delta G_{(overall\ reaction)}$, can be represented by:

$$\Delta G_{(overall\ reaction)} = \Delta G_{products} - \Delta G_{reactants} = \Delta G_{III} + \Delta G_{II} - \Delta G$$

However, the free energy of mixing of sulphides and oxides, i.e. Steps I and III, are normally small compared with ΔG (oxides $\rightarrow$ sulphides) and can therefore be neglected. Thus, under equilibrium conditions, and using electrically equivalent ion fractions,

$$\Delta G^{\ominus}_{(overall\ reaction)} = N'_{Ca^{2+}} \, \Delta G^{\ominus}_{[CaO \rightarrow CaS]} + N'_{Na^+} \, \Delta G^{\ominus}_{[Na_2O \rightarrow Na_2S]}$$

and $\quad -RT \ln K = -N'_{Ca^{2+}} (RT \ln K_{[CaO \rightarrow Ca_2S]}) - N'_{Na^+} (RT \ln K_{[Na_2O \rightarrow Na_2S]})$

or $\quad \ln K = N'_{Ca^{2+}} \ln K_{[CaO \rightarrow CaS]} + N'_{Na^+} \ln K_{[Na_2O \rightarrow Na_2S]} \tag{5.5}$

Applying equation (5.4),

$$\ln K'g(f) = N'_{Ca^{2+}} \ln K_{Ca} + N'_{Na^+} \ln K_{Na}$$

$$\ln K' + \ln g(f) = N'_{Ca^{2+}} \ln K_{Ca} + N'_{Na^+} \ln K_{Na}$$

$$\ln K' = N'_{Ca^{2+}} \ln K_{Ca} + N'_{Na^+} \ln K_{Na} - \ln g(f)$$

or $\quad \ln K' = \Sigma N'_i + \ln K_i - \ln g(f) \quad \text{(Flood's theory)} \tag{5.6}$

An example of the application of Flood's theory (equation (5.6)) can be seen in the phosphorous–oxygen equilibrium between metal and slag in iron and steelmaking using the following two-step reaction:

Step I

$$2P_{(m)} + 5O_{(m)} = P_2O_{5\,(s)}; \quad K_p = \frac{a_{P_2O_{5(s)}}}{a_{O_{(m)}}^5 \cdot a_{P_{(m)}}^2} \tag{5.7}$$

where $\Delta G^{\ominus} = -705\,600 + 557T$ (J) which is positive at $1600\,^{\circ}C$. Assuming infinite dilution and Henry's law, i.e. standard state = 1 wt%, for metal solution and Raoult's law for P in the slag

$$\Delta G^{\ominus} = -RT \ln K_p = -RT \ln \frac{\gamma_{P_2O_{5(s)}} N_{P_2O_{3(s)}}}{a_{O_{(m)}}^5 \, a_{P_{(m)}}^2} \tag{5.8}$$

Thus, for dephosorisation, $a_{P_2O_5}$ must be very low, i.e. the acidic P_2O_5 must be neutralised with a basic slag, producing PO_4^{3-} ions in the slag.

Step II Reaction of P_2O_5 with basic oxide MO to the slag.

$$P_2O_{5(s)} + 3MO_{(s)} = 3MO.P_2O_{5(s)} \text{ or } M_3(PO_4)_2, \text{ or } 3M^{2+} + 2(PO_4)^{3-}$$

Hence, combining steps (I) and (II):

$$2P_{(m)} + 5O_{(m)} + 3MO_{(s)} = 3MO.P_2O_{5(s)} \tag{5.9}$$

or in the form of an ionic equation:

$$2P_{(m)} + 5O_{(m)} + 3O_{(s)}^{2-} = 2(PO_4)_{(s)}^{3-}$$

where, according to Flood and Temkin:

$$K = \frac{N_{(PO_4)^{3-}}^2}{a_{P_{(m)}}^2 \, a_{O_{(m)}}^5 \, N_{O_{(s)}^{2-}}^3}$$

K is strongly influenced by the cationic composition of the slag and, according to Flood, can be expressed as a function of the electrical equivalent cationic fraction $N'_{M^{2+}}$ of each cation (Ca^{2+}, Mn^{2+}, Fe^{2+}, Mg^{2+}) or mole fraction of basic oxides N_{MO}, by

$$\log K' = \Sigma N'_{M^{2+}} \log K_M - \ln g(f) = \Sigma N_{MO} \log K_M - \ln g(f)$$

where K_M is the equilibrium constant established for each element present in the basic oxides, e.g. Ca, Mn, Mg, Fe, etc. for the slag–metal reaction under consideration, taking the general form of reaction (5.9). Plotting $\log K'$ versus N' for each element in the slag allows the value for K_M to be determined from the slope of the line as shown in *Fig. 5.18*. Also, considering reaction (5.7), and assuming ideal solution in the metal phase, i.e. $f_P = 1 = f_0$:

$$K_P = \frac{a_{P_2O_5}}{a_P^2 \cdot a_O^5} = \frac{N_{P_2O_5}}{(wt\%P)^2} \frac{\gamma_{P_2O_5}}{(wt\%O)^5} = \gamma_{P_2O_5} K'_P$$

Thus, $\log \gamma_{P_2O_5} = \log K_P - \log K'_P$

But $\log K'_P = N'_{M^{2+}} \log K_M$

for each cation in the slag. Hence,

$$\log \gamma_{P_2O_5} = - \Sigma N'_{M^{2+}} \log K_M + \log K_P$$

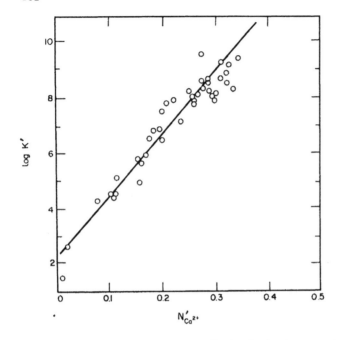

Figure 5.18 Graphical determination of log K_{Ca} by plotting the apparent equilibrium constant ($\log K'$) against the electrical equivalent ion fraction of Ca^{2+}, $N'_{Ca^{2+}}$ (from Coudurier et al.[20], p. 261)

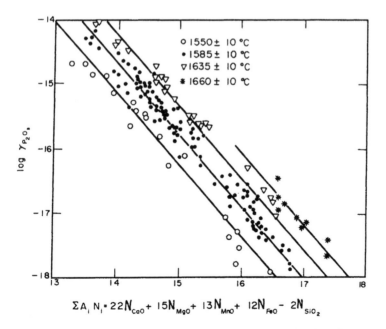

Figure 5.19 The effect of slag chemistry and temperature on $\gamma P_2 O_5$ (from Turkdogan and Pearson[8])

or
$$\log \gamma_{P_2O_5} = -\Sigma N_{MO} \log K_M + \log K_P$$

Plotting the calculated values of $\gamma_{P_2O_5}$ for typical steelmaking slags against $N_{MO} \log K_M$ for each oxide, as shown in *Fig. 5.19*, it is possible to provide information on the ability of each of the basic oxides to be an efficient dephosphoriser. One such approach produced the following relationship[8]:

$$\log \gamma_{P_2O_5} = -1.12(22N_{CaO} + 15N_{MgO} + 13N_{MnO} + 12N_{FeO} - 2N_{SiO_2}$$

$$- \frac{42\,000}{T} + 23.58$$

giving the general relationship:

$$\log \gamma_{P_2O_5} = -1.12 \Sigma N_{MO} \log K_{MO} - 4200/T + 23.58$$

These data indicate that CaO exerts the largest influence on $\gamma_{P_2O_5}$ and is the most effective dephosphoriser, i.e.

$$2P_{(m)} + 5O_{(m)} + 3CaO = Ca_3(PO_4)_2 ; \; \Delta G^{\ominus} \sim -1428\,760 + 632.6T\,J$$

FeO plays a double role in dephosphorisation, i.e. as an oxidant and basic oxide. Although FeO is a basic oxide forming $Fe_3(PO_4)_2$ it also provides $O_{(m)}$ to the metal. But CaO is a stronger dephosphoriser. Therefore increasing FeO dilutes the slag of CaO which reduces dephosphorisation. Hence, up to 16–20% FeO, the oxidising effect is dominant. Above 20% FeO, the dilution of CaO is dominant, resulting in a decrease in the partitioning of P, as seen in *Fig. 5.20*.

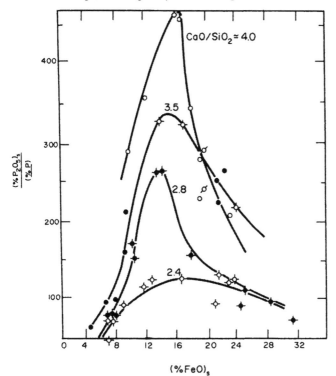

Figure 5.20 Influence of FeO and basicity on the partitioning of phosphorus in steel (from Balaiaiva *et al.*[21])

5.7.1.4 Masson's theory[9,10]

Masson's theory is used to calculate the activity of a basic oxide in a silicate slag, taking into account that more complex anions than SiO_4^{4-} are usually present. Slags are complex solutions containing polymeric silicate anions, the degree of polymerisation being controlled by the character and quantity of the basic oxides present. Highly basic slags contain mainly SiO_4^{4-}, while less basic slags contain SiO_4^{4-}, $Si_2O_7^{6-}$, $Si_3O_{10}^{8-}$, etc. Consider the equilibria between these complex anions:

(a) $SiO_4^{4-} + SiO_4^{4-} = Si_2O_7^{6-} + O^{2-}$; $K_a = \dfrac{N_{Si_2O_7^{6-}} \cdot N_{O^2}}{N_{SiO_4^{4-}} \cdot N_{SiO_4^4}}$

(b) $Si_2O_7^{6-} + SiO_4^{4-} = Si^3O_{10}^{8-} + O^{2-}$; $K_b = \dfrac{N_{Si_3O_{10}^{8-}} \cdot N_{O^{2-}}}{N_{Si_2O_7^{6-}} \cdot N_{SiO_4^{4-}}}$

(c) $Si_3O_{10}^{8-} + SiO_4^{4-} = Si_4O_{13}^{10-} + O^{2-}$; $K_c = \dfrac{N_{Si_4O_{13}^{10-}} \cdot N_{O^{2-}}}{N_{Si_3O_{10}^{8-}} \cdot N_{SiO_4^{4-}}}$

Where K_a, K_b and K_c are the corresponding equilibrium constants. Each of these reactions consists of the addition of a silica tetrahedron to the existing chain and the mole fractions of each anion can be given by:

$$N_{Si_2O_7^{6-}} = \frac{K_a N_{SiO_4^{4-}}}{N_{O^{2-}}} N_{SiO_4^{4-}}$$

$$N_{Si_3O_{10}^{8-}} = \frac{K_b N_{Si_2O_7^{6-}}}{N_{O^{2-}}} N_{SiO_4^{4-}} = K_b K_a \left(\frac{N_{SiO_4^{4-}}}{N_{O^{2-}}}\right)^2 N_{SiO_4^{4-}}$$

$$N_{Si_4O_{13}^{10-}} = \frac{K_c N_{Si_3O_{10}^{8-}}}{N_{O^{2-}}} N_{SiO_4^{4-}} = K_c K_b K_a \left(\frac{N_{SiO_4^{4-}}}{N_{O^{2-}}}\right)^3 N_{SiO_4^{4-}}$$

The sum of the silicate anions is therefore given by:

$$\Sigma N_{(silicate\ ions)} = N_{SiO_4^{4-}} + N_{Si_2O_7^{6-}} + N_{Si_3O_{10}^{8-}} + N_{Si_4O_{13}^{10-}} + N_{Si_nO_{(3n+1)}^{2(n+1)-}}$$

$$= N_{SiO_4^{4-}}\left[1 + K_a\left(\frac{N_{SiO_4^{4-}}}{N_{O^{2-}}}\right) + K_a K_b\left(\frac{N_{SiO_4^{4-}}}{N_{O^{2-}}}\right)^2 + K_a K_b K_c\left(\frac{N_{SiO_4^{4-}}}{N_{O^{2-}}}\right)^3\right]$$

But the sum of the series $\Sigma(1 + x + x^2 + \ldots x^n) = 1/(1-x)$ when $x < 1$. Hence, substituting

$$K \frac{N_{SiO_4^{4-}}}{N_{O^{2-}}} = x$$

where K is the composite equilibrium constant:

$$\Sigma N_{(silicate\ ions)} = \frac{N_{SiO_4^{4-}}}{1 - K\left(\dfrac{N_{SiO_4^{4-}}}{N_{O^{2-}}}\right)} \tag{5.10}$$

The silica content of the slag can be found by chemical analysis and the mole fraction of SiO_2, N_{SiO_2}, can be related to $N_{O^{2-}}$ as given by

$$N_{SiO_2} = \frac{n_{SiO_2}}{n_{\text{total oxides}}}$$

i.e. $\quad N_{SiO_2} = \dfrac{n_{SiO_2}}{n_{SiO_2} + n_{MO\,(\text{free})} + n_{MO\,(\text{as silicates})}}$

However, before this can be calculated, it is necessary to know the equivalent number of moles of SiO_2 required to produce each of the complex silicate anions, i.e. $Si_2O_7^{6-}$, $Si_3O_{10}^{8-}$, $Si_4O_{13}^{10-}$ as given below:

(i) $\quad n_{SiO_2}$ as $SiO_4^{4-} = n_{SiO_4^{4-}}$ since $SiO_2 + 2O^{2-} = SiO_4^{4-}$

(ii) $\quad n_{SiO_2}$ as $Si_2O_7^{6-} = 2n_{Si_2O_7^{6-}}$ since $2\,SiO_2 + 3O^{2-} = Si_2O_7^{6-}$

(iii) $\quad n_{SiO_2}$ as $Si_3O_{10}^{8-} = 3n_{Si_3O_{10}^{8-}}$ since $3SiO_2 + 4O^{2-} = Si_3O_{10}^{8-}$

(iv) $\quad n_{SiO_2}$ as $Si_4O_{13}^{10-} = 4n_{Si_4O_{13}^{10-}}$ since $4SiO_2 + 5O^{2-} = Si_4O_{13}^{10-}$

Thus

$$\Sigma n_{SiO_2} = n_{SiO_4^{4-}} + 2n_{Si_2O_7^{6-}} + 3n_{Si_3O_{10}^{8-}} + 4n_{Si_4O_{13}^{10-}} + \ldots$$

And the total number of oxides, i.e. SiO_2 + basic oxides, can be determined from the same four equations, i.e.

(i) $\quad$ each SiO_4^{4-} ion produces a total of $3n$ oxides

(ii) $\quad$ each $Si_2O_7^{6-}$ ion produces a total of $5n$ oxides

(iii) $\quad$ each $Si_3O_{10}^{8-}$ ion produces a total of $7n$ oxides

(iv) $\quad$ each $Si_4O_{13}^{10-}$ ion produces a total of $9n$ oxides

Thus $n_{\text{oxides}} = n_{O^{2-}} + 3n_{SiO_4^{4-}} + 5n_{Si_2O_7^{6-}} + 7n_{Si_3O_{10}^{8-}} + 9n_{Si_4O_{13}^{10-}}$

and converting these values to mole fractions:

$$N_{SiO_2} = \frac{N_{SiO_4^{4-}} + 2N_{Si_2O_7^{6-}} + 3N_{Si_3O_{10}^{8-}} + 4N_{Si_4O_{13}^{10-}}}{N_{O_2} + 3N_{SiO_4^{4-}} + 5N_{Si_2O_7^{6-}} + 7N_{Si_3O_{10}^{8-}}} \tag{5.11}$$

When there are only silicate and O^{2-} ions present as anions, Temkin's theory gives:

$$N_{\text{silicate ions}} = 1 - N_{O^{2-}}$$

and $N_{SiO_4^{4-}}$ can be calculated from (5.10) and (5.11) above as a function of K and $N_{O^{2-}}$, i.e.

$$\Sigma_{\text{silicate ions}} = \frac{N_{SiO_4^{4-}}}{1 - K\left(\dfrac{N_{SiO_4^{4-}}}{N_{O^{2-}}}\right)} \tag{5.12}$$

Thus,
$$N_{SiO_4^{4-}} = \Sigma N_{silicate\ ions} \left[1 - K\left(\frac{N_{SiO_4^{4-}}}{N_{O^{2-}}}\right)\right]$$

$$= (1 - N_{O^{2-}})\left[1 - K\left(\frac{N_{SiO_4^{4-}}}{N_{O^{2-}}}\right)\right]$$

$$= \frac{1 - N_{O^{2-}}}{1 + K\left(\dfrac{1 - N_{O^{2-}}}{N_{O^{2-}}}\right)}$$

i.e.
$$N_{SiO_4^{4-}} = \frac{1 - N_{O^{2-}}}{1 - K\left(1 - \dfrac{1}{N_{O^{2-}}}\right)} \tag{5.13}$$

The fractions of the other silicate ions can be calculated from reactions (a), (b) and (c), and assuming a composite K for K_a, K_b, K_c gives:

$$N_{Si_n O_{(3n+1)}^{2(n+1)-}} = K\left(\frac{N_{SiO_4^{4-}}}{N_{O^{2-}}}\right)^{n-1} N_{SiO_4^{4-}} \tag{5.14}$$

Combining equations (5.12), (5.13) and (5.14) and assuming that the sum of the series:

$$1 + 2x + 3x^2 + \ldots = \frac{1}{(1 - x)^2}$$

and the sum of the series:

$$3 + 5x + 7x^2 + 9x^3 + \ldots = \frac{3 - x}{(1 - x)^2}$$

where $x = N_{SiO_4^{4-}}$:

$$N_{SiO_2} = \frac{1}{3 - K + \dfrac{N_{O^{2-}}}{1 - N_{O^{2-}}} + \dfrac{K(K - 1)}{N_{O^{2-}}/(1 - N_{O^{2-}}) + K}}$$

This solution allows $N_{O^{2-}}$ to be calculated knowing K and using Temkin's relationship:

$$N_{O^{2-}} = \frac{a_{MO}}{N_{M^{2+}}} \quad \text{i.e. } a_{MO} = N_{M^{2-}} + N_{O^{2-}}$$

Therefore, the activity of a basic oxide MO (a_{MO}) can be calculated knowing $N_{O^{2-}}$, $N_{silicate\ ions}$ and N_{SiO_2} and the characterisation of the slag can be determined.

Such values calculated by this method for CaO, FeO, MnO, etc. in silicate slags agree well with experimental data. However, the theory is not applicable to all slag systems since:

(i) constants, K, for initial silicates $Si_2 O_7^{6-}$, $Si_3 O_{10}^{8-}$ are different from those of the more complex silicates.

(ii) Masson's theory takes account only of linear chains (or branching chains in other theories) and cyclic forms may exist in high polymers.

(iii) Masson's theory, as with Temkin and Flood, assumes ideal behaviour for anions and cations separately. Some ion interaction takes place in real solutions.

5.7.2 Molecular theory[11]

The molecular theory assumes ideal behaviour for all molecules present in the slag. Simple oxides associate to form complex molecules or remain as non-combined or free compounds rather than ions, e.g.

$$CaO + Al_2O_3 = CaAl_2O_4$$

$$2CaO + SiO_2 = Ca_2SiO_4$$

$$4CaO + P_2O_5 = Ca_4P_2O_9$$

Each component in the slag may be in either the combined or free state. It is, therefore, necessary to consider all possible molecules that may exist in the slag. Using the example of a ternary slag containing MnO, CaO, SiO$_2$, the molecules that could exist include: MnO, CaO, $CaSiO_3$, Ca_2SiO_4, $MnSiO_3$, Mn_2SiO_4, $Ca_2Si_2O_6$, $Ca_4Si_2O_8$. Since there are eight possible compounds, eight equations are needed to determine the mole fraction (N) of each:

(1) $MnO + SiO_2 = MnSiO_3$; $K_1 = N_{MnSiO_3}/N_{MnO} \cdot N_{SiO_2}$

(2) $2MnO + SiO_2 = Mn_2SiO_4$; $K_2 = N_{Mn_2SiO_4}/N_{MnO}^2 \cdot N_{SiO_2}$

(3) $CaO + SiO_2 = CaSiO_3$; $K_3 = N_{CaSiO_3}/N_{CaO} \cdot N_{SiO_2}$

(4) $2CaO + SiO_2 = Ca_2SiO_4$; $K_4 = N_{Ca_2SiO_4}/N_{CaO}^2 \cdot N_{SiO_2}$

(5) $2CaO + 2SiO_2 = Ca_2Si_2O_6$; $K_5 = N_{Ca_2Si_2O_6}/N_{CaO}^2 \cdot N_{SiO_2}^2$

(6) $4CaO + 2SiO_2 = Ca_4Si_2O_8$; $K_6 = N_{Ca_4Si_2O_8}/N_{CaO}^4 \cdot N_{SiO_2}^2$

(7) $Ca_4Si_2O_8 = Ca_2Si_2O_6 + 2CaO$; $K_7 = N_{Ca_2Si_2O_6} \cdot N_{CaO}^2/N_{Ca_4Si_2O_8}$

(8) $Ca_4Si_2O_8 = 2(Ca_2Si_2O_4)$; $K_8 = N_{Ca_2Si_2O_4}^2/N_{Ca_4Si_2O_8}$

It is necessary to determine the equilibrium constant for each of these reactions in terms of the mole fractions of the species taking part in the reaction, and assuming ideal behaviour. The number of mols of CaO, SiO$_2$, MnO in the slag are given by:

$$n_{CaO} = n_{CaO(free)} + n_{CaSiO_3} + 2n_{Ca_2SiO_3} + 2n_{Ca_2Si_2O_6} + 4n_{Ca_4Si_2O_8}$$

$$n_{SiO_2} = n_{SiO_2(free)} + n_{MnSiO_3} + n_{Mn_2SiO_4} + n_{CaSiO_3} + n_{CaSiO_4}$$
$$+ 2n_{Ca_2Si_2O_4} + 2n_{Ca_4Si_2O_8}$$

and

$$n_{MnO} = n_{MnO(free)} + n_{MnSiO_3} + 2n_{Mn_2SiO_4}$$

and are obtained by chemical analysis of the slag. The mole fractions of the other compounds can then be deduced knowing the equilibrium constants.

This method is useful in particular cases but ideal behaviour assumption in slags is not always valid. Also, the actual molecules formed and the equilibrium constants which relate the mole fractions of these molecules are not often known with any degree of accuracy.

An example of the application of the molecular theory to a complex slag is that used by Winkler and Chipman[12] in the dephosphorisation of steel. The slag composition was as follows:

Component	CaO	SiO_2	FeO	Fe_2O_3	MgO	P_2O_5	MnO
Weight %	20.78	20.50	38.86	4.98	10.51	1.67	2.51
Mole fraction	0.232	0.215	0.339	0.0195	0.165	0.0075	0.022

Assuming that the species present in the slag are free CaO, $Ca_4P_2O_9$, $CaFe_2O_4$, $Ca_2Si_2O_6$, $Ca_4Si_2O_8$, FeO, MgO and MnO, and that MgO and MnO are included in CaO-containing compounds, the following equilibria can be considered:

(1) $\quad 4CaO + P_2O_5 = Ca_4P_2O_9$; $K_1 = N_{Ca_4P_2O_9}/N_{CaO}^4 \cdot N_{P_2O_5}$

(2) $\quad CaO + Fe_2O_3 = CaFe_2O_4$; $K_2 = N_{CaFe_2O_4}/N_{CaO} \cdot N_{Fe_2O_3}$

(3) $\quad 2CaO + 2SiO_2 = Ca_2Si_2O_6$; $K_3 = N_{Ca_2Si_2O_6}/N_{CaO}^2 \cdot N_{SiO_2}^2$

(4) $\quad 4CaO + 2SiO_2 = Ca_4Si_2O_8$; $K_4 = N_{Ca_4Si_2O_8}/N_{CaO}^4 \cdot N_{SiO_2}^2$

It was assumed that (i) the reactions for the formation of $Ca_4P_2O_9$ and $CaFe_2O_4$ are complete, (ii) the following equilibrium existed for the silicates:

(5) $\quad (Ca_4Si_2O_8) = (Ca_2Si_2O_6) + 2(CaO)_{free}$: $K_5 = 1 \times 10^{-2}$

MgO and MnO are included with CaO. In order to characterise the slag, the mole fractions of each of the six free components need to be determined. Therefore, six equations are needed which describe the equilibria that exist amongst these components in the slag. One of these is given, i.e. K_5; the others can be determined from a consideration of the mass balance of those components in a similar fashion to that used for ionic theories. Therefore, considering the number of moles, n, of each component present in the slag:

(i) $\quad$ FeO: $n_{FeO} = 0.339$

(ii) $\quad Fe_2O_3$: $CaO + Fe_2O_3 = CaFe_2O_4$: $n_{CaFe_2O_4} = 0.0195 = n_{Fe_2O_3}$

(iii) $\quad P_2O_5$: $4CaO + P_2O_5 = Ca_4P_2O_4$: $n_{Ca_4P_2O_9} = 0.0075 = n_{P_2O_5}$

(iv) $\quad$ CaO, MgO, MnO: $\Sigma(CaO + MgO + MnO)$

$$= n_{\left(\begin{smallmatrix} CaO \\ MnO \\ MgO \end{smallmatrix}\right)_{free}} + 2n_{Ca_2Si_2O_6} + 4n_{Ca_4Si_2O_8} + 4n_{Ca_4P_2O_9} + n_{CaFe_2O_4}$$

i.e. $\qquad 0.419 = n_{(CaO)_{free}} + 2n_{Ca_2Si_2O_6} + 4n_{Ca_4Si_2O_8} + 0.0495$

Therefore $\qquad 0.3695 = n_{\left(\begin{smallmatrix} CaO \\ MgO \\ MnO \end{smallmatrix}\right)_{free}} + 2n_{Ca_2Si_2O_6} + 4n_{Ca_4Si_2O_8}$ $\qquad$ (5.15)

(v) $\quad$ Since there is no free SiO_2

$$n_{SiO_2} = 2n_{Ca_2Si_2O_6} + 2n_{Ca_4Si_2O_8} = 0.215 \qquad (5.16)$$

(vi) The sixth equation is that given for reaction (5) assuming ideal solution.

$$K_5 \ = \ 10^{-2} \ = \ N^2_{CaO(free)} \ \frac{N_{Ca_2Si_2O_6}}{N_{Ca_4Si_2O_8}}$$

$$= \ \frac{n^2_{CaO(free)}}{\Sigma n^2_{oxides}} \ \frac{n_{Ca_2Si_2O_6}}{n_{Ca_4Si_2O_8}} \ \frac{\Sigma n_{oxides}}{\Sigma n_{oxides}}$$

Therefore

$$10^{-2} \ = \left(\frac{n_{CaO(free)}}{\Sigma n_{oxides}} \right)^2 \ \frac{n_{Ca_2Si_2O_6}}{n_{Ca_4Si_2O_8}}$$

But

$$\Sigma n_{oxides} \ = \ n_{CaO(free)} \ + \ n_{FeO} \ + \ n_{Ca_2Si_2O_6} \ + \ n_{Ca_4Si_2O_8} \ + \ n_{CaFe_2O_4} \ + \ n_{Ca_4P_2O_9}$$

Therefore $\quad \Sigma n_{oxides} \ = \ n_{CaO(free)} \ + \ 0.339 \ + \ \dfrac{0.215}{2} \ + \ 0.0195 \ + \ 0.0075$

$$= \ n_{CaO(free)} \ + \ 0.4735$$

Thus $\qquad 0.01 \ = \left(\dfrac{n_{CaO(free)}}{n_{CaO(free)} \ + \ 0.473} \right)^2 \ \dfrac{n_{Ca_2Si_2O_6}}{n_{Ca_4Si_2O_8}}$ $\qquad$ (5.17)

In the three equations (5.15), (5.16) and (5.17) there are three unknowns and therefore a solution can be determined. Solving for these three unknowns gives:

$$n_{Ca_2Si_2O_8} \ = \ 0.052$$
$$n_{Ca_2Si_2O_6} \ = \ 0.056$$
$$n_{CaO(free)} \ = \ 0.0505$$

Finally, the molar fractions of the actual species are related to the number of moles by:

$$N_i \ = \ n_i / \Sigma n$$

Hence:

$$n_{FeO} \ = \ 0.339; \ N_{FeO} \ = \ 0.647$$
$$n_{CaFe_2O_3} \ = \ 0.0195; \ N_{CaFe_2O_3} \ = \ 0.037$$
$$n_{Ca_4P_2O_9} \ = \ 0.0075; \ N_{Ca_4P_2O_9} \ = \ 0.014$$
$$n_{Ca_2Si_2O_8} \ = \ 0.052; \ N_{Ca_2SiO_8} \ = \ 0.099$$
$$n_{Ca_2Si_2O_6} \ = \ 0.056; \ N_{Ca_2Si_2O_6} \ = \ 0.107$$
$$n_{CaO(free)} \ = \ 0.0505; \ N_{CaO(free)} \ = \ 0.096$$
$$\Sigma n \ = \ 0.5245; \ \Sigma N \ = \ 1.00$$

Characterising the slag as determined above, it is now possible to determine the oxygen and phosphorus contents of the steel in equilibrium with the slag at $1600\,^\circ$C given the following additional data:

$$\log(\%O)_{sat} = \frac{6320}{T} + 2.734$$

and

$$4(CaO)_s + 2(P)_{1\%} + 5(O)_{1\%} = (Ca_4P_2O_9)_{(s)}; \log K = \frac{71\,667}{T} - 28.73$$

where (s) denotes slag, 1% denotes less than 1wt% oxygen in liquid iron solution, i.e. dilute solution, and (sat) indicates oxygen saturation in the liquid iron. The amount of oxygen dissolved in iron in equilibrium with pure FeO can be calculated from:

$$\log(\%O)_{sat} = \frac{6320}{1873} + 2.734 = -0.640 \text{ and } (\%O)_{sat} = 0.23 \text{ wt\%}$$

In this steelmaking unit the oxygen used to oxidise P to P_2O_5 is from the FeO. Hence at oxygen saturation in the iron the value of $a_{FeO} = 1$. Using this assumption the value of a_{Fe} when oxygen does not saturate the liquid iron is given by

$$a_{FeO} = \%O/\%O_{(sat)}$$

Also since ideal solution is assumed between oxygen and iron:

$$a_{FeO} = N_{FeO} = 0.647$$

Hence, $\%O_{(in\,Fe)} = 0.23 \times 0.647 = 0.149 \text{ wt\%}$

Examination of the equilibrium constant for the formation of $Ca_4P_2O_{9\,(s)}$ at $1000\,^\circ$C gives:

$$\log K = \frac{71\,667}{1873} - 28.73 = 9.55 \text{ and } K = 3.55 \times 10^9$$

Therefore, again assuming ideal solutions for both slag and liquid iron

$$3.55 \times 10^9 = \frac{N_{Ca_4P_2O_9}}{(CaO(free)^4 \cdot (\%O)^5_{(m)} \cdot (P)^2_{(m)}} = \frac{0.014}{(0.096)^4 (0.149)^5 (P)^2_{(m)}}$$

giving a phosphorus content in the steel of 0.025 wt%. Assuming 100 kg of slag per 1000 kg steel:

Amount of P in the slag = (wt% P_2O_5 × 62/142 × 100) kg =

1.67/100 × 62/142 × 100 = 0.73 kg

Amount of P in the steel = (0.025/100 × 1000) kg = 0.25 kg

Therefore the total P content distributed between slag and steel = 0.98 kg. If the appropriate slag refining conditions, i.e. basic oxidising slag, had not been used most of this phosphorus would have partitioned into the steel giving a phosphorus content of 0.98/1000 × 100 wt% = 0.098 wt% P in steel.

5.7.3 Summary of slag theories

A summary of the molecular and ionic slag theories is given below in note form:

Molecular
- assumes all species exist in molecular form and all molecular species are known
- equilibria between molecular species in slag are needed and the relationships between each molecular species must be known
- assumes ideal behaviour of all molecules in slag.

Temkin
- assumes slag solution is completely dissociated into ions, no interaction between ions of the same charge and complete randomness, i.e. two ideal solutions, of cations and anions

$$a_{RO} = N_{R^{2+}} N_{O^{2-}}$$

Flood
- based partly on Temkin and considers equilibrium between elements dissolved in metal phase and its ions in the slag, i.e. partitioning of elements between metal and slag
- assumes ideal behaviour in slag, non-ideal in metal
- uses electrically equivalent ion fraction, N'

$$\text{i.e. } N' = \frac{2n_{B^{2+}}}{n_{A^+} + 2n_{B^{2+}} + 3n_{C^{3+}} + 4n_{D^{4+}} \dots}$$

- development of Flood's theory to give

$$\ln K' = \Sigma N'_{A^{2+}} \ln K_n + \ln g(f)$$

Masson's theory
- assumes slags are complex solutions containing polymeric silicate anions, e.g. $Si_nO_{(3n+1)}^{(2n+2)-}$
- degree of polymerisation is dependent on character and quantity of the basic oxides present
- assumes ideal solution in slag and non-ideal in metal
- establishes relationships between

$$\left.\begin{array}{l} N_{\text{(silicate ions)}} \\ \\ N_{SiO_2} \end{array}\right] \quad \begin{array}{l} \text{with complex values between the} \\ \text{equilibrium constants and } N_{O^{2-}} \end{array}$$

Temkin's, Masson's and Flood's theories apply only to basic slags. Masson's and Flood's theories are more applicable to practice although neither have provided complete correlations with experimentally determined thermodynamic data. Masson's theory is more useful in slags in which complex silicates are present.

REFERENCES

1 Richardson, F.D. 1953 *The Physical Chemistry of Melts*, p. 63. London: Inst. Min. Met.
2 Herasymenko, P. 1938 *Trans. Faraday Soc.*, **304**, 124.
3 Herasymenko, P. and Speight, G.E. 1950 *J.I.S.I.*, **166**, 169 and 269.
4 Temkin, M. 1945 *Acta Physicochimica URSS*, **20**, 411.
5 Flood, H. and Muan, A. 1950 *Acta Chem. Scand.*, **4**, 359.

6   Flood, H. and Grjotheim, K. 1952 *J.I.S.I.*, **171**, 64.
7   Flood, H., Forland, T. and Grjotheim, K. 1953 *Physical Chemistry of Metals*, p. 46. London: Inst. Min. Met.
8   Turkdogan, E.T. and Pearson, J. 1953 *J.I.S.I.*, **175**, 398.
9   Masson, C.R. 1965 *Proc. R. Soc.*, A287, 201.
10 Whiteway, S.G., Smith, I.B. and Masson, C.R. 1970 *Can. J. Chem.*, 48, 33.
11 Schenck, H. 1945 *Physico-chemistry of Steelmaking* (trans.), p. 455. London: BISRA.
12 Winkler, T.B. and Chipman, J. 1946 *Trans. AIME*, **167**, 111.
13 Bodsworth, C. and Bell, H.B. 1972 *Physical Chemistry of Iron and Steel Manufacture*, 2nd edn, London: Longman.
14 Richardson, F.D. 1952 *Chemistry and Industry*, p. 50. London:
15 Bowen, N.L. *et al.* 1933 *Am. J. Sci.*, 5th series, **26**, 204.
16 Maddocks, W.R. 1935 *J. Iron St. Inst., Carnegie Schol. Mem.*, **24**, 64.
17 Greig, P. 1927 *Am. J. Sci.*, 5th series, **13**, 41.
18 Rankin, D. and Wright, J. 1915 *Am. J. Sci.*, 4th series, **39**, 52.
19 Oelson, W. and Maetz, 1941 *Mitt. K. - Wilh. Inst. Eisenforsch.*, **23**, 195.
20 Coudurier, L., Hopkins, D.W. and Wilkomirsky, I. 1978 *Fundamentals of Metallurgical Processes*, Oxford: Pergamon.
21 Balajaiva, K., Quarrel, A.G. and Vajragupta, P. 1946 *J.I.S.I.*, **153**, 115–145.
22 Hatch, G.G. and Chipman, J. 1949 *Trans. AIME*, **196**, 309–313.
23 Bishop, H.L., Lander, H.N., Grant, N.J. and Chipman, J. 1956 *Trans. AIME*, **206**, 862.

FURTHER READING

Turkdogan, E.T. 1983 *Physicochemical Properties of Molten Slags and Glasses*, London: The Metals Society.

For self-test questions on this chapter, see Appendix 5.

6

Aqueous Metal Solutions and Electrochemistry

Electrochemistry is the study of the relationship between electricity and chemical behaviour and may be divided into two broad overlapping areas: (a) ions in solution and conductivity—*ionics*; and (b) production of electricity by an *electrochemical* reaction and the reverse process, that is, production of an electrochemical reaction by an applied electrical force, i.e. *electrolytic* reaction or electrolysis. The study of both these electrochemical processes is called *electrodics*.

6.1 Ionics I: Ions in solution

6.1.1 Electrolytes

In a metal an electric current is constituted by a flow of *electrons*—the 'positive' electric current flows in the opposite direction to electron flow—but in non-metallic liquids (solutions and melts) and to a limited extent in solids,e.g. AgCl(s), a flow (or migration) of *ions* constitutes an electric current. In both of these cases passage of electricity usually results in chemical decomposition and the process is known as electrolysis. Substances which conduct electricity by ion migration are known as electrolytes. These are acids, bases and salts. A salt is composed entirely of ions arranged in fixed positions in a lattice (Section 1.3.1.5) but in solution or in the molten state the ions are mobile and can flow under the influence of a potential difference.

6.1.2 Aqueous solutions of electrolytes

When a salt dissolves in water, energy is required because the ions move far apart and work is done against mutual attractions. Energy is provided largely by the process of hydration by which water molecules bond to the ions (see *Figs 6.1(a)* and *(b)*) to form hydrated ions. The extent of hydration depends on the size and charge on the ion.

A small highly charged ion will clearly be strongly hydrated, e.g. K^+ (ionic radius $= 0.13$ nm) is much less strongly hydrated than Al^{3+} (ionic radius $= 0.05$ nm). Ions in aqueous solution are normally written Li^+(aq) etc but '(aq)' will usually be omitted for clarity. The hydrated ions of opposite charge show little tendency to attract in dilute solution because the high permittivity[*] of water reduces the force between charges. The force, F, between two charges, Q_1 and Q_2, is given by[1],

$$F = \frac{Q_1 Q_2}{4\pi l^2 \epsilon}$$

---

[*]The permittivity is a parameter which allows for the fact that the magnitude of the force between two charges a certain distance apart varies with the medium in which the charges are situated.

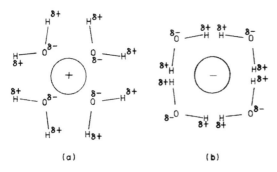

Figure 6.1 Arrangement of H_2O molecules in: (a) a hydrated cation; (b) a hydrated anion. These figures are not intended to be an exact true representation of any known structure

where l is the distance between the charges and ϵ is the permittivity (relative to free space) of the medium. For water the relative permittivity is 78.54 at 298 K, so that the force of attraction is approximately 1/80 of that in a vacuum. Also, water molecules associated with ions (hydration sheaths) inhibit ionic association. At high concentrations, however, some of the ions associate to form ion pairs, e.g. for magnesium sulphate ion pairs form thus:

$$Mg^{2+} + SO_4^{2-} \rightleftharpoons Mg^{2+}SO_4^{2-}$$

This phenomenon explains in part the reduction in the conductivity of electrolyte solutions at high concentrations (see Section 6.3.3) since ion pairs are non-conducting.

6.1.3 Ionic activity

As with other two-component systems, deviations from ideal behaviour occur in all but extremely dilute solutions. Mixtures in which one component is a solute are ideal if Henry's law is obeyed (see Section 4.3.6) and deviations from ideality are considered with reference to Henry's law.

For a particular ion, approach to ideality occurs as concentration tends to zero and interaction is then only with solvent and not with other ions. For electrolytes in solution, the standard state is generally taken as an ideal solution of 1 mol dm^{-3} or 1 mol kg^{-1} (1 molal).

6.1.4 Mean ionic activity

Since an ion cannot exist separately from its counter-ion* all measurements of activities for electrolyte solutions will inevitably be of mean activities for the two ions. Consider a solution of a 1:1 electrolyte†, A^+B^-. According to Section 2.2.3.6 the free energies for the two ions, G_{A^+} and G_{B^-}, are given by equations (6.1a) and (6.1b) respectively:

*Ion of opposite charge.

†That is an electrolyte containing univalent anions and cations in the ratio 1:1.

$$G_{A^+} = G_{A^+}^{\ominus} + nRT \ln a_{A^+} \qquad (6.1a)$$

and

$$G_{B^-} = G_{B^-}^{\ominus} + nRT \ln a_{B^-} \qquad (6.1b)$$

where $G_{A^+}^{\ominus}$ and $G_{B^-}^{\ominus}$ are the standard free energies, a_{A^+} and a_{B^-} are the activities of A^+ and B^- respectively and n is the number of moles. The mean free energy is given by:

$$\frac{G_{A^+} + G_{B^-}}{2} = \frac{G_{A^+}^{\ominus} + G_{B^-}^{\ominus}}{2} + nRT \ln a_{\pm} \qquad (6.2a)$$

where $a_{\pm}$ is the mean activity of A^+ and B^-. Now, from equations (6.1a) and (6.ab),

$$\frac{G_{A^+} + G_{B^-}}{2} = \frac{G_{A^+}^{\ominus} + G_{B^-}^{\ominus}}{2} + \frac{nRT}{2} \ln(a_{A^+} a_{B^-}) \qquad (6.2b)$$

A comparison of equations (6.2a) and (6.2b) shows that:

$$nRT \ln a_{\pm} = \tfrac{1}{2} nRT \ln(a_{A^+} a_{B^-})$$

Hence,

$$a_{\pm} = (a_{A^+} a_{B^-})^{\frac{1}{2}} \qquad (6.3)$$

In general, for an electrolyte of the type $X_p^{q+} Y_q^{p-}$ the mean ionic activity is given by:

$$a_{\pm} = (a_{X^{q+}}^p \, a_{Y^{p+}}^q)^{1/(p+q)} \qquad (6.4)$$

Thus for $MgCl_2$:

$$a_{\pm} = (a_{Mg^{2+}} \, a_{Cl^-}^2)^{1/3}$$

This argument can now be extended to include the activity coefficient, f, since

$$a_{A^+} = f_{A^+}[A^+]$$

where $[A^+]$ is concentration of A^+ and f_{A^+} is the activity coefficient of cations. It is easily shown that for a 1:1 electrolyte (A^+B^-)

$$f_{\pm} = (f_{A^+} f_{B^-})^{\frac{1}{2}} \qquad (6.4a)$$

where $f_{\pm}$ is the mean ionic activity coefficient. Some mean ionic activities are shown in *Table 6.1*.

Table 6.1 MEAN ACTIVITY COEFFICIENTS (TO TWO DECIMAL PLACES) OF SELECTED ELECTROLYTES AT 298 K IN AQUEOUS SOLUTION. FOR FURTHER DATA CONSULT ROBINSON AND STOKES[5] AND THE *HANDBOOK OF PHYSICS AND CHEMISTRY*[20]

Electrolyte	Molality/mol kg^{-1}			
	0.01	0.1	1.0	2.0
HCl	0.90	0.80	0.81	1.0
NaCl	0.90	0.78	0.66	0.67
NaOH	0.90	0.77	0.68	0.70
KNO$_3$	0.90	0.74	0.44	0.33
BaCl$_2$	0.72	0.51	0.40	–
CuSO$_4$	0.41–0.44	0.15	0.04	–
ZnSO$_4$	0.39	0.15	0.04	0.04
AlCl$_3$	–	0.34	0.54	–

6.1.5 Debye–Hückel theory

By considering the behaviour of completely dissociated ions in solution and making a number of assumptions, Debye and Hückel[2,3] obtained a theoretical expression for the mean ionic activity coefficient in terms of electronic charge, temperature, ionic strength and other constants. The ionic strength, I, is defined by

$$I = \frac{1}{2} \sum_i c_i(z_i)^2 \tag{6.5}$$

where c_i is the concentration of ion i and z_i is the charge of ion i. For example, for 0.01 molal NaBr in water,

$$I = \frac{1}{2}[0.01(+1)^2 + 0.01(-1)^2]$$
$$= 0.01 \text{ mol kg}^{-1}$$

For 0.02 molal aqueous Na$_2$SO$_4$ it can easily be shown that $I = 0.06$ mol kg^{-1}. The Debye–Hückel expression is given by:

$$\log_{10} f_\pm = -\frac{A|z_+||z_-|I^{\frac{1}{2}}}{1 + BaI^{\frac{1}{2}}} \tag{6.6}$$

where A and B, the Debye–Hückel constants, are given in terms of the Avogadro number, the electronic charge, the Boltzmann constant, temperature and permittivity of the solvent; z_+ is the cationic charge and z_- is the anionic charge. The term, a, the ion size parameter, is not accurately known. It seems, however, that a value for the factor Ba of 1 mol$^{-\frac{1}{2}}$ kg$^{\frac{1}{2}}$ can be used. In dilute solution, equation (6.6) simplifies to:

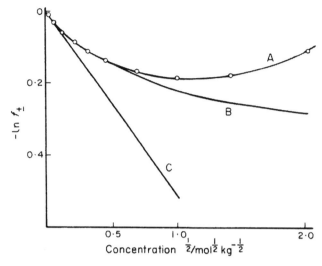

Figure 6.2 Variation of −ln (mean ionic activity coefficient) with (molality)$^{1/2}$ for aqueous solutions of sodium chloride at 298 K. (A) experimental data; (B) equation (6.6); (C) equation (6.6a)

$$\log_{10} f_{\pm} = -A|z_{+}||z_{-}|I^{1/2} \tag{6.6a}$$

$$\approx \log_{10} y_{\pm} \text{ (aqueous solutions at 298 K)*}$$

which is known as the Debye–Hückel limiting law. The limits of concentrations for which equations (6.6) and (6.6a) are valid for aqueous NaCl at 298 K are shown in *Fig. 6.2*. Typically equation (6.6) is valid up to concentrations of about* 0.02 mol kg^{-1} while the limiting law is valid up to concentrations of less than 0.001 mol kg^{-1}.

Other expressions have been developed[3] which agree with experimental results up to much higher concentrations.

6.1.6 Measurement of mean ionic activity

One method for finding mean ionic activities is from emf measurements[3,4,5]. In Section 6.3.12 it will be shown that the electrode potential, E_M, of a metal is given by:

$$E_M = E_M^{\ominus} + \frac{RT}{nF} \ln a_{M^{n+}} \tag{6.21}$$

where $E_M^{\ominus}$ is the standard electrode potential, n is the number of electrons taking part in the reaction, $a_{M^{n+}}$ is the activity of M^{n+} and a_M is assumed to be unity. As explained above, $a_{\pm}$, the mean ionic activity should be used but this point is taken so much for granted that expressions such as $a_{M^{n+}}$ will be used constantly in the text. For example, for zinc in $ZnSO_4$:

*$y_{\pm}$ is the mean *molar* activity coefficient and $f_{\pm}$ is the mean *molal* activity coefficient.

$$E_{Zn} = E_{Zn}^{\ominus} + \frac{RT}{nF} \ln a_{Zn^{2+}}$$

Thus, knowing $E_{Zn}^{\ominus}$, $a_{Zn^{2+}}$ can be found directly by measuring the electrode potential of zinc under appropriate conditions (Section 6.3.7). More sophisticated calculations are needed if $E_{Zn}^{\ominus}$ is not assumed.

Other methods for determining activities are also available. They can be determined, for example, from measurements of solubility[3] and freezing point depression[5].

6.1.7 Acids and bases

According to the Brønsted–Lowry theory acids are proton (H^+ ion) donors. For example, hydrogen chloride is a covalent gas but in aqueous solution it reacts as follows:

$$HCl + H_2O \longrightarrow H_3O^+ + Cl^-$$

The proton, H^+, produced by HCl is so small that it bonds strongly with at least one molecule of water†:

$$H^+ + H_2O \longrightarrow H_3O^+$$

In very dilute solution, acids such as HCl, HBr, HNO_3 and $HClO_4$ are considered completely dissociated but above about 10^{-3} mol dm^{-3} association (recombination of the dissociated species) can occur at 298 K. Acids such as those mentioned above which are completely or nearly completely dissociated in solution are *strong acids*. Some acids are only partially dissociated even in very dilute solutions. These are known as *weak acids* and a typical weak acid is ethanoic (acetic) acid. It is only 1.3% dissociated in a 0.1 mol dm^{-3} solution and 13% dissociated in a 10^{-3} mol dm^{-3} solution at 298 K. The dissociation of the general weak acid, HA, may be represented by the following equilibrium:

$$HA \rightleftharpoons H^+ + A^-$$

A *base* is a proton acceptor. Oxides such as CuO and hydroxides such as Ca(OH)$_2$ are bases and their reaction with H^+ is most simply represented by the following ionic equations:

$$O^{2-} + 2H^+ \longrightarrow H_2O$$

and

$$OH^- + H^+ \longrightarrow H_2O$$

---

*Depends on value of a used.

†However, for simplicity, H^+ rather than H_3O^+ will be used throughout.

Soluble bases such as KOH are also known as *alkalis* though alkaline solutions can be produced, not only by dissolving bases but also salts of weak acids and strong bases, e.g. aqueous Na_2CO_3, since their hydrolysis yields OH^- ions:

$$CO_3^{2-} + H_2O \rightleftharpoons HCO_3^- + OH^-$$

The bases mentioned above are *strong*. A good example of a weak base is an aqueous solution of ammonia which dissociates as follows:

$$NH_3 + H_2O \rightleftharpoons NH_4^+ + OH^-$$

Heys[6] and Bell[7] give fuller accounts of the behaviour of acids and bases.

6.1.8 The pH scale

The H^+ concentration of an aqueous solution is a measure of the acidity of a solution but for most solutions the values of concentration (e.g. 3.7×10^{-5} mol dm^{-3}) provide a cumbersome scale. Hence the pH scale is used. The term pH is defined as follows:

$$pH = -\log_{10}([H^+]/\text{mol dm}^{-3} \, y_\pm) \tag{6.7}$$

but for many purposes, the simplified expression $pH = -\log_{10}a_{H^+}$ or even $pH = -\log_{10}([H^+]/\text{mol dm}^{-3})$ is adequate. These expressions give a scale of values whic for most solutions, lies between 0 and 14. Strictly speaking, activity rather than concentration of H^+ is used.

Example

The pH of 7.0×10^{-4} mol dm^{-3} HCl is calculated as follows. Assuming complete dissociation and a unit activity coefficient:

$$[H^+] = 7.0 \times 10^{-4} \text{ mol dm}^{-3}$$

$$\therefore \quad pH = -\log_{10}(7 \times 10^{-4})$$

$$= 3.15$$

Water itself dissociates as follows:

$$H_2O \rightleftharpoons H^+ + OH^-$$

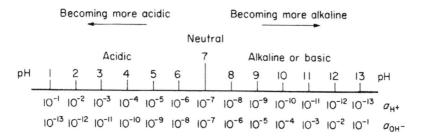

Figure 6.3 pH scale at 298 K

for which the equilibrium constant (known as the ionic product, K_w) is given by:

$$K_w = [H^+][OH^-]$$

assuming unit activity for H_2O and unit activity coefficients for H^+ and OH^- ions. At 298 K, $K_w = 1.0 \times 10^{-14}$ mol^2 dm^{-6}. Thus for pure water,

$$[H^+] = [OH^-] = 10^{-7} \text{ mol dm}^{-3} \quad \text{and} \quad pH = 7.0$$

Hence at 298 K, a pH of 7.0 represents neutrality. In acidic solutions, $[H^+] > [OH^-]$ and pH < 7. In basic solutions, $[H^+] < [OH^-]$ and pH > 7. Part of the pH scale is shown in *Fig. 6.3*.

6.1.9 Buffer solutions

A buffer solution has a pH which is hardly affected by contamination with small amounts of acid and base. It can be used for standardising a pH meter because of its reliability. It can also be used for adjusting and maintaining the pH of solutions, a point which is particularly important in electroplating (Section 6.4.10). Some of the simpler buffers consist of a weak acid or base and a salt of that acid or base. For example, an acid buffer may be made by mixing ethanoic acid and sodium ethanoate. An alkaline buffer is made by mixing NH_4Cl and aqueous NH_3. Simple equations are available[6] for determining the pH of simple buffers. Other mixtures or even single substances such as potassium hydrogen phthalate can be used and buffer tablets and powders for making buffer solutions are commercially available.

The behaviour of a simple buffer will be explained here using the example of a general weak acid HA and its salt MA. The acid dissociates as follows:

$$HA \rightleftharpoons H^+ + A^-$$

Both $[HA]$ and $[A^-]$ are large compared with $[H^+]$. The HA concentration is large because the acid is weak and $[A^-]$ is large due to the addition of MA. Thus limited additions or removals of H^+ have very little effect comparatively speaking on $[HA]$ and $[A^-]$. Most of any H^+ added reacts with a little of the A^- to form HA and a small proportion of HA can dissociate to replenish losses in H^+ without affecting the equilibrium position. Thus $[H^+]$ is maintained more or less constant.

6.1.10 pH measurement

Although a number of methods are available for pH determination, for example, colorimetry[8], the pH meter is the most widely used for accurate pH measurement. It consists of two electrodes, most commonly the glass electrode and the saturated calomel electrode although other electrode systems can also be used[8,9,10]. Both of these electrodes are described in Section 6.3.6. The former has a potential which is a function of the pH of the solution into which it is dipped, while the latter is a reference electrode which has a fixed and reproducible potential at a given temperature. The potential difference between the two electrodes is thus dependent on pH. A voltmeter incorporating a pH scale measures the potential difference. A temperature correction must usually be made, either manually or automatically by connecting a resistance thermometer which dips into the solution. Most pH meters are first standardised using one or two buffer solutions. Both mains and the more portable battery operated pH meters are available. Generally speaking an accuracy of 0.1 in pH is expected, although some instruments are more accurate.

Great care should be taken of the electrodes, and they should be kept immersed in water to prevent them drying out. The saturated calomel electrode must be checked to ensure that excess KCl is present. Potter[9] and Wilson[10] describe pH meters in more detail. The pH meter is used for checking electrolyte solutions in electrowinning and electroplating, for pH control in froth flotation, in corrosion control and in hydrometallurgical processes such as gaseous reduction (Section 7.3.2).

6.2 Ionics II: Electrolytic conduction

The electrical resistance (R) of a metal is related to the length (l) and the cross-sectional area (A) according to the expression[1]

$$R = \rho l/A \qquad (6.8)$$

where ρ is the resistivity of the metal (the resistance of a piece of metal of unit cross section and unit length).

For a solution or melt, electrons arrive or leave via electrodes, so that the term A becomes the mean surface area of the electrodes and the term l is the distance between the electrodes. It is more usual when referring to electrolytes to use the terms conductance (reciprocal resistance) and conductivity (reciprocal resistivity). Thus, equation (6.8) becomes:

$$\text{conductance} = \frac{1}{R} = \frac{\kappa A}{l} \qquad (6.9)$$

where $\kappa = 1/\rho$, the conductivity.

6.2.1 Measurement of conductivity

The resistance of a solution is measured in a conductivity cell (*Fig. 6.4*) which forms part of a Wheatstone bridge circuit*. A detailed description of the theory and use of the Wheatstone bridge is given by Nelkon and Parker[1], Findlay[11] and Tyler[12]. It is important to use a high frequency alternating current (e.g. 1000 Hz) to minimise polarisation effects (see Section 9.1.5). The conductivity is now calculated from equation (6.9) using a known value of l/A, the cell constant, which is found experimentally from conductivity measurements of solutions of known conductivity rather than direct measurement of the cell dimensions. This is obviously difficult to do accurately and there is no guarantee that the cell behaves according to the dimensional values of A and l.

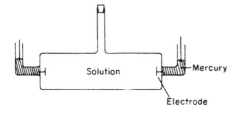

Figure 6.4 One type of conductivity cell

6.2.2 Molar conductivity

The current carried by the electrolyte depends on the rate of flow of ions through the solution which in turn is proportional to: velocity of ions (ionic mobility) $\times$ concentration of ions. Thus, as concentration rises, so does the current at a given voltage, and also the conductivity. In order to eliminate this purely mechanical effect on conductivity so that chemical effects can be considered, the term molar conductivity (Λ) is used. This is given by:

$$\Lambda = \kappa/c \qquad (6.10)$$

where c is electrolyte concentration. The units of κ are $\Omega^{-1}\ cm^{-1}$ or $\Omega^{-1}\ m^{-1}$ and the units of Λ are $\Omega^{-1}\ cm^2\ mol^{-1}$ or $\Omega^{-1}\ m^2\ mol^{-1}$. The ohm, Ω, is the unit of resistance. The preferred SI unit for Ω^{-1} is the siemens, S.

*The arrangement is sometimes known as a conductivity bridge.

Example

The conductivity of a 0.01 mol dm^{-3} NaI solution is 1.19×10^{-3} S cm^{-1}: calculate the molar conductivity.

In this case

$$\Lambda = \frac{1.19 \times 10^{-3} \text{ S cm}^{-1}}{0.01 \times 10^{-3} \text{ mol cm}^{-3}}$$

$$= 119 \text{ S cm}^2 \text{ mol}^{-1}$$

6.2.3 Variation of molar conductivity with concentration

(a) Weak electrolytes

As a weak electrolyte is only partially dissociated into ions, only a fraction of the solute is actually conducting, but the fraction increases as the solution is diluted.

Thus, the molar conductivity, given by equation (6.10), increases with dilution since κ falls less rapidly than c as the solution is diluted. The molar conductivity tends to a maximum (Λ_0) as the concentration tends to zero and dissociation tends to completion.

The variation of molar conductivity with concentration is shown in *Fig. 6.5*. It can be seen that a value for Λ_0 cannot accurately be obtained from extrapolation but it can be calculated[3] from molar conductivities for strong electrolytes.

(b) Strong electrolytes

The molar conductivities of strong electrolytes in solution also decrease with increasing concentration (*Fig. 6.6*) but the absolute molar conductivities are greater (cf. *Fig. 6.5*). than for similar concentrations of weak electrolytes. Extrapolation yields Λ_0. There are two main reasons for decreasing molar conductivities on increasing concentration. Firstly, there is ion-pair formation (see Section 6.1.2) which causes a reduction in the proportion of solute which is capable of conduction. Secondly, there is a decrease in ionic mobility so that for example, a tenfold increase in concentration would lead to something less than a tenfold rate of ion flow. This is explained in terms of two effects: the *asymmetry effect* and the *electrophoretic effect*.

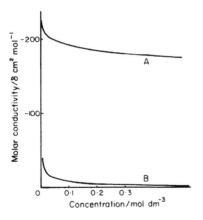

Figure 6.5 Variation of molar conductivity with concentration for: (A) a strong electrolyte; (B) a weak electrolyte

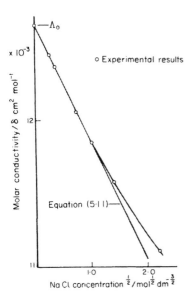

Figure 6.6 Variation of molar conductivity with (concentration)$^{1/2}$ for NaCl in aqueous solution at 298 K

(i) The asymmetry effect

An ion in solution is surrounded by an ionic atmosphere composed of ions of opposite charge or counter-ions. As the central ion moves, it is hindered both by the attraction of the ionic atmosphere and by repulsion as it moves into ionic atmospheres of the same charge.

(ii) The electrophoretic effect

Ions moving through a solution tend to carry solvent molecules with them and ions of opposite charge moving in the opposite direction are thus hindered by the solvent flow. The theories of conductivity were developed by Onsager and others and are discussed in more detail by Robbins[3], and Robinson and Stokes[5].

The theories lead, for dilute solutions, to the Onsager limiting law, which in its simplest form (for a 1:1 electrolyte) is

$$\Lambda = \Lambda_0 - (A + B\Lambda_0)\sqrt{c} \qquad (6.11)$$

where A and B are constants and Λ is the molar conductivity at concentration, c. The

Table 6.2 SELECTED MOLAR CONDUCTIVITIES IN AQUEOUS SOLUTION AT INFINITE DILUTION

Solute	Molar conductivity/S cm^2 mol^{-1}
HCl	426.2
NaCl	126.5
NaOH	248
KNO$_3$	145.0
BaCl$_2$	280
ZnSO$_4$	266

terms A and B are functions of temperature, permittivity and viscosity of the solvent. For several electrolytes, plots of Λ versus $\sqrt{c}$ give straight lines up to concentrations of about 0.2 mol dm^{-3} (see *Fig. 6.6*). Extrapolation to $\sqrt{c} = 0$ gives values of Λ_0. Some typical molar conductivities at infinite dilution (zero concentration) are given in *Table 6.2*.

6.2.4 Kohlrausch's law of independent migration

According to Kohlrausch's law of independent migration, each ion contributes to the total conductance—a conductance which is independent of other ions. Thus for any 1:1 electrolyte, A^+B^-, the molar conductivity is given by:

$$\Lambda = \lambda_{A^+} + \lambda_{B^-} \tag{6.12}$$

where λ_{A^+} and λ_{B^-} are the molar conductivities of the ions A^+ and B^- respectively*. This arises because both cations and anions independently contribute to the total current. Although of opposite charge, they are, of course, moving in opposite directions, so the currents due to each can be simply added. Ionic conductivities are determined via transport numbers (see Section 6.2.5). Some typical values are quoted in *Table 6.3*.

Table 6.3 SELECTED MOLAR CONDUCTIVITIES OF IONS IN AQUEOUS SOLUTION AT INFINITE DILUTION AND 298 K. FOR FURTHER DATA CONSULT ROBINSON AND STOKES[5] AND THE *HANDBOOK OF PHYSICS AND CHEMISTRY*[20]

Cations	Molar conductivity/ S cm² mol⁻¹	Anions	Molar conductivity/ S cm² mol⁻¹
H^+	349.8	OH^-	198.3
Na^+	50.1	F^-	55.4
K^+	73.5	Cl^-	76.4
Ag^+	61.9	Br^-	78.1
NH_4^+	73.4	I^-	76.8
$\frac{1}{2}Mg^{2+}$	53.1	NO_3^-	71.4
$\frac{1}{2}Ca^{2+}$	59.5	$\frac{1}{2}CO_3^{2-}$	69.3
$\frac{1}{2}Zn^{2+}$	53	$\frac{1}{2}SO_4^{2-}$	80
$\frac{1}{2}Cu^{2+}$	53.6–56.6		
$\frac{1}{3}Fe^{3+}$	68		

It is interesting to note that H^+ and OH^- have particularly high values. This has been explained by the so-called Grotthuss chain theory. According to this theory, a proton (H^+) does not itself need to travel through the solution but the effect is achieved by a transfer of bonds as shown in *Fig. 6.7*—a process which is much more rapid. The same theory applies to OH^-.

6.2.5 Transport numbers

The transport number of an ion is defined as the fraction of the total current carried by that ion. For a cation, the transport number, t_+, is given by:

$$t_+ = i_+/I \tag{6.13}$$

where i_+ is the current carried by cations and I is the total current. Similarly, for anions,

$$t_- = i_-/I$$

*In general for the electrolyte $X_p^{q+} Y_q^{p-}$,

$$\Lambda = p\lambda_{X^{q+}} + q\lambda_{Y^{p-}}$$

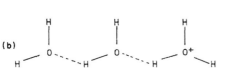

Figure 6.7 Grotthuss chain theory: (a) the transfer of electrons so that there is an interchange of covalent and hydrogen bonds; (b) the result—a transfer of charge

and it follows that $t_+ + t_- = 1$. The transport numbers of cation and anion in a solution are not necessarily equal as it is likely that they will have different mobilities (see Section 6.2.2). Since conductivity is proportional to current, i.e. $\Lambda \propto I$, and $\lambda \propto i$, then,

$$\lambda_+ \propto i_+ \quad \text{and} \quad \lambda_- \propto i_-$$

Therefore,

$$\frac{\lambda_+}{\Lambda} = \frac{i_+}{I} = t_+$$

and, similarly for anions,

$$\frac{\lambda_-}{\Lambda} = \frac{i_-}{I} = t_-$$

Thus, ionic conductivities can easily be calculated using transport numbers.

Two important methods for determination of transport numbers are the Hittorf method and the moving boundary method (*Fig. 6.8*). The latter is outlined here as an

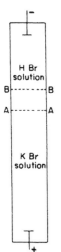

Figure 6.8 The basis of the moving boundary method for the determination of transport numbers. The boundary A–A moves to a new position, B–B

example and both methods are described in more detail by Robbins[3], Moore[2] and Findlay[11]. A capillary tube contains two solutions having a common ion, e.g. HBr solution and KBr solution. There is a distinct boundary between the two layers which can be observed using the sharp change in refractive index or by the use of an indicator. A potential difference is maintained across the ends of the tube and as a current flows, the H^+ ions migrate faster than the K^+ ions resulting in a movement of the boundary. From the quantity of electricity passed, the concentration of HBr and the distance moved by the boundary, it is possible to determine the transport numbers of H^+ and Br^-. Some typical transport numbers are shown in *Table 6.4*.

Table 6.4 SELECTED CATION TRANSPORT NUMBERS FOR 0.1 mol dm^{-3} AQUEOUS SOLUTIONS AT 298 K

Electrolyte	Cation transport number
HCl	0.83
NaCl	0.38
KCl	0.49
$BaCl_2$	0.43
$AgNO_3$	0.47
$CaCl_2$	0.41

6.2.6 Conduction in fused salts

It is useful to extend the discussion of conductivity to fused salts (melts) because of their importance in the electrolytic extraction of metals (Section 7.4.2). Molten salts have much higher conductivities than equivalent aqueous systems. For example, the conductivity of molten KCl at 800 $^\circ$C is about 20 times greater than that of a 1 mol dm^{-3} solution at 25 $^\circ$C. In the absence of a solvent the molar conductivity will be given by:

$$\Lambda = \kappa V_{mol}$$

where V_{mol} is the molar volume, that is, the volume of one mole of salt.

There are technical problems when measuring conductivities of fused salts. For example, traces of oxygen and water must be excluded from the melt and for accuracy, measurements are made over a range of frequencies of the alternating current applied to the bridge. The molar conductivity is very sensitive to temperature, so thermostatting is especially important. The variation of molar conductivity with temperature has been expressed in the form[12]

$$\Lambda = A_0 \exp(-E_\Lambda/RT)$$

where A_0 is a constant, and E_Λ is an activation energy (see Section 3.5). This relationship is valid for a number of salts over a range of approximately 200 K. For other substances, however, E_Λ is itself a function of temperature and plots of log Λ versus $1/T$ show slight curvature.

For further information concerning conduction and other features of fused salts consult references[3,13,14,15].

6.3 Electrodics I

6.3.1 Electrode potentials

Consider a metal, M, dipping into an aqueous solution of its own ions M^{n+}(aq). The following equilibrium is established:

$$M \rightleftharpoons M^{n+}(\text{aq}) + ne \qquad (6.I)$$

which leaves the metal with a negative charge. The M^{n+}(aq) ions are attracted to the metal (*Fig. 6.9*) forming an electric double layer and there is thus a potential difference

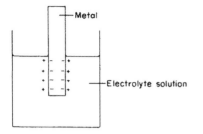

Metal

Electrolyte solution

Figure 6.9 A simplified electric double layer

between metal and solution: this shall be referred to as the *absolute electrode potential*. The absolute electrode potential clearly depends on the equilibrium position of reaction (6.I) and this is affected by the following factors:

(i)   the metal, its purity, crystallographic and phase structure, surface condition etc.;
(ii) activity of M^{n+} in solution;
(iii) temperature.

The effect of (ii) and (iii) on the electrode potential will be dealt with in Section 6.3.12, while the condition of the metal electrode is discussed below.

6.3.2 The metal electrode

The position of equilibrium for a reaction is related to ΔG and while entropy changes cannot be ignored, it is still very useful to consider the enthalpy contribution to reaction (6.I). Now, the reaction

$$M(s) \longrightarrow M^{n+}(aq) + ne$$

is made up of three processes[*]:

(a) sublimation $M(s) \longrightarrow M(g)$
(b) ionisation $M(g) \longrightarrow M^{n+}(g) + ne$
(c) hydration $M^{n+}(g) + aq \longrightarrow M^{n+}(aq)$

The enthalpy changes for sublimation, ΔH_a, and for ionisation (ionisation energy), ΔH_b, are positive and metals with low values of ΔH_b (for example, alkali metals) will tend to have low values of ΔG and hence high absolute electrode potentials.

However, hydration energy is negative, so small, highly charged ions (e.g. Al^{3+}), i.e. those which are highly hydrated, will also tend to have high absolute electrode potentials. Thus ionisation and hydration have similar periodic trends but cancel each other out to some extent. In group I, for example, the lithium ion, Li^+, is most hydrated but has the highest ionisation energy[16], and no predictions about its electrode potential relative to other elements can be made without more complete information. In general, however, group I metals have the highest absolute electrode potentials because of their low ionisation energies followed by groups II and III but ΔH_a, ΔH_b and ΔH_c, enthalpy of hydration need to be considered to predict relative electrode potentials within each group. In fact, lithium has the highest absolute electrode potential (see electrochemical

---

[*]It should be made clear that the three processes mentioned do not occur in discrete stages in a given order but simultaneously. The overall reaction involves the break-up of the solid metal lattice to give entities which move about independently and at random as atoms do in the gas phase. The *effect* is, therefore, the same as if the metal were heated until it vaporised, i.e. until it sublimed and a similar enthalpy change is involved here. Ionisation and hydration must also occur since the products are hydrated ions.

series, *Table 6.5*), presumably because of the importance of its hydration energy. Lithium is followed by potassium, caesium, calcium and then sodium.

This explains the unpredictable order of metals in the electrochemical series and shows that it is not strictly speaking a 'reactivity' series though some correlation with reactivity is possible. The term 'reactivity' is vague, it depends to some extent on ionisation energy and kinetic factors and is slightly subjective. The electrode potential is a function of the enthalpy and the entropy changes associated with reaction (6.I) as measured under certain conditions (see below).

The electrode potential depends also on the presence and concentration of alloying elements in the metal electrode and on its microstructure etc., all of which are important in the consideration of corrosion (Section 9.1) and other electrochemical processes.

6.3.3 Comparison of electrode potentials—electrochemical cells

The absolute electrode potential cannot be measured. Any attempt to connect the solution to a voltmeter involves dipping another metal into the solution and will thus alter the system. Apart from this, it is a principle of electrostatics that a potential difference cannot be defined between two points which are in different phases. It is possible, however, to compare electrode potentials by making an electrochemical cell which, incidently, can be made to produce electricity (batteries, accumulators etc.).

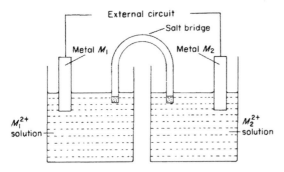

Figure 6.10 A galvanic cell

A simple electrochemical (or galvanic) cell is shown in *Fig. 6.10*. It consists of two half-cells: metal electrodes, M_1 and M_2, dipping into aqueous solutions of their ions (M_1^{2+} and M_2^{2+} respectively). To complete the circuit, the electrodes are connected via the external circuit but there must also be an electrical connection between the two solutions without allowing them to mix. This can be done using a salt bridge as shown in *Fig. 6.10*, though other arrangements are also possible. A porous barrier as shown in

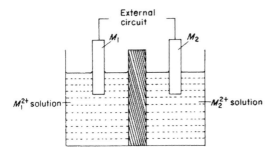

Figure 6.11 A galvanic cell with porous barrier

Fig 6.11 can also be used but is less satisfactory. This is because cations flow from solution of M_1^{2+} to solution of M_2^{2+} through the porous barrier and anions flow in the reverse direction (see cell mechanism, Section 6.3.8). As the rates of flow are unlikely to be equal (see transport number, Section 6.2.5), a potential difference builds up between the solutions: *the liquid junction potential*[3]. The salt bridge largely overcomes this problem. It is composed of a concentrated solution of KCl or KNO_3, so most of the current is carried by K^+ and Cl^- (or NO_3^-) ions which migrate at similar velocities.

6.3.4 Diagrammatic representation of cells

Cells may be represented simply in the following way: for the cell in *Fig. 6.10*:

$$M_1(s)|M_1^{2+}(aq), 1 \text{ mol dm}^{-3} \| M_2^{2+}(aq), 1 \text{ mol dm}^{-3}|M_2(s)$$

i.e.,

half-cell 1 $\|$ half-cell 2

or

electrode 1 | electrolyte 1 $\|$ electrolyte 2 | electrode 2

A single vertical line, |, represents a phase boundary, i.e. between solid M_1 and the solution M_1^{2+}(aq). A double vertical line, $\|$, represents a salt bridge while a porous barrier can be represented thus: $\vdots$ or |. According to the *IUPAC*[*] *convention*[17, 18], for a positive emf the half-cell with the highest absolute electrode potential is written on the left. Electron flow is then left to right, and the corresponding cell reaction is

$$M_1(s) + M_2^{2+} \longrightarrow M_1^{2+} + M_2(s)$$

6.3.5 The standard electrode potential

Although in principle the electrode potentials of any pair of half-cells can be compared, the most useful way of obtaining consistent relative electrode potentials is to compare the electrode potential of all half-cells with the same reference electrode. This is the standard hydrogen electrode[†]:

$$Pt(s)|H_2 \ (1 \text{ atm})|H^+, \quad a_{H^+} = 1$$

Pure hydrogen at one atmosphere pressure is saturated with water vapour and passed over an electrode of platinised platinum[3]. This dips into an acid solution of unit ionic activity. The whole cell, composed of the half-cell of unknown electrode potential and the standard hydrogen electrode is maintained at the required temperature.

The emf of the cell gives the electrode potential of the half-cell relative to the standard hydrogen electrode. The relative electrode potential[‡], E_M, is numerically equal to the emf of the above-mentioned cell and in accordance with the IUPAC convention is given by:

$$E_M = \phi_{H_2} - \phi_M \tag{6.14}$$

where ϕ_{H_2} and ϕ_M are the *absolute electrode potentials* of the standard hydrogen electrode and the metal respectively. Thus for a metal with an absolute electrode potential greater than hydrogen, E_M is negative while for a metal which has an absolute electrode potential less than hydrogen, E_M will be positive. The sign of the electrode potential indicates the direction of flow of electrons—from metal to hydrogen if negative and from hydrogen to metal if positive: the electrode potential is a vector quantity.

---

[*]International Union of Pure and Applied Chemistry, Stockholm, 1953.

[†]A hypothetical electrode: in practice measurements are carried out at a series of low H^+ activities and the results extrapolated.

[‡]. . . or simply the 'electrode potential' as it will be called from this point in the text.

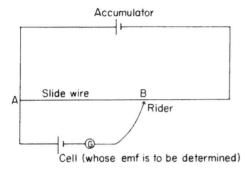

Figure 6.12 Basic potentiometers circuit; G is the galvanometer

The emf is measured using a potentiometer; a simplified circuit diagram is shown in *Fig. 6.12*. The basis of the method is that the emf is measured without drawing any current and this is done using an equal and opposite emf which is under thermodynamically reversible conditions. A rider on the slide wire is moved until there is zero deflection on the galvanometer. The voltage drop across AB of the slide wire will then equal the emf of the cell. The slide wire is first calibrated using a standard cell. The electrode potential of a cell under standard conditions (unit activity of metal, and unit mean ionic activity of electrolyte solution) is known as the standard electrode potential, $E_M^{\ominus}$.

Table 6.5 SELECTED STANDARD ELECTRODE POTENTIALS AT 298 K IN AQUEOUS SOLUTION. FOR FURTHER DATA SEE APPENDIX 2

Electrode reaction	$E_M^{\ominus}$ *or* $(\phi_{H_2} - \phi_M)/V$
$Li^+ + e \rightleftharpoons Li$	−3.05
$K^+ + e \rightleftharpoons K$	−2.93
$Cs^+ + e \rightleftharpoons Cs$	−2.92
$Ca^{2+} + 2e \rightleftharpoons Ca$	−2.87
$Na^+ + e \rightleftharpoons Na$	−2.71
$Mg^{2+} + 2e \rightleftharpoons Mg$	−2.37
$Al^{3+} + 3e \rightleftharpoons Al$	−1.66
$Zn^{2+} + 2e \rightleftharpoons Zn$	−0.76
$Fe^{2+} + 2e \rightleftharpoons Fe$	−0.44
$Pb^{2+} + 2e \rightleftharpoons Pb$	−0.13
$H^+ + e \rightleftharpoons \frac{1}{2}H_2 (g)$	0.00
$Cu^{2+} + 2e \rightleftharpoons Cu$	0.34
$Ag^+ + e \rightleftharpoons Ag$	0.80
$Hg^{2+} + 2e \rightleftharpoons Hg$	0.85

Some typical values are shown in *Table 6.5*. Note that $E_M^{\ominus}$ for H_2 is zero since all other electrode potentials are measured by comparison with the hydrogen electrode.

6.3.6 Reference electrodes

There are practical difficulties in setting up the standard hydrogen electrode, so other half-cells of known and reproducible electrode potentials (reference electrodes[3,19]) were devised and these can be used for reference purposes. The standard electrode potential of a metal is then easily calculated from the potential difference between that of the half-cell and the reference electrode.

Example

Calculate the standard electrode potential of zinc from the following data: the emf of a cell comprising a reference electrode and a zinc half-cell under standard conditions is 1.005 V. The electrode potential of reference electrode $(E_{ref}) = 0.242$ V

$$emf = E_{ref} - E_{Zn}^{\ominus}$$

(to be consistent with equation (6.14)). Hence,

$$E_{Zn}^{\ominus} = 0.242 - 1.005 = -0.763 \text{ V}$$

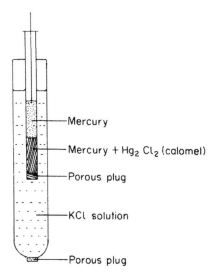

Mercury

Mercury + Hg$_2$Cl$_2$ (calomel)

Porous plug

KCl solution

Porous plug

Figure 6.13 The calomel electrode

One of the most commonly used reference electrodes is the saturated calomel electrode (*Fig. 6.13*):

$$\text{Hg(l)} \mid \text{Hg}_2\text{Cl}_2\text{(s)} \mid \text{KCl saturated, (aq)}$$

6.3.7 Indicator electrodes

These are half-cells used for determining the activities of suitable ions, e.g. H^+, in solution. Their electrode potentials are dependent on ionic activity; the emf between the electrode and a reference electrode can then be used to measure ionic activity. An example is the Ag|AgCl electrode (a silver wire coated with solid silver chloride) which can be used to measure chloride activity. The reaction is:

$$\text{AgCl(s)} + e \rightleftharpoons \text{Ag(s)} + \text{Cl}^-$$

Another example is the glass electrode (*Fig. 6.14*) which, incidentally, also makes use of the Ag|AgCl electrode. It is represented as:

$$\text{Ag} \mid \text{AgCl(s)} \mid \text{HCl(aq)} \mid \text{glass}$$

The potential difference across the glass membrane depends on the pH of the solution into which the electrode dips. Applying the Nernst equation (Section 6.3.12):

$$E_{glass} = E_{glass}^{\ominus} + \frac{RT}{F} \ln \left(\frac{1}{a_{H^+}}\right); \quad n = 1$$

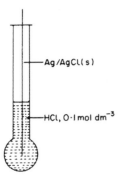

Figure 5.14 The glass electrode

Hence,

$$E_{glass} = E^{\ominus}_{glass} + \frac{RT}{F} \times 2.303 \text{ pH}$$

The glass electrode is used in the pH meter (Section 6.1.10).

6.3.8 Cell mechanism

Consider the cell in *Fig. 6.10* where M_1 is zinc and M_2 is copper and the two are connected via the external circuit by an electrically conducting (metal) wire. When the cell gives current, electrons flow from zinc (which has the higher absolute electrode potential* ($\phi_{Zn} > \phi_{Cu}$) to copper, i.e. $E_{Zn} < E_{Cu}$ (consult equation (6.14)). The zinc electrode is called the *anode* which goes through the following oxidation (electron donation or metal dissolution) reaction:

$$Zn \longrightarrow Zn^{2+} + 2e$$

i.e. oxidation (anodic reaction), and the copper electrode is called the *cathode* which goes through the following reduction (electron consumption or metal deposition) reaction:

$$Cu^{2+} + 2e \longrightarrow Cu$$

i.e. reduction (cathodic reaction). Adding the cathodic and anodic reactions together and cancelling the two electrons from each side gives the overall cell reaction:

$$\overbrace{Cu^{2+} + \underbrace{Zn \longrightarrow Cu}_{\text{anodic reaction}} + Zn^{2+}}^{\text{cathodic reaction}}$$

Since cations are produced in the zinc half-cell and anions in the copper half-cell, there must be a flow of ions between the two solutions: anions from the Cu^{2+} solution to the Zn^{2+} solution and cations in the reverse direction (see also Section 6.3.3). The current in the external circuit and that flowing through the salt bridge are equal. As the concentration of Zn^{2+} increases, the electrode potential, E_{Zn}, of zinc increases (see equation (6.19)) while the electrode potential of copper, E_{Cu}, decreases as the concentration of Cu^{2+} decreases. This can be shown using equation (6.19). Hence, the cell

*A higher *absolute* electrode potential, ϕ_M, results in a more negative electrode potential, E_M or $(\phi_{H_2} - \phi_M)$, and therefore a greater ability to give up electrons via the anodic reaction, i.e. $M \xrightarrow{H_2} M^{2+} + 2e$. This must be balanced by an equal consumption of electrons which occurs at the electrode which has a lower absolute electrode potential—the two electrodes being connected by a material which conducts electrons.

emf decreases until it reaches zero (see Section 6.3.13) when equilibrium is established. In the above case this occurs at 298 K when

$$\frac{a_{Cu^{2+}}}{a_{Zn^{2+}}} \approx 5 \times 10^{-38}$$

that is, at complete removal of copper ions. For cells in which the systems have closer $E_M^{\ominus}$ values, e.g. $Fe|Fe^{2+}$ (-0.440 V) and $Cd|Cd^{2+}$ (-0.403 V), equilibrium is established at a much higher ratio of activities. In this case when

$$\frac{a_{Cd^{2+}}}{a_{Fe^{2+}}} \approx 5.5 \times 10^{-2}$$

Thus, if $a_{Cd^{2+}} < 5.5 \times 10^{-2} \, a_{Fe^{2+}}$ the electron flow will be in the reverse direction, that is Cd to Fe. Under these conditions cadmium will be the anode and iron the cathode (see Section 6.3.15).

6.3.9   Concentration cell

A potential difference also exists between two half-cells if the electrode systems are the same but the electrolyte (or electrode) activities are different, for example

$$Zn(s)|Zn^{2+}, 0.1 \text{ mol dm}^{-3} \| Zn^{2+}, 1 \text{ mol dm}^{-3} |Zn(s)$$

This is an example of a concentration cell. A concentration cell is set up when iron is immersed in differentially aerated water. The half-cell reaction is

$$\frac{1}{2}O_2 + 2e + H_2O \longrightarrow 2OH^-$$

and a current flows between those parts of the iron exposed to different concentrations of dissolved oxygen and corrosion occurs (see Section 9.1.2.2).

A further example is the cell which is set up when an alloy, e.g. 60/40 brass, is immersed in an electrolyte solution. Within the microstructure of the alloy there are grains of α and β phases, each having a different electrode potential.

6.3.10   Redox potentials

For most of the electrode systems discussed so far, an electrode potential is produced when an element is in contact with a solution of its own ions. The reaction (6.I) is oxidation (left to right) and reduction (right to left). An electrode potential can also be developed at a platinum wire (an inert electrode*) which dips into a solution containing two different ions of the same element, e.g. Fe^{2+} and Fe^{3+}. The reaction is:

$$Fe^{3+} + e \rightleftharpoons Fe^{2+}$$

and the electrode potential is defined and measured in the same way as for metal–metal ion systems. Systems involving oxidation and reduction of the ions of the same element are known as reduction–oxidation or simply *redox* systems. Strictly speaking all the systems discussed so far are redox systems, although where the electrode itself is a reactant, the term electrode potential is still used. In other cases, the term redox potential applies. The electrochemical series can be expanded to include all redox systems and is known as the redox series (Appendix 2).

---

*An inert electrode is one which does not take part in the electrode reactions but which will conduct electrons and allow other electrode reactions to take place on its surface.

6.3.11 Cell thermodynamics

Consider the cell which is used for measuring the standard electrode potential of metal, M:

$$M(s)\,|\,M^{n+}, a_{M^{n+}} = 1\,\|\,H^+, a_{H^+} = 1\,|\,Pt(s)\,|\,H_2(g),\,1\text{ atm}$$

The reactions which give rise to absolute standard electrode potentials, $\phi_M^\ominus$ and $\phi_{H_2}^\ominus$ are respectively,

$$M \longrightarrow M^{n+}(aq) + ne \qquad\qquad (6.\text{I})$$

and

$$\tfrac{1}{2}H_2(g) \longrightarrow H^+ + e \qquad\qquad (6.\text{II})$$

but the standard electrode potential of M is given by equation (6.14), that is

$$E_M^\ominus = \phi_{H_2}^\ominus - \phi_M^\ominus$$

Thus the reaction which corresponds to $E_M^\ominus$, is given by

$$n[\text{reaction (6.II)}] - [\text{reaction (6.I)}]$$

which gives:

$$M^{n+} + \tfrac{1}{2}nH_2(g) \longrightarrow M + nH^+ \qquad\qquad (6.\text{III})$$

that is the reduction of M^{n+} by H_2. This is consistent with the IUPAC convention which gives the electrode potential *always* assuming the reduction (cathodic) reaction, i.e. $M^{n+} + 2e \to M$.

Consider now the transfer of a single electron* from the hydrogen electrode to the metal under reversible conditions. The work done by an electron in passing through a potential difference, E, is given by eE, where e is the electronic charge. In this case, the work done by the system is $eE_M^\ominus$. (In this example, $\phi_M^\ominus > \phi_{H_2}^\ominus$, so $E_M^\ominus$ is negative and the work done is negative, i.e. work must be done to transfer the electron). For an Avogadro number, N, of electrons, the work done by the system is

$$NeE_M^\ominus = FE_M^\ominus$$

where F, the Faraday constant, is the total charge on an Avogadro number of electrons, i.e. 96 500 coulombs. For 1 mole of reaction (6.III), that is for 1 mole of M, nN electrons are transferred and the total work done by the system is $nFE_M^\ominus$. The work done *on* the system is, therefore, $-nFE_M^\ominus$ and as the system is under reversible conditions, this is the maximum or reversible work (see Section 2.2.1.1).

Now the maximum work done on a system (apart from pV work—Section 2.2.1.1) is the Gibbs free energy change, ΔG. Thus

$$\Delta G_M^\ominus = -nFE_M^\ominus \qquad\qquad (6.15)$$

where $\Delta G_M^\ominus$ is the standard free energy change for reaction (6.III). In general then:

$$\Delta G = -nFE \qquad\qquad (6.15a)$$

where E is the electrode potential, n is the number of electrons transferred from the oxidised state to the reduced state per atom or ion, and ΔG is the free energy change per mole for the *reduction* of the oxidised state with H_2 as described above.

---

*In effect, transfer of an infinitesimal amount of charge.

6.3.12 The Nernst equation

According to the van't Hoff isotherm for reaction (6.III), per mole,

$$\Delta G = \Delta G^{\ominus} + RT \ln \left(\frac{a_H^n + a_M}{p_{H_2}^{n/2} \, a_{M^{n+}}} \right) \tag{6.16}$$

Since the hydrogen electrode is under standard conditions, $p_{H_2} = 1$ atm and $a_{H^+} = 1$, equation (6.16) becomes:

$$\Delta G = \Delta G^{\ominus} + RT \ln \left(\frac{a_M}{a_{M^{n+}}} \right) \tag{6.17}$$

If metal, M, is pure, then $a_M = 1$, and equation (6.17) can be simplified still further to give:

$$\Delta G = \Delta G^{\ominus} + RT \ln \left(\frac{1}{a_{M^{n+}}} \right) \tag{6.18}$$

Now substituting for ΔG and $\Delta G^{\ominus}$ from equations (6.15a) and (6.15), this gives:

$$-nFE_M = -nFE_M^{\ominus} + RT \ln \left(\frac{1}{a_{M^{n+}}} \right) = -nFE^{\ominus} - RT \ln(a_{M^{n+}})$$

Dividing by $-nF$:

$$E = E_M^{\ominus} + \frac{RT}{nF} \ln(a_{M^{n+}}) \tag{6.19}$$

This equation can also be written generally as:

$$E_M = E_M^{\ominus} + \frac{RT}{nF} \ln \left(\frac{a_{reacts}}{a_{prods}} \right)$$

which is known as the Nernst equation and allows the electrode potential of any half-cell to be calculated knowing $a_{M^{n+}}$, $E_M^{\ominus}$ and n.

Example

Calculate the electrode potential of zinc (E_{Zn}) at 298 K when the activity of Zn^{2+} is 0.01. The standard electrode potential of zinc is -0.763 V at 298 K, $R = 8.314$ J K^{-1} mol^{-1}, $F = 96\,500$ C. The metal is pure.

$$E_{Zn} = -0.763 + \frac{8.314 \times 298}{2 \times 96\,500} \ln 0.01$$

Thus, $E_{Zn} = -0.822$ V.

The Nernst equation also applies to other types of redox systems. For example for the reaction,

$$Fe^{3+} + e = Fe^{2+}$$

the equation becomes

$$E_{Fe^{2+}/Fe^{3+}} = E_{Fe^{2+}/Fe^{3+}}^{\ominus} + \frac{RT}{F} \ln \left(\frac{a_{Fe^{3+}}}{a_{Fe^{2+}}} \right)$$

6.3.13 Calculation of emf (cell potential)

Consider the cell:

$$M_1(s)|M_1^+, a_1 \, \| M_2^+, a_2 |M_2(s)$$

where M_1 and M_2 are different metals (assumed to be univalent for simplicity) and a_1 and a_2 are the mean molar activities of the two solutions. For M_1/M_1^+, the electrode potential is E_{M_1} and for M_2/M_2^+ it is E_{M_2}. $E_{M_2} > E_{M_1}$, that is M_1 is higher than M_2 in the electrochemical series. The reactions corresponding to these electrode potentials, as mentioned earlier, in each case have been determined assuming the reduction (cathodic) reaction, i.e.

$$M_1^+ + \tfrac{1}{2}H_2 \longrightarrow M_1 + H^+ \tag{6.IV}$$

and

$$M_2^+ + \tfrac{1}{2}H_2 \longrightarrow M_2 + H^+ \tag{6.V}$$

The overall cell reaction is

$$M_2^+ + M_1 \longrightarrow M_2 + M_1^+$$

which is given by

$$[\text{reaction (6.V)}] - [\text{reaction (6.IV)}]$$

Thus, the emf (E_{cell}) is given by:

$$E_{cell} = E_{M_2} - E_{M_1} \tag{6.20}$$

and E_{cell} will then always be positive. This is consistent with the negative free energy change for the overall cell reaction, ΔG_{cell}, which is given by:

$$\Delta G_{cell} = -nFE_{cell} \tag{6.15b}$$

This equation may be easily obtained by combining equation (6.15a) for reaction (6.IV) and (6.V).

Example

Calculate the emf of the following cell:

$$Zn(s)\,|\,Zn^{2+}\,\|\,Cu^{2+}\,|\,Cu(s)$$

given that $E_{Zn} = -0.76$ V and $E_{Cu} = 0.40$ V

$$E_{cell} = E_{Cu} - E_{Zn}$$
$$= 0.40 - (-0.76)\ \text{V}$$
$$= 1.16\ \text{V}$$

If only standard values are known, the actual electrode potentials must first be calculated using the Nernst equation assuming the reduction (cathodic) reaction for each. The more positive potential will, therefore, be the more cathodic electrode since it will have the more negative ΔG value for the reaction $M_2^{2+} + 2e = M_2$ and will therefore function as a cathode while the less positive potential will function as an anode, i.e. $M_1 = M_1^{2+} + 2e$. Hence, employing the same basis used to determine equation (6.20), $E_{cell} = E_c - E_a$, where the subscripts c and a stand for cathode and anode respectively. This basis is also used in electrolytic cells as well as galvanic cells.

6.3.14 Concentration cell potentials

Consider the concentration cell

$$M(s)\,|\,M^+,\,a_1\,\|\,M^+,\,a_2\,|\,M(s)$$

where $a_1 < a_2$. The electrode potentials of the half-cells are given by:

$$E_1 = E_M^{\ominus} + \frac{RT}{F} \ln a_1$$

and

$$E_2 = E_M^{\ominus} + \frac{RT}{F} \ln a_2$$

Since $a_1 < a_2$, E_2 will be more positive than E_1, and

$$E_{cell} = E_2 - E_1 = \frac{RT}{F} \ln \left(\frac{a_2}{a_1} \right) \tag{6.21}$$

A similar equation is obtained where the electrodes are of the same metal but have different activities.

Example

Calculate the emf at 298 K between two points A and B on a piece of iron dipping into aerated water if p_{O_2} (at A) : p_{O_2} (at B) = 10. The half-cell reaction (reduction) is:

$$\tfrac{1}{2}O_2 + H_2O + 2e \rightleftharpoons 2OH^-$$

for which,

$$E = E_{O_2/OH^-}^{\ominus} + \frac{RT}{2F} \ln \left(\frac{p_{O_2}^{\frac{1}{2}}}{a_{OH^-}^2} \right)$$

(assuming $a_{H_2O} = 1$). Applying equation (6.21), and assuming a_{OH^-} to be constant throughout the solution,

$$E_{cell} = \frac{RT}{2F} \ln \left(\frac{p_{O_2} \text{ at A}}{p_{O_2} \text{ at B}} \right)^{\frac{1}{2}} = \frac{8.314 \times 298}{2 \times 96\,540} \ln 10^{\frac{1}{2}}$$

Hence, $E_{cell} = 1.48 \times 10^{-2}$ V.

6.3.15 Application of the redox series

In the redox series, systems are listed in order of decreasing reducing ability. For example, metals such as Na and Mg are good reducing agents whereas Cl^- and F^- are very poor. The oxidising ability of the oxidised species such as Na^+, Mg^{2+}, Cl_2 and F_2 obviously increase down the series. The series can thus be used, with caution, to determine the direction of redox reactions. For example, consider the systems:

$$Mg^{2+} + 2e \rightleftharpoons Mg \qquad E^{\ominus} = -2.37 \text{ V}$$

and

$$Cu^{2+} + 2e \rightleftharpoons Cu \qquad E^{\ominus} = 0.337 \text{ V}$$

The first system is reducing with respect to the second and it is possible to predict that magnesium, if added to copper (II) sulphate solution, would reduce Cu^{2+} to Cu and would itself be oxidised to Mg^{2+}.

Care must be exercised, however, in making these predictions since order in the redox series has come about through electrochemical measurements (i.e. the reactions take place in an electrochemical cell) and the conditions are different when substances are mixed together. Also, the redox series refers to standard conditions and the order in the series does not always apply in other conditions. Consider again the systems Fe/Fe^{2+} and Cd/Cd^{2+} (Section 6.3.8). Under standard conditions, Fe reduces Cd^{2+} to Cd, but when $a_{Cd^{2+}} < 5.5 \times 10^{-2} a_{Fe^{2+}}$, Cd reduces Fe^{2+} to Fe.

Another problem is that some metals have several cations so that two or more systems must be taken into account. Consider, for example, the effect of adding iron to $CoCl_2$ solution. The following products are possible: Co, Co^{2+}, Co^{3+}, Fe^{2+}, Fe^{3+} and Fe. Investigation of the redox series (Appendix 2) shows that Fe can be expected to reduce Co^{2+} to Co as $E^\ominus_{Fe/Fe^{2+}} = -0.44$ V and $E^\ominus_{Co/Co^{2+}} = -0.28$ V but Fe^{2+}, not Fe^{3+}, is the product as Fe^{2+}/Fe^{3+} is *below* Co/Co^{2+} ($E^\ominus_{Fe^{2+}/Fe^{3+}} = 0.77$ V). Oxidation of Co^{2+} to Co^{3+} is also not possible ($E^\ominus_{Co^{2+}/Co^{3+}} = 1.82$ V).

An important application of this type of replacement reaction is in the hydro-metallurgical extraction process known as cementation (Section 7.3.2).

6.3.16 Kinetic considerations

Using redox potentials it is possible, as described above, to predict the direction of redox reactions, that is the thermodynamic feasibility of a reaction. However, redox potentials give no information about the rates of reaction. Some redox reactions are slow due to a high activation energy resulting from the need for covalent bonds to be broken, for example in reactions involving ClO_3^-, NH_3 and SO_4^{2-}. Corrosion (i.e. oxidation) of a metal is inhibited by protective coatings (anodising, etc.). Redox reactions involving the production of H_2 or O_2 are also slower because of the high activation energies involved in gas production (see H_2 overpotential, Section 9.1.5).

6.3.17 Variation of redox potential with pH

For redox reactions involving H^+ or OH^-, the redox potential obviously depends on pH and even for those reactions apparently not involving H^+ or OH^- ions, pH may need to be considered, e.g. formation of an insoluble metal hydroxide or passivating film on the metal (electrode) surface (see Section 9.1.4). Consider for example the reduction of Ni^{2+} in solution with H_2:

$$Ni^{2+} + H_2 \longrightarrow Ni + 2H^+$$

For the half-cell reaction

$$Ni^{2+} + 2e \longrightarrow Ni$$

where

$$E_{Ni} = E^\ominus_{Ni} + \frac{RT}{2F} \ln a_{Ni^{2+}}$$

the electrode potential, E_{Ni}, is independent of pH, and for the other half-cell reaction,

$$H^+ + e \longrightarrow \tfrac{1}{2}H_2$$

(note the *reduction* reaction is used to determine E_{H_2})

$$E_{H_2} = E^\ominus_{H_2} + \frac{RT}{F} \ln a_{H^+} \text{ at } p_{H_2} = 1 \text{ atm}$$
$$= \frac{RT}{F} \times 2.303 \log_{10} a_{H^+}$$
$$= -\frac{RT}{F} \times 2.303 \text{ pH}$$

(as pH $= -\log_{10} a_{H^+}$). Thus, the electrode potential, E_{H_2}, decreases as the pH of the solution increases. The variation of E_{Ni} and E_{H_2} have been plotted in *Fig. 6.15* which forms part of the so-called Pourbaix diagram for the Ni/H_2O system (Section 9.1.4). It can be seen that the lines cross at about pH 4.2. Below this pH, $E_{Ni} < E_{H_2}$ and Ni reduces H^+ to H_2. Above pH 4.2, it becomes possible to reduce Ni^{2+} with H_2 at 1 atm. Near pH 6, there is a break in the line as $Ni(OH)_2$ is now the thermodynamically more stable reaction product (rather than Ni^{2+}) and therefore the electrode potential for the system

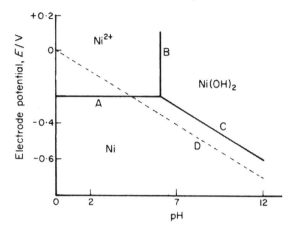

Figure 6.15 Pourbaix diagram (part) for the system Ni–H_2): (A) $Ni^{2+} + 2e \rightleftharpoons Ni$; (B) $Ni^{2+} + 2OH^- \rightleftharpoons Ni(OH)_2$; (C) $Ni(OH)_2 + 2e + 2H^+ \rightleftharpoons Ni + H_2O$; (D) $H^+ + e \rightleftharpoons \frac{1}{2}H_2$, $p_{H_2} = 1$ atm. All activities are unity (except a_{H^+})

$$Ni(OH)_2 + 2H^+ + 2e \longrightarrow Ni + 2H_2O$$

is plotted. In this case the electrode potential is pH dependent. The slope is the same as for the H_2/H^+ line, that is $-(RT/F) \times 2.303$.

It is possible to extend diagrams such as shown in *Fig. 6.15* to other M^{2+} activities, other M_2 pressures and indeed to other systems, and such diagrams can be used directly to predict conditions for selective extraction of metals (Section 7.3.2) and, as mentioned previously, in the construction of Pourbaix (E–pH) diagrams which are used for studying electrodeposition and corrosion processes.

6.4 Electrolysis

6.4.1 Basic considerations

The fundamental definition of electrolysis is the decomposition of a liquid (aqueous or molten) ionic compound by the passage of an electric current. An electrolytic cell is shown diagrammatically in *Fig. 6.16*. The cathode is the electrode at which electrons are consumed and the anode is the electrode at which electrons are produced. In this respect the electrolytic cell is exactly the same as the electrochemical (corrosion) cell considered in Chapter 9 but the electrode polarities are reversed (see Appendix 3).

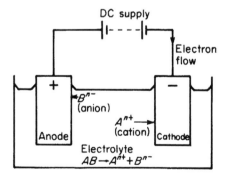

Figure 6.16 Simple electrolytic cell

Thus the cathode reactions for electrolysis may be considered to be the same as those given by reactions (9.II) to (9.VIII). The anode reaction in corrosion, however, is always metal dissolution (reaction (9.I)). In electrolysis this reaction is only one of a number of possibilities including gas evolution, gas adsorption and metal ion oxidation by the reverse of reactions (9.V), (9.II) and (9.VII) respectively. Some examples of simple electrolytic processes should make the situation clearer.

In the electrolysis of *molten* sodium chloride above 804 $°C$ only two species of ion exist, namely Na^+ cations and Cl^- anions. (In practice the operating temperature is reduced to 600 $°C$ by addition of other chlorides but these ions do not interfere with the basic process.) Therefore the reaction at the cathode is

$$Na^+ + e \longrightarrow Na$$

i.e. reduction, and molten sodium (melting point 98 $°C$) is produced. The reaction at the anode is

$$2Cl^- \longrightarrow Cl_2 + 2e$$

i.e. oxidation, and chlorine gas is produced and collected as a valuable by-product.

The electrolysis of dilute sulphuric acid involves not only H^+ and SO_4^{2-} ions but also OH^- ions from the water which is also present. Thus, at the cathode, hydrogen gas is evolved in accordance with reaction (9.II). However, both SO_4^{2-} and OH^- ions are attracted to the anode. Because the discharge potential (Section 6.4.9) for hydroxyl ions is much lower than for sulphate ions the former are able to lose their electrons more easily. The anodic reaction is therefore

$$OH^- \longrightarrow OH + e$$

and hydroxyl ions are immediately replaced in accordance with the equilibrium

$$H_2O \rightleftharpoons H^+ + OH^-$$

The OH radical is unstable and therefore a more accurate representation of the anodic reaction is:

$$4OH^- \longrightarrow 4OH + 4e$$

followed by

$$4OH \longrightarrow 2H_2O + O_2$$

or, as an overall reaction

$$4OH^- \longrightarrow 2H_2O + O_2 + 4e$$

(see reaction (9.V)) and oxygen gas is evolved at the anode. Therefore, for every four moles of electrons flowing round the circuit two moles of hydrogen and one mole of oxygen are produced, i.e. the electrolysis of dilute sulphuric acid is equivalent to the decomposition of water by:

$$2H_2O \longrightarrow 2H_2 + O_2$$

In the electrolysis of aqueous copper sulphate solution the *ionisation* reactions are:

$$CuSO_4 \longrightarrow Cu^{2+} + SO_4^{2-}$$
$$H_2O \longrightarrow H^+ + OH^-$$

The cathodic reaction is the discharge of copper ions (reaction 9.VI)). The anode reaction, however, depends on the electrode material. If the anode is 'inert', e.g. platinum or carbon, oxygen gas is evolved by hydroxyl ion discharge as for dilute sulphuric acid. If the anode is copper the dissolution of copper is more favourable than ion discharge and hence Cu^{2+} ions are produced (reaction 9.I)). This is the basis of electrorefining* of copper (Section 7.4).

The concentration of ions present also affects the nature of the electrode product. Consider the ionisation reactions for aqueous sodium chloride solution:

$$NaCl \longrightarrow Na^+ + Cl^-$$

$$H_2O \longrightarrow H^+ + OH^-$$

The cathode product, hydrogen, is the same irrespective of concentration since sodium is too reactive to be deposited as a metal from aqueous solution (sodium may be deposited as an amalgam at a mercury cathode). In dilute solution oxygen is evolved at the anode with traces of chlorine. In concentrated solution the discharge potential of Cl^- ions is such that mainly chlorine gas is evolved.

6.4.2 Faraday's laws of electrolysis

Faraday's first law states that the mass (m) of any substance discharged (deposited or dissolved) at an electrode is proportional to the quantity of electricity passed which is measured by coulombs and 1 coulomb (C) is equivalent to 1 ampere second (A s). Hence:

$$m \propto It$$

or

$$m = \frac{WIt}{nF} \tag{6.22}$$

where m is the mass (g) discharged of atoms of relative atomic mass W by the passage of a current I (A) for a time t(s), n is the valency of the atom and F the Faraday constant (96 500 C). Sometimes equation (6.22) is expressed as

$$m = ZIt/F$$

where Z is the chemical or electrochemical equivalent, W/n. An element which has more than one valency state has a corresponding number of Z values.

Faraday's second law may be simply expressed in that 1 mole of ions of any substance is discharged by the valency number of Faradays to produce 1 mole of atoms of that substance. Thus, 1 mole of copper (63.5 g) is produced from 1 mole of Cu^{2+} ions by 2 Faradays ($Cu^{2+} + 2e \longrightarrow Cu$). Similarly, 1 mole of aluminium (27 g) requires 3 Faradays to discharge 1 mole of Al^{3+} ions. Care needs to be taken when evolution of diatomic gases is considered. One mole of hydrogen *atoms* is produced from 1 mole of hydrogen ions (H^+) by 1 Faraday, but 1 mole of hydrogen *gas* (hydrogen molecules, H_2) requires 2 Faradays since

$$2H^+ + 2e \longrightarrow H_2$$

It should be noted that 1 Faraday is equivalent to 1 mole of electrons.

6.4.3 Current efficiency

Equation (6.22) gives the theoretical maximum mass of substance which can be discharged by a given current in a given time. Current efficiency refers to the cathode product which in practice is less than theoretically predicted. This parameter is defined as

$$\text{current efficiency} = \frac{\text{actual mass discharged}}{\text{theoretical mass discharged}} \times 100\% \tag{6.23}$$

---

*Electrorefining is the electrolytic *refining* of a metal used as an *anode* in an electrolytic cell whereas electrowinning is the electrolytic *extraction* (winning) of a metal from an *electrolyte* containing the metal ions. Electrowinning may use inert or soluble anodes.

Example

Determine the current efficiency of a chromium plating cell in which a current of 10 A flowing for 90 minutes deposits 0.9 g of chromium from chromic acid electrolyte (CrO_3) comprising Cr^{6+} ions. The relative atomic mass of chromium is 52.

Using Faraday's first law to find theoretical mass deposited:

$$m = \frac{52 \times 10 \times (90 \times 60)}{6 \times 96\,500} \text{ g}$$

$$= 4.85 \text{ g}$$

Therefore

$$\text{current efficiency} = \frac{0.9}{4.85} \times 100\%$$

$$= 18.6\%$$

In general, chromium plating from chromic acid has a current efficiency of between 12% and 20%. Some processes such as aluminium refining operate at current efficiencies in excess of 99%. There are many reasons for the difference between theoretical and actual yields some of which depend on operating conditions. For example, loss of current by electrical leakage and conversion to heat may be contributory factors. There may be recombination of electrode products unless these are well separated, e.g. $Mg + Cl_2 \longrightarrow MgCl_2$, or simultaneous unproductive unwanted electrode reactions may occur, e.g. hydrogen evolution or $Fe^{2+}|Fe^{3+}$ redox reactions. The electrode product may react with the environment, e.g. sodium metal in moist air, or undergo chemical or physical reactions with the electrodes or the electrolyte, e.g. re-solution of cathode deposit.

Although current efficiency is usually taken as a measure of the efficiency of cathodic deposition, it is also possible to define anodic current efficiency and this may be defined in the same way as equation (6.23). For the situation where a metal anode dissolves chemically in the electrolyte, the weight loss of the anode is greater than theoretically predicted and a current efficiency of greater than 100% is achieved, which can never be the case at the cathode. A high anodic current efficiency has less significance on the overall process than a high cathodic current efficiency. However, a low anodic current efficiency may be significant (see Section 6.4.9).

6.4.4 Energy efficiency

This parameter is usually expressed as the power (W) required to produce 1 kg of product and has units kW h kg^{-1}. Copper refining has an approximate energy efficiency of 0.2 kW h kg^{-1} whereas the figure for aluminium refining (Section 7.4.2) is 20 kW h kg^{-1}, i.e. less energy is used in refining 1 kg of copper than for the same mass of aluminium.

Sometimes energy efficiency may be calculated as:

$$\text{energy efficiency } (\%) = \text{current efficiency } (\%) \times \frac{\text{theoretical cell voltage}}{\text{applied voltage}} \qquad (6.24)$$

Theoretical cell voltage will be considered in Section 6.4.6. One use of equation (6.24) is to study the interrelationship between energy efficiency and current efficiency. If the applied voltage could be kept constant and the current efficiency increased, the energy efficiency, measured as a percentage, increases, i.e. more current is usefully used decreasing the overall power (current $\times$ voltage) needed. If the current efficiency is maintained constant and the applied voltage altered the situation is rather more complicated. Increasing the voltage increases the power input but that part of the applied voltage performing useful work may not change in proportion. This is because part of the applied voltage is needed to overcome polarisation and resistance effects in the cell (Section 6.4.8) and this fraction may markedly increase. Power input may be reduced by reducing the applied voltage but there is a minimum voltage below which electrolysis does not proceed at a significant rate (Section 6.4.8). Hence optimum conditions in practice are achieved by a balance between current efficiency and energy efficiency. Even though aluminium refining has a relatively poor energy efficiency there is no alternative low energy route. Energy is not only consumed by electrolysis but also by heat required to keep the electrolyte molten. The only way to improve energy efficiency in this case is to use as large a plant as possible. The most important variable in electrolysis, however, is input current since this determines the output of the cell as implied by Faraday's first law (equation (6.22)).

6.4.5 Current density

This parameter is the current per unit area of electrode and has units $A\ m^{-2}$, but is often quoted as $A\ cm^{-2}$. For example in copper refining the current density is $220\ A\ m^{-2}$ and for aluminium refining $15\ 000\ A\ m^{-2}$ are required. Normally the measurable geometric area of the electrode is satisfactory for determining current density but in some electroplating processes a rough surface greatly reduces the expected current density. Both anodic and cathodic current densities may be considered but they need not necessarily be equal.

Activation and concentration polarisation effects are directly dependent on current density (Section 9.1.5.4) and therefore it may be desirable to reduce current density. If the current were reduced then output would also be reduced, if the electrode area were increased current density would decrease without affecting the applied current. In some cases a high current density is desirable. For example, gas cleaning by promoting cathodic gas evolution. Also in electroplating the current density used is as high as possible without producing 'burning' (dark, powdery deposits). The highest current density at which 'good' plating is achieved is called the limiting current density but this term is not usually the same as i_L in Section 9.1.5.4. Current density also affects the covering power and throwing power of plated metal on a substrate.

6.4.6 Theoretical cell voltage

The voltage, E_{cell}, is the minimum theoretical voltage required to electrolyse a given electrolyte. In principle as soon as this voltage is achieved between an anode and a cathode, electrolysis, i.e. discharge of ions, occurs. Within this topic area it is necessary to distinguish between aqueous electrolytes and fused salt electrolytes.

In aqueous electrolysis the theoretical cell voltage is dependent on the reversible electrode potentials of each electrode. The Nernst equation (Section 6.3.12) and electrochemical series (Section 6.3.3) are used to calculate E_{cell} (Section 6.3.13). The most simple example of an aqueous electrolyte is water (acidified) (see Section 9.1.4).

At pH 7, $E_{r,H_2} = -0.413$ V and $E_{r,O_2} = +0.817$. Therefore

$$E_{cell} = E_c - E_a = -(0.413) - (0.817) = -1.23\ V$$

If platinum electrodes dipping into acidified water are connected together directly no current flows and no emf exists between them. If an external current is applied oxygen is evolved at the anode and hydrogen at the cathode only when the potential difference between the electrodes has been increased to a theoretical minimum of 1.23 V. Another method of calculating E_{cell} is by using free energy data, and the equation $\Delta G = -nFE_{cell}$ (Section 6.3.11).

$$2H_2O \longrightarrow 2H_2 + O_2 \qquad \Delta G_{298}^{\ominus} = +474\ kJ$$

$n = 4$ since four electrons are transferred:

$$4H^+ + 4e \longrightarrow 2H_2 \quad \text{and} \quad 4OH^- \longrightarrow 2H_2O + O_2 + 4e$$

Therefore

$$E_{cell} = -\frac{474\ 000}{4 \times 96\ 500}\ V$$

$$= -1.23\ V$$

The signs of both ΔG and E_{cell} should be noted. The positive value of ΔG indicates the non-spontaneous direction of reaction, i.e. the reaction as written above should normally go from right to left. Therefore the negative value of E_{cell} indicates that this voltage must be overcome in order to produce hydrogen and oxygen. Hence applying 1.23 V produces no reaction but applying fractionally more than 1.23 V results in electrolysis.

Calculation of E_{cell} in this way is particularly important when considering fused salt electrolytes where $E^{\ominus}$ data are not as readily available as thermodynamic data (see Section 7.4).

An interesting variation in electrolytic processes is the electrorefining of nickel from nickel sulphide. The nickel sulphide from the matte (Section 7.2.10.2) is cast into anodes weighing about 250 kg each. The sulphide is regarded as Ni_3S_2 at this stage rather than NiS and contains about 75% nickel, 20% sulphur, 3% copper, 0.5% cobalt, 0.5% iron with small amounts of lead, zinc, selenium and precious metals. A diaphragm cell is used for electrolysis to separate the impure anolyte from the catholyte (see Section 7.4). Cobalt, iron and copper ions present in the anolyte would deposit at the cathode (Section 6.4.9) unless removed. The anolyte is therefore pumped away to be purified and then returned to the cathode compartment. In addition to a diaphragm (which increases electrical resistance) between the compartments the anodes are placed in very large bags to collect the anode slime. The electrolyte used consists principally of nickel sulphate. The anodic reaction for electrolysis is

$$Ni_3S_2 \longrightarrow 3Ni^{2+} + 2S + 6e$$

The sulphur produced is collected as part of the *anode slime*. Nickel ions in the purified catholyte are deposited on the pure nickel cathode according to $3Ni^{2+} + 6e \longrightarrow 3Ni$. The overall reaction is therefore,

$$Ni_3S_2 \longrightarrow 3Ni + 2S \quad \Delta G_{298}^{\ominus} = +203\ kJ$$

Since $n = 6$, $E_{cell} = -0.35$ V.

6.4.7 Polarisation and overpotential

These topics are considered in detail in Section 9.1.5.4 so that here the implications only will be emphasised. $E^{\ominus}$ for Ni plating from aqueous $NiSO_4$ solution is -0.25 V but plating does not begin until the cathode potential is about -0.45 V. Since the overpotential, η, is defined by equation (9.6) as

$$\eta = E_p - E_r \quad (E_p = -0.45\ V, E_r = -0.25\ V)$$

Therefore $\eta_c = -0.20$ V, which is the amount by which the cathode potential must be made more negative before nickel can plate out. Overpotential is defined by equation (9.8):

$$\eta = \eta^A + \eta^C + \eta^R$$

Each type of polarisation acts *against* the applied voltage, e.g. the applied voltage must include an extra contribution with E_{cell} to overcome the activation energy for an electrode reaction. In addition, local electrode concentration and resistive effects set up 'back emf's' as indicated by η^c and η^R respectively which must also be overcome by the applied voltage. These polarisation effects take place at both cathode and anode and the applied voltage, E_{app}, may be defined as:

$$E_{app} = E_{cell} + \eta_a + \eta_c \qquad (6.25)$$

6.4.8 Decomposition voltage

The decomposition voltage is the minimum voltage at which appreciable electrolysis occurs. This is best explained by reference to *Fig. 6.17*. The graph of current against voltage shows that as the voltage increases there is little change in current. From Faraday's first law (equation (6.22)) this implies little product. Suddenly, however, a small change of voltage produces a marked increase in current (product). Extrapolating

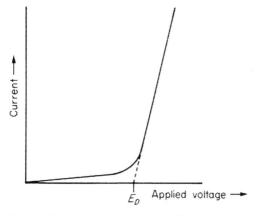

Figure 6.17 Graph of current against voltage for electrolysis showing decomposition voltage, E_D

as shown gives the decomposition potential E_D, which is the minimum applied voltage necessary to produce a significant yield and takes account of resistance in the external circuit, E_{cont} (contacts, busbars, etc.) and electrolyte resistance, E_{sol}. Therefore the decomposition voltage may be defined as

$$E_D = E_{cell} + \eta_a + \eta_c + E_{res} \qquad (6.26)$$

where

$$E_{res} = E_{cont} + E_{sol}$$

Electrolytic decomposition of acidified water has been shown to take place at a theoretical applied voltage of 1.23 V. However, polarisation and, to a lesser extent, resistive effects contribute a further 0.5 V to this figure in practice.

For nickel electrorefining (of Ni_3S_2 anodes) $E_{cell} = -0.35$ V, $\eta_a = 1.1$ V, $\eta_c = 0.2$ V, $E_{cont} = 0.3$ V, $E_{sol} = 0.9$ V. Therefore $E_D = 2.85$ V. Note that negative signs are ignored because *all* component voltages in E_D must be overcome by the applied voltage. In general, voltages for aqueous electrolysis are less than for fused salt electrolysis, extra power being required in the latter case to keep the electrolyte molten in addition to overcoming the higher resistance of molten electrolytes. However, voltages cannot be consider in isolation from current requirements. For example, copper extraction using a copper sulphate electrolyte requires about 2 V but the current is about 20 000 A, depending on electrode area (current density 200 A m^{-2}). The current efficiency is 80–90% to make the energy efficiency about 2 kW h kg^{-1}. A typical cell for aluminium extraction require about 40 000 A and 5 V and with a current efficiency of about 85% achieves an energy efficiency of about 18 kW h kg^{-1}.

Where electrolytically soluble anodes are used for aqueous electrorefining E_{cell} is approximately zero (see Section 7.4). Therefore the voltage requirement is reduced considerably. For example, a typical industrial cell for copper refining requires 15 000 A (current density 220 A m^{-2}) and 0.2–0.3 V and operates at a current efficiency of about 95% to give an energy efficiency of about 0.2 kW h kg^{-1} refined copper. Aluminium refining requires 6–8 V and 16 000 A (current density 15 000 A m^{-2}) operating at a current efficiency of greater than 99% to give an energy efficiency of about 20 kW h kg^{-1} refined aluminium. Electrowinning is generally a much greater energy consuming process than electrorefining–aluminium is an exception.

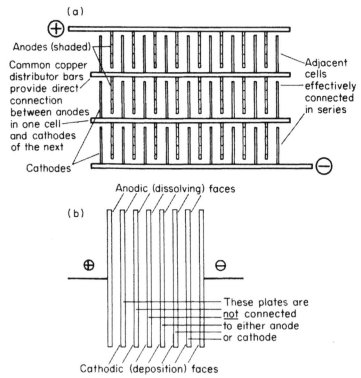

Figure 6.18 Schematic plan view of electrorefining systems: (a) monopolar or parallel system; and (b) bipolar or series system

The different cells by which electrorefining can be achieved are shown in *Fig. 6.18*. The monopolar (multiple or parallel) system is the most popular of the two and is used for refining copper, gold, silver and lead. The bipolar (series) system is effectively a number of cells in series and used mainly for copper. The current efficiency of the monopolar system is 90–98% whereas that for the bipolar system is 70–75%. However, the energy efficiency of the former is about 0.30 kW h kg^{-1} and that for the latter is much smaller, 0.13 kW h kg^{-1}. This lower energy requirement is a consequence of the simpler electrical system.

The bipolar system produces less pure cathodes because of residual impurities which always remain; the cathodes being removed before all the impure material has been refined. An added advantage of the monopolar system is that it can refine metals of relatively low purity. Nowadays modern electrorefining plants for copper are exclusively monopolar.

6.4.9 Discharge potential

This relates to a single electrode and is the minimum potential which must exist between the electrolyte and an electrode to bring about a continuous discharge of an ion. It may be defined as (for that particular electrode reaction):

$$\text{discharge potential} = E_r + \eta$$

Consider discharge of zinc ions at a zinc electrode where $E_r = -0.76$ V and $\eta = -0.20$ V giving a discharge potential of -0.96 V. Consider discharge of hydrogen gas at a zinc electrode under the same conditions. $E_r = 0.0$ V and η^A, the hydrogen overpotential, is -1.13 V. Therefore the discharge potential for hydrogen is -1.13 V. If the applied voltage is such that the electrode potential is between the discharge potentials of zinc and hydrogen, zinc will be deposited without hydrogen discharge. (The absolute values of η used here are of course subject to operating conditions (current density, pH and type of plating bath). Nevertheless the importance of overpotential is clear.) Hydrogen will only be formed with the zinc when the electrode potential is more negative than -1.13 V. Thus it is possible to deposit zinc from aqueous solution without hydrogen evolution even though zinc has a more negative electrode potential than hydrogen in the electrochemical series (Appendix 2). High hydrogen overpotentials* also permit manganese, chromium, iron, cadmium, cobalt, nickel, tin and lead to be deposited from aqueous electrolytes; an important point since the process costs compared with those for fused salt electrolysis are lower. Reactive (base) elements for which the discharge potentials are more negative than the hydrogen discharge potentials, would produce hydrogen rather than metal in aqueous solution, e.g. electrolysis of NaCl solution (Section 6.4.1). In this latter case any sodium, even if it were formed, would immediately react with the water present and dissolve. Therefore deposition of metals such as calcium, sodium, magnesium, beryllium, aluminium and niobium requires fused salt electrolytes.

The concept of discharge potential is useful in explaining the importance of impurity removal in electrolysis. For example if, in the previous example, there were impurities present with discharge potentials more positive than zinc, these impurities would deposit with the zinc, thereby reducing the current efficiency. Consider electrorefining of nickel from Ni_3S_2 (Section 6.4.6), $E_r = -0.25$ V and $\eta = -0.2$ V. Therefore the discharge potential is -0.45 V. Consider the discharge potentials for iron and cobalt on the nickel cathode to be -0.48 V and -0.40 V respectively. Cobalt would be deposited because it has a *more positive*† discharge potential than nickel. In fact some iron would also be deposited because the electrode potential would be more negative than -0.48 V in practice. Hence the need for a diaphragm cell and purification of the anolyte which together help to produce a current efficiency of greater than 95%. The anode potential in this process is about $+1.2$ V. Therefore any metal for which the discharge (dissolution) potential is *more negative*† than this value will dissolve. Assuming that the discharge potentials for deposition are the same as those for dissolution then iron, cobalt and nickel will dissolve at the anode. This assumption implies that the overpotentials (positive) are the same at anode and cathode which is unlikely in practice. Also there may be some contribution to the reversible electrode potential for nickel as calculated from the Nernst equation due to the different activities of nickel ions in the anolyte and catholyte. However, suitable choice of operating variables ensures discharge potentials are less than the operating anode potential. Since $E^{\ominus}$ for $S + 2e = S^{2-}$ is -0.48 V, for sulphur to precipitate at the anode, i.e. $S^{2-} = S + 2e$, the anode potential of this reaction will be $+0.48$ V. Hence formation of sulphur will be favoured at the applied anode potential even allowing for the

---

*The more modern view is that the exchange current density for hydrogen evolution is low, which produces a lower rate of discharge of H^+ ions than metal ions.

†Calculated from Nernst equation using IUPAC convention.

unknown overvoltage contribution. The discharge potentials of the precious metals are more positive than 1.2 V and therefore these metals do not dissolve but collect as an anode slime. At the anode potential used decomposition of water occurs and oxygen is evolved by the reverse of reaction (9.IV), i.e. $2H_2O \rightarrow O_2 + 4H^+ + 4e$, which also makes the solution more acidic. This effect reduces tne anode efficiency with the result that nickel is deposited from the catholyte faster than it is dissolved in the anolyte. This nickel deficiency is corrected during anolyte purification.

There is another way to approach the problem of which ion species is discharged or produced at an electrode. Application of the Nernst equation to the equilibria $Co^{2+} + 2e = Co$; $S + 2e = S^{2-}$, etc. at anode and cathode at the potentials used indicates whether solution or discharge takes place. Thus for cobalt for example the Nernst equation becomes

$$E = -0.28 + \frac{RT}{2F} \ln \left(\frac{a_{Co^{2+}}}{a_{Co}} \right)$$

where $a_{Co^{2+}}$ refers to activity of Co in solution and a_{Co} to activity in the solid electrode. The value of E is either $+1.2$ V or -0.45 V depending on choice of anode or cathode respectively for consideration. Deductions may be made simply by determining the value of the activity ratio $a_{Co^{2+}}/a_{Co}$, an absolute value of one of these activities is not required.

It may be shown that at the anode $a_{Co^{2+}} \gg a_{Co}$ ($a_{Co^{2+}}/a_{Co} = 1.4 \times 10^{50}$ at 25 °C) and therefore dissolution of cobalt is strongly favoured. Similarly, at the cathode $a_{Co} \gg a_{Co^{2+}}$ ($a_{Co^{2+}}/a_{Co} = 1.7 \times 10^{-6}$ at 25 °C), implying that discharge of Co^{2+} ions could take place. Note that for sulphur the activity ratio is $a_s/a_{s^{2-}}$ and at the anode $a_s \gg a_{s^{2-}}$ ($a_s/a_{s^{2-}} = 8.5 \times 10^{56}$ at 25 °C) which implies that, in effect, the sulphur does not dissolve (therefore consideration of the cathode reaction in this case becomes unnecessary).

The reader may wish to attempt other problems of this type and it is suggested that the Nernst equation is applied to the $Fe^{2+} + 2e = Fe$ equilibrium ($E^{\ominus} = -0.44$ V) during the electrorefining of copper ($E_c \simeq E_a = +0.28$ V). It is left for the reader to show that although iron is dissolved at the anode Fe^{2+} ions are not discharged at the cathode (unlike electrowinning of nickel from Ni_3S_2) and therefore a diaphragm cell is not required for copper refining. The reader may also care to deduce that silver $Ag^+ + e = Ag$ ($E^{\ominus} = +0.79$ V, $n = 1$) is not dissolved at the anode.

6.4.10   Electroplating

Electrowinning, electrorefining and electroplating all depend on the same basic principles of electrolysis. Electroplating is carried out to serve one or both of the functions of decoration and corrosion protection. The extent to which each of these is important depends on the application. For example, zinc and cadmium on steel have essentially protective functions. A very common electrodeposit which combines both functions is chromium.

A schematic comparison of corrosion resistance of chromium plating is given in *Fig. 6.19*. A bright nickel layer is required underneath the thin (0.3 µm minimum) chromium layer for two reasons. First, to achieve levelling because chromium has no levelling action and, secondly, to produce a bright finish to which colour, lustre and tarnish resistance are provided by a chromium layer. However, bright nickel and chromium are both cathodic to steel and corrosion can occur as shown. Bright nickel has a poorer resistance than dull nickel. The latter cannot be used alone without seriously affecting the lustre of the subsequent chromium layer. Therefore a duplex nickel layer is used and the resulting coating has enhanced corrosion resistance, as shown in *Fig. 6.19(b)*. Corrosion resistance may be improved further by the use of 'microcracked' chromium (*Fig. 6.19(c)*). Chromium cracks when more than about 0.5 µm thick but the plating process is controlled to give a specific crack pattern of

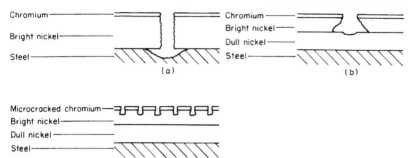

Figure 6.19 The effect of different types of chromium + nickel plating on corrosion of steel: (a) rapid corrosion of steel beneath cathodic layer once pit has penetrated nickel; (b) the presence of dull nickel reduces pit penetration; (c) microcracked chromium spreads the corrosion attack

30–80 cracks per millimetre. The effect of the cracks is to increase the exposed relatively anodic nickel area beneath and thus reduce the current density, i.e. the corrosion rate. The microcracked chromium layer is usually more than 0.8 μm thick and its use means that the nickel layer may be thinner. Incorporating an undercoat of copper between the nickel and the substrate allows the thickness of the nickel layer to be reduced still further. The actual thickness of the layers depends on component application and is specified in BS1224. For engineering purposes the chromium layer must be resistant to wear. To achieve this the chromium layer may be between 12 μm and 0.4 mm thick depending on application but 0.13 mm is generally considered thick enough for average use. This relatively thick chromium layer is usually called 'hard' chromium (800–900 VPN) but in fact it is no harder physically than chromium used for decorative purposes. The thickness and evenness of the deposit are extremely important as are adherence of the deposit to the substrate and usually the decorative effect.

Electrolytes, for electroplating, usually fall into the categories acidic or alkaline, although sometimes near neutral plating baths of pH 5 to 8 are also classified. Acid baths are usually based on sulphates (e.g. for nickel, cobalt, zinc, tin and copper), chlorides (e.g. for nickel, iron and zinc) or fluoborates (e.g. for nickel, copper, lead and tin–lead alloys). The well known Watts bath for electroplating nickel comprises sulphate and chloride. The advantage of fluoborates is that for most metals they are very soluble and the activity of metal is greater. These activity and solubility factors permit the use of higher current densities than, for example, with sulphate solutions. The main disadvantage of fluoborates is their high cost.

Alkaline electrolytes contain complex anions which contain the metal to be deposited. The most important complexing agent is cyanide and cyanide solutions are used to plate copper, cadmium, gold, silver, zinc and indium. Although many other metals such as nickel also form cyano-complexes the electrodeposit obtained is not as good as that obtained by other electrolytes. There is also the potential danger of producing highly toxic HCN gas. Other alkaline electrolytes make use of other complexes, e.g. stannate $Sn(OH)_6^{2-}$ for tin, pyrophosphate $(P_2O_7)_2^{6-}$ for copper, dichromate $Cr_2O_7^{2-}$ (and other Cr complex anions) for chromium. It is of interest to note that all these ions are negatively charged yet metal ions must be discharged at the cathode. Considering a copper cyanide complex $Cu(CN)_4^{3-}$ the copper plating reaction may be written:

$$Cu(CN)_4^{3-} + e \longrightarrow Cu + 4CN^-$$

Copper in this complex ion has valency 1. It is not clear whether this reaction takes place directly by a dipole mechanism (i.e. the complex has an electropositive and an electronegative end) or by an intermediate dissociation reaction giving $Cu^+ + 4CN^-$ ions in the electrode double layer.

Cyanide solutions generally have better macrothrowing power but poorer microthrowing power than acid solutions. Macrothrowing power may be defined as the ability to give a cathodic coating of uniform thickness over an undulating surface, crevices and protuberances. Microthrowing power is the ability to deposit into small surface discontinuities of less than 0.25 mm width. Concentration polarisation is the key factor in determining micro- and macrothrowing power. If, for a cyanide solution, the temperature were increased and/or agitation increased and/or free CN^- ion concentration decreased, the effect of concentration polarisation would be reduced. This would increase the microthrowing power but reduce the macrothrowing power. Similarly, if the solution had a high concentration of copper cyano-complex ions the effect of local cation depletion and consequent concentration polarisation is reduced.

The presence of complex ions is also very important when alloy deposition is considered. For example, brass plating from cyanide solution is achieved by the discharge of $Cu(CN)_4^{3-}$ and $Zn(OH)_4^{2-}$ (rather than $Zn(CN)_4^{2-}$) ions. The reversible electrode potentials of copper and zinc in acid solutions of unit activity are $+0.34$ V and -0.76 V respectively and the corresponding potential–current density curves are shown in *Fig. 6.20(a)*. Clearly there is no potential which could be applied at which both metals could be simultaneously deposited from these acid solutions. All the copper would be deposited first and then the zinc. The effect on the E–i curves using the cyanide solution is shown in *Fig. 6.20(b)*. Thus, simultaneous co-deposition is now possible. The reason for this behaviour is that the formation of complexes reduces the activity of copper ions to about 10^{-18} g ions/l although the concentration of copper in solution is about 1 mol l^{-1}. The copper ion is Cu^+ in the complex and $E^{\ominus}$ for this ion is $+0.55$ V.

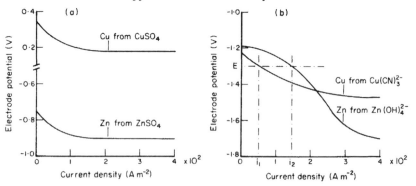

Figure 6.20 Electrode potential–current density (polarisation) curves for copper and zinc ions: (a) from sulphate electrolytes; and (b) from cyanide solutions

The zinc complex ion is relatively weak, however, and the activity of Zn^{2+} ions in solution is not as markedly affected.

It should be noted that at a given potential the corresponding values of i_1 and i_2 give values of the current flowing, the cathode area being the same in each case. Therefore it is possible to determine the masses of each component deposited in a given time by using Faraday's first law (equation (6.22)). Correct choice of working cathode potential ensures the deposition of the correct proportion of each element.

The more reactive metals such as tantalum and niobium require fused salt electrolytes for electroplating but such processes are only pursued by highly specialised companies. Organic electrolytes have also been used to electrodeposit the more reactive elements but only plating aluminium from diethyl ether containing $AlCl_3$ has any commercial significance.

Metal ions in solution as simple or complex ions tend to be poor conductors of electricity and aqueous electrolytes usually have a stray acid or alkali component to increase conductivity. Many plating solutions need to operate over fairly narrow ranges of pH. The effect of pH on plating, e.g. of nickel, may be considered by consulting the appropriate Pourbaix diagram (*Fig. 6.15*). If the solution is too alkaline $Ni(OH)_2$ may be precipitated with the nickel. If the solution is too acidic hydrogen evolution may be too rapid. Therefore use is made of a boric acid–borate buffer solution to maintain the pH within the range 5 to 6. Other variables include temperature and agitation. Increasing either of these parameters reduces the concentration polarisation. Thus, most plating processes operate at about $50\,°C$ and some are also agitated using compressed air. Since plating processes operate at a limiting current density governed by concentration polarisation, reducing polarisation at a given potential enables a higher current density to be used. However, local increases in current density above the optimum level may produce a powdery, 'burned' deposit.

Addition agents are added to aqueous electrolytes for specific purposes. For example, impurity precipitation is brought about by addition of sodium polysulphide to zinc cyanide baths. Wetting agents (surfactants or depitters) may be added to promote bubble removal from the cathode, e.g. in nickel and chromium plating. If a bubble adhered to a cathode metal would not be able to deposit in that region and discontinuities would result. Organic levelling agents may be added to improve the microthrowing power, facilitating deposition in the microscopic ridges rather than on the peaks. Brightening agents may also be added. Where the substrate is bright, e.g. bright nickel, a brightener is not required as a thin electrodeposited layer on top maintains the brightness. In general brightening agents are organic compounds based on N or S groups, such as amines or sulphides. The concentration of brightener is somewhat critical since too much may permit deposition of it which makes the bright deposit brittle. The principal purpose of electrolytic brightening is to remove the need for a post-plating polishing operation. A common feature of all these addition agents is that they are consumed during plating and the electrolyte must be replenished either continuously or at intervals. The mechanisms by which the more complex of these agents function has not been established.

The mechanism of deposition at a cathode is often called electrocrystallisation. Once the ion has arrived at the double layer the process of crystallisation begins by the ion releasing ligands, i.e. water molecules ('solvation sheath') or complexing ions. The metal ion is adsorbed at the cathode surface and discharged to become an 'ad-atom'. The ad-atom then diffuses across the cathode surface until it becomes incorporated in the metal lattice at a surface irregularity. This procedure is similar to that envisaged for oxide formation (Section 9.2). Thus a monatomic layer is formed which will have grain boundaries. Subsequent depositon tends to occur by epitaxial growth, i.e. the atomic lattice of the substrate is maintained, where possible. This process requires matching atomic layers across the substrate–deposit interface and is called coherence. Clearly coherence implies strong bonding and therefore good adhesion between the metal and the deposited coating. Adherence is also affected by cleaning treatments prior to electrodeposition. If the atomic lattices of substrate and deposit are very different, or impurities are present, or high growth rates (high current densities) are maintained epitaxial growth is prevented. The initial deposited monolayer may still be coherent but the growth pattern favours the lattice of the deposited metal, indeed at high current densities even this may not be possible. In the case of electroforming coherence is not required. Metal is deposited electrolytically on a mandrel of the desired shape which may easily be stripped at the end of the process. Many deposits are formed with residual tensile or compressive stresses. With more ductile deposits these present no problem but in the extreme case cracking or peeling may occur. Stresses may occur by the presence of impurities in the deposit or mismatch of lattice parameters. Sometimes stress-reducing addition agents may be added to the electrolyte to overcome the problem. Anodes in electroplating are of the same two basic types used in electrowinning, namely, soluble (e.g. copper) or inert (e.g. lead or lead alloy for chromium plating). Simple arrangements of anode and cathode are satisfactory for simple shapes but where the component to be electroplated has a more complex shape an auxiliary anode or a bipolar anode must be used (*Fig. 6.21*).

This procedure is necessary because throwing power effects, outlined previously, would otherwise produce a deposit at A and B which was much thicker than that at C. For some solutions the current density at the anode must be lower than that at the cathode. The surface area of the anode is therefore increased by increasing the perimetal length. This may be done by using anodes of oval cross section or, in more extreme cases, fluting the anode surface.

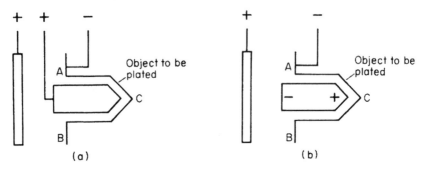

Figure 6.21 Principle of improving throwing power by using supplementary anodes: (a) auxiliary anode connected to DC supply; and (b) bipolar anode (not connected to DC supply)

Electroless plating, or autocatalytic plating, deposits a metal coating on a substrate by a controlled chemical reduction catalysed by the metal or alloy being deposited (ASTM B 374). Nickel, platinum, silver, copper and gold have been deposited by this process. Because of the chemical rather than electrical nature of the process, use of a suitable reducing agent enables metals to be deposited on plastics.

In order to produce a suitable 'keyed' (roughened) surface for the coating a preliminary etching process is required. Autocatalytic plating should not be confused with immersion (displacement or cementation) plating. In this case a base metal is dipped into a solution containing more noble metal ions and the standard displacement reaction ensues. In general such deposits have no practical value since they tend to be non-adherent and powdery but they do have commercial significance for tin.

6.5 Partition (distribution): of solutes between immiscible phases

When a solute, e.g. iodine, is shaken with two immiscible solvents, water and trichloromethane (chloroform), it dissolves to some extent in both solvents. The process is

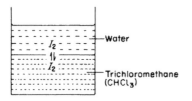

Figure 6.22 Distribution of I_2 between water and CHCl$_3$

known as distribution or partition. At a fixed temperature equilibrium is eventually established (see *Fig. 6.22*) and there is no further change in the solute concentration in either solvent. The equilibrium can be written:

$$X \, (\text{in} \, S_1) \, \rightleftharpoons \, X \, (\text{in} \, S_2)$$

where X is the solute and S_1 and S_2 are the solvents. For this equilibrium, the equilibrium constant, K_{dist}, known as the distribution or partition coefficient or distribution ratio is given by:

$$K_{dist} = \frac{a_{X \text{ in } S_2}}{a_{X \text{ in } S_1}} \approx \frac{[X \text{ in } S_2]}{[X \text{ in } S_1]}$$

where a refers to activity.

It has been found that the ratio of the concentrations of solute in the two solvents is indeed constant over a range of solute concentrations at a given temperature. If the solute exists in different molecular forms in the two solvents, this will of course not be true. For example, ethanoic (acetic) acid dimerises in ethoxyethane (ether) but not in water, i.e.

$$2CH_3COOH(\text{in water}) \rightleftharpoons (CH_3COOH)_2(\text{in ethoxyethane})$$

Then,

$$K_{dist} = \frac{[(CH_3COOH)_2 \text{ in ethoxyethane}]}{[CH_3COOH \text{ in water}]^2}$$

Distribution also occurs in liquid metal. For example, the distribution of Mn, C, S, Si and Cr between liquid iron and liquid slag in steelmaking (Section 7.2.10.6) and distribution ratios are available for several systems over a range of temperatures[21]. They can be used to determine the activities of elements in equilibrium in the two phases.

Example

The partition ratio for manganese between iron and slag (a_{Mn}/a_{MnO}), is 8.8 at $1600\,^{\circ}C$[21]. For a certain slag the manganese activity at $1600\,^{\circ}C$ is 0.1 (measured as the activity of MnO as liquid slags are ionic—see Sections 4.4.2 and 6.1.4). Calculate the activity of manganese in iron which is in equilibrium with the slag assuming unit carbon activity (carbon-saturated iron) and 1 atm CO pressure at $1600\,^{\circ}C$. Thus, for the reaction $MnO_{(slag)} + C_{(Fe)} \rightleftharpoons Mn_{(Fe)} + CO$

$$K_{dist} = \frac{a_{Mn}p_{CO}}{a_{MnO}a_C}$$

since $p_{CO} = 1$ atm and $a_C = 1$

$$K_{dist} = \frac{a_{Mn}}{a_{MnO}} = 8.8$$

and

$$a_{Mn} = 8.8 \times 0.1 = 0.88$$

6.5.1 Extraction

The equilibrium between iodine in water and chloroform is also established if pure trichloromethane is shaken with an aqueous solution of iodine. Some of the iodine transfers to the trichloromethane. The iodine is said to have been extracted with trichloromethane. This technique is called *solvent extraction* when applied to aqueous solutions and is particularly useful when a solute is to be obtained from a rather dilute solution. Using a small volume of a more efficient solvent a more concentrated solution is obtained. This is the situation for the iodine–water–trichloromethane system and also applies to some metallurgical systems. For example, zinc is extracted with di-2-ethyl hexyl phosphoric acid (see Section 7.3.2) from an aqueous solution. This technique can also be applied to liquid metal solutions, e.g. in the Parkes process[22] for the removal of silver from lead. The lead–silver alloy is melted with zinc in which the silver is more soluble and below a certain temperature liquid zinc and liquid lead are immiscible (see Section 7.2.10.3). After separation the zinc is distilled off leaving the silver. For each application it can readily be shown that extraction with a number of small volumes of solvent is more efficient than one large volume[6]. In fact, continuous circulation of fresh solvent is the most efficient procedure.

The leaching of ores (Section 7.3.1) is another example of a distribution system—distribution taking place between solvent (leachant) and ore.

Partition and activity

The distribution ratio can be utilised in determining the Raoultian activities for certain systems. The distribution ratio is given in terms of Henrian activities of the dilute solute in each solvent. If, however, Raoultian activities of the solvent are used it can be shown that the solute activities are the same in both solvents when equilibrium is established. The chemical potential of X (μ_X) * in the two solvents is given according to Section 2.2.3.6 by:

$$\mu_{X \text{ in } S_1} = \mu_X^\ominus + RT \ln a_{X \text{ in } S_1}$$

and

$$\mu_{X \text{ in } S_2} = \mu_X^\ominus + RT \ln a_{X \text{ in } S_2}$$

(cf. equations (6.1a) and (6.1b). At equilibrium $\mu_{X \text{ in } S_1} = \mu_{X \text{ in } S_2}$. Hence

$$a_{X \text{ in } S_1} = a_{X \text{ in } S_2}$$

For some systems, the activity coefficient of the solute in one solvent is considered to be unity, i.e. ideal solution. Thus, in this case,

$$a_{X \text{ in } S_1} = N_{X \text{ in } S_1} = a_{X \text{ in } S_2} \tag{6.27}$$

and the activity of X in S_1 is found by measuring the mole fraction, N, of X in S_2 when equilibrium between X in S_1 and S_2 has been established. This seems to be the case when aluminium distributes between silver and iron[23] at 1600 °C. The aluminium–silver liquid metal solution is considered ideal. Equation (6.27) also applies to a solid in equilibrium with a saturated solution of that solid in another solid. For example, pure carbon $(a_C = 1)$ is in equilibrium with a saturated solution of carbon in iron. At 1540 °C saturated iron contains[23] 5.2% C, and carbon activity in this solution is thus unity.

6.6   Ion exchange

Ion exchange is used for separating cations or anions in solution. Ion exchange materials are naturally occurring silicates or synthetic resins—cross-linked polymers (e.g. polystyrene), onto which ions are attached. A cation exchange resin has *anions* attached (e.g. SO_3^-) and allows exchange and hence separation of *cations*†. Similarly an anion exchange resin can be used for the separation of anions. Separation of ions is carried out by pouring a solution of ions through a tube or column containing the ion-exchange material in the form of beads (say 60–100 mesh) suspended in a suitable solution.

A cation exchange resin is often converted to the acid form $(R-X^-H^+)$ by mixing with an acid before use. In the presence of metal ions, the resin exchanges cations:

---

*μ_X is the partial molar free energy (chemical potential) of X in solution $(\bar{G}_X)$. At equilibrium the free energy of X in each solvent is equal.

†This is its counter-ion.

$$R-X^-H^+ + Na^+ \longrightarrow R-X^-Na^+ + H^+$$

and the metal ions thus become more or less strongly attached to the resin. Each ion can then be eluted (washed off) in turn using suitable solvents such as citric acid solution which complexes the rare earths. Those ions least strongly attached will be removed first and the others later.

Ion exchange forms the basis of ion exchange chromatography used for quantitative and qualitative analysis[24], for water softening (Ca^{2+} ions are exchanged for Na^+ ions), for the deionising of water (anions *and* cations exchanged for OH^- and H^+ respectively, which react to give water) and for the extraction of certain metals (Section 7.3.3).

6.7 Adsorption

6.7.1 Gibbs adsorption isotherm

Consider a solution containing solute X. The region close to the liquid–air interface (surface) where the solute concentration differs from that in the bulk is called the surface phase (*Fig. 6.23*). Let the number of moles of X in the surface phase *in excess*

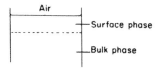

Figure 6.23 The surface phase

of the number in the same volume of bulk phase be n_X. Since the thickness of the surface phase is comparatively very small, it can be regarded as a plane. Let the surface area of this plane be A. The term *surface excess concentration*, Γ, can now be defined and given by:

$$\Gamma = n_X/A$$

The term Γ can be positive or negative as there can be an excess or a deficiency of X at the surface.

The first law of thermodynamics may be represented as:

$$dU = dq - dw \tag{6.28}$$

(cf. equation (2.10)*). For the surface phase,

$$dw = pdV - \gamma dA \tag{6.29}$$

where γ is surface tension. According to the second law, $G = H - TS$, where $H = U + pV$ (Section 2.2.1.1) thus

$$G = U + pdV - TS$$

which fully differentiated becomes,

$$dG = dU + pdV + Vdp - SdT - TdS \tag{6.30}$$

Substituting equations (6.28) and (6.29) into equation (6.30) gives,

$$dG = dq - pdV + \gamma dA + pdV + Vdp - SdT - TdS$$

Since $dq = TdS$, see equation (2.16)*,

$$dG = Vdp - SdT + \gamma dA$$

---

*It is acceptable to compare these equations bearing in mind that d indicates a small change whereas Δ indicates a large change.

and for an open system containing the single solute X

$$dG = Vdp - SdT + \gamma dA + \mu_X dn_X$$

where μ_X is the chemical potential of X in solution. At constant temperature and pressure,

$$dG = \gamma dA + \mu_X dn_X \qquad (6.31)$$

Integrating equation (6.31) gives

$$G = \gamma A + \mu_X n_X$$

Differentiating gives

$$dG = \gamma dA + A d\gamma + \mu dn_X + n_X d\mu_X \qquad (6.32)$$

Comparison of equations (6.31) and (6.32) shows that

$$A d\gamma + n_X d\mu_X = 0$$

Therefore,

$$d\gamma = -\frac{n_X}{A} d\mu_X$$

Substituting Γ for n_X/A, gives

$$d\gamma = -\Gamma d\mu_X \qquad (6.33)$$

But

$$d\mu_X = RT \, d(\ln a_X)$$

Substituting in equation (6.33) for $d\mu_X$, produces the Gibbs adsorption isotherm:

$$d\gamma = -RT\Gamma d(\ln a_X) \qquad (6.34)$$

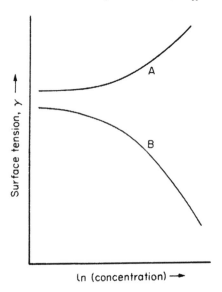

Figure 6.24 Variation of surface tension with ln (concentration) for two types of solute (see text). The two curves are not necessarily on the same scale

or for dilute solutions

$$d\gamma = -RT\Gamma d(\ln[X]) \tag{6.35}$$

The variation of surface tension with ln(concentration) is of two basic types (see *Fig. 6.24*).

Type A. The surface tension increases slightly with concentration. The slope of the γ versus ln $[X]$ plot is positive which means that Γ is negative. That is, there is a depletion of X in the surface phase. This type of behaviour is shown by electrolyte solutions in which the solute–solvent attraction is stronger than the solvent–solvent attraction.

Type B. Surface tension decreases with increasing concentration. The slope of the γ versus ln $[X]$ plot is negative and Γ is positive. This means that solute molecules tend to migrate to the surface. Substances of this type are called surface active agents (surfactants) and are composed of molecules of the type RX where R is a hydrophobic (water-hating) part such as a hydrocarbon chain and X is hydrophilic (water-liking) and is ionic or polar. Molecules of RX at the surface are orientated with the hydrophobic group located away from the water and the hydrophilic group in the water (see *Fig. 6.25*). An example of RX is butan-1-ol.

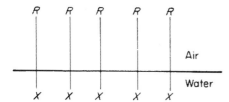

Figure 6.25 Disposition of a surface active agent RX at the surface of an aqueous solution. R = alkyl chain (hydrophobic), X is polar or ionic, e.g. $-COO^-$ (hydrophilic)

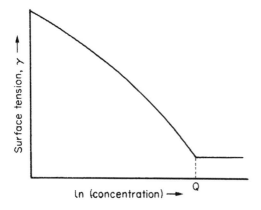

Figure 6.26 Variation of surface tension with ln (concentration) for a strongly adsorbed surface active agent. The point Q is the critical micelle concentration

Some substances are strongly adsorbed on the surface (*Fig. 6.26*). For example, laurylamine hydrochloride $(CH_3(CH_2)_{11} NH_3^+ Cl^-)$ and sodium dodecyl sulphate $(CH_3(CH_2)_{11} SO_4^- Na^+)$ at a certain concentration, completely cover the surface forming a monomolecular layer (monolayer). The surface excess concentration increases

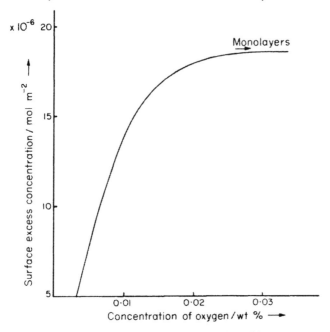

Figure 6.27   Variation of surface excess concentration with concentration for oxygen in iron at 1550 °C (after Kozakevitch[27])

with concentration reaching a maximum when monolayers are formed (*Fig. 6.27*). There is also a discontinuity in the γ versus $\ln[X]$ curve, at point Q in *Fig. 6.26*, when this happens and thereafter increasing concentration causes molecules of solute to associate to form globules, called *micelles*. For a solute containing 16 carbon atoms, for example, there are about 60 molecules in each micelle. The concentration at which micelles begin to form is the *critical micelle concentration* and for sodium dodecyl sulphate this occurs at about 0.01 mol dm^{-3} at 20 °C. Shaw[25] discusses micelles in more detail.

Surface active agents are made use of in froth flotation (Section 7.1.3), and in the study of surface tension of liquid metals (Section 4.2).

The phenomenon of surface activity is not restricted to aqueous solution but also occurs in liquid metals. The effect of the concentration of various solutes (e.g. phosphorus, sulphur) on the surface tension has been measured for liquid iron at 1500 °C. This, and work on similar systems (liquid copper, aluminium etc.), is reviewed by Richardson[26]. The variation of surface excess concentration with bulk concentration for the Fe-O and Fe-S systems at 1500 °C is also described. It seems that for the Fe-O system monolayers are formed when the bulk concentration of oxygen reaches[27] 0.04–0.05 wt%. Surface active agents have interesting effects on reaction rates in liquid metal systems. For example, the rate of loss of oxygen from liquid copper seems to be increased as a result of the surface activity of oxygen[26]. The latter is strongly adsorbed on the surface from which it evaporates. As it does so, fresh oxygen replaces it from the bulk phase. This process causes turbulence at the surface creating an increase in the effective surface area and hence an increased rate of loss of oxygen.

The rate of removal of nitrogen from liquid iron has been shown[28] to be decreased by the presence of oxygen. Kozakevitch and Urban[29] have shown that the absorption of nitrogen by liquid iron is similarly affected (*Fig. 6.28*). The effect is attributed to the physical blocking of oxygen atoms at the surface reducing the rate of loss of nitrogen

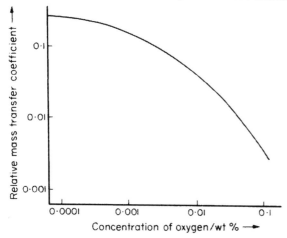

Figure 6.28 Variation of relative mass transfer coefficient with oxygen concentration for absorption of nitrogen by iron at 1550 °C (after Kozakevitch and Urbain[29])

atoms from the liquid. If this is so, the rate of reaction will be proportional to the fraction of uncovered surface, which can be shown to be proportional to the reciprocal of the oxygen concentration when the fraction of surface covered is nearly unity. Thus, the rate should be proportional to $1/[O]$. Turkdogen[30] has shown that this is indeed true for this system. Richardson[26] reviews the effects of surface active agents on the rates of various gas–metal reactions.

6.7.2 Adsorption of gases

When a solid is exposed to a gas, adsorption of the gas can take place. That is, the molecules of gas (the adsorbate) become more or less strongly attached to the surface atoms of the solid*. The bonding between gas and solid can be of the van der Waals type and the process is then called *physical adsorption* or *physisorption* or the bonding can be more chemical in nature and the process is then called chemical adsorption or *chemisorption*.

Physisorption

As the pressure of gas is increased the amount of gas adsorbed increases, rapidly at first, then more slowly as the surface becomes covered with adsorbed molecules. Eventually a monolayer forms. The variation of amount adsorbed plotted against gas pressure at a given temperature (the adsorption isotherm) for a system such as N_2 on silica is shown in *Fig. 6.29*. It should be noted that p/p_0 rather than pressure, p, is plotted on the x-axis. The saturated vapour pressure, p_0, of the gas is the pressure at which liquefaction occurs and p/p_0 is more useful than p because of the relationship between physisorption and liquefaction. Both are due to van der Waals forces. After monolayer formation (point M, *Fig. 6.29*), further layers of adsorbate form leading ultimately to liquefaction at $p/p_0 = 1$. Physisorption tends to occur at low temperatures and is reduced at higher temperatures. It is usually reversible.

---

*A discussion of adsorption on microporous solids will not be included here, but is usually included in more detailed accounts of adsorption.

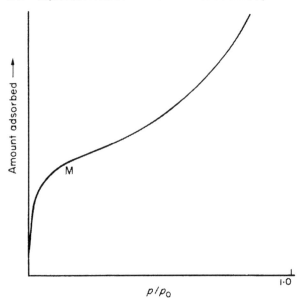

Figure 6.29 Adsorption isotherm for simple physisorption

Chemisorption

As with physisorption, the amount of gas adsorbed increases with increasing gas pressure (*Fig. 6.30*), but after the formation of monolayers no further adsorption occurs because the bonding between gas and solid is much stronger than the forces (van der Waals) between gas molecules. Thus, at pressures where monolayers form there is little tendency for multilayer formation. Unlike physisorption, chemisorption occurs at higher temperatures and is not always reversible. For example, H_2 is chemisorbed on nickel reversibly but attempts to remove O_2 chemisorbed on charcoal results in the production of CO.

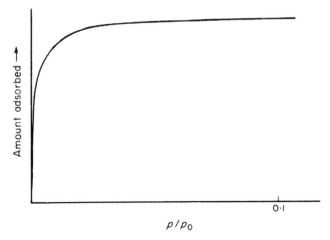

Figure 6.30 Adsorption isotherm for chemisorption

Other types of adsorption isotherms have also been identified[2] but are outside the scope of this book.

Theory of adsorption

There are three theories of adsorption in common use: the Langmuir theory, the Freundlich theory and the BET (Brunauer, Emmett and Teller) theory; these are fully discussed by Shaw[25]. Only the Langmuir theory of *monolayer formation* will be outlined here.

Langmuir theory (1916)

The Langmuir theory is based on the following assumptions:

(i)   There are a fixed number of adsorption sites and each site can hold one molecule (atom) of adsorbate.
(ii) The enthalpy change which accompanies adsorption is the same for all sites and is independent of the number of sites occupied.
(iii) There is no interaction between adsorbed molecules.

The rate of adsorption at a fixed temperature is proportional to the gas pressure, p, and the fraction of the surface which is still unoccupied. The fraction of surface covered by adsorbed molecules is θ, thus rate of adsorption is given by:

$$k_a p(1 - \theta)$$

where k_a is the rate constant for adsorption (see Section 3.2). The rate of desorption at the same temperature for the same system depends only on the fraction of surface covered, that is

$$\text{rate} = k_d \theta$$

where k_d is the rate constant for desorption. At equilibrium, the rate of adsorption and desorption are equal. Therefore,

$$k_a p(1 - \theta) = k_d \theta$$

Rearranging, this becomes:

$$\theta = \frac{bp}{1 + bp} \tag{6.36}$$

where $b = k_a/k_d$. Since the amount, x, adsorbed for a given sample or per unit mass or surface area of solid is proportional to θ, another form of this equation is also used:

$$x = \frac{ap}{1 + bp} \tag{6.37}$$

where $a = \text{constant} \times b$. It can easily be shown that,

$$\frac{1}{\theta} = 1 + \frac{1}{bp}$$

and

$$\frac{p}{x} = \frac{1}{a} + \frac{bp}{a}$$

and plots of $1/\theta$ versus $1/p$ and p/x versus p give straight lines for many systems where monolayers are formed showing that in these cases the Langmuir theory agrees with experimental data. The other theories mentioned above are generally used when there is multilayer formation. The theories of adsorption discussed above have also been applied to the adsorption on solids of solutes from solution.

REFERENCES

1. Nelkon, M. and Parker, P. 1982 *Advanced Level Physics*, 5th edn. London: Heinemann Educational.
2. Moore, W.J. 1972 *Physical Chemistry*. London: Longman.
3. Robbins, J. 1972 *Ions in Solution*, Vol. 2: *An Introduction to Electrochemistry*. London: Oxford University Press.
4. Salzberg, H.W., Morrow, J.I., Cohen, S.R. and Green, M.E. 1970 *Physical Chemistry, A Modern Laboratory Course*. New York: Academic Press.
5. Robinson, R.A. and Stokes, R.H. 1970 *Electrolyte Solutions*. London: Butterworths.
6. Heys, H.L. 1985 *Physical Chemistry*, 5th edn. London: Harrap.
7. Bell, R.P. 1971 *Acids and Bases: Their Quantitative Behaviour*. London: Chapman and Hall.
8. Mattock, G. 1961 *pH Measurement and Titrations*. London: Heywood and Co.
9. Potter, E.C. 1961 *Electrochemistry*. London: Cleaver-Hume Press.
10. Wilson, A. 1970 *pH Meters*. London: Kogan Page.
11. *Findlay's Practical Physical Chemistry*, 1973, revised B.P. Levitt. London: Longman.
12. Tyler, F. 1977 *A Laboratory Manual of Physics*, 5th edn. London: Edward Arnold.
13. Bockris, J. O'M. and Reddy, A.K.N. 1970 *Modern Electrochemistry*. London: MacDonald.
14. Delimarskii, Yu. K. and Markov, B.F. 1961 *Electrochemistry of Fused Salts*. Sigma.
15. Laity, R.W. 1962 *J. Chem. Educ.*, 39, 67.
16. Sienko, M.J. and Plane, R.A. 1963 *Physical Inorganic Chemistry*. New York: Benjamin.
17. Denaro, A.R. 1960 *School Sci. Rev.*, 42, 389.
18. Denaro, A.R. 1960 *School Sci. Rev.*, 43, 135.
19. MacInnes, D.A. 1961 *The Principles of Electrochemistry*. New York: Dover.
20. *Handbook of Physics and Chemistry* 1987–8, 68th edn, ed. R.C. Weast. Cleveland: Chemical Rubber Co.
21. Bodsworth, C. and Bell, H.B. 1972 *Physical Chemistry of Iron and Steel Manufacture*. London: Longman.
22. Alcock, C.B. 1976 *Principles of Pyrometallurgy*, p.316. London: Academic Press.
23. Biswas, A.K. and Bashforth, G.R. 1962 *The Physical Chemistry of Metallurgical Processes*. London: Chapman and Hall.
24. Browning, D.R. 1973 *Chromatography*. London: Harrap.
25. Shaw, D.J. 1981 *Introduction to Colloid and Surface Chemistry*, 3rd edn. London: Butterworths.
26. Richardson, F.D. 1974 *Physical Chemistry of Melts in Metallurgy*, Vol. 2. London: Academic Press.
27. Kozakevitch, P. 1968 *Surface Phenomena in Metals, Soc. Chem. Ind. Monogr.* 28, p.223.
28. Pehlke, R.D. and Elliott, J.F. 1963 *Trans. AIME*, 227, 844.
29. Kozakevitch, P. and Urbain, G. 1963 *Mémoires Scientifique de la Revue Metallurg.*, 60, 143.
30. Turkdogen, E.T. 1970 *Basic Oxygen Steelmaking, AIME Metallurg. Soc. Monogr.*, p.115.

FURTHER READING FOR SECTION 6.4

Biswas, A.K. and Davenport, W.G. 1979 *Extractive Metallurgy of Copper*. Oxford: Pergamon Press.
Boldt Jnr J.R. 1967 *The Winning of Nickel*. London: Methuen.
Gilchrist, J.D. 1979 *Extraction Metallurgy*. Oxford: Pergamon Press.
Lowenheim, F.A. 1978 *Electroplating*. New York: McGraw-Hill.
Parker, R.H. 1977 *An Introduction to Chemical Metallurgy*. Oxford: Pergamon Press.
Rosenqvist, T. 1983 *Principles of Extractive Metallurgy*. McGraw-Hill Kogakusha.
The Canning Handbook on Electroplating 1982 London: W. Canning & Co. Ltd.

For self-test questions on this chapter, see Appendix 5.

7

Metal Extraction Processes

Metals occur in the earth's crust in the following chemical states:

(i)   oxides, e.g. Fe_2O_3, TiO_2, Cu_2O, SnO_2;
(ii)   sulphides, e.g. PbS, ZnS, Cu_2S, Ni_2S_3, HgS;
(iii) oxysalts, such as silicates, sulphates, titanates, carbonates, e.g. $FeCO_3$, $ZrSiO_3$;

and to a lesser extent in other forms such as in 'native' (elemental) form or as arsenides, e.g. $PtAs_2$.

By far the most important groups with respect to quantity and occurrence are sulphides and oxides. Apart from the precious metals (Au, Ag, Pt), metals rarely occur in the uncombined or 'native' form. This is due to the reactive nature of metals which combine with the environment producing such compounds as oxides and sulphides. Precious metals are least reactive as can be seen from the appropriate $\Delta G^{\ominus}-T$ diagrams (*Figs 2.11* and *2.13*).

Metal ores are concentrations of the above metal compounds associated with other unwanted minerals (*gangue*) such as silicates. It is therefore necessary to separate the metal-bearing component (*value*) from the unwanted gangue prior to the extraction process. If this is not done the subsequent extraction of the metal will be less efficient, more costly and it will be more difficult to produce the metal in a state of high purity.

Once the metal-bearing constituent and the gangue have been separated the *concentrate* produced can be subjected to the extraction process. There are three main routes that can be used:

(i)   pyrometallurgy—incorporates smelting, converting and fire refining of the metal concentrate;
(ii)   hydrometallurgy—provides the metal in the form of an aqueous solution followed by subsequent precipitation of the metal;
(iii) electrometallurgy—uses electrolysis to extract the metal. Electrowinning is the extraction of the metal from the electrolyte while electrorefining is the refining of the impure metal which is in the form of the anode.

The choice of extraction route will largely depend on the cost per tonne of metal extracted which, in turn, will depend on the type of ore, availability and cost of fuel (coal, oil, natural gas or electricity), production quantity and rate and the required metal purity. Unless cheap hydroelectric power is available or the metal is highly reactive, such as aluminium, electrometallurgy is an expensive route but it usually provides the metal in an extremely pure state of about 99.9+ %. Electrorefining is often used as a final refining process after pyrometallurgical extraction. Hydrometallurgy tends to be slower than pyrometallurgical extraction and reagent costs tend to be high but it is ideal for extracting metals from lean ores. Again, a final electrorefining process is often adopted. Owing to the abundance and relatively low cost of fossil fuels (oil, coal, coke and natural gas) and the fact that pyrometallurgy is more adaptable than hydrometallurgy and electrometallurgy to high production rates, pyrometallurgical processes provide the main routes for the extraction of metals. For this reason the major portion of the present chapter will be spent on pyrometallurgical extraction.

Thermodynamics and reaction kinetics are used to determine the most suitable process steps, heat balances (enthalpy) and reaction rates but material balances, that is the calculation of the required amount of input raw materials to give the necessary production rates, must also be deduced. Often, material and heat balances give valuable information regarding process efficiency and indicate areas where improvements can be made.

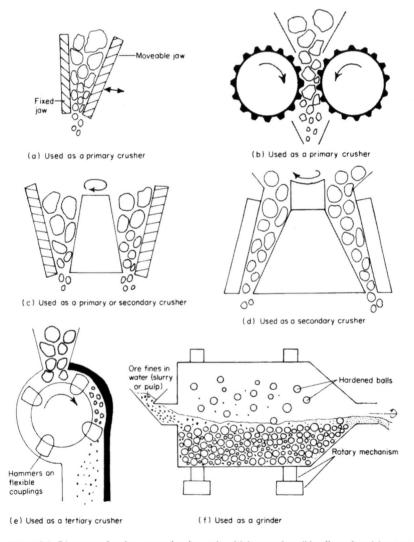

Figure 7.1 Diagrams of various comminution units: (a) jaw crusher; (b) roll crusher; (c) gyratory crusher; (d) cone crusher; (e) hammer mill; (f) ball mill

7.1 Ore preparation and mineral processing

The metallic ore must be concentrated with respect to the metal to be extracted to facilitate the subsequent extraction process. This is achieved by subjecting the ore to a series of comminution (crushing and grinding), screening and separation operations[1,24] with the aim of producing a mineral concentrate in which the element to be extracted is present to approximately 25 wt% or more.

7.1.1 Comminution processes

The comminution processes involve crushing the ore into fine particles to aid separation of the metal-bearing components (value) from the unwanted gangue. The reduction ratios (entry size: exit size) vary usually between 3:1 to 4:1. Thus, depending on the final size required there are a series of crushing operations using primary, secondary and tertiary crushers, with a final grinding or milling operation if a very fine particle size is required, e.g. 6 mm for sintering or less for froth flotation separation or pelletising. Owing to the fact that some fines are produced with each crushing operation irrespective of the reduction ratio a screening operation is used between each crusher, removing the fines which would not be further reduced in subsequent crushing or grinding operations. Various types of crushers and milling operations are shown in *Fig. 7.1*.

7.1.2 Classification processes

The first stage of separation of value from gangue is by classification which separates different particles due to their different rates of travel through a fluid (normally water). Therefore, the particles will be separated due to their different densities, sizes and shapes. Normally, the lighter gangue particles will be collected as 'tailings' in an over-

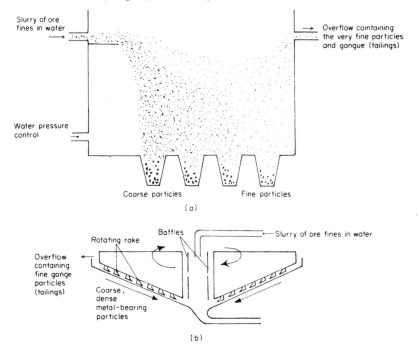

Figure 7.2 Diagrams of: (a) simple sluice box classifier and (b) a bowl or rake classifier

flow from the classifier while the denser metal-bearing particles will be collected as an underflow from the bottom of the classifier. Both portions will exist as a pulp of minerals in the water. A prolonged dwelling time in a fluid is required to separate the mineral particles from the water in the pulp when the ore has been finely ground. This is normally provided by a 'thickener' which is also used to reclaim the fluid from the pulp produced after grinding or to concentrate the solids in the pulp. Two types of classifier are shown diagrammatically in *Fig. 7.2*.

7.1.3 Separation processes

As classification is only a rough separation process further separation of the different mineral particles is required. This is achieved using the different physical properties of the various particles present, such as magnetic, electrical conductivity and surface properties. By far the most sensitive and most important separation technique is that of froth flotation which uses the difference in surface free energy of the particles (Section 4.2). Air bubbles are introduced into a cell containing the different mineral particles in water (pulp) and the fluid stirred to increase bubble formation (*Fig. 7.3*). The main aim

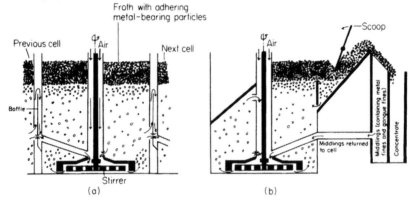

Figure 7.3 Diagram of a typical froth flotation cell: (a) front view; (b) side view

of the process is to achieve a selective adsorption of the required metal-bearing particles onto the air bubbles which rise to the top of the cell liquid effecting a separation from the gangue particles which fall to the bottom of the cell. This separation is achieved by producing conditions in the liquid which lower the interfacial energy between the required metal-bearing particles and the air bubbles, thereby providing 'wetting' of the air bubbles by the particles. Owing to the small differences in surface free energies between different particles certain reagents must be added to the cell liquid to ensure separation. These are called frothers, collectors and modifying agents[1] (activators, depressants, dispersants, regulators):

(i) Frothers are added to stabilise the air bubbles in the liquid. The life of an air bubble in water is about 0.01 s. Adding terpinol, a common frother, lowers the interfacial energy between the air bubble and the liquid allowing most of the air bubbles to rise without bursting, even on reaching the surface of the liquid. The bubbles with the adhering mineral particles can then be scooped off and the mineral separated.

(ii) Collectors offer selective adsorption between a certain mineral particle and the air bubble thereby lowering the interfacial energy and providing separation. Typical collectors include xanthates, thiocarbonates, thiocarbomates, dithiophosphates, used to float sulphide particles; fatty acids and soaps used to separate oxides; and amines used to separate silicates.

(iii)   Modifying agents intensify the action of the collector by either decreasing (activators) or increasing (depressants) the interfacial energy between a mineral particle and the air bubble, thereby increasing the selectivity of the process. Regulators control the pH of the solution preventing precipitation of compounds in the liquid which might seriously affect the separation efficiency. The addition of reagents to the cell liquid is carried out before or during the flotation process and is termed 'conditioning'. Only a few parts per million of total reagents are required to provide efficient froth flotation but most mineral particles must be ground between 200×10^{-3} mm and 5×10^{-3} mm and diluted to between 25–45% with water (pulp). Flotation cells are usually arranged in series with the liquid continuously passing from one cell to the next. The size of each cell depends on the total flotation time required and the space and handling arrangements available.

7.1.4   Agglomeration processes

Agglomeration is the reforming of fine particles into larger lumps and is usually conducted if:

(i)   the separation requirements of an ore necessitate grinding and froth flotation and the particles produced are too fine to be added directly to a reverberatory or blast furnace;

(ii)   a rich ore is too dense to be efficiently reduced due to poor contact between the reducing gas and the solid ore. Grinding the dense ore and reforming it into a more porous material of appropriate sized lumps improves the efficiency of the subsequent smelting process.

The main methods of agglomerating ore fines are sintering and nodulising conducted at elevated temperatures such that partial or incipient melting occurs, and pelletising and briquetting conducted at room temperature usually employing a binder. Pellets are subsequently fired at an elevated temperature to improve their mechanical properties.

7.1.4.1   Sintering

Ore fines of less than or equal to 6 mm diameter are thoroughly mixed with about 5–6% coal dust (heat producer) and about 10–12% water (improves permeability) and fed onto the grate of a sinterstrand (*Fig. 6.4*). The sinter bed of ore, and coal is passed under an ignition hood, fired by oil or natural gas, which ignites the coal in the bed.

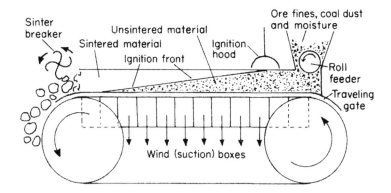

Figure 7.4 Diagram of a sinterstrand using downdraughting

The heat or ignition front is sucked down through the bed as it passes over several suction or wind boxes. The aim is to complete the sintering by the last but one wind box; the last wind box being used for cooling the sinter. The sinter is finally broken up into conveniently sized lumps.

The sintering process occurs in two stages:

(i) Calcination during which decomposition and drying take place removing H_2O, CO_2, SO_2 and also removing any volatile metals, e.g. Cd, As. These may be recovered with appropriate gas cleaning plant from the flue gases.

(ii) Incipient fusion during which the surface of the ore fines partially melts and the liquid is drawn into the interstices between the ore fines by capillary action. This increases the strength of the sinter.

Sintering is used for agglomeration of oxide and sulphide concentrates. The main difference is that in sintering sulphide ores updraughting (air blowing) rather than downdraughting (air being sucked through the bed) is used after ignition so that sulphur dioxide can be collected more easily and reprocessed to elemental sulphur or sulphuric acid. Sintering sulphide ores also provides a roasting operation in which the metal sulphides are converted to oxides.

7.1.4.2 Nodulising

Ore fines, coal dust and moisture are again thoroughly mixed and fed into a rotating drum at an elevated temperature—similar to that used for sintering. The rotating action and high temperature result in the formation of nodules in the same way that snowballs are produced by rolling snow flakes. However, due to the poor control over temperature, the lower melting point constituents of the concentrate tend to fuse resulting in irregularly shaped nodules of low permeability. For this reason nodulising is normally only used for agglomerating metallic fines produced from smelting, polishing and machining operations in the production of secondary precious metals (Section 7.2.2.7) where low melting point gangue constituents are not present.

7.1.4.3 Pelletising

Ore fines of less than about 0.20 mm are unsuitable for sintering due to the danger of complete fusion of such a small particle. These ore fines are thoroughly mixed with 10% water to improve permeability and a binder such a bentonite, lime, salts or organic material, and fed into an inclined rotating drum or disc. A nucleus of a number of fines is produced due to the rotating action, the particles being bonded together by the binder and water moving into the crevices between the particles due to capillary attraction. On further rotation each nucleus grows forming a pellet by collecting more fines in a similar 'snowballing action' described for nodulising. A slight pressure is also required but this is usually produced by the weight of the particles on each other. The final pellet size is controlled by the residence time in the drum which in turn is controlled by the rate of rotation of the drum or disc, the length or diameter of the drum or disc, the angle of inclination and any pressure applied. There is an optimum size of pellet for reduction in a blast furnace resulting from the best compromise between high permeability (large pellets) and high surface area (small pellets) which in turn is dependent on the blast furnace practice and composition of the concentrate. Iron oxide pellets are generally controlled between 10 to 25 mm in diameter.

The green pellet thus produced would be too soft to withstand the load of the charge above it in a blast furnace shaft and therefore needs to be hardened by firing on a travelling grate machine, similar to the sinterstrand. Updraughting is used since downdraughting would cause crushing of the soft green pellets. Alternatively, the green pellets may be fired in a small shaft furnace.

7.1.4.4 Briquetting

Ore fines are mixed with a similar binder material as that used for pelletising and mechanically pressed at room temperature. The briquetting may be subsequently hardened in a tunnel kiln. In some cases a binder is not necessary. Hot briquetting is also being used to a limited extent, mainly in the reprocessing of metal-bearing particles in flue dusts. Owing to the excessive wear on the press, briquetting tends to be an expensive process and has limited application.

The choice of pellet or sinter charges in blast furnace reduction processes depends on the composition of the ore, the permeability of the agglomerate and the total processing costs.

Generally pellets offer better reducibility than sinter during reduction smelting in a blast furnace due to the more favourable distribution of porosity in the pellets. The majority of the pores are concentrated at the surface of the pellet and therefore provide very high areas of contact initially at the lower temperature. When the pellet has been reduced at its surface the higher temperatures available lower down the blast furnace stack somewhat compensate for the reduced area of contact compared with sinter which has a more uniform distribution of porosity. Sintering provides favourable chemical effects such as the elimination of CO_2 and sulphur as SO_2 but sintering has a very low thermal efficiency due to the large amounts of air needed to be heated up and drawn through the sinter bed (air has a high specific heat–Section 2.1.4.1). Also, because of the finer particles required for pelletising, the preceding separation process will be more efficient, usually resulting in a higher metal content than in sinter. However, in both cases the increased comminution and agglomeration costs have to be justified by an increase in the production of metal.

The addition of a flux material in the sinter or pellet mix will provide a self-fluxing agglomerate.

7.2 Pyrometallurgical extraction processes

Metal sulphides, carbonates, hydrates and basic carbonates are not normally convenient starting materials for the extraction of the metal. Most metal sulphides are not reduced with carbon or hydrogen, the most common and readily available reducing agents, since CS_2 and H_2S have less negative values for $\Delta G^{\ominus}$ of formation (*Fig. 2.13*) than most metal sulphides. For instance,

$$2Cu_2S + C \longrightarrow 4Cu + CS_2 \qquad \Delta G^{\ominus}_{1100°C} = +160 \text{ kJ mol}^{-1}$$
$$Cu_2S + H_2 \longrightarrow 2Cu + H_2S \qquad \Delta G^{\ominus}_{1100°C} = +130 \text{ kJ mol}^{-1}$$

At the same time sulphides, carbonates, hydrates and basic carbonates are not soluble in water thereby rendering them unsuitable for a hydrometallurgical extraction route involving leaching and precipitation.

It is, therefore, more convenient to convert these ore concentrates into oxides which are readily reduced with carbon or hydrogen, or into sulphates which are soluble in water or sulphuric acid.

7.2.1 Drying and calcination

The simplest initial treatment to remove water which is physically combined with the concentrate is by heating to above the boiling point of water (drying). The evaporation of water is endothermic ($+44$ kJ mol^{-1}) and thus requires heat to bring the mineral to the appropriate temperature and for the evaporation process.

Calcination involves the chemical decomposition of the mineral and is achieved by heating to above the mineral's decomposition temperature (T_D) (Section 2.2.3.10) or by reducing the partial pressure of the gaseous product (p_{H_2O}, p_{CO_2}) below that of its equilibrium partial pressure for a certain constant temperature. For example,

$$CaCO_3 \longrightarrow CaO + CO_2 \qquad T_D = 900\,^{\circ}C \text{ (under standard thermodynamic conditions)}$$

Calcination is mainly used to remove water, CO_2 and other gases which are chemically bound in metal hydrates and carbonates as these minerals have relatively low decomposition temperatures.

Drying and calcination are conducted in rotary kilns, shaft furnaces or fluidised bed furnaces.

7.2.2 Roasting of metal concentrates

The most important roasting reactions are those concerning metal sulphide concentrates and involve chemical combination with the roasting atmosphere. Possible reactions include:

$$2MS + 3O_2 \longrightarrow 2MO + 2SO_2 \quad \text{(dead roast)} \qquad \text{where } K = \frac{a_{MO}^2 \, p_{SO_2}^2}{a_{MS}^2 \, p_{O_2}^3}$$

$$MS + 2O_2 \longrightarrow MSO_4 \quad \text{(sulphating roast)} \qquad \text{where } K = \frac{a_{MSO_4}}{a_{MS} \, p_{O_2}^2}$$

$$MS + O_2 \longrightarrow M + SO_2 \quad \text{(reduction roast)} \qquad \text{where } K = \frac{a_M \, p_{SO_2}}{a_{MS} \, p_{O_2}}$$

Other equilibria which need to be taken into account include:

$$\tfrac{1}{2}S_2 + O_2 \longrightarrow SO_2 \quad \text{and} \quad SO_2 + \tfrac{1}{2}O_2 \longrightarrow SO_3$$

Thus, when p_{SO_2} is large and p_{O_2} is low, p_{S_2} becomes large. Also when p_{O_2} and p_{SO_2} are large p_{SO_3} becomes large which is the required condition for sulphating roasting; the sequence of reactions being,

$$MS + \tfrac{3}{2}O_2 \longrightarrow MO + SO_2$$
$$SO_2 + \tfrac{1}{2}O_2 \longrightarrow SO_3$$
$$MO + SO_3 \longrightarrow MSO_4$$

giving the overall sulphating reaction,

$$MS + 2O_2 \longrightarrow MSO_4$$

If the metal forms several sulphides, oxides, sulphates and basic sulphates, e.g. M_2S, M_2O_3, $M_2(SO_4)_3$, MSO_4, xMO, further equilibria must be considered. By examination of the equilibrium constant for each of these roasting reactions it is possible to determine the values of p_{O_2}, and p_{SO_2} at which each of the roasting products (calcine) is in equilibrium with the metal sulphide at a constant temperature. Increasing or decreasing the p_{O_2} or p_{SO_2} results in either the calcine or metal sulphide being stable. Further changes in p_{O_2} and p_{SO_2} may produce other roasting reactions. Kellog[2] has used this criterion to construct thermodynamic phase diagrams for the roasting reactions at a constant temperature (*Fig. 7.5*). Reactions which involve both SO_2 and O_2 are seen to have a diagonal line since the equilibrium phases produced will depend on the partial pressure of both gases. The roasting of a metal sulphide to a sulphate or a basic sulphate will produce a vertical reaction line since only O_2 will take part in the reaction. The 'Kellog diagram' provides the *predominance areas* for each phase within which p_{O_2} and p_{SO_2} can be varied without altering the roasting product or calcine. It should be noted that if roasting is carried out in air the sum of the partial pressure of O_2 and SO_2 is about 0.2 atm, i.e. $p_{O_2} + p_{SO_2} = 0.2$ atm. (Compare with Section 7.3.2.1.)

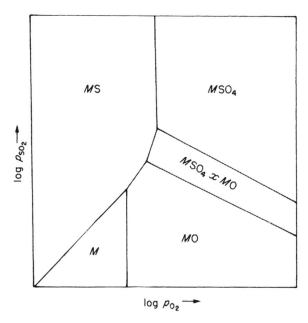

Figure 7.5 Hypothetical thermodynamic phase diagram for the roasting of a metal sulphide concentrate at a constant temperature. (Adapted from *Fig. 8.8* of *Principles of Extractive Metallurgy* by T. Rosenqvist, © 1974, McGraw-Hill Inc. Used with permission of McGraw-Hill Book Co.)

The ore concentrate will undoubtedly contain other metal sulphides each with their own phase predominance areas dependent upon p_{O_2}, p_{SO_2} and temperature. By examination of the appropriate diagrams it is possible to obtain information which will enable the roasting operator to select the most appropriate roasting conditions for each particular concentrate. In this way selective roasting of one metal sulphide to its oxide may be achieved while another metal present in the concentrate remains as the sulphide.

One of the most common roasting operations is that of *partial roasting* in which the metal concentrate containing several metal sulphides is roasted to preferentially oxidise the impurity metal sulphides and to reduce the level of sulphur present. The metal oxides thus formed can be separated and collected in an appropriate slag on subsequent smelting. Thus partial roasting is conducted on the more noble metal sulphides from which impurity sulphides are readily oxidised. The $\Delta G^{\ominus}-T$ relationships for the more common sulphide roasting reactions are given in *Fig. 7.6*. The reaction lines lower down the diagram indicate those metal sulphides which are more readily converted to an oxide.

A dead roast is used when the metal oxide is to be reduced by carbon or hydrogen. A sulphating roast is used when the metal sulphate is subsequently leached with a dilute sulphuric acid solution. Metal sulphates decompose at low temperatures therefore sulphating roasting is normally conducted at about 600–800 °C, i.e. below the corresponding decomposition temperature, with a restricted amount of air while dead roasting is conducted at 800–900 °C with excess air, i.e. a high p_{O_2}/p_{SO_2} ratio. Differential sulphating roasting is possible by operating at a temperature which will decompose one sulphate but not another. Thus, roasting a Ni_3S_2–FeS concentrate (*Fig. 7.6*) at 850 °C will produce $NiSO_4$ and Fe_2O_3 since $Fe_2(SO_4)_3$ will decompose at this temperature, i.e. $Fe_2(SO_4)_3 \longrightarrow Fe_2O_3 + 3SO_3$. Reduction roasts are generally rare since this reaction

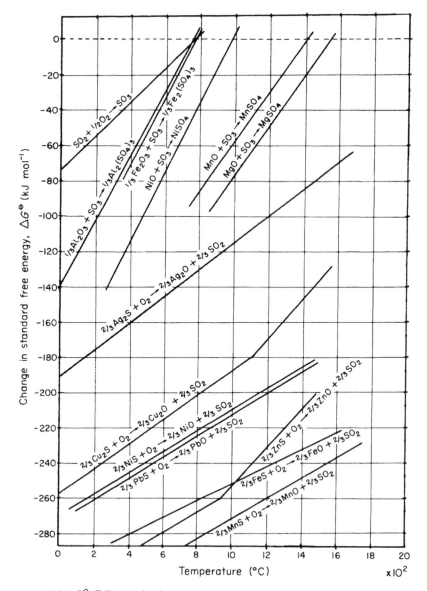

Figure 7.6 $\Delta G^{\ominus}$-T diagram for the more important roasting reactions

usually requires very low p_{O_2} values and high temperatures demanded by the thermodynamic considerations. Other roasting reactions include:

(i)   Chloridising roast which is generally used for the conversion of a reactive metal such as Ti, Zr, U, which forms extremely stable oxides, to a less stable chloride or other halide. The halide is relatively easy to reduce with another element which forms a more stable halide.

In the production of titanium from TiO_2-containing concentrates, the TiO_2 is subjected to a chloridising roast in the presence of carbon at 500 °C:

$$TiO_2 + C + 2Cl_2 = TiCl_4 + CO_2 \quad \Delta G^{\ominus}_{500°C} = -295 \text{ kJ}$$

Since $TiCl_4$ is thermodynamically less stable than TiO_2, carbon must be added to make the $\Delta G^{\ominus}$ for the reaction more negative. This is the case in most chloridising roast operations, i.e.

$$TiO_2 + 2Cl_2 \longrightarrow TiCl_4 + O_2 \quad \Delta G^{\ominus}_{500°C} = +97 \text{ kJ}$$

$TiCl_4$ gas is subsequently reduced with magnesium (Kroll process) at about 850 °C as $MgCl_2$ is more stable (more negative $\Delta G^{\ominus}$ of formation) than $TiCl_4$.

$$\underset{\text{(gas)}}{TiCl_4} + \underset{\text{(liquid)}}{2Mg} \longrightarrow \underset{\text{(solid sponge)}}{Ti} + \underset{\text{(liquid)}}{2MgCl_2}$$

Sodium and calcium have also been used to reduce $TiCl_4$. Liquid $MgCl_2$ and excess Mg are removed by vacuum distillation. Iodides (ZrI_4) and fluorides (UF_4) are also used in extraction of the more reactive metals; iodides are decomposed at relatively low temperatures while fluorides may be reduced with Na or Ca as above.

(ii) Volatilising roasts remove volatile impurity elements and oxides such as Cd, As_2O_3, Sb_2O_3, ZnO. These may be recovered from the process fume using bag filters.

(iii) Magnetising roast using controlled reduction of haematite (Fe_2O_3) to magnetite (Fe_3O_4) which can be subsequently magnetically separated from the gangue.

Drying and calcination will also take place as the temperature of the ore concentrate is increased in the initial stages of the roasting operation.

The roasting reactions are gas–solid reactions and therefore rely on the diffusion of oxygen into and sulphur dioxide out of each concentrate particle. Thus, if the metal sulphide forms several oxides, concentric layers of different oxides are produced with the possibility of some unreacted sulphide at the centre. As sulphates occupy larger

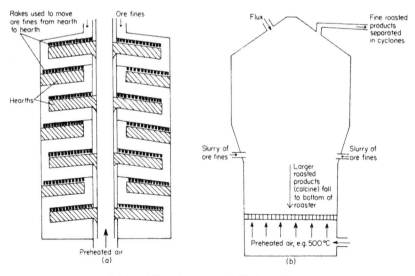

Figure 7.7 Diagrams of: (a) multihearth roaster and (b) fluo-solids roaster

volumes than oxides, sulphating roasting is slower than oxide formation due to the reduced diffusion rate through the more dense sulphate layers, usually resulting in a relatively small amount of sulphate in the roast product or calcine. As a draught is required to assist diffusion rates, the sinterstrand is an ideal roasting unit. The multi-hearth roaster is also a common roasting unit in which the ore fines are passed from a series of higher hearths to lower hearths against a counter-current flow of the hot air; most of the roast reactions occurring as the ore fines pass between the hearths.

Reducing the particle size (less than 6 mm) improves the gas–solid contact and increases throughput. This principle is used in the modern flash roasters in which pre-heated ore particles are injected through a burner with air, and fluidised bed roasters in which the fine ore particles are suspended in the roasting gas. A development which incorporates both these principles is the fluo-solids roaster in which air and ore fires are injected into the side of a reactor and fluidised by an upward draught of preheated $(500\,^\circ C)$ air.

It has been claimed[3] that the fluo-solids roaster offers certain advantages over the multihearth unit for the partial roasting of Cu_2S ores. Such advantages include greater control over sulphur elimination, less space and operating labour required, it is more amenable to automation and a higher quality matte from the reverberatory furnace is produced. These roasting units are shown diagrammatically in *Fig. 7.7*.

7.2.3   Smelting

Smelting is mainly a process of melting and separation of the charge into two immiscible liquid layers, i.e. liquid slag and liquid *matte* or liquid metal. Mattes are mixed molten sulphides of heavy metals and will therefore be produced on smelting metal sulphide concentrates, normally under neutral or slightly reducing conditions. A third immiscible liquid layer may be present if arsenic and antimony are present in the concentrate by the formation of a *speiss* which is a mixture of molten arsenides and antimonides of heavy metals – particularly iron, cobalt and nickel.

Metal oxide concentrates are smelted under highly reducing conditions with carbon or some other suitable reducing agent thereby producing a liquid slag–liquid metal separation. In both cases smelting provides a liquid–liquid separation process which may be controlled so that a considerable amount of the impurities will separate to and be collected in the slag, leaving the matte or metal richer in the metal required to be extracted. Matte smelting is therefore a concentrating stage in the overall extraction of a metal from its sulphide ore while metal oxide smelting is the main extraction process producing an impure metal which must be subsequently refined. Needless to say that in both matte and metal oxide smelting choice of slag composition to give the optimum balance of basicity and fluidity is extremely important so that maximum removal of impurities takes place. Metal sulphides have lower melting points than metal oxides, therefore, matte smelting can be conducted at lower temperatures than metal oxide smelting. Reverberatory furnaces are used where a neutral or only a slightly reducing condition is required for the smelting operation. Thus, matte smelting is nor-mally carried out in reverberatory furnaces (*Fig. 7.8(a)*) which are so shaped to allow the combustion products of the fuel (oil, gas, pulverised coal) to be reflected or 'reverber-ated' from the inclined furnace roof on to the charge. The long, wide hearth allows good slag–matte separation, but the process tends to be slow. Electric arc furnaces are also used to a lesser extent (*Fig. 7.8(b)*) mainly when higher temperatures $(1500\,^\circ C+)$ are required. The modern flash smelting units (*Fig. 7.8(d)*) provide an improvement in the production rate by incorporating the principles of flash roasting with smelting. Flash smelting injects finely ground ore together with oxygen or preheated air and a flux into a smelting unit similar to a reverberatory furnace.

The ore fines are superheated in the furnace atmosphere above the slag and the fine particle size produces high areas of contact between the roasting gas, flux and concen-trate. The oxidised product subsequently settles to provide a liquid slag–liquid matte separation. The concentrate composition is normally of a low grade; the impurities

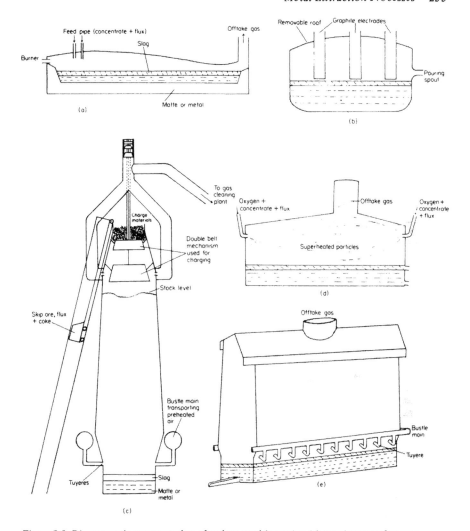

Figure 7.8 Diagrammatic representation of various smelting units: (a) reverberatory furnace; (b) direct electric arc furnace; (c) circular blast furnace; (d) flash smelter furnace; (e) rectangular blast furnace

providing sufficient exothermic reactions during the oxidising process to negate the need for using a fuel. The process is then said to be *autogenous* or self-supporting with respect to heat.

Where the concentrate is in a lumpy form and/or a high reduction potential is required a blast furnace (*Fig. 7.8(c)*) is used for smelting. The blast furnace is an ideal unit for metal oxide smelting with coke when the lump ore is variable in size and composition, although maximum efficiency is achieved using a properly graded charge. Preheated air is introduced through tuyeres at the base of the shaft or stack of the blast furnace. The air reacts with coke to produce CO which acts as the reducing gas. In this instance the coke is used as an indirect reducing agent since the coke itself does not perform the reduction but is converted to CO. The CO reduces the metal oxide to the metal and is itself oxidised to CO_2, i.e.

$$MO + CO \longrightarrow M + CO_2$$

Lower down the stack towards the hearth direct reduction between the coke and the metal oxide increases. The CO/CO_2 ratio is used as a major control parameter in blast furnace smelting of metal oxides. The combustion of the coke also provides the necessary heat, while the high strength of coke at temperatures up to $2000\,^{\circ}C$ allows full support of the charge in the stack of the blast furnace.

A blast furnace of circular cross section is used for metal oxide smelting with tuyeres evenly spaced around the base of the stack, just above the slag. This arrangement produces temperatures sufficient to melt the high melting point oxides as they approach the tuyeres. The ore, coke and flux material are charged to the furnace at the top of the stack or shaft. The charge is dried and subsequently reduced as it descends. This provides a counter current technique in which the reducing gases rise through the descending ore, creating excellent contact and favourable reaction rates.

For the lower boiling point metals such as Pb and Zn, a blast furnace of rectangular cross section is used with tuyeres placed only along the two long sides (*Fig. 7.8(e)*). This produces a less severe build-up in temperature at the tuyeres and reduces the danger of metal loss through volatilisation of the low boiling point metals. When the blast furnace is used for matte smelting of metal sulphides only sufficient coke is charged to provide the necessary furnace atmosphere and heat since a reduction reaction is not required. The success of the matte smelting operation is judged by the *fall* and *grade* of the matte:

$$\text{fall of matte} = \frac{\text{weight of matte}}{\text{total weight of materials charged}} \times 100\%$$

$$\text{grade of matte} = \frac{\text{weight of metal (to be extracted) in matte}}{\text{total weight of matte}} \times 100\%$$

Some metal oxide smelting, e.g. PbO, is also conducted with coke or anthracite in reverberatory furnaces. Air blowing may also be used if a sulphide concentrate is charged. To aid the rate of the smelting operation, a temperature is chosen which provides a free-running slag. This is often considerably greater than the minimum reduction temperature for metal oxide smelting or the melting point of the metal or matte produced. Reaction rates are also improved in hearth furnace smelting by mechanical rabbling of the melt.

Since electric arc smelting is used where high temperatures (in excess of $1500\,^{\circ}C$) are required, the main application of electric smelting is therefore in the production of ferro alloys, Ni_3S_2 smelting, and in ironmaking areas where electricity is cheap and production requirements are too low for the use of a blast furnace process. For the larger furnace capacities six graphite electrodes may be used.

The product of matte smelting provides the metal to be extracted in the form of a concentrated metal sulphide which needs to be converted to the metal, whereas the product of metal oxide smelting provides the metal to be extracted in the form of an impure metal which needs to be refined. As a general rule the more noble metals, i.e. higher than the $\frac{2}{3}NiS + O_2 \rightarrow \frac{2}{3}NiO + \frac{2}{3}SO_2$ reaction line in *Fig. 7.6*, are normally subjected to a partial roasting–matte smelting–converting route since the more reactive impurity metals may be preferentially oxidised during partial roasting and separated to the slag on subsequent smelting.

7.2.4   Matte converting

Converting of the sulphide matte is achieved by blowing air and/or oxygen through the liquid matte effecting preferential oxidation of the more reactive impurity metal sulphides, e.g. MeS, to their respective oxides which are collected in an appropriate slag:

$$2MeS \; + \; 3O_2 \; \longrightarrow \; MeO \; + \; 2SO_2$$
$$\begin{pmatrix} \text{impurity} \\ \text{sulphide} \end{pmatrix} \qquad\qquad \begin{pmatrix} \text{impurity} \\ \text{oxide} \end{pmatrix}$$

This is normally conducted in a horizontal converter having tuyeres along one side (*Fig. 7.13*). The air blowing is controlled to convert the remaining more noble metal (less stable metal oxide) to the required metal, i.e.

$$MS \; + \; O_2 \; \longrightarrow M \; + \; SO_2$$

Owing to the larger volume of MS present compared with impurity metal sulphides, oxidation of MS to MO will take place initially followed by a *mutual reduction reaction* between MS and MO:

$$MS \; + \; 2MO \longrightarrow 3M \; + \; SO_2$$

In practice this is achieved by adding more MS to the converter or lowering p_{O_2}. Thus, oxidation of a metal sulphide to its metal is only possible where the mutual reduction reaction has a negative free energy change. This is the case for copper and lead above about $900\,^{\circ}C$ and more noble metals such as Au and Ag but not for iron. Nickel is an intermediate case in which the nickel sulphide may be only partially converted to metal (see Section 7.2.10.2).

7.2.5 Reduction of metal oxides

Metal oxides may be reduced to the metal by carbon, carbon monoxide, hydrogen or other metals which form more stable oxides (more negative ΔG for oxide formation) than the metal oxide to be reduced. If the oxides and the reduced phases are assumed to be pure solids and MO is the hypothetical metal oxide required to be reduced, the reaction lines:

$$2M(s) \; + \; O_2(g) \; \longrightarrow \; 2MO(s) \qquad\qquad\qquad\qquad\qquad (7.\text{I})$$

$$C(s) \; + \; O_2(g) \; \longrightarrow \; CO_2(g) \qquad \Delta H^{\ominus}_{298} \; = \; -397\,\text{kJ} \qquad (7.\text{II})$$

$$2C(s) \; + \; O_2(g) \; \longrightarrow \; 2CO(g) \qquad \Delta H^{\ominus}_{298} \; = \; -222\,\text{kJ} \qquad (7.\text{III})$$

$$2CO(g) \; + \; O_2(g) \; \longrightarrow \; 2CO_2(g) \qquad \Delta H^{\ominus}_{298} \; = \; -572\,\text{kJ} \qquad (7.\text{IV})$$

$$2H_2(g) \; + \; O_2(g) \; \longrightarrow \; 2H_2O(g) \qquad \Delta H^{\ominus}_{298} \; = \; -486\,\text{kJ} \qquad (7.\text{V})$$

on the $\Delta G^{\ominus}-T$ diagram (*Fig. 2.11*) can be used to assess the various reducing agents. From *Fig. 2.11* it can be seen that if a sufficiently high temperature is achieved any metal oxide may be reduced with carbon according to the reaction:

$$2MO(s) \; + \; C(s) \; = \; M(s) \; + \; CO_2(g) \qquad \text{below } 650\,^{\circ}C \qquad (7.\text{VI})$$

or

$$MO(s) \; + \; C(s) \; = \; M(s) \; + \; CO(g) \qquad \text{above } 650\,^{\circ}C \qquad (7.\text{VII})$$

Reaction (7.VI) is therefore only applicable to the reduction of the less stable, low melting point oxides such as PbO.

As the minimum reduction temperature (Section 2.2.3.10) rises above $1800\,^{\circ}C$ the cost of reduction with carbon increases substantially. This is due to the problem of providing refractory materials that can cope with these high temperatures and the increased reactivity of the metal with its environment resulting in increased contamination. For these reasons iron, manganese, chromium, tin, lead and zinc, are the main metal oxides reduced with carbon. Owing to the slow rates of reaction for (7.VI) and (7.VII) it is

likely that CO is first formed by reaction (7.III) and which then reduces the metal
oxide according to the reaction:

$$MO(s) + CO(g) \longrightarrow M(s) + CO_2(g) \qquad\qquad (7.VIII)$$

The CO_2 produced is simultaneously reduced with the carbon, i.e.

$$CO_2(g) + C(s) \longrightarrow 2CO(g) \qquad \Delta H_{298}^{\ominus} = +175 \text{ kJ} \qquad (7.IX)$$

Addition of reactions (7.VIII) and (7.IX) produces the overall reaction given in (7.VII).
This two-stage process offers an improvement in the reaction kinetics of the reduction
with carbon due to the gas–solid reactions of (7.VIII) and (7.IX) compared with the
solid–solid reaction of (7.VII). This point becomes more important at lower temperatures (below 700 °C).

By examination of the enthalpies for these equations it is seen that the amount of
heat needed to be added to or evolved from the extraction reaction will depend on the
$\Delta H^{\ominus}$ value for reaction (7.VIII). This may be found by reversing reaction (7.I), i.e.
dissociation of the metal oxide ($2MO \rightarrow 2M + O_2$), and adding reaction (7.IV). Metal
oxide dissociation is endothermic and will therefore be the controlling factor. However,
the large exothermic value of reaction (7.IV) produces exothermic reduction of the
lesser stable oxides when CO is used as the reducing agent. The reduction of metal
oxides with CO alone can be assessed by examining reactions (7.I) and (7.IV) in
Fig. 2.11.

Under standard thermodynamic conditions CO gas will reduce all the oxides above
the $6FeO + O_2 \rightarrow 2Fe_3O_4$ reaction line and, as stated above, is normally achieved with
an evolution of heat for the less stable oxides, e.g. Fe. Thus reduction of most metal
oxides with carbon is endothermic due to the smaller exothermic nature of reaction
(7.III) compared with reaction (7.IV) while reduction of FeO and less stable oxides
with CO is exothermic; the enthalpy values become more positive in each case with
increasing stability of the oxide.

Changing the partial pressures of CO, CO_2 and O_2 will alter the position of the reaction lines on the $\Delta G^{\ominus}$-T diagram as discussed in Sections 2.2.3.4 and 2.2.3.10. If the
CO/CO_2 ratio is increased the $2CO + O_2 \rightarrow 2CO_2$ reaction line is lowered (more negative ΔG) allowing other metal oxides, e.g. ZnO, to be reduced with CO. In the same
way reducing the partial pressure of CO will make the reaction line for $2C + O_2 = 2CO$
more negative thereby lowering the minimum reduction temperature of metal oxides
with carbon. This latter procedure is not normally carried out in practice due to the
cost of operating at low pressures.

Coke added to the blast furnace first reacts with the preheated air injected at the
base of the stack through the tuyeres to produce CO_2. At the very high temperatures
achieved at the tuyeres, thermodynamics indicates that CO should be formed but the
non-equilibrium condition and excess of oxygen result in the production of CO_2
which is subsequently reduced to CO by reaction with coke outside the tuyere zones
(see *Fig. 7.20*):

inside tuyere zones	$C + O_2 \longrightarrow CO_2$	$\Delta H_{298}^{\ominus} = -397 \text{ kJ}$	
outside tuyere zone	$CO_2 + C \longrightarrow 2CO$	$\Delta H_{298}^{\ominus} = +175 \text{ kJ}$	

Using an earlier discussion, reduction of metal oxides with CO is the most favourable
reaction at the lower temperatures (below 700 °C) in the blast furnace while above
this temperature reduction with C is thermodynamically favourable. However, due to
the high CO/CO_2 ratios, the $2CO + O_2 \rightarrow 2CO_2$ reaction line is lowered and the
improved reaction kinetics associated with the gas–solid reaction (7.VIII) compared
with the solid–solid reaction (7.VII), reduction of metal oxides with CO takes place at
higher temperatures than 700 °C in the blast furnace. The limit to reduction with CO
in the blast furnace is generally determined by the temperature at which slag begins to
form. At the onset of slag formation the surface of the metal oxide will be covered

with a layer of liquid slag which will prevent contact with carbon monoxide*. The only method available for reduction at this stage is for the solid coke to penetrate the slag layer and provide a solid coke–solid metal oxide reaction according to reaction (7.VII). Thus, although predominantly indirect reduction of metal oxides with coke takes place in the blast furnace there is also some direct reduction.

Reduction of metal oxides with hydrogen is of less importance industrially than with C or CO. The lower enthalpy value for the reaction $2H_2 + O_2 \rightarrow 2H_2O$ results in less exothermic reduction of metal oxides with hydrogen than with carbon monoxide; most metal oxides being reduced endothermically. Examination of reactions (7.II), (7.III) and (7.V) in *Fig. 2.11* indicates that H_2 is a better reducing agent of metal oxides than carbon at the lower temperature (below 650 °C). From the same thermodynamic reasoning CO *should* be a better reducing agent than H_2 below 800 °C and H_2 better than CO above 800 °C. However, due to its increased diffusivity, hydrogen is often superior to CO below 800 °C in the reduction of metal oxides[4] while, due to improved reaction kinetics, CO is a more effective reducing agent above 800 °C. In certain cases CO and H_2 are added as joint reducing agents.

Other reducing agents

In certain cases a metal oxide is reduced with another metal which forms a more stable oxide. These are called metallothermic reduction reactions and are normally used when the metal to be extracted forms stable carbides (Ti, Cr, Nb) on reduction with carbon. These reactions never go to completion, leaving some residual unreacted reducing agent in the final metal product together with some unreduced metal oxide in the oxide or slag phase. Metallothermic reductions are normally exothermic, the stronger the reducing agent the more exothermic the reaction. Certain of these reactions may be achieved without any initial heat supply. Si, Al and, occasionally, Mg are used as the reducing agents. The main application of this technique is in the production of low-carbon ferro alloys as in the aluminothermic production of ferrotitanium, ferrovanadium, ferroniobium and silicothermic production of ferrochromium in electric arc furnaces; scrap iron or iron ore being added in the required amount; e.g.

$$3MO + 2Al \longrightarrow 3M + Al_2O_3$$
$$2MO + Si \longrightarrow 2M + SiO_2$$

The reactions are normally highly exothermic. Vanadium has been extracted by reducing V_2O_5 with calcium in a sealed reaction vessel or 'bomb' operated with an inert atmosphere.

Thermal decomposition

Oxides of the more noble metals, e.g. Au, Ag, Hg, Pt, Pd, have low decomposition temperatures at normal atmospheric pressure and may be easily reduced to the metal by thermal decomposition:

$$2Ag_2O \longrightarrow 4Ag + O_2 \quad \text{above } 220\,°C$$
$$2PdO \longrightarrow 2Pd + O_2 \quad \text{above } 900\,°C$$

see *Fig. 2.11* and Section 2.2.3.10.

7.2.6 Fire refining

The impure metal from metal oxide smelting and sulphide matte converting may be refined by air and/or oxygen blowing using a preferential oxidation technique which produces oxides that are insoluble in the molten metal. Where a less severe oxidising condition is required, oxidising slags and fluxes can be used instead of air blowing.

*CO is insoluble in most liquid slags.

Preferential oxidation of an impurity element is only possible for impurity metals which form more stable oxides (more negative ΔG of oxide formation) than the metal to be refined. The degree of refinement achieved will also depend on the activities of the impurity metals present.

For example, fire refining a copper melt at $1250\,^\circ C$ where nickel is present as the main impurity can only be carried out until the activity of the nickel is reduced to a level that the ΔG for NiO formation equals that for Cu_2O formation. Beyond this point the ΔG for Cu_2O formation will be more negative than that for NiO formation and therefore copper will be preferentially oxidised and lost to the slag. This can be demonstrated by using the van't Hoff isotherm (vHI) for non-equilibrium conditions for the reactions:

$$4Cu_{(in\ melt)} + O_2 \longrightarrow 2Cu_2O_{(in\ slag)}$$
$$2Ni_{(in\ melt)} + O_2 \longrightarrow 2NiO_{(in\ slag)}$$

Thus, when

$$\Delta G_{NiO(1250\,^\circ C)} \leqslant \Delta G_{Cu_2O(1250\,^\circ C)}$$

and applying VHI to determine the limiting condition, i.e. the limiting a_{Ni}/a_{Cu} in the melt:

$$\Delta G^{\ominus}_{NiO} + RT \ln\left(\frac{a^2_{NiO(slag)}}{a^2_{Ni(melt)}p_{O_2}}\right) \leqslant \Delta G^{\ominus}_{Cu_2O} + RT \ln\left(\frac{a^2_{Cu_2O(slag)}}{a^4_{Cu(melt)}p_{O_2}}\right)$$

This will result in a minimum residual nickel which is dependent upon the equilibrium nickel activity under the oxidising conditions used.

Fire refining will provide at best a metal purity of about 99.5% and is only useful for Fe, Cu, Pb, and more noble metals, e.g. *cupellation* of a Pb–Ag melt in which Pb is preferentially oxidised to the slag and the remaining silver is electrorefined. Other preferential refining reagents include sulphur for removal of Cu as Cu_2S; chlorine for removal of Al and Mg as $AlCl_3$ and $MgCl_2$ respectively and is also used in the refining of gold (Miller process). Again, the process relies on preferential formation of insoluble compounds, e.g. $AuCl_3$, which can be separated from the melt in a slag or flux. Higher purity levels (99.9%+) may be achieved by electrorefining techniques (Sections 6.4 and 7.4).

7.2.7   Distillation

Low boiling point metals can be separated and refined from higher boiling point metals by distillation and subsequent condensation to the pure metal due to their difference in vapour pressures (see *Fig. 7.19*). Considering the evaporation of metal as:

$$M(l) = M(g)$$

thus, the standard free energy of evaporation will be given by Equation (2.30):

$$\Delta G^{\ominus} = -RT \ln \overset{\circ}{p}_M$$

(assuming $a_{M(l)} = 1$, i.e. pure liquid), where $\overset{\circ}{p}_M$ is the partial pressure of pure M at temperature T. This can be written

$$\ln \overset{\circ}{p}_M = -\Delta G^{\ominus}/RT$$
$$= -\frac{\Delta H^{\ominus} + T\Delta S^{\ominus}}{RT} \tag{7.1}$$

The $\Delta S^{\ominus}$ of evaporation of different metals is approximately constant but $\Delta H^{\ominus}$ of evaporation, or latent heat of evaporation (L_e) varies systematically within the periodic table such that Zn, Cd, Hg (group IIB) tend to be volatile, while other transition metals of Group VIA, i.e. Cr, Mo, W, have very low vapour pressures (see Section 1.4.2 and *Fig. 1.21*). Therefore, equation (7.1) may be rewritten:

$$\ln \overset{\circ}{p}_M = -\frac{L_e}{RT} + C \qquad (7.2)$$

Thus, increasing the temperature of a molten metal increases the vapour pressure; this relationship is shown in *Fig. 7.9*. It is important to note that equation (7.2) is essentially the Clausius–Clapeyron equation, which can be used to calculate the vapour pressure of a metal (see Section 2.2.3.9, i.e. equation (2.38)). If the vapour pressure of an alloying (solute) element in a liquid melt is required, equation (4.6) must be used once $\overset{\circ}{p}_M$ has been calculated using equation (2.38).

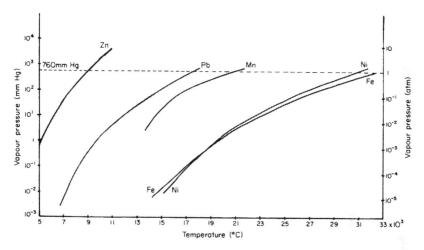

Figure 7.9 Plot of vapour pressure of various liquid metals against temperature. Note that the vapour pressure of the metal at its boiling point will be 760 mm Hg or 1 atm. (Vapour pressures calculated from data tables in *Metallurgical Thermochemistry* by O. Kubaschewski and E. Ll. Evans published by Pergamon Press, 1967, 4th revised edition.)

The range of metals that may be refined by distillation is mainly limited to those with boiling points less than 1000 °C, as above this temperature the practical problems increase considerably.

Using vacuum distillation extends the range of metals that can be refined in this way, since, under vacuum, metals which have a vapour pressure approximating to that of the vacuum, i.e. about 10^{-3} atm can be continuously distilled. *Table 7.1* lists the boiling points of various metals and their corresponding temperatures at which a vapour pressure of 10^{-3} atm is achieved. Thus under normal atmospheric pressure metals up to zinc may be refined by distillation; under vacuum distillation metals up to lead can be refined.

From equation (4.6) it can be seen that complete separation of metals using distillation is not possible since as the mole fraction of the metal in the melt is decreased so is its vapour pressure, eventually resulting in the evaporation of two or more metals whose vapour pressures are equal at the operating temperature. At this stage selective condensation may be used to aid separation. This involves the condensation of the less volatile component in the hottest part of the condenser and which can be redistilled. The more volatile component remains in the vapour and can be condensed in a cooler and more remote section of the condenser.

At constant temperature the maximum rate of evaporation is achieved when equilibrium is achieved between the number of molecules leaving the liquid metal surface

Table 7.1   MELTING AND BOILING POINTS OF SELECTED METALS

Metal	Melting point (°C)	Boiling point (°C)	Temperature at which vapour pressure is 10^{-3} atm (°C)
Mercury (Hg)	−39	357	121
Cadmium (Cd)	321	765	384
Sodium (Na)	98	892	429
Zinc (Zn)	420	907	477
Tellurium (Te)	450	990	509
Magnesium (Mg)	650	1105	608
Calcium (Ca)	850	1487	803
Antimony (Sb)	630	1635	877
Lead (Pb)	327	1740	953
Manganese (Mn)	1244	2095	1269
Aluminium (Al)	660	2500	1263
Silver (Ag)	961	2212	1334
Copper (Cu)	1083	2570	1603
Chromium (Cr)	1850	2620	1694
Tin (Sn)	232	2730	−
Nickel (Ni)	1453	2910	1780
Gold (Au)	1063	2970	1840
Iron (Fe)	1539	3070	1760
Molybdenum (Mo)	2600	4800	3060
Niobium (Nb)	2468	4927	−
Tungsten (W)	3380	5400	3940

and those condensing on the liquid metal surface. Under these conditions the rate of evaporation of component A in solution has been shown to be[5]:

$$\text{rate of evaporation} = 3.5 p_A \left(\frac{M}{T}\right)^{\frac{1}{2}}$$

where p_A is the vapour pressure of A in the molten metal, M the molecular weight of A and T the temperature of the melt in kelvin. It has been assumed that all molecules striking the surface condense. If this is not the case a further factor must be considered. In most metal evaporation processes the rate controlling step is the transfer of vapour away from the liquid metal surface rather than the rate of evaporation itself.

7.2.8   Halide metallurgy

The fact that metal halides (chlorides, iodides, fluorides, bromides) are normally formed with a smaller negative free energy change than their oxides (*Fig. 7.10*) can be utilised to extract those metals which form particularly stable oxides, e.g. Ti, Zr, Al, U, Mg. The stability of halides increases in the order: iodide; bromide; chloride; fluoride. The extraction of titanium via this route has already been described in Section 7.2.2 and the extraction of zirconium follows a similar route, i.e. chloridising roast to $ZrCl_4$ which is reduced with magnesium (Kroll process). Most metal halides are volatile, e.g. $SnCl_4$, $TiCl_4$, $AlCl_3$, or have relatively low decomposition temperatures, the latter point applies to iodides in particular: e.g.

$$ZrI_4 \xrightarrow{1400°C} Zr + 2I_2$$

This makes it possible to separate certain metal halides and also offers scope for purification by distillation.

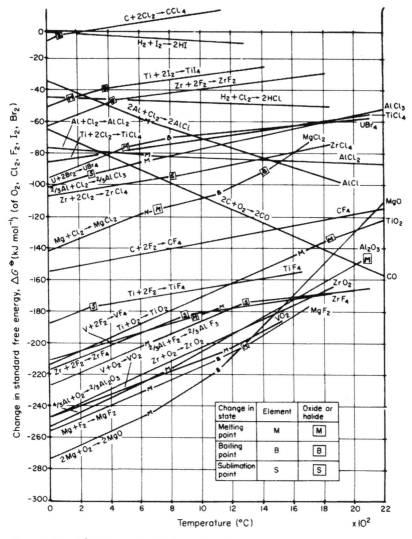

Figure 7.10 $\Delta G^{\ominus}$-T diagram for the formation of various metal halides and oxides

A further important point is that halides generally form more highly conducting melts (Section 6.2.6) with lower melting points than oxides. Thus metal halides may be used as molten salt electrolytes in the electrolytic extraction and refining of the more reactive metals such as Al, Mg, Na. These latter metals cannot be used in the form of aqueous electrolytes since hydrogen evolution at the cathode would occur at the expense of metal deposition. Owing to the fact that carbon tetrachloride (CCl_4) is less stable than most metal chlorides (*Fig. 7.10*) carbon cannot be used as a reducing agent for metal chlorides. On the other hand HCl becomes more stable as the temperature increases, therefore allowing hydrogen to be used as a reducing agent for several metal chlorides at high temperatures.

Certain metals form several chlorides, e.g., Fe, Nb, Al. The higher chlorides are easily reduced with hydrogen to lower chlorides while various disproportionate reactions may occur at certain temperatures allowing some interesting separations which could well provide future extraction routes, e.g.:

$$AlCl_3(g) + 2Al(l) \xrightarrow{1000\,^{\circ}C} 3AlCl(g)$$

followed by

$$3AlCl(g) \xrightarrow[\text{below } 800\,^{\circ}C]{\text{cooling}} 2Al(s) + AlCl_3(g)$$

also

$$5NbCl_3(s) \xrightarrow{\text{above } 800\,^{\circ}C} 2Nb(s) + 3NbCl_5(g)$$

The main disadvantage of using halides in extractive metallurgy is that many halides, apart from alkali metal halides, are readily hydrolysed: e.g.

$$MeCl_2 + H_2O \longrightarrow MeO + 2HCl$$

and must therefore be stored in dry places. Fluorides are generally less susceptible to this problem but tend to be more expensive. Halogens are costly reagents and are not cheaply regenerated in the extraction process. Halogens also cause considerable corrosion problems in the processing plants.

More stable metal halides of U, Ti, Be, Zr, can be reduced by a metallothermic reduction process using Al, Mg, Ca or Na as the reducing agent. For instance

$$UF_4 + 2Mg = U + 2MgF_2 \qquad \Delta H^{\ominus}_{298} = -375 \text{ kJ}$$

The heat produced is sufficient to fuse both products allowing separation of MgF_2 into an appropriate slag.

Anhydrous metal halides are prepared via three routes:

(a)   Halogenation of oxides is the most common method, in which metal oxides are mixed with a suitable reducing agent (coke, hydrogen or sulphur) and charged through the top of a shaft furnace. Chlorine is introduced at the bottom and the heat required to start the reaction is generated electrically in a coke bed. The reactions are normally exothermic and may be sufficient to maintain the process autogenously between 500 °C and 700 °C (see Section 7.2.2); e.g.

$$2MgO(s) + 2Cl_2(g) + C(s) = 2MgCl_2(l) + CO_2$$

It is important to avoid the production of oxyhalides, e.g. $VOCl_3$, which lead to oxygen contamination of the metal on subsequent reduction.

(b)   Halogenation of ferro alloys or carbides with chlorine can be conducted without the need of a reducing agent and without the danger of producing oxyhalides.

(c)   Crystallisation of the aqueous solution which has been produced by leaching with an appropriate acid such as hydrochloric acid. This may only be used for the preparation of alkali metal halides since other halides, as mentioned above, will hydrolyse in the presence of water.

Purification of the metal halides may be achieved by fractional distillation of volatile chlorides; reprecipitation and recrystallisation; solvent extraction or ion exchange separation (Section 7.3.3) of alkali metal halides; or by fractional reduction of a volatile chloride with hydrogen.

7.2.9 Continuous extraction processes

The ultimate goal of the extraction metallurgist is to be able to perform all the individual extraction processes in one unit with no or minimum fuel input, resulting in efficient separation of the refined metal from the slag and maximum recovery of any valuable by-products. In this way capital and operating costs will be at a minimum, material handling and fuel costs will be reduced and, since the operations will be conducted in a single unit, total automation of the process becomes easier to achieve. Thus, for the extraction of metal from sulphide concentrates the continuous extraction unit must be capable of roasting, smelting, converting and fire refining; while for metal oxide concentrates it must be capable of reducing the metal oxide followed by a fire refining oxidation treatment.

Some basic principles and requirements can therefore be listed in order to achieve this goal:

(i) one unit which is separated into roasting, smelting and converting or reducing and oxidising sections;
(ii) use of fine ore particles to provide high areas of contact between reactants and therefore high reaction rates;
(iii) use of oxygen to provide autogenous processing;
(iv) unit designed to provide movement of slag in opposite direction to that of the metal being extracted and therefore facilitate separation of the two;
(v) recovery of process fume to extract valuable metallics and production of chemicals and slag treatment (cleaning) to recover metallics;
(vi) jetting and lancing to provide turbulence to aid reaction kinetics and effect movement of slag in opposite direction to the refined metal.

Although these idealised requirements are not easily achievable, a number of pilot plant operations have been developed. In this respect several furnaces have been developed for the continuous extraction of metal from metal sulphide ores[6] while the spray steel making[7], CIP[8] and IRSID[9] processes for continuous refining have been operated either on a pilot plant or semi-commercial scale. A diagrammatic representation of the WORCRA unit is given in *Fig. 7.11*.

The sloping hearth and angled oxygen jetting used in the WORCRA furnace (*Fig. 7.11*) provide the counter current flow of slag with respect to the denser metal. The WORCRA furnace has also been used to process metal oxide ores on an experimental basis. The Q–S process in commercial operation in Canada for the extraction of lead from sulphidic ores is similar in design principles to the WORCRA process. In the IRSID process oxygen blowing of the molten iron in the reactor produces a slag–metal–gas emulsion with high areas of contact between the phases, providing high reaction

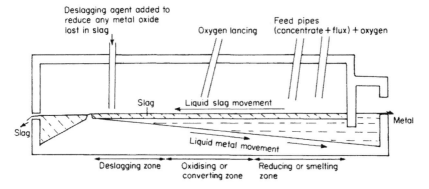

Figure 7.11 WORCRA continuous extraction unit (diagrammatic)

rates. The emulsion is allowed to overflow into a decanting vessel for slag–metal separation with final alloy adjustment conducted in a third container or 'nuanceur'. The main problem with these units is in the accurate control over the final metal composition.

7.2.10   Pyrometallurgical extraction procedures for selected metals

The extraction of the more common metals, i.e. Fe, Cu, Ni, Zn, Pb, Mg, is outlined below. The extraction of each metal will vary from plant to plant depending on the available raw materials and fuels, capacity and metal purity requirements, and political and sociological factors. The routes described are 'typical' and somewhat 'idealised' but will serve to explain the previously discussed theory. In each case it has been assumed that the starting point for the extraction process is the metal ore concentrate after appropriate ore preparation.

7.2.10.1   *Extraction of copper from copper sulphide concentrates*

Ore preparation and flotation normally provide a Cu_2S concentrate containing 25% Cu or more from an ore which 'as mined' may have contained as little as 0.5 to 2% Cu. The main impurity is iron and is often associated with nickel, gold and silver and traces of Zn, Se, As, Sb, Te, Co, Sn, Pb.

Partial roasting

Partial roasting of the concentrate is conducted in multihearth, flash or fluidised bed units. The main aim of the partial roasting process is to reduce the sulphur level but ensuring all of the copper is present as Cu_2S and part of the iron is retained as FeS. The reason for retaining some FeS in the calcine is to provide autogenous processing due to the large exothermic reaction between FeS and oxygen during the converting stage. Some preferential oxidisation of certain impurity metal sulphides is also encouraged in partial roasting, e.g.

$$ZnS + 3O_2 \longrightarrow 2ZnO + 2SO_2$$

These oxides will be collected in the slag on subsequent smelting.

Smelting

The calcine product is smelted with an acid (silica) slag in a reverberatory or flash smelting unit. A sulphide matte is produced containing Cu_2S–FeS plus precious metals (Ag, Au) and other impurity sulphides which were not oxidised during partial roasting. If arsenic and antimony are present a liquid speiss layer is formed. The basic FeO produced in the partial roasting operation reacts with the acid slag, $FeO + SiO_2 \rightarrow FeSiO_3$, which also collects other impurities such as amphoteric oxides (ZnO).

Converting

The sulphide matte is transferred to a converter (*Fig. 7.12*) through which is blown air and/or oxygen. The impurity metal sulphides are preferentially oxidised (consult *Fig. 7.6*); the oxidation of the remaining FeS which was deliberately retained during partial roasting provides sufficient heat to make the converting process autogenous. That is

$$2FeS + 3O_2 \quad 2FeO + 2SO_2 \quad \Delta H^{\ominus}_{298} = -932 \text{ kJ} \qquad (7.X)$$
$$\text{(matte)} \qquad \text{(slag)}$$

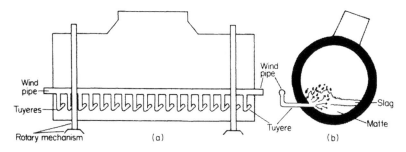

Figure 7.12 Horizontal converter (diagrammatic): (a) front view; (b) side view

Since FeO is more stable (more negative $\Delta G^{\ominus}$ of formation) than Cu_2O, any Cu_2O produced during roasting or smelting and which may separate to the slag will be returned as Cu_2S to the matte on contact with FeS:

$$Cu_2O \; + \; FeS \longrightarrow Cu_2S \; + \; FeO$$
$$\text{(slag)} \quad \text{(matte)} \quad \text{(matte)} \quad \text{(slag)}$$

The turbulence produced on air/oxygen blowing during converting will aid the reaction rate and recovery of copper sulphide. Again an acid (silica) slag is used to react with the FeO and decrease its activity. This improves the removal of iron from the copper as can be seen by applying the van't Hoff isotherm for non-equilibrium conditions (equation (2.29)) to equation (7.X), i.e.

$$\Delta G_{FeO} \; = \; \Delta G^{\ominus}_{FeO} + RT \ln \left(\frac{a^2_{FeO(slag)} \, p^2_{SO_2}}{a^2_{FeS} \, p^3_{O_2}} \right)$$

Thus as a_{FeO} is decreased the ratio $(a^2_{FeO} p^2_{SO_2})/(a^2_{FeS} p^3_{O_2})$, becomes less than unity thereby making the RT term negative. This results in a more negative value of ΔG, increasing the driving force for the removal of iron to the slag. Other more stable sulphides will be removed in the same way. More matte is charged and the Cu_2S is converted to copper by the exothermic reaction (see Section 7.2.4):

$$Cu_2S \; + \; O_2 \longrightarrow 2Cu \; + \; SO_2 \qquad \Delta H^{\ominus}_{298} \; = \; -212 \text{ kJ}$$

This reaction is only possible because the mutual reduction reaction $(Cu_2S + 2Cu_2O \rightarrow 6Cu + SO_2)$ has a negative ΔG value at $1200\,^{\circ}C$. The copper content after converting is approximately 98% and is called *blister copper* due to the bubbling of sulphur dioxide through the melt producing a blister effect. The precious metals, as would be expected from their free energies of oxidation, are not oxidised and are collected in the blister copper.

Fire refining

The blister copper containing the precious metals and small amounts of other impurity elements is subjected to controlled air blowing in a hearth furnace with possible use of oxidising slags and fluxes. The more reactive metals are oxidised and removed to the slag until the copper begins to oxidise. At this stage any copper oxide formed is reduced by *poling* the melt with tree trunks or bubbling gaseous NH_3 or gaseous hydrocarbons through the melt. The hydrocarbons released from burning the trees produces carbon monoxide and hydrogen which reduce the copper oxide, returning the copper to the melt from the slag. Gaseous ammonia, which dissociates to hydrogen and nitrogen, and gaseous hydrocarbons have the same effect, i.e.

$$Cu_2O + CO \longrightarrow 2Cu + CO_2$$
$$\text{(slag)} \qquad\qquad \text{(melt)}$$

$$Cu_2O + H_2 \longrightarrow 2Cu + H_2O$$
$$\text{(slag)} \qquad\qquad \text{(melt)}$$

There is now a problem of reducing other oxides which have similar thermodynamic properties to copper and which have separated to the slag as impurity oxides. Therefore, careful alternating of controlled oxidation (air blowing) and reduction (poling) is carried out to provide maximum copper yield with a minimum impurity level. Final deoxidation may be achieved by addition of a phosphorus (15%)–copper alloy, removing the oxygen as P_2O_5. Alternatively, if residual phosphorus is to be avoided due to its presence reducing the electrical conductivity of copper, lithium or carbon may be used as the deoxidant.

Fire refining will produce a copper purity of about 99.5%. If a higher purity is required electrorefining must be used. A flow sheet of the process is given in *Fig. 7.13*.

7.2.10.2 *Extraction of nickel from nickel sulphide concentrates*

Nickel sulphide ore with a typical nickel content of 1–3% is processed to provide a nickel sulphide concentrate containing some copper and iron as sulphides and smaller amounts of Co, Ag, Pt as sulphides, some arsenides and a siliceous gangue. There are several pyrometallurgical and hydrometallurgical extraction techniques used but the one outlined here will be that used by the International Nickel Company[10]. Owing to the similar thermodynamic characteristics of nickel and copper sulphides the major problem is concerned with their separation. The initial process steps of partial roasting, smelting and converting are similar to those used in the extraction of copper from copper sulphide ore, eventually providing nickel and copper sulphides in a form that can lead to efficient separation of the two. Again, a certain amount of iron sulphide is retained in the calcine after partial roasting to aid autogenous converting of the nickel sulphide–copper sulphide matte. It is not possible to convert nickel sulphide to nickel as was achieved for copper sulphide converting to copper due to the mutual reduction reaction between nickel sulphide and nickel oxide having a positive standard free energy change at $1200\,^\circ C$:

$$4NiO + Ni_3S_2 \longrightarrow 7Ni + 2SO_2 \qquad \Delta G^{\ominus}_{1200\,^\circ C} = +21 \text{ kJ}$$

However, small amounts of nickel sulphide and copper sulphide are converted to provide a nickel–copper alloy containing the majority of the precious metals present in

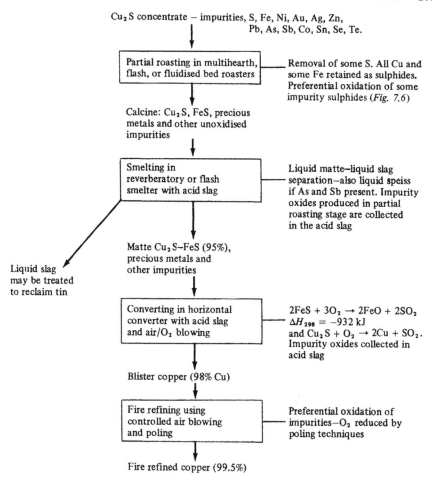

Cu₂S concentrate – impurities, S, Fe, Ni, Au, Ag, Zn,
Pb, As, Sb, Co, Sn, Se, Te.

| Partial roasting in multihearth, flash, or fluidised bed roasters | — Removal of some S. All Cu and some Fe retained as sulphides. Preferential oxidation of some impurity sulphides (*Fig. 7.6*) |

Calcine: Cu₂S, FeS, precious metals and other unoxidised impurities

| Smelting in reverberatory or flash smelter with acid slag | — Liquid matte–liquid slag separation–also liquid speiss if As and Sb present. Impurity oxides produced in partial roasting stage are collected in the acid slag |

Matte Cu₂S–FeS (95%), precious metals and other impurities

Liquid slag may be treated to reclaim tin

| Converting in horizontal converter with acid slag and air/O₂ blowing | — $2FeS + 3O_2 \rightarrow 2FeO + 2SO_2$ $\Delta H_{298} = -932$ kJ and $Cu_2S + O_2 \rightarrow 2Cu + SO_2$. Impurity oxides collected in acid slag |

Blister copper (98% Cu)

| Fire refining using controlled air blowing and poling | — Preferential oxidation of impurities–O₂ reduced by poling techniques |

Fire refined copper (99.5%)

Figure 7.13 Flow sheet for pyrometallurgical extraction of copper from a Cu₂S concentrate

the concentrate. Thus, the main products of the matte converting operation are nickel and copper sulphides, which form a matte, and a small amount of nickel–copper alloy, which contains the precious metals. The FeS present in the matte is converted to iron oxide providing heat for the process as in copper matte converting. Again, an acid silica slag is used for the same reasons as in copper converting.

The three products Ni_3S_2, Cu_2S, Ni–Cu alloy are subsequently cooled slowly through the range 925 °C to 370 °C, which allows first Cu_2S, then the Ni–Cu alloy and finally the Ni_3S_2 to solidify in large grains. These are separated by first crushing and grinding the matte followed by magnetic separation of the Ni–Cu alloy and froth flotation separation of the remaining Ni_3S_2 and Cu_2S. The Cu_2S is sold to a copper extractor and the Ni_3S_2 is treated in various ways to extract the nickel. The nickel sulphide may be cast into anodes and directly electrorefined into nickel using a diaphragm (see Section 6.4.6) to separate the anolyte (electrolyte around the anode) and the catholyte (the electrolyte around the cathode). If this were not done impurities (Cu, Co and Fe), being close to nickel in the electrochemical series, would result in contamination of the cathode nickel. Sulphur collects as part of the anode slime together with any precious metals that may be present. Alternatively, the Ni_3S_2 matte is sintered to produce nickel oxide which is subsequently reduced with carbon or hydrogen to provide an impure nickel:

$$Ni_3S_2 + 7O_2 \xrightarrow{750\,°C} 3NiO + 2SO_2$$

$$NiO + \begin{cases} C \\ H_2 \end{cases} \xrightarrow{400\,°C} Ni + \begin{cases} CO \\ H_2O \end{cases}$$

The crude nickel is then refined by the *carbonyl process* which relies on the fact that nickel produces a volatile carbonyl, $Ni(CO)_4$, on reaction with carbon monoxide at 50 °C and which can be dissociated in a volatiliser to provide pure nickel (99.9%) pellets on increasing the temperature to 180 °C. Other impurities such as Fe, Co form carbonyls ($Fe(CO)_5$, $Co_2(CO)_8$) but the rate of formation of $Fe(CO)_5$ is low and $Co_2(CO)_8$ has low volatility, leaving Fe, Co and precious metals as a residue in the volatiliser:

$$Ni + 4CO \xrightarrow{50\,°C} Ni(CO)_4 \xrightarrow{180\,°C} Ni + 4CO$$

Alternatively, the nickel oxide produced on sintering the nickel sulphide can be roasted with producer gas ($CO + H_2$) at 400 °C, which is volatilised as $Ni(CO)_4$ on exposure to excess carbon monoxide at 50 °C. The overall reaction being:

$$NiO + H_2 + 4CO \xrightarrow{50\,°C} Ni(CO)_4 + H_2O$$

$$Ni(CO)_4 \xrightarrow{180\,°C} Ni + 4CO$$

The pure nickel pellets grow relatively slowly. This may be increased by 'seeding' the volatilising chamber by addition of pure nickel shot which provides centres on which the nickel can nucleate and grow. A flow chart of the process is given in *Fig. 7.14*.

7.2.10.3 *Extraction of lead and zinc from PbS-ZnS concentrates in the Imperial smelting furnace*

PbS and ZnS are often found together in as mined ores. Before the Imperial Smelting Furnace (ISF)[11] was developed, expensive froth flotation separation techniques were required to separate the sulphides which were subsequently smelted individually after dead roasting. Lead oxide can be smelted with coke and flux in a blast furnace producing a crude lead bullion while zinc oxide can be smelted with coke in horizontal or vertical retorts. Extraction of lead from PbO in the blast furnace relies on reduction with CO. By examination of the $\Delta G^{\ominus}$–T diagram (*Fig. 2.11*) at 500 °C reduction with CO is thermodynamically more favourable than reduction with coke: i.e.

$$PbO + CO = Pb + CO_2 \qquad \Delta G^{\ominus}_{500\,°C} = -147 \text{ kJ mol}^{-1}$$
$$PbO + C = Pb + CO \qquad \Delta G^{\ominus}_{500\,°C} = -76 \text{ kJ mol}^{-1}$$

since reduction with CO provides a greater thermodynamic driving force. At the same time a solid–gas (PbO–CO) reaction will be kinetically more favourable than a solid–solid (PbO–C) reaction. On the other hand, reacting ZnO with coke will produce a reduction reaction but only above 950 °C (see *Fig. 2.11*):

$$ZnO(s) + C(s) = Zn(g) + CO_2(g) \qquad \Delta G^{\ominus}_{1000\,°C} = -29 \text{ kJ mol}^{-1}$$

The reaction between ZnO and CO at 1000 °C will have a positive standard free energy change and will therefore only be possible if the partial pressure of CO is high, that is the CO/CO_2 ratio is large, i.e.

$$ZnO(s) + CO(g) = Zn(g) + CO_2(g) \qquad \Delta G^{\ominus}_{1000\,°C} = +76 \text{ kJ}$$

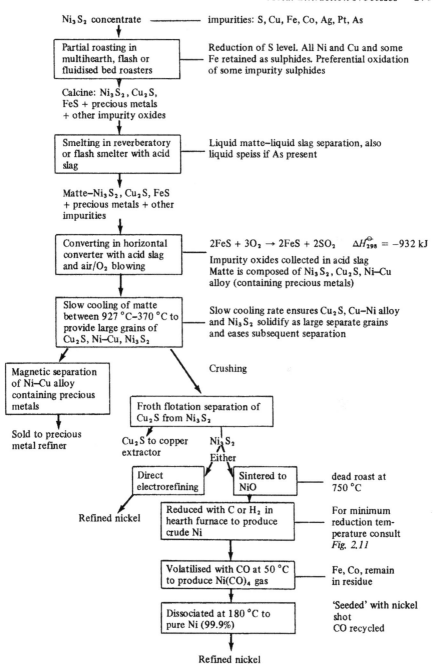

Ni₃S₂ concentrate ——————— impurities: S, Cu, Fe, Co, Ag, Pt, As

| Partial roasting in multihearth, flash or fluidised bed roasters | Reduction of S level. All Ni and Cu and some Fe retained as sulphides. Preferential oxidation of some impurity sulphides |

Calcine: Ni₃S₂, Cu₂S, FeS + precious metals + other impurity oxides

| Smelting in reverberatory or flash smelter with acid slag | Liquid matte–liquid slag separation, also liquid speiss if As present |

Matte–Ni₃S₂, Cu₂S, FeS + precious metals + other impurities

| Converting in horizontal converter with acid slag and air/O₂ blowing | $2FeS + 3O_2 \rightarrow 2FeS + 2SO_2$ $\Delta H^{\ominus}_{298} = -932$ kJ
Impurity oxides collected in acid slag. Matte is composed of Ni₃S₂, Cu₂S, Ni–Cu alloy (containing precious metals) |

| Slow cooling of matte between 927 °C–370 °C to provide large grains of Cu₂S, Ni–Cu, Ni₃S₂ | Slow cooling rate ensures Cu₂S, Cu–Ni alloy and Ni₃S₂ solidify as large separate grains and eases subsequent separation |

Crushing

| Magnetic separation of Ni–Cu alloy containing precious metals |

Sold to precious metal refiner

| Froth flotation separation of Cu₂S from Ni₃S₂ |

Cu₂S to copper extractor

Ni₃S₂

Either

| Direct electrorefining |

| Sintered to NiO | dead roast at 750 °C |

Refined nickel

| Reduced with C or H₂ in hearth furnace to produce crude Ni | For minimum reduction temperature consult *Fig. 2.11* |

| Volatilised with CO at 50 °C to produce Ni(CO)₄ gas | Fe, Co, remain in residue |

| Dissociated at 180 °C to pure Ni (99.9%) | 'Seeded' with nickel shot
CO recycled |

Refined nickel

Figure 7.14 Flow sheet for the pyrometallurgical extraction of nickel from Ni₃S₂ concentrate

The equilibrium CO/CO_2 ratio for this reaction at 1000 $^\circ$C is 50 (Section 2.2.3.10.3) thus for ZnO to be reduced by CO the actual CO/CO_2 ratio must be greater than 50. The zinc produced at this temperature is in gaseous form and will react with any oxidising gas present (O_2 or CO_2) to revert to ZnO if the temperature falls below 950 $^\circ$C; ZnO being more stable than CO_2 at temperatures below 950 $^\circ$C.

Therefore the zinc vapour produced must condense in a reducing atmosphere. For this reason zinc oxide reduction with coke is conducted in horizontal or vertical retorts which are heated *externally* thereby preventing oxidation of CO to CO_2. This limitation produces a low thermal efficiency and means that zinc has to be produced in small horizontal retorts placed in batches (*Fig. 7.15(a)*). The zinc vapour is condensed

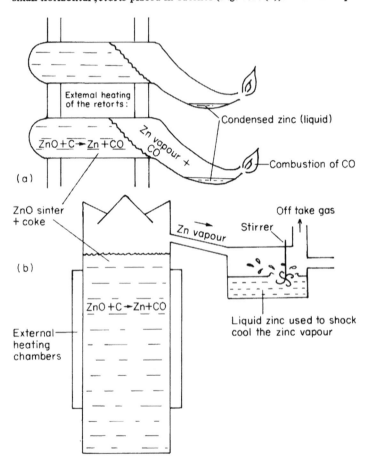

Figure 7.15 Diagrammatic representation of: (a) horizontal retort; and (b) vertical retort; for ZnO reduction with carbon

and collected at a controlled distance from the reaction chamber. Externally heated vertical retorts (*Fig. 7.15(b)*) can be operated on a continuous basis with zinc oxide briquettes and coke being charged at the top of the retort and the zinc vapour being removed and rapidly condensed in a liquid zinc bath at the top of the reactor. This provides an increased throughput compared with the horizontal retort.

The development of the ISF provides the simultaneous reduction of a PbO–ZnO mixture (produced by sintering the PbS–ZnS concentrate) in a rectangular blast furnace. A liquid lead bullion is collected in the hearth of the furnace and zinc vapour is rapidly condensed or 'shock cooled' on passing from the top of the furnace (*Fig. 7.16*). The shock cooling is achieved by spraying liquid lead at 600 °C on to the zinc vapour as it leaves the top of the blast furnace shaft and prevents the reversion of zinc to zinc oxide; the condensed zinc liquid being soluble in the liquid lead at this temperature. As the liquid metal mixture cools zinc solubility decreases until at 400 °C only 2.0% of

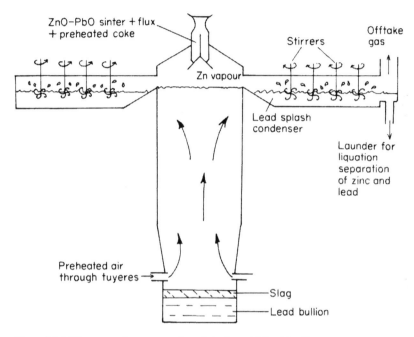

Figure 7.16 Diagrammatic representation of the Imperial Smelting Furnace

zinc remains in the lead; the liquid zinc floating above the denser liquid lead. This is a *liquation* separation process which is made possible due to the change in solubility between the two molten metals on cooling*. The PbO–ZnO sinter mix is charged with preheated coke and a limestone flux at the top of the blast furnace and preheated (850–950 °C) air is injected through tuyeres at the bottom of the stack.

The reduction reactions occur in the stack of the furnace in three distinct regions (*Fig. 7.17*). The bottom third of the stack is the zinc reduction zone where zinc oxide is reduced with coke:

$$ZnO(s) + C(s) \longrightarrow Zn(g) + CO(g)$$

This is the hottest part of the stack. There is also a small amount of reduction of ZnO with CO. The top third of the stack is the lead reduction zone where lead oxide is reduced with carbon monoxide:

$$PbO(s) + CO(g) = Pb(l) + CO_2(g) \qquad \Delta H^{\ominus}_{298} = -572 \text{ kJ mol}^{-1}$$

This reaction will take place with no further coke requirement since it is exothermic; the CO being produced by the reduction of zinc oxide and the combustion of coke and

*The reader should consult the Pb–Zn phase diagram.

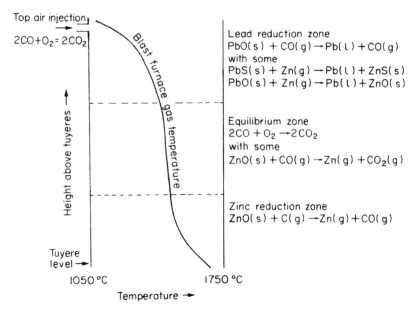

Figure 7.17 The three regions or zones in the Imperial Smelting Furnace (diagrammatic)

air lower down the stack. This is a considerable factor in the overall economics of the process. The middle zone of the stack is termed the equilibrium zone.

A controlled injection of preheated air at the top of the stack is made to maintain the temperature above 1000 °C and prevent reversion of zinc gas to zinc oxide. Although the air reacts with carbon monoxide producing carbon dioxide the large exothermic nature of the reaction ensures that the temperature is maintained above the reversion (950 °C) temperature.

$$2CO + O_2 \longrightarrow 2CO_2 \qquad \Delta H_{298}^{\ominus} = -572 \text{ kJ mol}^{-1}$$

The ISF is completely computer controlled with fully automatic charging cutting labour requirements to a minimum. The main impurities in the zinc are lead and cadmium, which can be substantially removed from the zinc by reflux distillation. The first fraction of zinc distilled will contain most of the cadmium, the last fraction will contain 1–2% lead. Further distillation of the zinc will produce a high grade zinc. A double reflux distillation column (*Fig. 7.18*) is usually employed: the impure zinc is introduced in the centre of the first column; zinc and cadmium evaporate; condensation of any lead present in the vapour takes place as it ascends the column; the enriched Zn–Cd vapour is allowed to condense and is transferred to the second column in which the temperature is lower than that in the first; this allows the cadmium to evaporate leaving an enriched (99.9%) zinc. The cadmium is recovered by condensation. A Zn–Pb alloy is removed from the bottom of the first column which can be refined by further distillation. Distillation of zinc is an endothermic process and is normally only competitive with electrorefining of zinc in areas with a cheap supply.

The main impurities in the lead are copper, precious metals, antimony, tin and arsenic, although the presence of arsenic and antimony may react with any iron present to form a separate speiss layer. As, Sb, Sn, Cd and Bi will be removed from the concentrate as volatiles to some extent during sintering, smelting, converting and refining. These can be collected from the process fumes which may be treated to extract the metals (see Section 8.2). Any cobalt and nickel present will also be collected in the speiss layer. The hot lead bullion is cooled to just above its freezing point (327 °C) when the less soluble impurities As, Sb, Sn, Fe will separate out in a liquation process. Copper is removed at this point as Cu_2S by the addition of sulphur, i.e. *drossing*.

Metals Sb, Sn, As which form more stable oxides than PbO can be preferentially oxidised with air blowing or litharge or oxidising flux, e.g. $NaNO_3$ additions at 750 °C. Gold and silver can be removed by the addition of zinc at 480 °C. Au and Ag produce intermetallic compounds with zinc, e.g. Ag_2Zn_3, which separate out into a dross. This

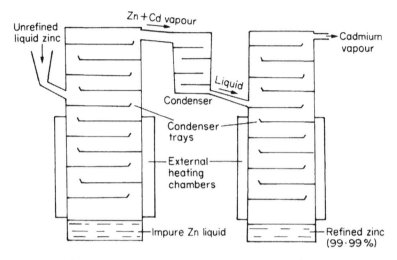

Figure 7.18 Double distillation or reflux unit used to refine zinc. Second reflux unit is operated at a lower temperature than the first unit

is the Parkes process in which silver is removed from liquid lead by a partitioning process; the distribution ratio of Ag in Zn to Ag in Pb being the governing factor (see Section 6.5.1). The zinc is removed by vacuum distillation and the remaining zinc containing a small amount of lead is refined by preferential oxidation of the lead as PbO (cupellation). If bismuth is present as an impurity, electrorefining of lead using an acid lead fluosilicate electrolyte is the main refining method used in which the bismuth is collected in the anode slime.

7.2.10.4 Extraction of magnesium from magnesium oxide concentrates

Magnesium occurs naturally in the earth's crust as magnesite ($MgCO_3$), dolomite ($MgCO_3.CaCO_3$) and brucite ($Mg(OH)_2$) and in sea water (0.13%) as $MgCl_2$.

Although the major tonnage of magnesium production is by electrolysis of fused magnesium chloride (see Section 7.4.2) some thermal reduction of MgO is carried out using silicon (metallothermic) or carbon (carbothermic) as the reducing agent.

Consultation of the $\Delta G^{\ominus}-T$ diagram for oxide formation (*Fig. 2.11*) indicates that MgO reduction with silicon is impossible due to the positive $\Delta G^{\ominus}$ value for the reaction at 1200 °C.

$$2MgO(s) + Si(s) = 2Mg(g) + SiO_2(s) \qquad \Delta G^{\ominus}_{1200°C} = +244 \text{ kJ mol}^{-1}$$

The above reaction assumes that the reactants and products are in their standard thermodynamic states, i.e. unit activity (pure) if solid or liquid and a partial pressure of 1 atm if in gaseous form. By manipulating these activities and partial pressures and applying the van't Hoff isotherm for non-equilibrium conditions (equation (2.29)) it is possible to achieve a negative ΔG value for the above reaction and therefore provide reduction of MgO with Si. Applying equation (2.29) to the above reaction at $1200\,°C$:

$$\Delta G_{1200°C} = \Delta G^{\ominus}_{1200°C} + RT \ln \left(\frac{p^2_{Mg} a_{SiO_2}}{a^2_{MgO} a_{Si}} \right)$$

$$= +244\,000 + 8.3 \times 1473 \ln \left(\frac{p^2_{Mg} a_{SiO_2}}{a^2_{MgO} a_{Si}} \right) \; J\,mol^{-1}$$

For the reduction reaction to occur at $1200\,°C$, $\Delta G_{1200°C} < 0$ (i.e. ΔG negative). Thus

$$8.3 \times 1473 \ln \left(\frac{p^2_{Mg} a_{SiO_2}}{a^2_{MgO} a_{Si}} \right) < -244\,000$$

and so

$$\ln \left(\frac{p^2_{Mg} a_{SiO_2}}{a^2_{MgO} a_{Si}} \right) < \frac{-244\,000}{8.3 \times 1473}$$

This condition can be readily achieved by lowering the partial pressure of the magnesium vapour produced using vacuum techniques while the activity of SiO_2 is lowered by additions of lime (CaO) to the slag which collects the silica, i.e. formation of a basic slag. In practice the lime is added in the form of calcined dolomite MgO.CaO (Pidgeon process). The silicon is added as high grade ferrosilicon which is cheaper to produce than pure silicon and also improves fluidity by lowering the melting point of the slag. The reaction is conducted in an externally heated Ni–Cr retort similar to the horizontal retort used in zinc extraction.

$$2MgO.CaO(s) + Fe–Si(s) \longrightarrow 2Mg(g) + Ca_2FeSiO_4$$

Again, several retorts are placed in batches. Since magnesium is the only gaseous product there is no danger of a reversion reaction to MgO during cooling.

By consultation of the $\Delta G^{\ominus}$–T relationships (*Fig. 2.11*) for the carbothermic reduction of MgO the minimum reduction temperature is $1850\,°C$:

$$MgO(s) + C(g) \longrightarrow Mg(g) + CO(g)$$

If the magnesium vapour cools below $1850\,°C$ the above reaction will reverse. Therefore, the Mg vapour must be rapidly 'shock cooled' with hydrogen or natural gas to prevent considerable reversion to MgO. The magnesium vapour condenses as a fine dust covered with oxide and which easily catches fire on exposure to air (pyrophoric). For this reason the magnesium dust is refined by vacuum sublimation which makes this process more expensive than the Pidgeon process or the electrolytic process.

7.2.10.5   Reduction of iron oxides

Iron occurs naturally as haematite (Fe_2O_3), magnetite (Fe_3O_4) and as the leaner limonite ($Fe_2O_3.3H_2O$) and siderite ($FeCO_3$) ores. The iron content varies from 20% to 70%.

Blast furnace smelting

The main method of reduction of these ores is with coke in the blast furnace. The richer ores may be added directly to the furnace in lump form while the leaner ores generally require some concentration. Agglomeration of the ores or concentrates will produce improved reaction kinetics but the agglomeration process must be substantiated by sufficiently improved production rates. Iron ore, coke and limestone as a flux are charged at the top of the furnace through a bell mechanism which helps to control

Table 7.2 TYPICAL CHARGE QUANTITIES PER TONNE OF HOT PIG IRON PRODUCED IN THE BLAST FURNACE SMELTING OF IRON ORES

Charge material	Tonnes per tonne of iron produced
Iron ore (50% Fe)	~2.0 depending on Fe content
Coke	0.35–0.6 (0.35 for agglomerated ore used with fuel injection at the tuyeres)
Limestone	~0.4 depending on Fe content
Preheated air through tuyeres	3.0 (~3.12 × 10⁶ m³)

the conditions at the top of the stack (*Fig 7.8(c)*). *Table 7.2* indicates the normal charge quantities.

As discussed in Section 7.2.5 reduction of iron oxides with CO (indirect reduction) rather than with coke (direct reduction) predominates until the onset of slag formation at about $1100\,^\circ C$. The series of reactions can be represented as:

Indirect reduction or stack reactions

$$3Fe_2O_3 + CO \rightarrow 2Fe_3O_4 + CO_2 \quad \text{at } 400\text{–}600\,^\circ C$$

$$Fe_3O_4 + CO \rightarrow 3FeO + CO_2 \quad \text{at } 600\text{–}800\,^\circ C \text{ (FeO unstable below } 570\,^\circ C. \text{ See } Fig. \text{ 7.20)}$$

$$2CO \rightarrow C + CO_2 \quad \text{at } 540\text{–}650\,^\circ C \text{ (Boudouard or carbon deposition reaction)}$$

$$FeO + CO \rightarrow Fe + CO_2 \quad \text{at } 800\text{–}1100\,^\circ C$$

$$CaCO_3 \rightarrow CaO + CO_2 \quad \text{at } 900\,^\circ C$$

Onset of slag formation at about $1100\,^\circ C$:

Direct reduction or hearth reactions at 1200–1800 $^\circ C$

$$P_2O_5 + 5C \rightarrow 2P(\text{in Fe}) + 5CO$$

$$MnO + C \rightarrow Mn(\text{in Fe}) + CO$$

$$SiO_2 + 2C \rightarrow Si(\text{in Fe}) + CO$$

$$S(\text{in ore} + \text{coke}) + CaO + C \rightarrow CaS(\text{in slag}) + CO$$

$$C(\text{in coke}) \rightarrow C(\text{in Fe})$$

$$FeO + C \rightarrow Fe + CO$$

The above reactions should be considered together with temperature distributions within the stack (*Fig 7.19*). These isotherms may be levelled out to some extent by using a graded burden size or an agglomerate (sinter or pellet) charge. *Fig. 7.20* shows the $\Delta G^{\ominus} - T$ relationships for the relevant reactions in ironmaking and steelmaking. Using *Fig. 7.20* it is possible to consider the theoretical thermodynamic relationships that occur when a pocket of preheated air enters the blast furnace via the tuyeres and reacts alternately with coke and iron oxides (FeO) represented in the diagram by an arrowed line. The preheated air first reacts with coke in the tuyere zone to produce CO_2 which then reacts with more coke outside the tuyere zone to produce CO by the overall reaction:

$$2C + O_2 \longrightarrow 2CO$$

If the CO then comes into contact with a lump of iron ore, e.g. FeO, it is reduced, i.e.

$$FeO + CO \longrightarrow Fe + CO_2$$

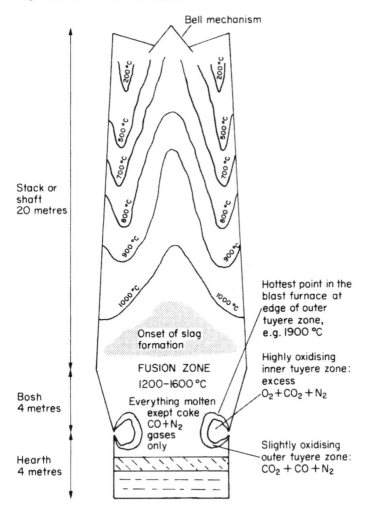

Figure 7.19 Diagrammatic representation of the distribution of temperature of the blast furnace gas for the reduction of iron ore in an ungraded burden charge

Thus, the pocket of blast furnace gas (arrowed line) alternates between the $2C + O_2 \rightarrow 2CO$ line and the $2Fe + O_2 \rightarrow 2FeO$ line and tries to reach equilibrium with these reactions. Due to the conditions in the blast furnace, equilibrium is never achieved but the blast furnace gas composition can be considered to alternate between these two reaction lines as it ascends the stack. At the lower temperatures higher up the stack the gas composition contains more CO than would be expected due to the blast furnace being well away from equilibrium conditions; the rapid ascent of the blast furnace gas flushing the CO to the top of the stack. At 545 °C to 650 °C the *Boudouard* or *carbon deposition* reaction $2CO \rightarrow 2C + CO_2$ is at an optimum. The reaction is thought to require a catalyst—possibly an iron–refractory interface. The carbon is deposited as a fine powder which may provide some direct reduction of the iron oxides. This is likely to be only a small amount. Owing to the onset of slag formation just above

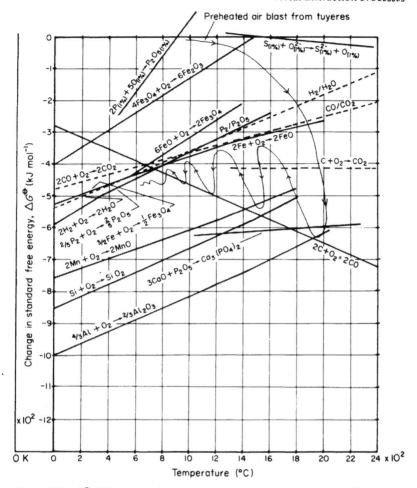

Figure 7.20 $\Delta G^{\ominus}$-T diagram for the important reactions concerned with ironmaking and steelmaking

the tuyeres any further reduction of iron oxide below this position in the stack is mainly by direct reduction with carbon. The pick up of Si, S, Mn, can be controlled to some extent by slag basicity and temperature on examination of the thermodynamic relationships involved.

Control of silicon in pig iron

The $\Delta G^{\ominus}$-T relationship[*] for the reduction of SiO_2 with carbon is given as[12]

$$SiO_2 + 2C \longrightarrow Si(\text{in Fe}) + 2CO$$
$$\Delta G^{\ominus} = +59\,3570 - 396T \text{ J mol}^{-1} \qquad (7.XI)$$

Thus, increasing hearth temperature will make $\Delta G^{\ominus}$ more negative favouring the forward reaction and increasing silicon pick up in the pig iron. At the same time SiO_2 is

[*]The $\Delta G^{\ominus}$-T relationship is normally given in the form $\Delta G^{\ominus} = \pm A \mp BT$ where A is $\Delta H^{\ominus}$ and B is $\Delta S^{\ominus}$, both of which can be positive or negative (Section 2.2.2.1).

an acidic oxide. Therefore, using a basic (CaO) slag the activity of SiO_2 (a_{SiO_2}) is reduced due to the reaction:

$$SiO_2 + 3CaO \longrightarrow Ca_2SiO_4 + \text{excess CaO}$$

If the van't Hoff isotherm for non-equilibrium conditions (equation (2.29)) is applied to reaction (7.XI):

$$\Delta G = \Delta G^\ominus + RT \ln \left(\frac{a_{Si(in\ Fe)} p_{CO}^2}{a_{SiO_2} a_C^2} \right)$$

using a basic slag reduces a_{SiO_2} which makes the ln term positive, thereby making ΔG more positive and favouring the reverse reaction,

$$Si(in\ Fe) + 2CO \longrightarrow SiO_2 + 2C$$

Therefore, silicon pick-up in the iron is lowered using a basic slag and increased using an acidic slag. In the same way the distribution of other elements between the slag and the pig iron can be deduced.

The corresponding relationship for MnO (basic oxide) reduction with carbon is[13]:

$$MnO + C = Mn(in\ Fe) + CO \quad \Delta G^\ominus = +290\ 300 - 173T \text{ J mol}^{-1}$$

Thus, Mn pick up is favoured by a high hearth temperature and a basic slag. The removal of sulphur to the slag relies on the formation of a more stable sulphide, e.g. CaS, than FeS which will separate into and react with the slag. This is best achieved by employing a basic (CaO) slag. Therefore, sulphur removal depends on the reaction,

$$FeS(in\ Fe) + CaO \longrightarrow CaS(in\ slag) + FeO$$

which is more usually written as the ionic equation[14]:

$$S(Fe) + O^{2-}(in\ slag) \longrightarrow S^{2-}(in\ slag) + O(in\ Fe)$$
$$\Delta G^\ominus = +72\ 000 - 38T \text{ J mol}^{-1} \quad (7.XII)$$

Thus, increasing hearth temperature will make $\Delta G^\ominus$ more negative favouring the removal of sulphur from the iron to the slag. Since each reaction tries to reach equilibrium, examination of the equilibrium constant will indicate the conditions which will favour sulphur removal to the slag.

The equilibrium constant (K) for reaction (7.XII) is:

$$K = \frac{a_{S^{2-}(in\ slag)} a_{O(in\ Fe)}}{a_{S(in\ Fe)} a_{O^{2-}(in\ slag)}}$$

from which the ratio of sulphur in the slag to sulphur in the iron is seen to be:

$$\frac{a_{S^{2-}(in\ slag)}}{a_{S(in\ Fe)}} = K \frac{a_{O^{2-}(in\ slag)}}{a_{O(in\ Fe)}} \tag{7.3}$$

To achieve maximum sulphur removal to the slag this ratio needs to be as large as possible which, from equation (7.3), requires the ratio $a_{O^{2-}(in\ slag)}/a_{O(in\ Fe)}$ to be as large as possible in order to maintain K at its constant value. It is clear then that the oxygen ion activity of the slag must be high and the oxygen potential of the pig iron must be low to provide a high removal of sulphur to the slag. The oxygen ion activity of the slag is increased by using a lime addition to the charge, i.e. basic slag formation:

$$CaO \longrightarrow Ca^{2+} + O^{2-}$$

while control of the oxygen potential of the pig iron is more difficult. Adjustment of the $a_{O(in\ Fe)}$ value is used to a much greater extent in removal of sulphur during steel-making which will be discussed later.

Thus, sulphur removal is favoured by a high temperature and basic slag, although it may be noted that oxygen solubility in Fe increases as the melt temperature increases. This latter point becomes important in steelmaking processes where highly oxidising conditions prevail.

Use of blast preheat, oxygen enrichment of the blast, injection of hydrocarbons and steam into the blast furnace and increasing the pressure of the blast furnace gas at the top of the stack (high top pressure) have resulted in considerable improvements in production rate and furnace efficiency. The main aim of the blast furnace manager is to achieve a *rapid* but *regular* charge descent.

Blast preheat and oxygen enrichment

Preheating the air blast means that less coke need be burnt to supply the heat. The blast furnace reduction of iron oxide requires extremely expensive high quality coke and any savings in the use of this material will lead to considerable cost improvements. The air is preheated by passing through a heated refractory network in hot blast (Cowper) stoves. Enriching the preheated air with oxygen will also reduce the coke rate (amount of coke consumed per tonne of iron produced) in the same way as preheated air. However, replacing some of the air with oxygen produces a reduction in blast furnace gas volume which can be passed up the stack. Hence, the amount of heat transferred from the ascending gas to the charge is decreased which reduces the amount of indirect reduction that can take place and therefore necessitates an equivalent increase in the amount of direct reduction in the hearth. As discussed earlier in the chapter, direct reduction is endothermic while indirect reduction is exothermic. Thus, oxygen enrichment is kept well below 9%.

It is cheaper to use an increased air preheat than oxygen enrichment but the preheat temperature is limited to the refractory properties in the hot blast stoves. Typical preheat temperatures vary in the range 900–1200 °C.

Hydrocarbon and steam injection

Injection of hydrocarbons into the stack of the blast furnace results in their combustion to provide two reducing gases, carbon monoxide and hydrogen, although some carbon dioxide and steam will also be produced. For example,

$$2CH_4 + O_2 \longrightarrow 4H_2 + 2CO \qquad \Delta H^{\ominus}_{298} = -71.4 \text{ kJ}$$

As discussed earlier (Section 7.2.5), hydrogen is a better reducing agent for metal oxides than carbon monoxide below 800 °C while CO is a better reducing agent than H_2 above 800 °C. The combination of these two reducing gases provides an increase in the efficiency of the indirect reduction reactions. However, the exothermic value of the combustion of the hydrocarbon is less than that for coke,

$$2C + O_2 \longrightarrow 2CO \qquad \Delta H^{\ominus}_{298} = -222 \text{ kJ}$$

(see reaction 7.III), and an increase in blast preheat is required to maintain hearth temperature. The hydrocarbons most commonly used are: (i) natural gas; (ii) atomised oil; and (iii) pulverised coal. All of these are injected into the furnace via the tuyeres, although a $CO + H_2$ mixture has also been injected just above the tuyeres. In most cases an increase in the amount of indirect reduction and reduction in coke rate has been produced. The injection of steam through the tuyeres also results in H_2 and CO formation but with an endothermic reaction with coke:

$$H_2O + C \longrightarrow H_2 + CO \qquad \Delta H^{\ominus}_{298} = +132 \text{ kJ} \qquad (7.XIV)$$

Therefore, steam injection will quickly lower the hearth temperature and is used in conjunction with a high blast preheat and/or oxygen enrichment to rapidly and efficiently control the hearth temperature. Typical values for steam injection vary between 1 to 4% enrichment of the blast, again resulting in an increased amount of indirect reduction, reduced coke rate and increased production.

High top pressure

Increasing the pressure of the blast furnace gas at the top of the stack will provide a smaller increase of the gas at the tuyere level. In this way the pressure difference between the gas at the top of the stack and that at the tuyeres will be reduced. Reducing the pressure difference in the stack will slow down the rate of ascent of the gases, increasing their residence time in the stack and increasing their contact with the iron oxides. This produces an increase in the efficiency of coke combustion and pig iron production rate and is represented diagrammatically in *Fig. 7.21*. Typical top pressures vary from 50.6 to 289.6 kN m^{-2} (0.5 to 2.8 atm) although pressures up to 361.8 kN m^{-2} (3.57 atm) have been tried.

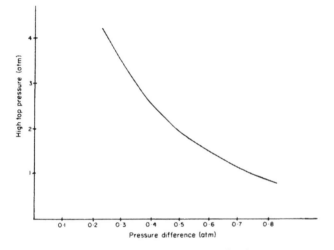

Figure 7.21 Variation of pressure difference of blast furnace gas at tuyeres to that at the top of the stack using different high top pressures. (Based on data presented in *Blast Furnace Theory and Practice* edited by J.H. Strassburger, published by Gordon and Breach Science Publishers for American Institute of Mining, Metallurgical and Petroleum Engineers Inc.)

Blast furnace practice using the above techniques has resulted in production rates in excess of 10 000 tonnes per day producing an impure pig iron with 3.5–4%C, 1–2% Si, about 1% Mn, 0.2–2.5% P and 0.04–0.2%S. However, blast furnace production has certain disadvantages:

(i)   Control over pig iron composition is generally poor with composition varying from tap to tap.
(ii) The large production rates may not be required, especially for developing countries whose steel outlets are limited.
(iii) High capital and operating costs are incurred.

Alternatives to blast furnace production of iron include:

(i)   Smelting in an electric arc furnace where cheap electricity is available.

(ii)   Low shaft furnace smelting similar to the bottom third of the blast furnace and using 9% oxygen enrichment. The smaller stack allows the weaker, cheaper carbon forms such as coals and lignites to be used as the reducing agent.

Both these processes rely mainly on direct reduction reactions with only a small amount of indirect reduction.

Direct reduction (DR) and direct smelting (DS) processes

The most important alternatives to the blast furnace are the *direct reduction* (DR) and *direct smelting* (DS) *processes*. In the DR processes carbon in the form of coal dust, coke breeze, char or anthracite or, alternatively, carbon monoxide and hydrogen are added to the reduction unit resulting in direct conversion of oxide to steel in one step. Therefore, even though CO and H_2 may be used as the reducing agents they are being used directly without any intermediate reactions; carbon monoxide and hydrogen being formed by combustion of hydrocarbons with air or steam prior to their use as reducing agents.

The processes are almost totally direct reduction processes with very little, if any, indirect reduction reactions taking place. Solid reducing agents (coal dust, coke breeze, char or anthracite) are used in rotary kiln reduction of pelletised iron ore; CO and H_2 are used in static bed reactors, low shaft furnaces (similar to the bottom third of the blast furnace) or fluidised bed furnaces in the reduction of pelletised iron ore. Diagrammatic representations of these processes are given in *Fig. 7.22*. The operating temperatures are in the range 1000–1200 °C except for the fluidised bed process using hydrogen as the main reducing agent. In this case a lower temperature, 750 °C, can be used

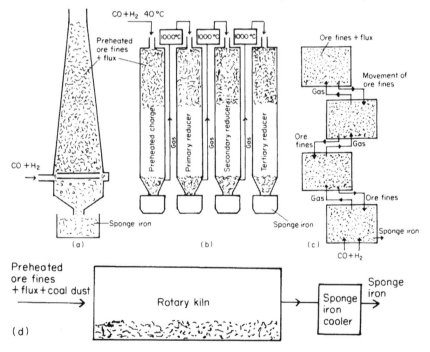

Figure 7.22 Diagrammatic representation of direct reduction processes for iron ore: (a) gaseous reduction in a shaft furnace; (b) gaseous reduction in static bed retorts; (c) gaseous reduction in fluidised beds; (d) solid reduction in a rotary kiln

due to the increased reducing efficiency of hydrogen below 800 °C and the improved heat transfer of the fluidised bed. The direct reduction processes are closer to equilibrium conditions than the blast furnace reduction of iron oxide and although these processes produce a solid or sponge iron with a lower purity than that of pig iron the compositional control is superior. Sponge iron is used mainly as a feedstock for electric arc steelmaking.

There are several radically new direct smelting (DS) processes which are emerging as contenders to both the blast furnace (BF) and the so-called direct reduction (DR) processes for ironmaking and steelmaking. These DS processes have the following salient features:

— coal used as the reductant;
— minimum iron ore processing (pelletising and sintering);
— preferable direct use of concentrate;
— minimum ancillary plant and material handling;
— low capital investment;
— molten iron product that can be readily processed into required steel composition;
— maximum continuous processing;
— maximum pollution control; and
— minimum energy consumption and application of cogenerative principles.

Direct smelting (DS) processes, unlike DR processes, produce a liquid iron product similar to the pig iron produced in the blast furnace, while the coupling of a pre-reduction stage with a final reduction smelting stage and the incorporation of cogenerative principles with respect to maximum energy utilisation appear to be common design features. The SKF plasma smelt process (*Fig. 7.23*) is one such plasma DS process incorporating prereduction in fluidised beds and final reduction in a low shaft furnace heated with a plasma arc heater through which is injected process off gas, coal and iron ore fines. The coke filled low shaft furnace serves as a gas reformer with the excess sensible and chemical energy from the process being used to generate extra electricity.

The KR process exemplifies the same principles applied to a total fossil fuel DS process in which prereduction is achieved in a shaft furnace using similar technology to the MIDREX process (*Fig. 7.22(a)*). The prereduced product is subsequently finally refined by being continuously fed into a converter similar to a steel converter described in Section 7.2.10.6 and *Fig. 7.24(c)*.

7.2.10.6   Steelmaking

Whereas ironmaking is a reduction process producing an impure iron containing carbon, silicon, manganese, phosphorus and sulphur as the main impurities, steelmaking is an oxidising (fire-refining) process in which the impurities in the pig iron are lowered to acceptable levels depending on the required steel specification.

The relevant steelmaking oxidation reaction lines are reproduced in the $\Delta G^{\ominus}-T$ diagram in *Fig. 7.20*. From this it would be expected that at the steelmaking temperature of about 1600 °C preferential oxidation of C, Si, Mn would take place before excessive oxidation of iron. The impurity elements in the pig iron oxidise during steelmaking with exothermic reactions[15]:

$$2P(1\% \text{ in Fe}) + 5O(\text{in Fe}) \longrightarrow P_2O_5(\text{in slag}) \qquad \Delta H = -1200 \text{ kJ} \qquad (7.XV)$$

$$Si(1\% \text{ in Fe}) + 2O(\text{in Fe}) \longrightarrow SiO_2(\text{in slag}) \qquad \Delta H = -745 \text{ kJ} \qquad (7.XVI)$$

$$Mn(1\% \text{ in Fe}) + O(\text{in Fe}) \longrightarrow MnO(\text{in slag}) \qquad \Delta H = -399 \text{ kJ} \qquad (7.XVII)$$

$$C(1\% \text{ in Fe}) + O(\text{in Fe}) \longrightarrow CO(\text{gas}) \qquad \Delta H = -133 \text{ kJ} \qquad (7.XVIII)$$

Under steelmaking conditions P_2O_5 exists in the slag as a liquid and is not evolved as a gas. Atomic oxygen (O) is used in the above reactions rather than molecular oxygen

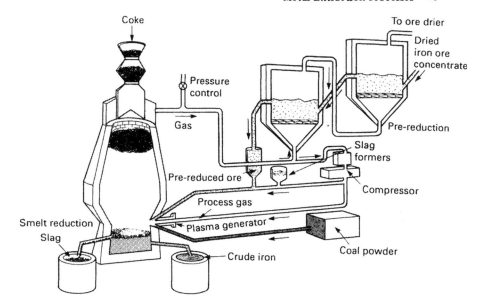

Figure 7.23 Schematic representation of the SKF plasma smelt process

(O_2) since it is assumed that oxygen is dissolved in the steel by dissociation of an oxide of iron at the slag–metal interface releasing atomic oxygen to the metal:

$$\underset{\text{(in slag)}}{Fe_2O_3} \longrightarrow \underset{\text{(in slag)}}{2FeO} + \underset{\text{(in Fe)}}{O} \qquad (7.XIX)$$

In oxygen blown processes, apart from direct reaction between the metaloids (Si, Mn, C, P) and molecular oxygen, iron is oxidised to FeO which either reacts with metaloids, e.g. $FeO + Mn \rightarrow Fe + MnO$, or is further oxidised to Fe_2O_3 which diffuses through the slag to release atomic oxygen to the melt by the dissociation reaction (reaction (7.XIX)) at the slag–metal interface. Therefore, slag fluidity, turbulence and slag–metal interfacial area are important factors in the supply of oxygen to the metal and hence on the production rates.

In practice, Si and Mn oxidise immediately but the oxidation of carbon is delayed due to the problems of producing a bubble of CO gas in the liquid steel. This is due to the very large interfacial energy between liquid steel and the CO gas increasing the critical nucleus bubble size required for nucleation:

$$r_c = \frac{2\gamma T_f}{L_f \Delta T} \quad \text{(see Section 4.3.11)} \qquad (7.4)$$

Under normal steelmaking conditions the critical radius size demanded for homogeneous nucleation of a CO bubble in liquid steel is too high and therefore the CO gas bubbles nucleate on various materials which reduce the value of γ, e.g. refractory crevices. This is called heterogeneous nucleation and accounts for the oxidation of carbon after silicon and manganese. The carbon–oxygen relationship is the most important in steelmaking and is shown in the graph in *Table 7.3* for the open hearth process. The removal rate (oxidation rate) of carbon is controlled during the steelmaking process so that the steel may be tapped as it approaches its required specification. This is called the *catch carbon* method. If the carbon is tapped below its required level, anthracite or pig iron is added to the steel in the ladle to bring the carbon up to specification. This is generally considered bad practice since it can lead to an increase in inclusions and a considerable lowering of the metal temperature. The transfer of oxygen across the slag–metal and gas–metal interfaces is generally considered to be rate controlling. The oxidation of carbon provides evolution of CO (reaction (7.XVIII)) which is insoluble in the slag and evolved into the furnace atmosphere giving the effect of a boiling action, i.e. the *carbon boil reaction*. The turbulence created by this reaction is an important factor in improving the reaction kinetics and, therefore, the production rates in the open hearth and electric arc steelmaking processes. The removal of phosphorus and sulphur is not so straight forward. Both these elements if present to any significant level will result in a deterioration of the mechanical properties of the steel. Normally, they are lowered to a maximum of 0.04% but much lower levels may be required for the high quality steels. Preferential oxidation of phosphorus at $1600\,^{\circ}C$ is not possible when reactants and products are in their standard thermodynamic states.

The main requirement for achieving preferential oxidation of phosphorus during steelmaking is:

$$\Delta G_{P_2O_5,1600^{\circ}C} < \Delta G_{FeO,1600^{\circ}C} \qquad (7.5)$$

Therefore, applying the van't Hoff isotherm for non-equilibrium conditions to the phosphorus oxidation reaction (7.XV):

$$\Delta G_{P_2O_5,1600^{\circ}C} = \Delta G^{\ominus}_{P_2O_5,1600^{\circ}C} + RT \ln \frac{a^2_{P_2O_5(\text{in slag})}}{a^2_{P(\text{in Fe})}\, a^5_{O(\text{in Fe})}} \qquad (7.6)$$

indicates that if equation (7.5) is to be satisfied the variable term (second term on the right-hand side) in equation (7.6) must be made highly negative. This is achieved in practice by using a basic slag which lowers $a_{P_2O_5(\text{in slag})}$, i.e.

$$4CaO + P_2O_5 \longrightarrow Ca_3(PO_4)_2 + \text{excess CaO} \qquad (7.XX)$$

and by increasing the supply of oxygen to the metal ($a_{O(\text{in Fe})}$ – the oxygen potential of the metal) by increasing iron ore and mill scale (Fe_3O_4–Fe_2O_3) additions or using oxygen blowing.

Examination of the $\Delta G^{\ominus}$–T relationship for the phosphorus oxidation reaction[16],

$$2P(\text{in Fe}) + 5O(\text{in Fe}) \rightarrow P_2O_5(\text{in slag})$$
$$\Delta G^{\ominus} = -683\,000 + 580T \text{ J mol}^{-1} \qquad (7.XXI)$$

shows that decreasing the melt temperature makes $\Delta G^{\ominus}$ more negative, favouring the forward reaction and phosphorus removal to the slag. However, the melt temperature must be maintained at such a level to ensure adequate fluidity otherwise the reaction kinetics will be hindered leading to slow refining rates. Therefore, the conditions necessary for phosphorus removal during steelmaking are: (i) basic slag formation; (ii) high oxygen potential in the metal; and (iii) low metal temperature.

Sulphur removal during steelmaking is effected in the same way as in ironmaking in that an element is added (Ca in the form of CaO) to the melt which forms a more stable sulphide than FeS or MnS. However, steelmaking conditions offer much greater control over sulphur elimination than in the blast furnace. The conditions necessary for sulphur removal during steelmaking are: (i) basic slag formation; (ii) low oxygen potential in the metal; and (iii) high metal temperature.

With respect to (ii) maximum sulphur elimination is achieved using a reducing slag in which anthracite or ferrosilicon is used as the reducing agent. The metaloids (C, Si, Mn, P) in the iron are first oxidised using an oxidising slag which is replaced by a reducing slag to remove the sulphur. However, this technique can be used efficiently only in the electric arc steelmaking process. In the other processes the oxygen potential of the metal is lowered after first oxidising C, Si, Mn, P, by stopping the iron ore and mill scale additions or stopping or reducing the oxygen blowing. In this respect the electric arc process offers the highest removal of sulphur while the oxygen blown processes offer only limited sulphur removal.

The main control problem during steelmaking is to ensure the sulphur and phosphorus levels have been lowered to below their maximum allowed value (normally 0.04%) before the carbon has reached its required specification. In this way the catch carbon technique can be used in which the carbon removal rate in the final stages of refining is so controlled to facilitate tapping of the furnace at the required composition. As can be seen from the previous discussion, the only common feature that aids both phosphorus and sulphur removal is a basic slag. Phosphorus removal is favoured by a low melt temperature and high oxygen potential in the metal while sulphur removal is favoured by a high melt temperature and low oxygen potential in the metal. The steel bath conditions, therefore, need to be varied to provide maximum removal of each element. However, a highly basic slag will remove some phosphorus and sulphur irrespective of the melt temperature and oxygen potential.

The main requirement is, therefore, to effect the formation of a basic slag as early as possible in the steelmaking process. This is best achieved by using fine powdered lime additions and a relatively high slag temperature to aid dissolution of the lime in the slag at an early stage. The high lime slag will then readily accept phosphorus and sulphur. Once the carbon has been removed to the required level a deoxidant, e.g. ferrosilicon, may be added to the bath to remove sufficient oxygen from the melt to prevent any more decarburisation. Further deoxidant additions using ferrosilicon, ferromanganese and aluminium may be added to the steel after tapping into a ladle or in a vacuum degassing unit or to the ingot mould during teeming. The amount of deoxidant used will depend on the structure required in the solidified steel, i.e. rimmed, semi-killed or killed. Low to medium (up to about 4%) alloying additions are added to the steel in the ladle after deoxidation thus preventing their removal to the slag as oxides. A greater amount of alloying additions cannot effectively be added to the ladle steel without undue cooling and loss of compositional uniformity and, therefore, must be added in the initial charge which will result in some loss (oxidation) to the slag. In the production of alloy steels, especially high alloy steels, it is necessary to achieve maximum recovery of these expensive materials. This is achieved by consideration of the appropriate thermodynamic properties, e.g. variation of the equilibrium constant with temperature and relationship between the activities of the relevant species at a certain oxygen potential.

It would be useful here to use the example of the production of a high chromium alloy steel such as 18–8 (Cr–Ni) stainless steel. The main process problem is to facilitate the oxidation of carbon to very low limits, i.e. 0.05%, without simultaneous oxidation of the expensive chromium alloy. The required amount of chromium in the steel is melted with other charge materials in an electric arc furnace using an oxidising slag.

This results in the formation of oxides of chromium which are collected in the slag[17]:

$$2Cr(\text{in Fe}) + 3O(\text{in Fe}) \rightarrow Cr_2O_3(\text{in slag})$$
$$\Delta G^{\ominus} = -887\,400 + 393T \text{ J mol}^{-1} \quad (7.XXII)$$

Some recovery of the chromium from the slag can be made by increasing the melt temperature to $1850\,^{\circ}C$ by oxygen lancing. This makes $\Delta G^{\ominus}$ for reaction (7.XXII) more positive, favouring the reverse reaction and chromium recovery to the steel. A deoxidant is added, e.g. ferrosilicon, which will form a more stable oxide than Cr_2O_3 thereby effecting further recovery:

$$2Cr_2O_3(\text{in slag}) + 3Si \longrightarrow 4Cr(\text{in Fe}) + 3SiO_2(\text{in slag})$$

A further consideration is the relationship* between the activities of carbon $(a_{C(\text{in Fe})})$ and chromium $(a_{Cr(\text{in Fe})})$ in the steel and the partial pressure of carbon monoxide (p_{CO}) in the furnace atmosphere at a constant oxygen potential in the steel $(a_{O(\text{in Fe})})$:

$$a_{C(\text{in Fe})} \propto a_{Cr(\text{in Fe})}p_{CO} \quad (7.7)$$

Thus as $a_{C(\text{in Fe})}$ decreases $a_{Cr(\text{in Fe})}$ will decrease at a constant value of p_{CO}. However, reducing p_{CO} will allow decarburisation to take place while still maintaining a high chromium content, i.e., lowering p_{CO} at high temperatures makes ΔG_{C-CO} more negative than $\Delta G_{Cr-Cr_2O_3}$ resulting in preferential oxidation of C. This is achieved by first melting the appropriate alloys in an electric arc furnace and transferring the melt to an argon–oxygen decarburisation (AOD)[18] converter through the bottom (or side) of which is blown oxygen and an inert gas (argon). The flushing of the inert gas reduces the partial pressure of the carbon monoxide thereby allowing carbon to be oxidised while retaining chromium in accordance with equation (7.7). Stainless steels with less than 0.05%C are required for welding applications; higher carbon levels may produce weld decay. Oxygen–steam[19] bottom injection in the newer CLU process has also been used in the same way as O_2–Ar. The same thermodynamic considerations can be used for other alloying elements, e.g. Ni, Co, Mo, Mn, Si, Mg, Al, Ti, Nb, W. In general Ni, Co, Mo are added without much danger of removal to the slag since they form oxides which are less stable than FeO. Manganese and silicon having been removed initially, the charge materials need to be alloyed back by addition of their ferroalloys. If these are oxidised to the slag they can be recovered in a similar way to chromium using appropriate reducing conditions. Magnesium, aluminium, titanium, niobium and tungsten form very stable oxides and must be added to the ladle steel after deoxidation. Addition of the alloying elements during vacuum melting will provide high recovery rates except for those elements with high vapour pressures.

*This relationship is determined by considering the equilibrium constants for the reactions (7.XXII) and (7.XVIII) occurring at a constant $a_{O(\text{in Fe})}$:

$$a_{O(\text{in Fe})} = \frac{1}{K_{Cr}^{1/3}} \cdot \frac{a_{Cr_2O_3}^{1/3}}{a_{Cr}^{2/3}} = \frac{1}{K_C} \cdot \frac{p_{CO}}{a_{C(\text{in Fe})}}$$

Therefore

$$a_{C(\text{in Fe})} = \frac{K_{Cr}^{1/3}}{K_C} \cdot \frac{p_{CO}a_{Cr}^{2/3}}{a_{Cr_2O_3}^{1/3}}$$

Assuming $a_{Cr_2O_3} = 1$, then $a_{C(\text{in Fe})} = Ka_{Cr}^{2/3}p_{CO}$, where $K = K_{Cr}^{1/3}/K_C$.

Steelmaking units can be classified under the headings:

(i)   pneumatic or oxygen blown converter processes: (a) top blown processes (LD, LD–AC, Kaldo, Rotor); (b) bottom blown processes (Bessemer, AOD, CLU, BOP, QBOP, LWS);

(ii) electric arc process;

(iii) open hearth process.

Figure 7.24 provides diagrammatic illustrations of the major steelmaking units with the electric arc represented in *Fig. 7.8(b)*, while *Table 7.3* provides a brief comparison of these units.

The Bessemer process was the first steelmaking process developed which oxidised the metalloids by blowing air through the bottom of a converter. The recent developments of oxygen–hydrocarbon (BOP) and oxygen–hydrocarbon–flux (QBOP) blown through the bottom of a converter improve process control and extend the range of steels that can be produced. The turbulence produced by oxygen blowing improves the reaction kinetics of the process.

Top blown oxygen processes offer less turbulence than the bottom blown processes but, when coupled with powdered lime injection (LD–AC), provide good phosphorus removal. These processes are often called BOS (basic–oxygen steelmaking). However, due to the high oxygen potential of the metal, only moderate sulphur removal is possible in oxygen blown processes.

More recently combined top and bottom blowing is being used in which oxygen is blown from the top and either an active, i.e. oxygen, or inert gas, e.g. N_2 or Ar, is blown from the bottom. Inert gas bottom blowing increases turbulence, reaction kinetics and control over the refining reactions, thereby giving the best of both blowing operations.

Increased processing efficiency is gained by completing the major desulphurising, deoxidising, alloying and degassing operations in a ladle, i.e. ladle metallurgy, having used the steelmaking unit to oxidise the major impurities, e.g. C, Si, Mn, P, etc. Typical ladle metallurgy units are schematically presented in *Fig. 7.25*.

No fuel is required with the pneumatic processes due to the large exothermic reactions produced (reactions (7.XV)–(7.XVIII)), but a molten metal charge is necessary although up to 30% scrap can be added as a coolant.

Incorporating a rotating action (Rotor) and inclining the converter (Kaldo) improves slag–metal contact and aids thermal efficiency by recovering the heat from the refractories, especially when a secondary lance (Kaldo) is placed above the slag to burn the CO produced from the refining reactions:

$$2CO + O_2 \longrightarrow 2CO_2 \quad \Delta H^{\ominus}_{298} = -572 \text{ kJ}$$

However, these rotor processes are subject to excessive refractory wear and costs. The electric arc process is slower than pneumatic processes but provides excellent control over the refining reactions and is most used in the production of high quality and alloy steels. The electric arc furnace is an ideal remelter of ferrous scrap and is, therefore, used with a charge material of at least 50% scrap or cold pig iron. The widest range of steels of any steelmaking unit can be produced in the electric arc, ranging from plain carbon to high alloy steels.

The open hearth furnace is similar to a reverberatory furnace with a heat recuperative system and can operate with any charge materials ranging from 100% cold charge to 100% hot metal charge. The long, shallow hearth allows very high control over refining reactions but incurs a prohibitively low production rate and occupies a large amount of floor space. For these latter reasons the electric arc and oxygen blown processes are replacing the open hearth furnace, for scrap metal charges and hot metal charges respectively, although some improvement in refining rates has been provided by injecting oxygen through tuyeres below the slag–metal interface in the open hearth (SIP) process.

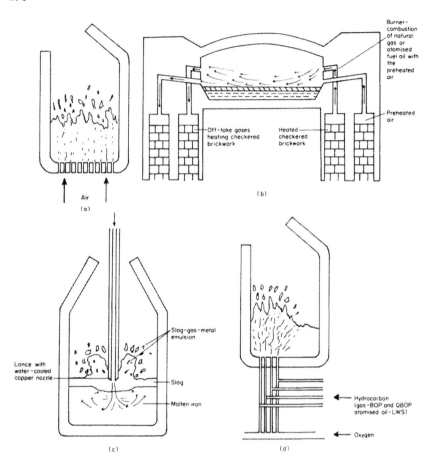

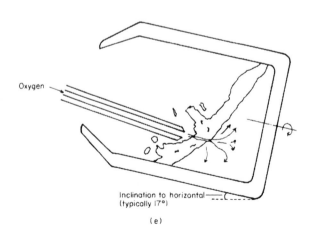

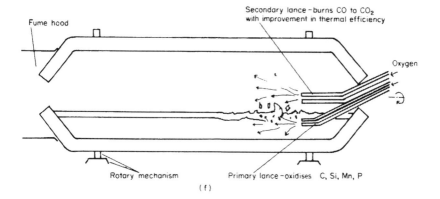

Figure 7.24 Diagrams of steelmaking units: (a) Bessemer; (b) openhearth; (c) top blown oxygen converter, BOS, e.g. LD and LD–AC; (d) bottom blown oxygen converter, e.g. QBOP; (e) Kaldo; (f) rotor

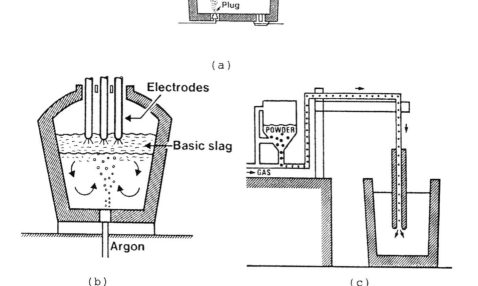

Figure 7.25 Ladle metallurgy equipment: (a) refining slag with bottom porous plug inert gas (Ar) stirring; (b) heated ladle refining unit with bottom porous plug, refining slag and slag-metal temperature control via electrodes; (c) deep ladle injection of calcium compounds for desulphurisation

Table 7.3 COMPARISON OF STEELMAKING UNITS

Unit	Fuel and oxygen supply	Charge materials	Production rates and range of steels produced	Remarks
(a) Open hearth furnace	Atomised oil or natural gas. Oxygen supply via iron ore and mill scale additions to the slag	Any charge composition ranging from 100% hot metal to 100% cold metal	100% molten metal–120 tonnes in 5–7 hours. 100% cold metal–120 tonnes in 10–12 hours. Furnaces up to 400 tonnes capacity have been used. Production of plain carbon steels and low to medium alloy steels, e.g. up to about 4% added to ladle steel after deoxidation	Long shallow bath with low turbulence provides high degree of control over refining reactions but slow production rates. Production rates may be increased by use of oxygen enrichment of the air during melting and oxygen injection of the melt during refining. High capital cost and floor space required. Relationship between carbon and oxygen is the most important in steelmaking. The rate controlling steps are the transfer of oxygen across the slag–metal interface for the open hearth and electric arc processes and also across the gas–metal interface in oxygen blown processes.

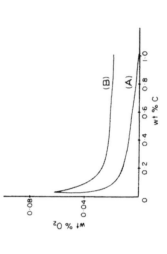

Carbon–oxygen relationship for: A, equilibrium conditions; and B, basic open hearth steelmaking. The LD C–O data are closer to A than B

(b) Electric arc	Electricity— 3 phase AC supply. Oxygen supply via iron ore and mill scale additions to the slag. Oxygen lancing may also be used during refining	100% cold charge and up to 50% molten metal	100% cold charge— 120 tonnes in 4 hours. Production of steels range from plain carbon to high alloy	Provides greatest degree of control over refining reactions, temperature, furnace atmosphere and slag conditions compared with all other steelmaking units. Excellent sulphur removal using reducing slag with basicity ratio of 4. Ideal unit for remelting scrap and producing high quality steels. Ultra-high power (UHP) furnaces (900 kV A with a transformer rating of 100 MV A) offers increased melting and refining rates compared with high power (HP) furnaces (400–600 kV A)
(c) Top blown oxygen converters. Also called BOS or BOF— basic oxygen steelmaking or basic oxygen furnace	Oxygen. Iron ore and mill scale additions to the slag may also be used	Molten pig iron, flux, up to 30% scrap as coolant	300 tonnes in 30–40 minutes Production of plain carbon and low to medium alloy steels, e.g. up to about 4% alloys added to ladle steel after tapping	Less turbulence than Bessemer, longer production time offering greater control over final composition (LD). Top blowing can be interrupted for sampling. Modification of injecting powdered lime with oxygen (LD–AC) improves phosphorus (good) and sulphur (moderate). Top blowing results in slag temperature being greater than melt temperature which provides early dissolution of CaO and formation of basic slag. Slag–metal–O_2 emulsion produced above slag—provides ideal oxidation of C, Si, Mn, P, but limited S removal as shown in the graph. Rotation (Rotor) and inclining (Kaldo) rotating vessel increases agitation of the melt and improves slag–metal contact but refractory costs tend to be excessive.

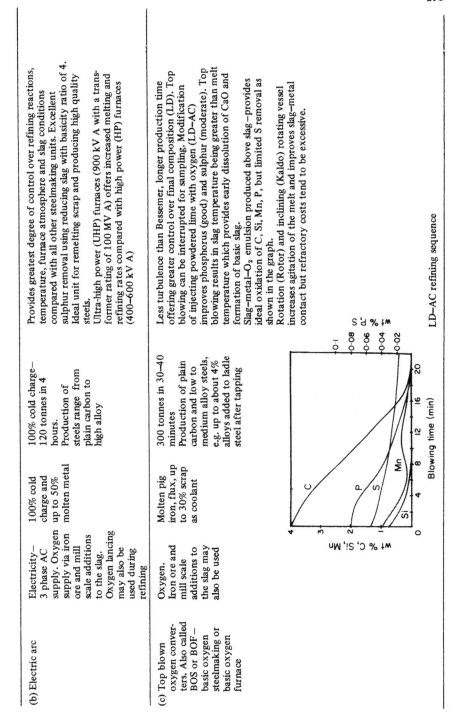

LD–AC refining sequence

Unit	Fuel and oxygen supply	Charge materials	Production rates and range of steels produced	Remarks
(d) Bessemer converter	Air	Molten pig iron, flux, up to 30% scrap as coolant	50 tonnes in 20 minutes. Production of low carbon steels only	The first bulk steelmaking process. High degree of turbulence and rapid reaction rates producing poor controllability. High phosphorus removal and poor sulphur removal due to high oxygen potential of the melt. Phosphorus removal after carbon due to low slag temperature not being able to dissolve CaO until late in process cycle. Thus, only suitable for production of low carbon steels. High build-up of N_2 in steel due to the use of air. Sequence of removal of elements is shown in graph. This was the forerunner of the modern top and bottom blown oxygen converters. Bessemer capacity limited to 50–60 tonnes due to mechanical problems involved with tilting (for charging and tapping) and bottom blowing.

Bessemer refining sequence

(e) Bottom blown oxygen converters	Oxygen–steam, oxygen–CO_2, oxygen–hydrocarbon. Iron ore and mill scale additions to the slag may also be used	Molten pig iron, flux, up to 30% scrap as coolant	200 tonnes in 30–40 minutes. Production of plain carbon and low to medium alloy steels, e.g. up to about 4% alloys	Use of oxygen provides increased slag temperature compared with Bessemer process. Allows dissolution of CaO in the slag at an early stage in the process cycle effecting P removal before C. Only moderate S removal due to high oxygen potential of melt. Excessive temperature build-up and bottom tuyere erosion is reduced by bottom blowing a hydrocarbon gas (BOP) or atomised oil (LWS) with the oxygen. As little as 2% hydrocarbon enrichment has been found to be effective while bottom injection of powdered lime (QBOP) has also been used to improve P and S removal.

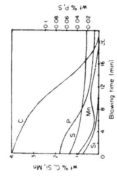

QBOP refining sequence

(f) Argon–oxygen decarburisation (AOD)	Argon (6%)–oxygen–N_2 may be used initially in place of Ar	Molten steel—initial melting usually conducted in electric arc	200 tonnes in 90 minutes. Production of high alloy steels and other alloys, e.g. Ni–Cr	AOD converter ideal for producing high Cr low C alloy steels after initial melting in an electric arc. Oxygen–steam bottom blowing (CLU) can be used in the same way as the AOD to produce high alloy steels

7.3   Hydrometallurgical extraction processes

Hydrometallurgical extraction processes involve dissolving (leaching) the metal value from the ore, concentrate or process waste in an aqueous solution (leaching reagent) and subsequently precipitating or isolating the metal required to be extracted. The ore mineral may not be soluble in water necessitating a pretreatment such as a sulphatising roast producing a water soluble metal sulphate. Addition of acids, alkalis or certain salts to the water may also aid dissolution of the required mineral. The leaching process will usually take several minerals into solution and it may be necessary to first purify the leached solution by precipitation of certain unwanted compounds or metals prior to the precipitation or isolation of the required metal. Precipitation results in a metal or group of metals being thrown out of solution in a solid form while isolation techniques result in the preferential selection of a metal ion or group of metal ions in a suitable solvent.

Precipitation may be effected by pH control, addition of certain chemicals, simply cooling the solution, or by chemical reduction or oxidation. Isolation techniques include reverse osmosis, ion exchange and solvent extraction.

Electrolytic separation techniques are sometimes classified as hydrometallurgical processes but these will be dealt with in a separate section.

7.3.1   Leaching

Leaching involves dissolving the required metal in the concentrate in a suitable reagent usually taking many other metals also present in the ore into solution at the same time. The leaching reagent must be cheap and recoverable from the solution to aid the total process economics. Some typical reagents are listed in *Table 7.4.*

Table 7.4 SOME COMMON LEACHING REAGENTS USED FOR VARIOUS METAL BEARING MINERALS

Mineral	*Leaching reagent*
Metal oxides	H_2SO_4
Metal sulphates	H_2SO_4 or even H_2O
Metal sulphides	$Fe_2(SO_4)_3$ solution
Cu/Ni minerals	NH_3 solutions, NH_4CO_3 −producing complex amines, e.g. $[Ni(NH_3)_6]^{2+}$
Al_2O_3	NaOH
Au, Ag	NaCN−producing complex cyanides, e.g. $Au(CN)_2^-$

The rate controlling steps in the leaching process are generally the transport of the reagent to and the transport of the leached solution away from the dissolution reaction site, rather than the dissolution reaction itself. Thus, the reaction kinetics of the leaching process are improved by increasing the diffusion rates of the reagent and leached solution through the pores of the mineral being leached; presenting the mineral in a fine particular form by prior crushing and grinding; providing increased agitation during leaching and increasing the temperature and pressure of the reaction. At the same time a high degree of crushing and grinding will aid any subsequent separation of the ore, thereby reducing the gangue content which may consume large quantities of the leaching reagent.

There are several different leaching methods used depending on such factors as required leaching rates, ore composition, subsequent isolation, precipitation or extraction process:

(i)   Heap leaching is used when ore lumps of less than 200 mm diameter are stacked in the open on drainage culverts and the leach reagent sprayed over the ore heaps. This is a slow process with low recovery rates.

(ii) Percolation leaching uses ore fines of about 6 mm placed in large tanks and percolated by a series of leach solutions increasing in concentration.

(iii) Agitation leaching uses ore fines of 0.4 mm diameter or less which would be unsuitable for percolation leaching. The fine ore particles are suspended in the leach solution and stirred mechanically or with compressed air jets. This is faster and more efficient than percolation leaching but is also more costly.

(iv) Pressure leaching at elevated temperatures is used to increase dissolution rates and to dissolve minerals which would be difficult to dissolve at atmospheric pressure. It also prevents dissolution of gangue components in certain cases. In this respect Ni, Co and Cu sulphides are dissolved in the presence of oxygen with ammonia solution between 71 and 87 °C at pressures of 7.1 to 10.5 atm ($709.3–1064$ kN m^{-2})[20]. The iron sulphide present in the ore is oxidised under these conditions and remains undissolved as ferric hydroxide ($Fe(OH)_3$). The solution containing the Ni, Co, and Cu as complex amines is first treated to precipitate the copper as Cu_2S by boiling at atmospheric pressure and after oxidation of the thiosulphates using air at 50 atm (5066.2 kN m^{-2}) pressure and 200 °C the nickel is precipitated by reduction with hydrogen.

The Bayer process treats impure bauxite ore with a strong caustic soda solution ($130–350$ g Na_2O/l) at 5–10 atm ($506.6–1013.2$ kN m^{-2}) pressure and between 150–170 °C. The aluminium is dissolved as the AlO_2^- anion,

$$Al_2O_3 \text{(bauxite)} + 2NaOH \xrightarrow[\text{8 atm}]{160\,^\circ\text{C}} 2NaAlO_2 + H_2O$$

leaving iron, silicon and titanium oxides undissolved. Hydrated alumina, $Al_2O_3.3H_2O$, is precipitated on cooling and 'seeding' with pure $Al_2O_3.3H_2O$ crystals. Subsequent calcination at 1200 °C provides the pure alumina which is used in the fused salt electrolytic extraction of aluminium (Section 7.4.2).

(v) Bacterial leaching has been used to increase the leaching rates of sulphides by a factor which ranges between 10 to 100. In this respect the bacterium thiobacyllus ferro-oxydans has been used in the leaching of copper sulphide ores containing iron. The bacterium oxidises Fe^{2+} to Fe^{3+} preventing the dissolution of iron and, therefore, increasing the dissolution rate of the copper.

The leached solution in each case will contain several unwanted minerals which may hinder the extraction of the required metal and it may be necessary to purify the leached solution by precipitating some of these unwanted minerals prior to the precipitation or isolation of the main metal.

Leach or *in situ* mining in which a leaching solution is flushed through an ore body, is likely to become more popular for the very lean ores which are difficult to mine, e.g. uranium ores.

7.3.2 Precipitation techniques

7.3.2.1 *Precipitation by pH and p_{O_2} control*

The precipitation of a metal from an aqueous solution is largely controlled by the pH and partial pressure of oxygen of the solution. Thus, the thermodynamic relationships between the metal ions, hydrogen ions and oxygen present in solution provide the hydrometallurgist with a considerable amount of knowledge. Rosenqvist[21] has explained this relationship by considering all the reactions between a hypothetical metal Me, H^+ ions and O_2 pressure, when the metal forms cations Me^{2+} and Me^{3+}, anions MeO_2^{2-} and MeO_2^-, and solid hydroxides $Me(OH)_2$ and $Me(OH)_3$. Therefore, the following reactions are possible:

$$2Me + 4H^+ + O_2 \longrightarrow 2Me^{2+} + 2H_2O \qquad (7.XXIII)$$

$$4Me^{2+} + 4H^+ + O_2 \longrightarrow 4Me^{3+} + 2H_2O \qquad (7.XXIV)$$

$$Me^{2+} + 2H_2O \longrightarrow Me(OH)_2 + 2H^+ \qquad (7.XXV)$$

$$Me^{3+} + 3H_2O \longrightarrow Me(OH)_3 + 3H^+ \qquad (7.XXVI)$$

$$Me(OH)_2 \longrightarrow MeO_2^{2-} + 2H^+ \qquad (7.XXVII)$$

$$Me(OH)_3 \longrightarrow MeO_2^- + H^+ + H_2O \qquad (7.XXVIII)$$

$$2Me + 2H_2O + O_2 \longrightarrow 2Me(OH)_2 \qquad (7.XXIX)$$

The equilibrium constant, K_1, for reaction (7.XXIII) is:

$$K_1 = \frac{a_{Me^{2+}}^2 \, a_{H_2O}^2}{a_{Me}^2 \, a_{H^+}^4 \, p_{O_2}}$$

Assuming H_2O and solid Me to have unit activities the above relationship can be rearranged to give:

$$a_{Me^{2+}}^2 = K_1 \, a_{H^+}^4 \, p_{O_2}$$

or in logarithmic form:

$$2 \log a_{Me^{2+}} = \log K_1 + 4 \log a_{H^+} + \log p_{O_2}$$

Since $pH = -\log a_{H^+}$, we can write

$$2 \log a_{Me^{2+}} = \log K_1 - 4pH + \log p_{O_2}$$

In the same way, similar relationships can be deduced for reactions (7.XXIV) to (7.XXIX) which give the values of p_{O_2} and pH in order to maintain an equilibrium condition between reactants and products. These relationships are given respectively below, assuming solids $Me(OH)_2$ and $Me(OH)_3$ have unit activities:

$$4 \log a_{Me^{2+}} = \log K_2 + 4 \log a_{Me^{3+}} - 4pH + \log p_{O_2}$$

$$\log a_{Me^{2+}} = -\log K_3 - 2pH$$

$$\log a_{Me^{3+}} = -\log K_4 - 3pH$$

$$\log a_{MeO_2^{2-}} = K_5 + 2pH$$

$$\log a_{MeO_2^-} = K_6 + pH$$

$$\log p_{O_2} = -\log K_7$$

It is, therefore, possible to determine the thermodynamically stable phases in certain aqueous solutions by plotting $\log p_{O_2}$ against pH. This is done in *Fig. 7.26* assuming unit activity for the metal ions in solution. This is similar to the 'Kellog' diagram given in *Fig. 7.5* and the Pourbaix diagram in *Figs 9.13* and *9.14*. Indeed they have each been plotted using the same thermodynamic principles.

The vertical lines in *Fig. 7.26* represent reactions which involve pH but not O_2 pressure (reactions (7.XXV)–(7.XXVIII)); the horizontal lines represent reactions which involve O_2 pressure but not pH (reaction (7.XXIX)), while diagonal lines represent reactions involving pH and O_2 pressure (reactions (7.XXIII) and (7.XXIV)).

From *Fig. 7.26*, the metallic phase Me is only stable at p_{O_2} values of less than 10^{-70} atm. At a low pH value increasing the O_2 pressure in the AA' direction brings Me into aqueous solution through reaction (7.XXIII). Further increase in the O_2 pressure oxidises Me^{2+} to Me^{3+} via reaction (7.XXIV). At a constant O_2 pressure,

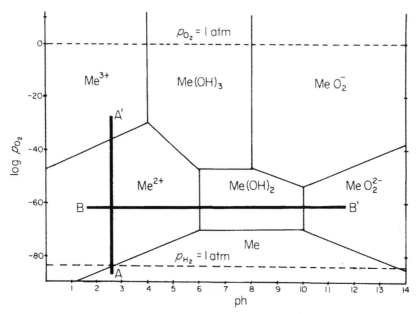

Figure 7.26 Predominance areas for the system Me–H_2O. (From *Principles of Extractive Metallurgy* by T. Rosenqvist, © 1974, McGraw-Hill Inc. Used with permission of McGraw-Hill Book Co.)

increasing the pH of the solution in the BB' direction results in precipitation of Me(OH)$_2$ initially by reaction (7.XXVI) and subsequent dissolution by the formation of MeO$_2^{2-}$ ions at higher pH values through the reaction (7.XXVIII). Different metals will produce different stable phases or 'predominance areas' allowing certain metals to be precipitated while others remain in solution by simply adjusting the O_2 pressure and pH of the solution. Activities of metal ions in solutions can be reduced by using complexing agents producing complex ions, e.g. Cu(NH$_3$)$_4^{2+}$. This will change the position of the line depending on the value of the relevant metal ion activity in reactions (7.XXIII) to (7.XXVIII). Hence, by careful use of these diagrams it is possible to precipitate unwanted minerals from solution retaining the metal to be extracted in the solution (purification). Alternatively, the metal to be extracted can be selectively precipitated. Chlorine may be used as an alternative oxidising agent to oxygen (see Section 6.3.15). In this case Cl_2 is commonly used in conjunction with a lime (CaO) addition to the aqueous solution in order to control the state of oxidation and pH respectively.

7.3.2.2 *Gaseous reduction*

Selective precipitation of metals can also be effected using electrochemical oxidation (electron loss) and reduction (electron gain). Gaseous reduction involves passing hydrogen or other suitable reducing gases through the leach solution resulting in selective precipitation (reduction) of metal ions by control of pH of solution and activity of metal ions in solution.

If two metal ions M^{2+} and Me^{2+} are present in solution they may both be precipitated by a reduction (cathodic) reaction, i.e.

$$M^{2+} + 2e \longrightarrow M \quad \text{and} \quad Me^{2+} + 2e \longrightarrow Me$$

Using the Nernst equation (Section 6.3.12) their respective electrode potentials are:

$$E_M = E_M^\ominus + \frac{RT}{2F} \ln a_{M^{2+}} \tag{7.8}$$

and

$$E_{Me} = E_{Me}^\ominus + \frac{RT}{2F} \ln a_{Me^{2+}} \tag{7.9}$$

At the same time the reducing gas, H_2, is oxidised: $\frac{1}{2}H_2 \longrightarrow H^+ + e$, where

$$E_H = E_H^\ominus + \frac{RT}{F} \ln \left(\frac{a_{H^+}}{p_{H_2}^{1/2}} \right)$$

$$= E_H^\ominus + \frac{RT}{F} \ln a_{H^+} - \frac{RT}{F} \ln p_{H_2}^{1/2}$$

Since $E_H^\ominus = 0$ and $pH = -\ln a_{H^+}$, we have

$$E_H = -\frac{2.303RT}{F} pH - \frac{RT}{F} \ln p_{H_2}^{1/2} \tag{7.10}$$

Examination of equations (7.8), (7.9) and (7.10) shows that E_M and E_{Me} at constant temperature vary only with the activity of their metal ions in solution, while E_H at constant temperature varies with the partial pressure of hydrogen and the pH of solution. Therefore, plotting E versus pH for each reaction at constant metal ion activity gives straight lines of zero slope for E_M and E_{Me} and a straight line of negative slope for E_H (*Fig. 7.27*).

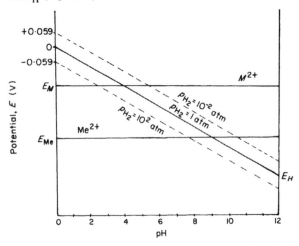

Figure 7.27 Potential–pH diagram used in selective precipitation (gaseous reduction)

Hydrogen will reduce M^{2+} only if E_M is more positive (cathodic) than E_H, i.e. if $E_H < E_M$, then $M^{2+} + H_2 \rightarrow M + 2H^+$ will take place. Using *Fig. 7.27*, M will be precipitated at a pH of 5 with a p_{H_2} of 1 atm. Increasing the pH to 10 will lead to subsequent precipitation of Me as the E_H line falls below E_{Me}. Using complexing agents will produce complexing of metal ions in solution, e.g. $Ni(NH_4)_3^{2+}$, which will alter the position of E_{Ni} and provide further selectivity. Changing p_{H_2} will also provide a further degree of manipulation. In this way several metals may be selectively precipitated from solution by simply increasing the pH.

7.3.2.3 Cementation

Cementation also involves electrochemical reduction (cathodic precipitation) of a metal ion from solution by passing the solution over a more anodic (less noble) metal. In this way copper has been precipitated from solution containing Cu^{2+} ions by passing the solution over scrap iron which dissolves anodically, i.e.

$$Cu^{2+} + 2e \longrightarrow Cu \qquad \text{cathodic precipitation}$$

$$Fe \longrightarrow Fe^{2+} + 2e \qquad \text{anodic dissolution}$$

Also, precious metals, Ag, Au, have been precipitated from solution by passage over zinc plates or on addition of zinc dust after leaching the metal with a cyanide solution, the Zn is removed by oxidation and the residue smelted with borax to produce a bullion.

Cementation is less selective than gaseous reduction precipitation since all the more noble (more cathodic) metals present in the solution will be precipitated until all the less noble or more anodic metal has been dissolved or is covered with the precipitated metal. In the latter case subsequent smelting can be carried out to extract the cemented metal from the more anodic metal.

7.3.3 Isolation techniques

7.3.3.1 Ion exchange

Ion exchangers are organic resins onto which either cations or anions have been loosely bonded. Passing the leach solution over the resin produces a selective exchange of one type of metal ion of the same charge and in the same quantity as those loosely bonded to the organic resin until equilibrium is reached (see Section 6.6). In this way the complex uranium anion $UO_2(SO_4)^{4-}$ has been isolated from its leach solution using a nitrate ion exchanger:

$$\underset{\text{(in solution)}}{UO_2(SO_4)^{4-}} + 4R'-NO_3^- = R'_4.UO_2(SO_4)_3^{4-} + 4NO_3^-$$

where R' is the organic resin (R) with an ionic group attached (in this case cationic) as described in Section 6.6, i.e. $R' = R-X^+$, such that NO_3^- ions are loosely bonded to X^+ ions. The extracted ion is recovered from the resin by stripping with a cheap solvent; extraction of the metal being completed by a reduction precipitation process such as electrodeposition. Zinc has been extracted using a strong acid cation exchanger[22], copper with a weak cation exchanger, chromium with a strong base cation exchanger and gold with an anion exchanger[23].

7.3.3.2 Solvent extraction

Certain organic solvents, e.g. ethers, esters and ketones, provide high solubility of some metallic compounds and in particular complex metal ions. The organic solvents are highly selective towards one metal ion and are immiscible in water and thus provide extraction or isolation of one metal ion from an aqueous solution containing many metal ions (see Section 6.1). Selectivity can increase with change in pH of solution. At the present time there is no adequate theory available for predicting which organic solvent could be used for which metal ion. Examples include the extraction of zinc and nickel from solution with di-2-ethyl hexylphosphoric acid or tertiary amines.

7.3.3.3 Reverse osmosis

Reverse osmosis produces a concentration of metal ions in solution from which metals may be subsequently extracted. Osmosis is the passage of water from a dilute to a more concentrated solution through a semipermeable membrane. This sets up an osmotic

Nickel concentrate containing Cr, Fe, Cu, Zn, as impurities

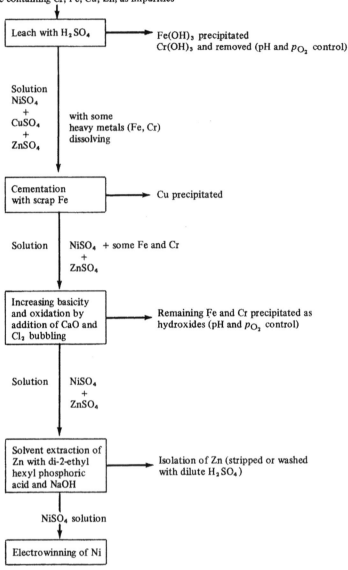

Figure 7.28 Hypothetical flow sheet for extraction of nickel from nickel concentrate

pressure. If a pressure in excess of the osmotic pressure is applied in the opposite direction water passes from the more concentrated (leached solution) to the dilute solution (water) leaving the metal ions in a more concentrated form. This is used increasingly in the removal of ions from electrochemical processing effluents allowing the recovery of the metal ions and the recycling of the associated water. These effluents are usually not highly ionised and therefore have low osmotic pressures. The process requires minimum fuel and is easily controlled. Cellulose acetate is the most popular membrane used. The use of some of these hydrometallurgical extraction techniques is shown in the flow sheet in *Fig. 7.28* in the extraction of various metals from a hypothetical nickel concentrate.

7.4 Electrometallurgical extraction processes

The theory of electrowinning and electrorefining was discussed in Section 6.4 and so it is only necessary here to explain the industrial processes for the electrometallurgical extraction of the major metals.

Electrowinning is used extensively for the extraction of reactive metals aluminium and magnesium from electrolytes of their fused salts and as an alternative to pyrometallurgical extraction of other metals such as copper and zinc from aqueous electrolytes. Electrorefining of impure metals using fused salts or aqueous electrolytes provides the metals in a very high state of purity. Impurity metals either collect around the anode (anode slimes contain Au, Ag) or go into solution in the electrolyte which necessitates removing the electrolyte from time to time and precipitating or isolating the impurities (purification). The anode slime also must be cleared periodically. These are roasted to preferentially oxidise the base metals and the precious metals melted to produce an Ag–Au alloy which is electrorefined in a nitrate electrolyte. The theoretical decomposition potential for electrowinning can be calculated (Section 6.4.6) using thermodynamic relationships. For example, the decomposition potential for the electrolytic production of magnesium using a $MgCl_2$ fused salt electrolyte at 750 °C will depend on the thermodynamics of the decomposition reaction:

$$MgCl_2(l) \longrightarrow Mg(l) + Cl_2(g) \qquad \Delta G^{\ominus}_{750°C} = +473 \text{ kJ}$$
$$\text{(electrolyte)} \quad \text{(at cathode)} \quad \text{(at anode)}$$

Using the relationship $\Delta G^{\ominus} = -nFE^{\ominus}$ for the above electrolytic cell reaction, we have

$$E^{\ominus}_{cell} = -\frac{473\ 000}{2 \times 96\ 500} = -2.46 \text{ V}$$

In practice higher voltages must be used than the theoretically determined emf in order to overcome polarisation and provide a satisfactory production rate.

Where there is danger of hydrogen reduction (evolution) occurring at the cathode at the expense of metal deposition or where impurity metals are anodically dissolved and may lead to codeposition on the cathode, special canvas diaphragms are used to separate the anode compartment of the cell from the cathode compartment.

In electrorefining, the value of the standard free energy change, $\Delta G^{\ominus}$, is zero as is the standard cell potential, $E^{\ominus}_{cell}$, e.g.

$$Cu(\text{anode}) \rightleftharpoons Cu(\text{cathode}) \qquad \Delta G^{\ominus} = 0$$

since the anodic dissolution of the copper:

$$Cu \longrightarrow Cu^{2+} + 2e \qquad E^{\ominus} = -0.34 \text{ V}$$

is balanced by the cathodic deposition of copper:

$$Cu^{2+} + 2e \longrightarrow Cu \qquad E^{\ominus} = +0.34 \text{ V}$$

If the impure copper anode has a copper activity, a_{Cu}, of 0.9 the emf for the electrorefining reaction at 25 °C would be:

$$E_{298} = E^{\ominus}_{cell} + \frac{RT}{nF} \ln \frac{a_{Cu(\text{impure})}}{a_{Cu(\text{pure})}}$$

But $E_{cell}^{\ominus} = 0$ and so

$$E_{298} = \frac{4.2 \times 298}{2 \times 96\,500}\ \ln 0.9$$

$$= -0.029\ \text{V}$$

Again higher voltages than 0.029 V are used in practice to overcome polarisation effects, e.g. 0.4 V with a current efficiency of 90 to 95% requiring about 0.4 kW h/kg of copper refined. Steam is used to maintain the copper sulphate electrolyte temperature at about 55 °C. Pourbaix diagrams (*Figs 6.15, 9.13, 9.14*) showing the stable phases which predominate ('predominance areas') for the metal at a certain potential and pH of electrolyte provide a useful guide in indicating the optimum conditions for electrolysis. Thus, the pH of the electrolyte and the applied potential should be chosen to provide the metal in suitable ionic form and not which precipitates the metal as a hydroxide or oxide.

7.4.1   Electrowinning and electrorefining of metals from aqueous solutions

The more noble metals such as Cu, Ni, Co, Cd, Fe, Zn, Mn can be extracted electrolytically from their leached aqueous sulphate solutions (see *Table 7.5*). Process control problems become more severe in the extraction of Zn, Mn and Fe due to the need to avoid hydrogen evolution and codeposition of more noble impurities, e.g. Sb, Cu, Co, at the cathode. For this reason impurities are removed from the solution prior to electrolysis by hydrometallurgical techniques such as cementation and ion exchange, while diaphragms can be used to restrict the movement of impurity ions or metal ions which exhibit variable oxidation states. These latter ions if unrestricted in movement could be alternately oxidised and reduced on passing between the anode and cathode and thereby substantially increase the energy requirements of the electrolytic cell. As it is, the energy requirements increase with the use of diaphragms. Usually, an anode is chosen which does not dissolve electrolytically during the process such as a Pb–Sb alloy in zinc extraction. However, a soluble anode, i.e. of the same metal being extracted, is used in some instances. In the case of an insoluble or inert anode the usual anodic reaction is that of oxygen evolution, i.e.

$$2OH^- \longrightarrow \tfrac{1}{2}O_2 + H_2O + 2e \qquad E^{\ominus} = -0.4\ \text{V}$$

The conventional electrode potential will be for the *reduction* reaction (Section 6.3.12), i.e.

$$\tfrac{1}{2}O_2 + H_2O + 2e \longrightarrow 2OH^- \qquad E^{\ominus} = +0.4\ \text{V}$$

and applying the Nernst equation at 25 °C (298 K):

$$E_{O_2} = +0.4 + \frac{8.3 \times 298}{2 \times 96\,500}\ \ln\left(\frac{p_{O_2}^{1/2}}{a_{OH^-}^2}\right)$$

$$= +0.4 + 0.015\ \log_{10}p_{O_2} - 0.059\ \log_{10}a_{OH^-}\ \text{volts}$$

but $\log_{10}a_{H^+} + \log_{10}a_{OH^-} = -14$ (see Section 6.1.8) and $-\log_{10}a_{H^+} = $ pH, thus $\log_{10}a_{OH^-} = $ pH $- 14$. Substituting this into the above electrode potential,

$$E_{O_2} = +0.4 + 0.015\ \log_{10}p_{O_2} - 0.059\ (\text{pH} - 14)\ \text{volts}$$

$$= +1.23 + 0.015\ \log_{10}p_{O_2} - 0.059\ \text{pH volts}$$

Thus in an acid solution where pH = 0 and $p_{O_2} = 1$ atm, $E_{O_2} = +1.23$ V. (The same value would apply if the reaction $\tfrac{1}{2}O_2 + 2H^+ + 2e \rightarrow H_2O$ was used.)

Table 7.5 DETAILS OF TYPICAL ELECTROWINNING CELLS FOR CERTAIN METALS

Metal	Typical electrolyte	Theoretical emf (V)	Typical emf used in practice (V)	Current efficiency (%)	Energy efficiency (kWh kg^{-1})	Remarks
Copper	6.5 to 8% $CuSO_4$ + 1.7 to 4.6% H_2SO_4	−0.89	1.8 to 2.5	80 to 90	2.0 to 3.0	Pb–Sb or Pb–Ag anodes. Presence of Fe^{2+} or Fe^{3+} ions reduces current efficiency
Zinc	10% $ZnSO_4$ + 0.5 to 22% H_2SO_4	−1.99	3.2 to 3.7	90 to 95	3.0 to 3.5	Temperature below 30 °C maintains a high H_2 overvoltage on hard rolled Al cathodes and makes process possible. pH about 7. Pb, Pb–Sb or Pb–Ag anodes
Manganese	13% $(NH_4)_2$ SO_4 + 0.01% SO_2 + 3% $MnSO_4$	−2.41	5 to 5.3	40 to 70	8.6 to 11.6	pH controlled between 8.2 to 8.4 to maintain high H_2 overvoltage. Diaphragm used at electrodes to keep impurity ions separate
Aluminium	3.5% Al_2O_3 in fused Na_3AlF_6 at 970 °C	−1.50	4.0 to 5.0	75 to 90	13 to 18	Carbon anodes with cathodes as cell lining (carbon)
Magnesium	7 to 15% $MgCl_2$ + 20% KCl + 20 to 40% $CaCl_2$ + 30 to 45% NaCl } fused salt mixture at 750 °C	−2.46	6 to 9	75 to 90	17.6 to 20	Carbon anodes and steel cathodes. Separation of anode and cathode compartments by use of ceramic curtains

The oxidation of OH^- ions will take place at the anode in preference to the oxidation of SO_4^{2-} ions since the former reaction has a more anodic (more negative) E value as calculated by the Nernst equation using the IUPAC convention (see Appendix 3 for an explanation of this point).

The cathodic reaction is that of metal deposition:

$$M^{2+} + 2e \longrightarrow M \qquad E_{M^{2+}} \text{ volts}$$

Hence, the theoretical cell e.m.f. is

$$E_{cell} = E_c - E_a = E_{M^{2+}} - (+1.23) \text{ V}$$
$$= E_{M^{2+}} - 1.23 \text{ V}$$

in an acid solution (see *Table 7.5* and Section 6.3.13).

Cathodes can be pure thin sheets of the same metal that is being extracted or of some suitable alloy which provides a high hydrogen overvoltage, if this is a problem, or improved electrode mechanical properties.

The spent electrolyte is often recirculated to the leaching plant after removal of impurities.

Electrorefining of gold, silver, nickel and in some cases lead using aqueous electrolytes follows the same general principles outlined earlier in the electrorefining of copper. Electrorefining of lead uses an electrolyte of silicofluoric acid, H_2SiF_6, since lead forms insoluble compounds with both sulphuric and hydrochloric acids. If bismuth is present as an impurity it remains undissolved in the anode slime.

Electrorefining of impure nickel containing copper, cobalt and iron as impurities provides problems due to their proximity to nickel in the electrochemical series. Diaphragms which separate the cell into anode and cathode chambers are used. Nickel, copper, cobalt and iron are anodically dissolved while the precious metals remain undissolved in the anode slime. The anolyte is removed from the cell, the iron and cobalt are precipitated as hydroxides by oxidation with air and chlorine respectively while copper is precipitated by cementation on nickel powder. The purified electrolyte is added to the cathode chamber where nickel is deposited. This complex process requires a voltage of 1.5–2.0 V. Direct electrorefining of nickel sulphide has also been developed using a similar diaphragm arrangement as above (Sections 6.4.3, 6.4.6, 6.4.7).

7.4.2 Electrowinning and electrorefining of aluminium and magnesium and the more reactive metals from fused salt electrolytes

All primary aluminium production is achieved by fused salt electrolysis. Purified bauxite from the Bayer process is added to molten cryolite (Na_3AlF_6) to about 1 to 8% together with minor additions of CaF_2 and AlF_3. The melting point of cryolite ($1010\,^\circ C$) is lowered by these additions so that the electrolytic process can be conducted at about $900–1000\,^\circ C$. The cell lining (*Fig. 7.29*) is of carbon which acts as the cathode while either prebaked or self-baking carbon anodes are dipped into the electrolyte and are consumed by oxidation to carbon dioxide. The carbon anodes, therefore, must be renewed either by replacing with prebaked anodes or by continuous feeding of carbon paste into a steel-lined chamber (self-baking type) above the electrolyte and which hardens as it descends towards the hot zone.

During electrolysis aluminium is deposited on the cathode and oxygen evolved at the anode, reacting with the carbon to produce CO_2 by the exothermic reaction:

$$O_2 + C \longrightarrow CO_2 \qquad \Delta H^\ominus_{298} = -397 \text{ kJ}$$

This aids the control of the electrolyte temperature but consumes the carbon anode. Impurities which are more noble (more cathodic) than aluminium such as iron and silicon will also deposit at the cathode unless removed from the alumina in the Bayer process. The cell reactions are not completely understood but a simplified presentation

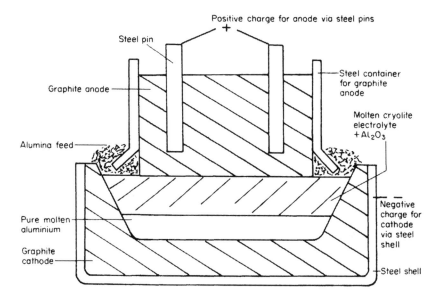

Figure 7.29 Electrolytic cell for aluminium extraction (diagrammatic)

of the cell reactions is given below. Dissociation of alumina present in the electrolyte may be represented as:

$$2Al_2O_3 \longrightarrow 3AlO_2^- + Al^{3+}$$

and the cathodic reaction as:

$$Al^{3+} + 3e \longrightarrow Al$$

and the anodic reaction as:

$$3AlO_2^- \longrightarrow 3Al + 3O_2 + 3e$$

Also at the anode:

$$3C(\text{electrode}) + 3O_2 = 3CO_2$$

Adding the anodic and cathodic reactions gives the overall cell reaction,

$$Al^{3+} + 3AlO_2^- + 3C \longrightarrow 4Al + 3CO_2$$

or more simply,

$$2Al_2O_3 + 3C \longrightarrow 4Al + CO_2 \qquad \Delta G_{1273}^{\ominus} = +1362 \text{ kJ}$$

(which can be calculated from the appropriate lines in *Fig. 2.11*). This requires a theoretical standard decomposition potential at 1000 °C of

$$E_{1273}^{\ominus} = -\frac{1\,362\,000}{nF} \quad \begin{matrix} = -4.7 \text{ Volts per 4 moles of Al} \\ = -1.176 \text{ Volts per mole of Al} \end{matrix}$$

where $n = 3$, $F = 96\,500$ J. If the activity of Al_2O_3 in the electrolyte is 0.08 the theoretical decomposition potential will be (assuming $a_C = a_{Al} = 1$ and $p_{CO_2} = 1$ atm)

$$E_{1273} = E_{1273}^{\ominus} + \frac{RT}{nF} \ln a_{Al_2 O_3}^2$$

$$= -1.176 + \frac{8.3 \times 1273}{3 \times 96\,500} \ln(0.08)^2$$

$$= -1.50 \text{ V}$$

The low current efficiency of 75 to 90% is thought to be due to anodic oxidation of the aluminium dissolved in the electrolyte. This needs to be further decomposed before any deposition can take place at the cathode. Therefore, a cell voltage of between 4 and 5 V is required to overcome this low efficiency and polarisation effects and at the same time provide a satisfactory production rate. This results in the large energy consumption (efficiency) of 13–18 kW h/kg of aluminium produced.

Electrorefining of aluminium is conducted at 700–900 °C using *three layer cell electrolysis* in which the impure molten aluminium anode, the barium–sodium–aluminium halide electrolyte and the pure molten aluminium form three distinct layers in an electrolytic cell. The impure molten aluminium anode, being the heaviest, lies at the bottom of the cell and the pure molten aluminium, being the lightest, cathodically deposits on a carbon electrode at the top of the cell; the molten halide electrolyte separating the anodic and cathodic aluminium. A current efficiency of 90–95% is produced using a cell voltage of 5 V. Therefore, three layer cell electrorefining of aluminium is expensive and is only used when high purity aluminium is demanded.

Electrowinning of magnesium (*Fig. 7.30*) is conducted at about 750 °C using a $MgCl_2$–$NaCl$–$CaCl_2$ electrolyte with carbon anodes dipped into the electrolyte in an iron pot and steel cathodes since magnesium does not alloy with iron. Magnesium is deposited at the cathode and chlorine evolved at the anode where there is some small consumption of the carbon if the electrolyte contains magnesium oxide, i.e.

$$2MgO + 2Cl_2 + C \longrightarrow MgCl_2 + CO_2$$

Magnesium, being lighter than the electrolyte, floats to the top of the cell after deposition on the steel cathodes and must be protected from reacting with the anode chlorine

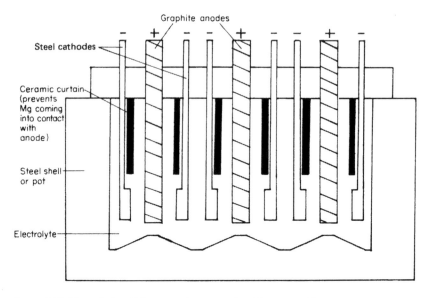

Figure 7.30 Electrolytic cell for magnesium extraction (diagrammatic)

gas by ceramic curtains which dip into the electrolyte and separate the anode and cathode compartments (see *Fig. 7.30*). As discussed earlier a theoretical decomposition voltage of 2.46 V is required but in practice up to 7 V are required to overcome polarisation and provide a reasonable production rate. The current efficiency is between 75–90% resulting in an energy efficiency of 18–20 kW h/kg of magnesium produced.

Electrowinning of sodium and calcium is similar to that of magnesium although a molten NaOH electrolyte may also be used for sodium. Some typical process details for electrowinning of Cu, Zn, Mn, Al and Mg are given in *Table 7.5*.

Electrowinning of titanium, zirconium and niobium is not usually conducted since these metals form ions of varying valency which will be alternately oxidised and reduced at the anode and cathode respectively. Their halides also have low solubility in salt mixtures and the high melting points of these metals would result in their cathodic deposition as a solid. It is more feasible to use electrorefining techniques for these metals. The impure or scrap metal anodes dissolve in their low valency state and the metal may be deposited on the cathode in powder form.

References

1. Pryor, E.J. 1965 *Mineral Processing*, 3rd edn, pp 471–90. Amsterdam: Elsevier.
2. Kellog, H.H. and Basu, S.K. 1960 *Trans. Met. Soc. AIME*, 218, 70–81.
3. 1973 *Metall. Mat. Tech.*, 5, (9) 442.
4. Bodsworth, C. and Bell, H.B. 1972 *Physical Chemistry of Iron and Steel Manufacture*, 2nd edn, p. 133. London: Longman.
5. St. Clair, H. 1958 *Vacuum Metallurgy*, ed. R.F. Bunshah, pp 298–305. New York: Reinhold.
6. Worner, H.K. 1968 *Proc. Symp. on Advances in Extractive Metallurgy*, pp 245–263. Institute of Mining Metallurgy.
7. Davies, D.R.G. *et al.* 1967 *J. Iron & Steel Industry*, 205, (7) 810–14.
8. 1969 BISRA *Annual Report*, p. 17.
9. Berthet, A. *et al.* 1969 *J. Iron & Steel Industry*, 207, (6) 790–7.
10. Boldt, J.R. and Queneau, P. 1967 *The Winning of Nickel*, pp 227–336. London: Methuen.
11. Morgan, S.W.K. and Woods, S.E. 1971 *Metall. Rev.*, review 156, pp 161–174 (incorporated in 1971 *Metals Mat.*, 5, (11)).
12. Bodsworth, C. and Bell, H.B. 1972 *Physical Chemistry of Iron and Steel Manufacture*, 2nd edn, p. 172. London: Longman.
13. Bodsworth, C. and Bell, H.B. 1972 *Physical Chemistry of Iron and Steel Manufacture*, 2nd edn, p. 165. London: Longman.
14. Bodsworth, C. and Bell, H.B. 1972 *Physical Chemistry of Iron and Steel Manufacture*, 2nd edn, p. 441. London: Longman.
15. Bodsworth, C. and Bell, H.B. 1972 *Physical Chemistry of Iron and Steel Manufacture*, 2nd edn, p. 196. London: Longman.
16. Bodsworth, C. and Bell, H.B. 1972 *Physical Chemistry of Iron and Steel Manufacture*, 2nd edn, p. 468. London: Longman.
17. Bodsworth, C. and Bell, H.B. 1972 *Physical Chemistry of Iron and Steel Manufacture*, 2nd edn, p. 281. London: Longman.
18. Leach, J.C.C. *et al.* 1978 *Ironmaking Steelmaking*, 5, (3), 108–20.
19. Norrman, T.O. and Johnsson, O. 1975 *Iron and Steelmaker*, 28
20. Boldt, J.R. and Queneau, P. 1967 *The Winning of Nickel*, pp 227–336. London: Methuen.
21. Rosenqvist, T. 1974 *Principles of Extractive Metallurgy*, p. 438. New York: McGraw-Hill.
22. Mokryshev, A.N. and Trushin, G.A. 1970 *Trans. Inst. Met. Obogarch. Akad. Nauk. Kaz. SSR*, 40, 44–53, 68–74.
23. Pritchard, E.J. and Peihoda, W.W. 1969 *Plating*, 56, (9) 1044–6.
24. Kelly, E.G. and Spottiswood, D.J. 1982 *Introduction to Mineral Processing*. New York: Wiley Interscience.

For self-test questions on this chapter, see Appendix 5.

8

Metal Melting and Recycling

8.1 Metal melting

Straight melting processes are conducted in air, vacuum or inert atmospheres and do not involve a refining operation.

Air melting is used for the less reactive metals (cast irons, steels, copper and nickel alloys) or the low melting point alloys (lead, aluminium, zinc, tin) where reaction with the furnace atmosphere is low, or limited due to a protective oxide skin formation.

Vacuum or inert atmosphere melting is used for high melting point or reactive metals (titanium and zirconium) which would otherwise react with the furnace environment. *Melting-refining* processes incorporate a refining operation on the molten charge usually by means of appropriate gas–metal and slag–metal reactions. The most common refining technique is the preferential oxidation of impurities by air or oxygen blowing or other oxidising agents, sulphur drossing to remove Cu, or addition of $NaNO_3$ and $NaOH$ to remove As, Sb and Sn from lead (Section 8.2.2.5). Operations which involve the use of two or three melting or refining units are called *duplex* or *triplex* processes respectively; the charge being melted in one unit and subsequently transferred to a second or third unit for refining, deoxidation, degassing, grain refining, alloy adjustment and temperature control. In this way the maximum efficient use can be made of each unit.

8.1.1 Physical and chemical considerations

Thermodynamic and reaction kinetic theories can be applied to melting and refining processes in the same way as they are applied to extraction processes and provide important information with respect to process control and economics. Metal loss due to volatilisation increases as its vapour pressure increases with increase in temperature (see Section 7.2.7). This becomes important in vacuum melting. If the vapour pressure of an element is above that of the vacuum there will be continuous loss of the element. The vapour pressure of an element can be calculated using the Clausius–Clapeyron equation (equation (2.39)).

Considerable loss of low boiling point metals can occur during air melting, e.g. Zn, Mg, Cd, Mn. *Table 7.1* gives the melting points and boiling points for certain metals and the temperature at which each metal exerts a vapour pressure of 10^{-3} atm. Difficulties in achieving complete solution and avoiding excessive metal loss due to volatilisation are experienced in melting metals which have large differences in their melting points. For example, mixing a low melting point metal (solvent) e.g. Zn, with a high melting point alloying metal (solute) e.g. Ni, creates problems in achieving complete solution. This can be overcome by producing an alloy of Zn and Ni which is richer in Ni than required and has an intermediate melting point between that of Ni and Zn. This intermediate alloy can be produced with little or no problems of solution of the two metals.

Addition of this alloy to the zinc melt in the required quantity will facilitate the melting of the specified Zn–Ni alloy composition. The intermediate Zn–Ni alloy is sometimes called a *temper* alloy. Fewer problems are experienced on adding a low melting point solute to a high melting point solvent metal providing its vapour pressure is not high.

8.1.2 Removal of gases from liquid metals

The solubility of gases in liquid metals was discussed in Section 4.4. The main problem gases are hydrogen (Cu, Ni, Fe, Mg, Al), nitrogen (Fe, Mg), oxygen (Cu, Ni, Mg, Fe, Al) and complex gases derived from sulphur, i.e. SO_2 (Cu, Ni) and carbon, i.e. CO and CO_2 (Fe, Ni). The presence of hydrocarbons and the dissociation of moisture present in the furnace atmosphere are the main sources of hydrogen pick-up:

$$H_2O \text{ (in atmosphere)} \longrightarrow 2H \text{ (in melt)} + O \text{ (in melt)}$$

Slags are permeable to moisture but not to hydrogen and, therefore, allow moisture to diffuse to the slag–metal interface resulting in hydrogen pick-up. Deoxidising the melt lowers the activity of oxygen, which allows a higher level of hydrogen pick-up in order to maintain the equilibrium condition. This can be seen by applying equation (2.29) to the above reaction:

$$\Delta G = \Delta G^\ominus + RT \ln \left(\frac{a^2_{H \text{ (in melt)}} \, a_{O \text{ (in melt)}}}{p_{H_2O \text{ (in furnace atmosphere)}}} \right)$$

Thus, deoxidation produces a decrease in the ratio $a^2_H a_O / p_{H_2O}$ eventually resulting in a negative variable term, $RT \ln(a^2_H a_O / p_{H_2O})$, and making ΔG more negative. This produces an increased driving force for the dissociation of moisture and an increase in the hydrogen level in the melt. This effect is shown in *Fig. 8.1*.

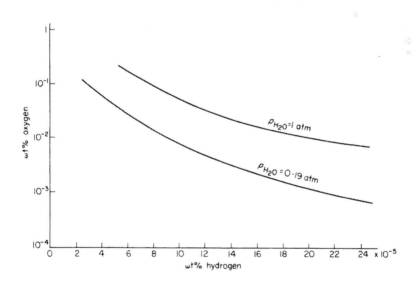

Figure 8.1 Relationship between equilibrium hydrogen and oxygen contents in liquid copper at 1250 °C

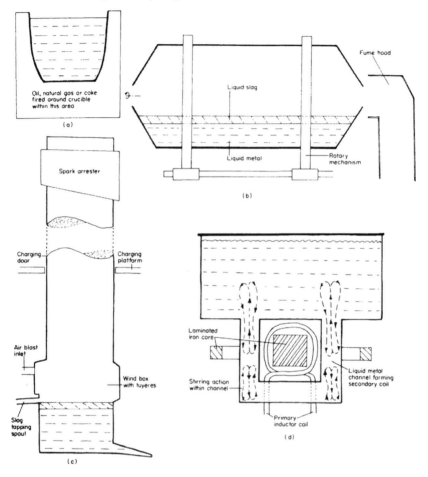

The main techniques used to lower hydrogen pick-up include:

(i)   Furnace atmosphere control. Melting with a slight excess of oxygen limits the
      hydrogen pick up as discussed above and also oxidises any hydrogen present in
      the furnace atmosphere. The metal is deoxidised prior to pouring.
(ii) Inert and active gas bubbling, e.g. N_2, Ar (see Section 4.4), Cl_2 (active, i.e. HCl
      (formed).
(iii) Vacuum treatment of the melt (see Section 4.4).

Vacuum degassing techniques must be used if H_2 is to be lowered to below 2 ppm.
The use of a slag will help to limit nitrogen pick-up from the furnace atmosphere while
nitrogen-bearing addition compounds, such as cyanamide and activated N_2 produced
from the arcing of the electrodes in electric arc melting will increase N_2 pick-up.
Nitrogen porosity may be controlled by:

(i)   addition of nitride-forming elements to the melt, e.g. Zr, Ti;
(ii) addition of alloying elements, e.g. Cr, Mn, V, which reduce the activity coefficient
      of nitrogen (f_N) in the melt (see Section 4.4—Sievert's law);

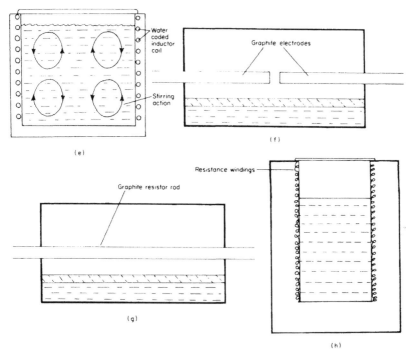

Figure 8.2 Diagrammatic representation of melting units: (a) crucible; (b) rotary furnace; (c) cupola; (d) channel induction furnace; (e) coreless induction furnace; (f) indirect electric arc furnace; (g) rod resistor furnace; (h) resistance furnace

(iii)   addition of austenite stabilisers to a ferrous melt since the solubility of N_2 is higher in austenite, γ, than other phases (see *Fig. 4.15*). N_2 in γ is greater than 0.02 wt%.

A certain level of AlN precipitation in ferrous alloys aids grain size control but excessive precipitation results in 'rock candy' fractures due to precipitation of the nitride at the grain boundaries.

8.1.3   Melting units

Melting units may be classified under fuel fired (oil, gas, coke or coal) furnaces and electric furnaces. Fuel fired units include crucible, reverberatory or hearth, rotary, converter and cupola or shaft furnaces. Electric units include induction (channel and coreless), arc (direct and indirect), resistor and resistance furnaces. Diagrammatic representations of most of these furnaces are given in *Fig. 8.2*, with the reverberatory furnace in *Fig. 7.8(a)*, direct electric arc in *Fig. 7.8(b)*, and converters in *Fig. 7.24*. Whereas oil must be atomised before burning in an oxidant (air or oxygen) and coke and coal may need to be crushed to a suitable size for packing around a crucible furnace or pulverised when used in a hearth or rotary furnace, gas can be combusted directly.

    In general, fuel fired furnaces, because of the requirement of combusting a fuel in the furnace atmosphere, provide only a limited amount of control over melting conditions, but can make efficient use of a slag to provide slag–metal refining. Therefore, where slag–metal refining reactions are required, fuel fired furnaces are normally used.

Table 8.1 SUMMARY OF MELTING UNITS

Melting unit	Fuel	Capacity (tonnes of Fe)	Melting rate (tonnes h^{-1} of Fe)	Temperature range (°C)	Range of metals melted
Crucible	Oil, gas, solid (coke) fired around crucible	Up to 1.0 (tilting) or 0.5 (pit)	0.4 to 0.75	200 to 1400	Al, Zn, brasses, some cast iron. Used mainly for low quality, low melting point metals. Poor thermal efficiency
Direct arc	Electricity, three phase. Alternating or direct current. Heat produced from arc and by induction in metal charge	Up to 200	2 to 40	1250 to 1750	Steel, cast iron, Cu and Ni alloys. Refining with use of appropriate slag composition. High thermal efficiency
Indirect arc	Electricity single phase. Alternating or direct current. Heat produced from arcing between two graphite electrodes above charge	Up to 2, but 0.25 more typical	0.25 to 1.0	1000 to 1750	Special cast irons, bronzes, gun metals, some Ni and Cu alloys. Metal–slag refining possible. Moderate to high thermal efficiency
Channel induction	Electricity. Alternating current. Heat produced by inducing current in molten charge in channel	Up to 25	0.5 to 2.0	600 to 1500	Al and low melting point alloys, brasses, some cast iron. Little or no refining possible. High thermal efficiency
Coreless induction	Electricity. Alternating current. Heat produced by inducing current in metal charge	Up to 50	0.5 to 3.0	600 to 1750	High melting point alloys, cast irons, Fe alloys–especially if expensive alloys used, e.g. Co, W, Ni, V, Cu. Little or no refining. High thermal efficiency
Resistor	Electricity. Alternating or direct current. Heat produced from high resistance rod (graphite) above charge	Up to 5.0	0.25 to 1.0	1000 to 1750	Al, Cu, Ni alloys, solders, type metals, special case irons and bronzes. No refining. Moderate thermal efficiency
Rotary	Oil, gas, solid (pulverised coal) fired above charge	0.5 to 5.0	0.1 to 5.0	800 to 1500	Al alloys, gun metals, brasses, bronzes, cast irons, steel. Metal–slag refining possible. Moderate thermal efficiency

Reverbera-tory	Oil, gas, solid (pulverised coal) fired above charge	5 to 200	2 to 20	600 to 1650	Al, Cu, Ni, alloys, steel. Metal–slag refining possible. Moderate thermal efficiency
Converter	Air or oxygen blown above or below molten charge. Heat produced from exo-thermic reactions between oxygen and the charge constituents	5 to 50 Up to 300 for bulk steelmaking	10 to 100	1650	High melting point metals. Steels in the main, some Cu and iron. Metal–slag refining possible. High thermal efficiency. Requires molten metal charge
Cupola	Coke charged to shaft combusted by injecting air through tuyeres at base of shaft	Continuous process	2 to 20	1300 to 1500	Cast irons, some Cu. Limited metal–slag refining. High thermal efficiency

Where a high degree of control over melting conditions, minimum contamination and metal loss are important electric furnaces are used; arc furnaces being more suitable than induction furnaces where high temperatures and some slag–metal refining are necessary. Induction furnaces are used mainly for straight melting and post melt treatments such as holding, alloying, grain refining, innoculation.

A summary of melting units is given in *Table 8.1*. For a more comprehensive comparison of melting units the reader should consult the references[1] and reading list at the end of the chapter.

8.1.4   Cast iron production

Cupola furnace melting is the most popular method of producing cast irons although electric induction melting has increased considerably over the past few years. This is especially due to the improved pollution control with induction melting. A small amount of cast iron melting is conducted in electric arc furnaces especially for high alloy irons and where there is a need for higher temperatures or a considerable amount of refining.

8.1.4.1   Cupola melting of cast irons

The cupola is a shaft furnace in which is placed the ferrous charge (pig iron, cast iron scrap, steel scrap), flux and coke in alternate layers; the main flux being limestone ($CaCO_3$). It may be operated on a continuous basis. The diameter of the cupola largely controls the output. A production rate of approximately 8.2 t m^{-2} of cross sectional area per hour (0.75 ton ft^{-2} h^{-1}) will normally result in the most efficient melting practice. A production rate of less than 6.6 t m^{-2} h^{-1} (0.6 ton ft^{-2} h^{-1}) results in underblowing while overblowing is produced with a production rate greater than 9.8 t m^{-2} h^{-1} (0.9 ton ft^{-2} h^{-1}). Both under- and overblowing will lead to low metal temperatures and an increased loss of metal and coke. These points are represented diagrammatically in *Fig. 8.3*.

Acid or basic melting practice, hot or cold air blast and the injection of oxygen, hydrocarbons (gas or oil) and graphite through the tuyeres may be used depending on the melting stock composition and the required cast iron analysis. The carbon content

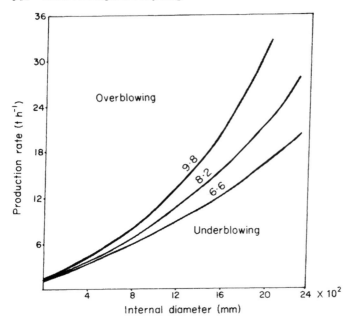

Figure 8.3 Relationship between metal production rate and internal diameter of the cupola. The curves are labelled with values of production rates in t m^{-2} h^{-1}

in the cast iron is derived from the amount of high carbon pig iron used in the charge materials and the transference of carbon from the coke. Sulphur (from coke and pig iron) and phosphorus must be kept below their maximum permissible limits—generally 0.1% S, and <0.1–0.9% P.

Melting with a basic slag and basic refractory lining improves carbon pick-up and aids sulphur, phosphorus and silicon removal from the iron (the thermodynamic explanation of this is given in Sections 7.2.10.5, 7.2.10.6). However, increased wear and cost of the basic lining compared with acid melting practice limits the use of basic cupola melting. Water cooling of the shell of the cupola helps to reduce refractory wear. The melt temperature increases as the air blast temperature increases. This produces an increased efficiency in carbon pick-up from the coke leading to a reduction in coke rate.

The hot blast may be provided by recuperation of the heat in the off-take gases or by external radiation or convector heating of the air. Increasing the melt temperature will also improve silicon retention in the iron and, when used with a basic slag, improve sulphur removal (see Section 7.2.10.5). Enrichment of the blast with oxygen leads to similar benefits as increasing the blast temperature. Oxygen enrichment is generally used to control the melt temperature during the initial melting period only, since preheating the air blast is cheaper to operate.

Injecting graphite particles through the tuyeres or by addition to the metal stream on tapping improves carbon pick-up and is especially useful for low carbon charges when increased amounts of steel scrap are used. However, the recovery of graphite tends to be erratic and there is a reduction in melt temperature when injection is employed.

The main variables in cupola melting, i.e. air blast rate, iron:carbon charge ratio, metal temperature and melting rate, can be plotted in the form of a *net diagram* (*Fig. 8.4*) for a cupola of a certain diameter. These diagrams provide the melting supervisor with

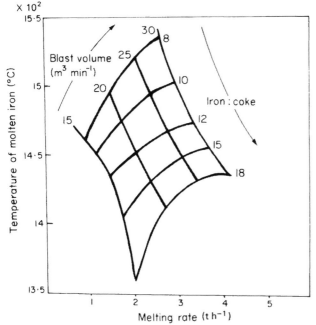

Figure 8.4 Simplified net diagram for a cupola of 550 mm internal diameter

important control information. The air blast is measured at standard temperature and pressure, i.e. 273 K and 760 mm Hg, and this must be taken into account.

8.1.4.2 Electric melting of cast irons

The coreless induction furnace provides a high degree of control over metal composition and temperature, furnace atmosphere and treatment time and is often used initially to melt a nodular (spheroidal graphite, SG) iron (Section 8.1.4.4) prior to desulphurisation and innoculation in a separate vessel. Electric arc melting of cast irons may be conducted in the direct arc furnace when higher temperatures are required, although as with cupola melting there is a greater tendency in the UK to use acid melting practice in the direct arc followed by a desulphurisation treatment in a second vessel.

There is greater use of arc furnace melting of cast iron in the USA than in the UK. Both the coreless induction furnace and the arc furnace are used to produce alloyed cast irons due to their high controllability and low metal losses. However, electric melting is unlikely to replace the cupola in the production of grey irons due to the high production rates required. Nevertheless, electric melting will continue to be used for the production of white irons owing to the fact that, in electric melting, silicon control is superior to that with cupola melting.

8.1.4.3 Desulphurisation and carburisation of iron

The technique of adding a carburiser and/or desulphurising agent to liquid irons as a post cupola or post electric melting practice allows cheaper raw materials to be used in the initial melting process. An example of this is the replacement of a large amount of pig iron by steel scrap in the initial cupola charge. Also, adjustment of the sulphur to carbon ratio is facilitated.

Calcium carbide, CaC_2, mixed into the molten iron and soda ash, Na_2CO_3, added to the ladle or launder are the most common desulphurising agents:

$$CaC_2 + S\text{(in Fe)} \longrightarrow CaS\text{(in slag)} + 2C\text{(in Fe)}$$

$$Na_2CO_3 + FeS \longrightarrow Na_2S\text{(in slag)} + FeO + CO_2$$

Coke breeze, coal dust or electrode scrap added to receiving furnaces or ladles and high purity carbon (graphite) added to the ladle or launder are the main carburising agents. Thorough mixing of the additions in the liquid iron is vitally important. This can be achieved by:

(i)   Blowing an inert gas or compressed air through a lance or porous plug (positioned in the bottom of the ladle) into the melt producing a vigorous stirring action (see *Fig. 7.25(a)*).

(ii) Injection of the carburising or desulphurising agent into the melt by means of a lance and an entrained carrier gas (see *Fig. 7.25(c)*).

(iii) Vibrating the treatment ladle.

(iv) Electric induction stirring of the melt.

(v)   Double ladling—transferring the liquid iron from one ladle to another. This is used only to a limited extent.

8.1.4.4   *Production of nodular or spheroidal graphite (SG) irons*

One of the significant developments in the production of cast irons is that of changing the shape of graphite from a flake to a spheroidal or nodular form. This provides improved mechanical properties. The melting sequence involves the addition of a spheroidising agent which allows the graphite to grow in a nodular rather than flake form, and an innoculant which provides nucleation centres for the nodules. The melting sequence for SG iron production involves:

(i)   Metal melting in a cupola, induction or arc furnace.

(ii) Desulphurisation as described above. This is essential since the spheroidising agents, Mg or Ce, subsequently added form stable sulphides.

(iii) Addition of spheroidising agents, Mg, Ce, Ni–Mg alloy or Ce–Mg–Si–Fe alloy, to the melt after transferring to a treatment ladle.

(iv) Innoculation of the melt with ferrosilicon. Impurities in the Fe–Si such as Al, Ba, Ca, provide centres on which the graphite nodules can nucleate. The innoculant is added to the stream of liquid iron on pouring into a ladle since their effectiveness *fades* with increasing the time between innoculation and pouring.

8.1.5   Steelmelting and refining

The theory and practice of steelmaking was thoroughly explained in Section 7.2.10.6 (i.e. the removal of C, S and P by slag basicity and oxygen potential control, deoxidation, alloying and degassing procedures) and needs no further explanation here. However, steelmelters rather than steelmakers generally produce much smaller quantities of steel and, therefore, the smaller electric arc furnace—5 tonnes to 35 tonnes—is the most common melting and refining unit. In the production of higher alloyed steels the coreless induction furnace is used where blending rather than refining is required and which offers a greater degree of control with minimum loss of alloying elements.

The high quality alloyed steels which demand strict control over the inclusion content are produced by the more modern processes of electroslag refining (ESR, *Fig. 8.5*), vacuum arc refining (VAR) and vacuum induction melting (VIM). Electroslag refining requires the steel charge to be in the form of an electrode which is in contact with slag (the second electrode) which is at a temperature above that of the melting point of the steel; the heat being generated by passing an AC or DC power supply between the steel and slag electrodes. The steel electrode melts and the droplets of metal fall through the slag which effects the refining process.

Thus, the choice of slag is important in that it must provide the required refining and electrical properties. In this respect calcium fluoride (CaF_2) is the most common slag constituent with additions of up to 30% alumina (Al_2O_3) to increase the slag's electrical resistance and up to 40% lime (CaO) and/or magnesia (MgO) added to provide the required basicity.

The ESR process results in a significant removal of inclusions and near-unidirectional solidification which, coupled with the melt–slag refining reactions, results in a considerable improvement in mechanical properties. However, the ESR process does not remove dissolved gases. VAR again uses the steel to be refined in the form of an electrode but within an evacuated environment and without a slag. This technique results in a considerable reduction in inclusion level and also dissolved gases.

VIM involves the melting of steel in an induction furnace under vacuum with a consequent lowering in the level of dissolved gases and some improvement in reduction of inclusions, especially if inert gas bubbling (rinsing) is used.

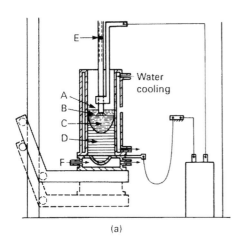

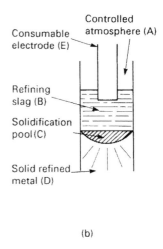

(a)   (b)

Figure 8.5 Schematic representation of Electroslag Refining (ESR): (a) refining unit; (b) electrode-slag–metal area. Vacuum Arc Refining (VAR) is similar except no slag (B) is used and the controlled atmosphere is a vacuum. ESR provides refining of elements (particularly sulphur) and structure and decreases the inclusion density. VAR provides refining of structure and lowers the hydrogen content and decreases the inclusion density. In (a), F is the water-cooled bottom plate.

8.1.6 Melting non-ferrous metals

Melting non-ferrous metals is usually a blending–remelting operation involving minimum refining. Therefore, electric and crucible furnaces are common melting units.

Reverberatory furnaces are used where some refining is required or the main charge constituent is bulky reclaimed scrap which requires larger melting units. Rotary furnaces offer a limited amount of refining but are ideal melting units for the smaller charge components such as borings, briquettes or fragmented scrap since these can be rapidly melted under a slag or flux on rotation of the vessel.

8.1.6.1 Nickel alloys

Melting nickel alloys is similar to melting steels. The charge material of pure nickel produced from electrorefined or carbonyl nickel and appropriate nickel scrap with maximum sulphur and lead contents of less than 0.01 wt% (lead is difficult to remove from the melt) is melted in a reverberatory or electric arc furnace. A thin lime slag cover is used and additions of carbon and nickel oxide are made to effect the carbon 'boil' reaction, producing carbon monoxide which helps to flush out hydrogen from the melt:

$$NiO + C \longrightarrow Ni + CO$$

A small addition of silicon is used to deoxidise the nickel melt and the slag made reducing by calcium–silicon and low sulphur charcoal additions. Alloying elements are added followed by final deoxidation with magnesium which has a high solubility in nickel and, due to the slightly lower melting point of nickel, is less prone to volatilisation than if used in steelmaking.

Chromium and copper additions are normally added after bath deoxidation with magnesium.

8.1.6.2 Copper alloys

Where some refining is required the charge materials comprising copper scrap and virgin copper are melted in a reverberatory or rotary furnace in a SO_2-free, oxidising atmosphere with an acid slag, or in an induction or crucible furnace where straight remelting is only needed. Cupolas have also been used for copper melting. Most impurity elements in the copper melt are preferentially oxidised to the slag.

Deoxidation is achieved with a Cu–15% P alloy or, if residual phosphorus is a problem with respect to conductivity properties, lithium or calcium boride (CaB_6) may be used. Alloying additions of Cr, Ag, Cd, Te, Zr, are made to the deoxidised molten copper while Al, Zn, Pb, Sn additions can be made prior to deoxidation. Indeed, alloying with aluminium and zinc may provide sufficient deoxidation so that further additions are not necessary. Extra zinc additions are normally needed prior to finishing the melt due to vaporisation loss of this element. After adding lead, the melt must be stirred before pouring to avoid this alloying element sinking to the bottom. Molten copper is susceptible to considerable dissolution of H_2, O_2 and SO_2 as shown diagrammatically in *Figs 8.1* and *8.6*. This is best minimised by operating with a controlled balance of oxidising–reducing conditions in the furnace and ensuring low sulphur levels in the charge.

As shown in Section 8.1.2, increasing O (in Cu) will lower H (in Cu) in order to maintain the equilibrium constant, K. Similarly, for SO_2 in the furnace atmosphere:

$$SO_2 \longrightarrow S \text{ (in Cu)} + 2O \text{ (in Cu)}$$

where

$$K = \frac{a_{S(\text{in Cu})} \, a^2_{O(\text{in Cu})}}{p_{SO_2 (\text{furnace atmosphere})}}$$

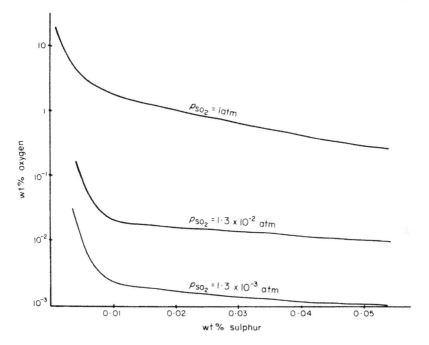

Figure 8.6 Relationship between equilibrium sulphur and oxygen contents in liquid copper at 1250 °C

therefore, increasing O(in Cu) will lower S(in Cu) in order to maintain the equilibrium constant. It is of interest to note that the same conclusion would have been made by adjusting a_O in the variable term of the van't Hoff isotherm for non-equilibrium conditions (see Section 8.1.3).

8.1.6.3 Aluminium alloys

Reverberatory or rotary furnaces are normally used to melt aluminium alloys, although crucible and induction furnaces may be used to melt small batches. Little refining is possible due to its high reactivity; melt blending being the main operating factor. However, magnesium can be removed to acceptable levels by bubbling Cl_2 gas through the melt or by additions of a halide flux, e.g. $NaCl-KCl-AlF_3$, i.e.

$$Mg(in\ Al) + Cl_2 \longrightarrow MgCl_2\ (dross)$$

or,

$$Mg(in\ Al) + NaCl(flux) \longrightarrow MgCl_2\ (dross) + Na(in\ Al)$$

If magnesium needs to be added it can be achieved by plunging the metal into the melt, thereby avoiding excessive oxidation loss. Mn, Ni, Ti and Cr are generally added in the form of temper alloys.

Injection of N_2 gas into the loop of a channel induction furnace or bubbled through the melt via a lance aids degassing and removal of inclusions. Alternatively, Cl_2-releasing compounds (e.g. hexochloroethane) can be added to the melt. Owing to the formation of a protective Al_2O_3 dross on the surface of the liquid metal fluxes need not be used.

8.1.6.4   Magnesium and zinc alloys

Induction or crucible furnaces are usually employed to melt magnesium alloys. Chloride- or fluoride-based fluxes are used to prevent oxidation and provide protection; the stock being dusted with flux prior to melting. Silicon refractories should be avoided as they can lead to silicon pick-up:

$$SiO_2 \text{ (refractory)} + 2Mg(melt) \longrightarrow 2MgO(dross) + Si(in\ Mg)$$

Zinc alloys are normally melted in crucible or induction furnaces or iron pots. A chloride-based flux, e.g. $ZnCl_2$, may be used for melt protection and cleaning. Volatilisation loss is the main problem.

8.2   Metal recycling

The finite limitations of the earth's mineral reserves have been emphasised on numerous occasions[2]. Thus, man must use his initiative and skill in developing alternative materials, seeking new ore deposits, extracting metals from leaner ores, exercising a high degree of conservation over metals which are presently readily available, and recycling metallic-bearing wastes.

Many tonnes of valuable metals are lost every year from such reclaimable sources as furnace flue dusts, slags and drosses, spent liquors from electrochemical plants, cleaning cloths and rags, effluents, refuse, spoil tips and tailings from mine wastes, spent catalysts and scrapped or obsolete components. Since the metal fraction of the metallic waste is often greater than that in primary ores where the metal is chemically combined (e.g. oxides, sulphides), extracting metals from scrap requires a much smaller energy output than when extracting metals from ores. A comparison of energy requirements and corresponding savings when producing metals from primary ores and scrap sources is given in *Table 8.2*.

Table 8.2 UNIT ENERGY FOR PRODUCTION OF PRIMARY AND SECONDARY METALS (AFTER KELLOG[3])

Metal	Energy requirements for extraction (GJ/t metal)		
	Primary from ore	Secondary from scrap	Energy saving
Magnesium	372	10	362
Aluminium	253	13	340
Nickel	150	16	134
Copper	116	19	97
Zinc	68	19	49
Steel	33	14	19
Lead	28	10	18

Recycled metal produced by the extraction and refining of metallic wastes is called *secondary metal* while metal produced from primary ores is called *primary metal*. There are two types of secondary metal producer: the secondary alloy producer who blends and refines the scrap to produce an alloy of required composition, and the secondary metal extractor who generally extracts one metal from the waste and sells any valuable by-products to other metal extractors or chemical processors. Both types of processor need to deal with a large range of sizes, shapes and compositions of the waste materials. In this respect the processing route and plant must be flexible in order to cope with a constantly changing input material.

8.2.1 Scrap processing

The majority of scrap processed is termed *new scrap* or *in process scrap*, and is mainly derived from manufacturers' and fabricators' waste. *Old scrap*, sometimes referred to as *post-consumer scrap*, largely originates from discarded or obsolete components and may be many years old, highly contaminated and difficult and costly to process. This latter source of metal may take up to twenty five years before being available for recycling.

Post-consumer recycle rates may be more indicative of scrap reclamation efficiency than total scrap recycle rates. A comparison of these two rates in Western Countries (Europe, North and South America, Africa, Japan, Asia, Australia and Oceanic Countries), is outlined in *Table 8.3*, where it is evident that the reprocessing of old scrap is very much less than for new scrap.

Table 8.3 COMPARISON OF POST-CONSUMER RECYCLE RATES AND SECONDARY METAL PRODUCTION FOR VARIOUS METALS IN WESTERN COUNTRIES[4]

Metal	Total consumption ($\times 10^3$ t)	Total secondary metal production		Post-consumer recycle rates (% of recycled metals)
		($\times 10^3$ t)	(% of total consumption)	
Aluminium	9375.7	2500.9	27.7	4
Zinc	4334.1	871.1	20.1	5
Copper	6193.4	3256.0	52.6	20
Lead	3039.4	1285.5	42.3	37
Tin	184.5	52.4	28.4	not available

Since modern society demands an increase in material wealth every year, recycling of metals cannot meet the increased demand, even if 100% scrap recycling was realised. Secondary metal production has increased over the past twenty years, but it has barely kept pace with increased consumption. Indeed, it is evident that in the last ten years, production of most secondary metals has fallen behind their rate of consumption.

The method and efficiency of recovering the metals from the remaining waste materials prior to extraction and refining play an important role in the efficiency of the overall recycling process. The valuable entrapped metallic particles present in process drosses and slags can be recovered by ball milling the drosses and grinding the slags. This fragments the brittle oxide materials (in which the metallics are locked) to a greater extent than the more ductile metals which can be separated on subsequent screening. Swarf and turnings after degreasing are shredded in a hammer mill and baled while light scrap such as sheet, tubing, and scrapped or faulty domestic utensils are simply baled. Heavy scrap such as girders, ingot moulds and railway lines need only to be cut into convenient sizes for charging to the melting furnace. Borings and fines are best briquetted, sintered or pelletised, although the cost of sintering and pelletising may be prohibitive. Large composite components such as washing machines and cars must be treated to separate the different metals present. To this end cryogenic processing (subjecting the components to very low temperatures, followed by shredding in a hammer mill) has been used to separate ferrous metals from non-ferrous metals. At very low temperatures the differences in ductility of metals becomes increased such that with liquid nitrogen ($-196\ ^\circ$C) treatment ferrous metals fragment to a greater extent than non-ferrous metals which can be separated on subsequent screening. One technique developed is to press the composite component into the form of a bale which is then subjected to a liquid nitrogen spray followed by crushing in a hammer mill and screening to separate the ferrous metals from the non-ferrous metals.

Crushed metals and non-metals have been separated by their different coefficients of restitution after being thrown against an angled plate. The metals and non-metals (plastics, wood, glass, etc.) rebound at different angles and distances, and can be collected in appropriate bins to give satisfactory separation of the two. The crushed and fragmented metals are normally separated due to their different magnetic, specific gravity, melting point or electrical conductivity properties.

8.2.2   Extraction of secondary metals

Scrap metal arising in solid form is normally subjected to a pyrometallurgical recycling route while that arising in liquid form, such as spent electrolytes from electrochemical processing of metals, is normally recycled by a hydrometallurgical route. Liquid effluent can be processed in two main ways to recover the associated metals:

(i)   Precipitation of a group of metals as a sludge, e.g. precipitation of the heavy metal hydroxides by addition of lime, followed by recovery of the individual metals using a pyrometallurgical or hydrometallurgical route.

(ii)   Extraction of a selected metal by application of ion exchange, solvent extraction, gaseous reduction or electrolysis.

Although there is a considerable overlapping of the hydrometallurgical and pyrometallurgical routes, the main industrial recovery techniques are based on pyrometallurgical extraction. Some typical recycling routes[4] are described for various metals in the remainder of this chapter.

8.2.2.1   Recycling of iron and steel

The theory and practice of cast iron production and steelmaking has been discussed in Sections 7.2.10.5, 7.2.10.6 and 8.1.4. There is considerable use of both cast iron scrap and steel scrap in the production of cast irons using cupola and electric melting and up to 100% steel scrap usage in electric arc and open hearth steelmaking processes, whereas up to 30% steel scrap may be used to lower the melt temperature and reduce materials' costs in converter steelmaking processes. The main problems in using scrap in steelmaking are the residual copper and tin levels which must be kept low to avoid serious reduction in mechanical properties. Higher levels can be tolerated in cast irons. Thermodynamically, Cu and Sn are relatively noble elements and continually build up in the steel.

8.2.2.2   Recycling of aluminium

The degree of refining of aluminium melts is severely limited. Only magnesium, calcium and sodium may be efficiently removed, as explained in Section 8.1.6.3. The secondary aluminium industry is, therefore, mainly concerned with the production of specification alloys, especially those used in the foundry industry. This results in a high degree of initial scrap preparation; by far the major emphasis being placed on the blending of the scrap to give a melt analysis of the required alloy composition. Therefore rapid instrumental analysis coupled with careful preparation of the charge (scrap aluminium with alloying and virgin aluminium additions) are required. To this end there is a growing use of computers. High density aluminium scrap is generally melted in reverberatory furnaces under low melting point fluxes, e.g. NaCl + KCl, with some fluorides and $CaCl_2$ to improve fluidity and separation of impurities from the melt to the slag. Some reverberatory furnaces are fitted with a well (*well furnace*), separated from the main furnace hearth by a refractory wall with two openings which allow molten metal to flow from the main furnace area to the well. The large scrap arisings are initially charged into the main hearth and on melting form a molten heel in the well. Subsequent charges are made to the well. Refining is generally restricted to removal of magnesium as a chloride (see Section 8.1.6.3). The liquid $MgCl_2$ rises through the melt and separates to the slag collecting other impurities, e.g. AlN, Al_2O_3, SiO_2, on the way. The removal of Cd, Pb, Bi, Sb, from the aluminium melt is achieved to some extent by forming an immiscible intermetallic compound by the addition of calcium.

Alloying can be effected via the addition of the element, e.g. lump silicon for Al–Si alloys, or an alloy, e.g. ferrosilicon for Al–Fe–Si alloys, or via the flux, e.g. titanofluorides, or borofluorides for titanium and boron alloys, which are particularly important in grain refining applications. 'Irony scrap', that is aluminium scrap which contains ferrous parts such as nuts and bolts, is melted in reverberatory furnaces with sloping hearths. The aluminium melts out (selective melting) at lower temperatures than the iron-bearing materials and separates to the lower end of the furnace. The aluminium is usually highly alloyed with zinc and other low melting point metals, while the residue is mainly ferrous metal which can be sold to the steelmakers. It is not currently possible to extract pure aluminium commercially although it has been achieved on the laboratory scale. The only successful method of removing Fe and Si from molten aluminium is by extremely expensive three layer cell electrolysis techniques (see Section 7.4.2).

8.2.2.3 Recycling of copper

Both pure copper and copper alloys are produced in large quantities from an extremely wide variety of copper containing waste materials by the secondary metal industry and follow the general techniques as described in the refining of copper in Section 8.2.10.1 with the obvious exception of matte smelting. The method adopted to produce pure copper largely depends upon the form and composition of the waste material. Medium to high grade scrap, i.e. greater than 40% Cu, is usually melted and refined with an acid slag in reverberatory or rotary furnaces while scrap with an average copper content of 20% is ideally smelted in a blast furnace with strict control over the CO/CO_2 ratio. The blast furnace copper (70–80% Cu) is subsequently transferred to a converter. The converter and reverberatory copper impurities, e.g. Fe, Zn, Ni, Sn, Pb, Si, Al, Mn, Mg, can be preferentially oxidised by air or oxygen blowing.

Tin is a valuable by-product and may be recovered in each unit either from the slag by critical slag control or from the fume on volatilisation by controlling the oxidising conditions. The refined copper is 'poled' with tree trunks or gas bubbling (NH_3, natural gas, propane) and cast into anodes for final electrorefining.

The copper-rich slags produced in the converter process can be recycled to the blast furnace. A typical flow sheet is given in *Fig. 8.7*. Secondary copper alloys are produced in a process route involving initial melting in a reverberatory or rotary furnace with subsequent air blowing in a converter to remove impurities. Final alloying additions can be made in an induction furnace.

8.2.2.4 Recycling of zinc

Zinc is mainly recycled as zinc chemicals such as zinc oxide, zinc chloride and zinc sulphate or as zinc–copper alloys. However, there is a large amount of recyclable zinc

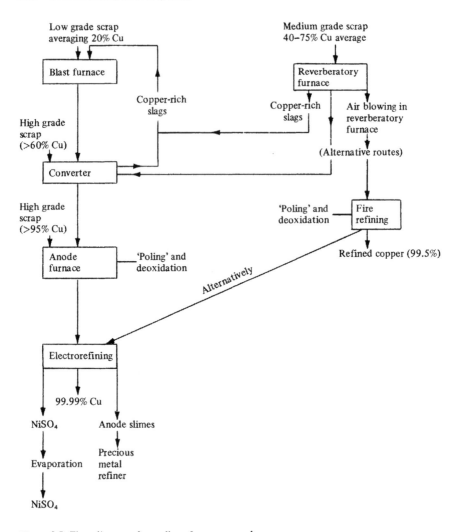

Figure 8.7 Flow diagram of recycling of scrap copper[4]

lost each year in the fumes from processing galvanised scrap due to the severe separation problems involved with this waste arising. Zinc dusts and zinc oxides collected from smelting and refining fumes and zinc drosses produced in melting zinc alloys are best recycled as pure zinc oxide or zinc dust or zinc chemicals. Zinc oxide is the most important zinc chemical, and is produced from zinc wastes by one of three processes:

(i)   Reduction of galvanisers' ash with coke or anthracite in a rotating kiln (Waeltz kiln). The zinc vapour is oxidised at the lower end of the kiln and the resulting zinc oxide collected from the fumes in a baghouse.

(ii) Melting metallic zinc scrap in a horizontal retort and the resulting zinc vapour oxidised to zinc oxide on contact with air.

(iii) Dissolving impure zinc oxide wastes in acid followed by solution purification, precipitation of zinc carbonate and calcination to ZnO. Zinc sulphate is made solely from zinc residues low in chloride content by leaching with sulphuric acid. This is followed by solution purification and cementation precipitation of copper, nickel, and cadmium by addition of zinc dust. Manganese and iron are removed by oxidation and the resulting solution evaporated to yield the monohydrate ($ZnSO_4.H_2O$).

Zinc chloride and zinc ammonium chloride are produced by treating zinc wastes with hydrochloric acid followed by solution purification.

A large proportion of zinc scrap is blended and melted to produce the appropriate zinc–copper alloys while a small amount (approximately 10%) is recycled as the refined metal.

8.2.2.5 Recycling of lead

High grade lead scrap can be satisfactorily melted and refined to final composition in a kettle furnace. Middle and low grade lead residues, lead-rich slurries after sintering are usually smelted in a reverberatory furnace, blast furnace or rotary hearth furnace.

Lead-rich materials and residues are normally contaminated with Cu, Sb, Sn, Al, As, Bi, Fe and S. Most of these contaminants may be removed by preferential oxidation using air blowing or oxidising fluxes to provide a crude metal. Elements which form high melting point compounds with the contaminating elements can also be added and by using suitable slag forming materials the compounds produced, being more stable than those formed between lead and the impurity elements, separate out into the slag, e.g. addition of Al to remove Fe, As, Cu, Sb. Lead residues containing more than 95% Pb are normally refined in cast iron kettle furnaces equipped with stirrers.

A high proportion of lead is used in the manufacture of storage batteries. Pure soft lead is used in the production of oxide pastes while antimonial lead (3–11% Sb) is used for grids, bridges and terminals.

The batteries are first drained of acid and the outer casing removed manually or by crushing followed by magnetic separation of the ferrous materials and density separation of the lead. The lead is subsequently smelted with coke and fluxes in a rotary, reverberatory or blast furnace. Copper is removed by addition of sulphur (drossing) producing Cu_2S as a dross, tin is removed as SnO_2 and $SnCl_2$ by addition of PbO_2 and $PbCl_2$ while arsenic and antimony are removed as Na_3AsO_4 and $NaSbO_3$ by addition of NaOH and $NaNO_3$ (Harris process); tin is also removed in this process as Na_2SnO_3. Each of these refining techniques is another example of preferential oxidation of the impurity element.

8.2.2.6 Recycling of tin

Owing to the volatile nature of tin compounds, secondary tin sources include airborne dusts from smelting operations, as well as chloride and oxide drosses from melting and tinning plants and normal solder, tin bronze and other alloy scrap arisings.

High grade oxidised residues are smelted with coke and soda-based fluxes in rotary furnaces. Lower grade tin residues are smelted with $CaO–SiO_2$ type slags in a blast furnace. The slags from the rotary furnace are also recycled to the blast furnace since they generally contain more tin than the blast furnace ($CaO–FeO–SiO_2$) slags. The rotary furnace smelting produces a crude high tin solder while the blast furnace produces a low tin solder. Both of these metal solders need refining since excess copper, antimony and zinc are normally present. Tin plate contains approximately 0.4% Sn which is presently being recovered from new tin plate scrap by a hydrometallurgical process involving dissolution of the tin in hot caustic soda solution containing sodium nitrate. The tin is recovered from the resulting stannate solution by electrolysis; the regenerated caustic soda solution being recycled to the dissolution stage. However, at present it is not possible to recover tin by this process from old or post-consumer tin plate, although this is being investigated. The detinned steel plate can be sold to steel-works if the residual tin content is low or to cast iron producers for melting in cupolas.

The tin from 'tin cans' (0.6–0.7% Sn) is currently not recovered although in the UK alone it would provide approximately 40% (2500 t/yr) of annual consumption. The only outlet for these tin cans is in cupola melting of cast iron but the tin content is too high for steelmaking. During cupola melting a large amount of tin is lost due to volatilisation and as entrained oxides of tin in the flue gases.

High iron-bearing tin slags are generally processed by a slag fuming technique to avoid the problem of Fe–Sn 'hard metal' alloy formation which would occur with normal smelting.

High zinc-bearing dusts can be treated with sulphuric acid to dissolve the zinc which is reprocessed as zinc vitriol (zinc sulphate); the tin residue being dried and smelted as described above.

8.2.2.7   Recycling of precious metals

Precious metals can be recovered from three main sources: solutions and electrochemical wastes; smeltable materials and slags; printed circuits and electronic scrap. Owing to their value rigorous recovery procedures are usually adopted in factories which manipulate, machine and fabricate these materials. Floor sweepings, water from hand washing and polishing cloths are all potential sources of secondary precious metals as well as the richer sources which include anode slimes from electrolytic plants, old medals and coins, brazing alloys and solders, jewellery and dental wastes and platinum catalytic reagents which are used in the chemical and petrochemical industry. In general, solid wastes follow a pyrometallurgical recovery route and liquids a hydrometrallurgical or electrometallurgical route with a final electrorefining process usually being adopted. However, intermediate products from one route may be passed into the other. There are three main process losses: slag, liquid or 'trade waste', and gaseous emissions which, because of their value and legislative requirements, must be kept to a minimum.

The cleaner, solid scrap sources are usually smelted in a blast furnace with litharge, coke and pyrites (FeS_2) which is added to provide a sulphide matte while coke provides the required reduction potential and heat. The precious metals are collected in a lead bullion created by the litharge addition and are also present in the sulphide matte to a lesser extent. The three liquids, i.e. slag, matte and bullion, are tapped at different levels from the blast furnace hearth. Fines containing precious metals are first nodulised before charging to the blast furnace while the larger items are crushed to aid charging.

Photographic film and paper, an important secondary source of silver, is incinerated before mixing with other constituents in the charge.

Richer scrap is smelted with coke and litharge in a reverberatory furnace, again resulting in a slag, matte and lead bullion. Silver is recovered from the bullion by preferential oxidation (cupellation) of impurities while gold recovery is achieved by chlorination (Miller process) with final electrorefining of each. Anode slimes from silver electrorefining are processed by leaching with H_2SO_4 or HNO_3. The Cu is cemented out with iron, and Ag and Rh subsequently cemented with Cu powder. The 'gold sand' left after leaching is treated electrolytically to reclaim fine Au, leaving the platinum group metals to be recovered from the spent electrolyte. Effluents from the plating industry are an important source of precious metals. Liquors containing Ag from the photographic processing operations are normally processed at source or by a local consortium and subsequently electrolytically refined.

Gold is also recovered from dilute solutions with an ion-exchange technique using a strong cation (OH^- or Cl^-) exchange resin:

$$R{-}OH^- + Au(CN)_2^- = R{-}Au(CN)_2^- + OH^-$$
$$R{-}Cl^- + Au(CN)_2^- = R{-}Au(CN)_2^- + Cl^-$$

Silver present in 'hypo' fixing solution after processing photographic film may be precipitated (cementation process) after passing the solution over steel wool. The resulting Fe–Ag sludge is subsequently smelted to recover the silver.

8.3 Fluxes used for non-ferrous metals

In addition to the use of fluxes to lower the melting point and aid solution of the slag components, a liquid or solid flux can be solely employed in non-ferrous metal melting and refining to aid dissolution of metal dross formed on the melt surface and provide protection, insulation and refining of the melt.

A protective flux is generally best applied as a low melting point fused salt usually around an eutectic composition which can easily and efficiently cover the molten metal surface. An *inspissating* flux is one which aids refining by coalescing impurities facilitating early and rapid removal to the dross or flux covering. Most non-ferrous fluxes are low melting point halides or salts, e.g. potassium fluoride, aluminium fluoride, sodium chloride, potassium chloride; fluorides being added to reduce the melting point of the chlorides.

As most halides are hygroscopic (accept water) they must be thoroughly dried to avoid hydrogen contamination of the melt.

Fluxes used to melt non-ferrous metals provide similar functions to those given by a slag and therefore the flux's specific gravity and acceptance of unwanted components and fluidity requirements are similar to those demanded by slags.

Fluxes may also be used to provide minor alloying additions and grain refining, e.g. modification of Al–Si alloys by using a NaCl flux and ensuring sufficient residual sodium is available for modification, i.e.

$$3NaCl_{(flux)} + Al_{(melt)} = AlCl_{3\ (in\ flux)} + 3Na_{(in\ Al)}$$

or using potassium titanofluorides and titanoborides as grain refiners, which react with carbon in the melt forming carbides of titanium and boron which act as nuclei in solidification. Sodium titanofluoride, titanoboride and silicofluoride fluxes may also be used to provide Ti, B, Si alloying although the extent of alloying in this way is limited to a few per cent only

8.3.1 Structures of flux materials (fused salts)

Fused salts are similar to solid salts in that they are composed of cations and anions which occupy two independent lattice structures and which are mutually interwoven. Cations are surrounded by anions and vice versa. The ions are free to move within each lattice and are more or less independent of each other. This is supported by the fact that molten salts are good electrical conductors and the current is carried by electrically charged ions.

The thermodynamics of fused salts becomes difficult to evaluate since the individual particles in the molten salt, being electrically charged ions, are impossible to measure. It is only possible to measure electrically neutral groups, that is the activity of ion pairs (see Section 6.1.4) where the pure salt is used as the standard state. For example, it is impossible to measure the activity of K^+ and Cl^- ions for a molten salt of KCl. Under this limitation the activity of the paired ions (KCl) is determined from statistical thermodynamics assuming ideal mixing, i.e.

$$\bar{S}_{Cl^-} = -R \ln N_{Cl^-} \qquad \bar{S}_{K^+} = -R \ln N_{K^+}$$

and

$$\bar{G}_{Cl^-} = -RT \ln N_{Cl^-} \qquad \bar{G}_{K^+} = -RT \ln N_{K^+}$$

from which it is possible to determine $\Delta \bar{S}_{KCl}$ and $\Delta \bar{G}_{KCl}$:

$$\Delta \bar{S}_{KCl} = -R(\ln N_{K^+} + \ln N_{Cl^-})$$

and

$$\Delta \bar{G}_{KCl} = RT(\ln N_{K^+} + \ln N_{Cl^-})$$

The reader is directed to Chapter 6 in particular Sections 6.1.4, 6.1.5, 6.1.6 and 6.2.6 for a more detailed examination of the activities of fused salts.

In a mixture of molten salts in which the bonding between one ionic pair is extremely strong, e.g. $Mg^{2+} + 2Cl^-$, a complex ion $(MgCl_4^{2-})$ may be produced. Thus a mixture of KCl and $MgCl_2$ molten salts may be regarded as $MgCl_4^{2-}$ anions, K^+ cations and additional Cl^- anions and Mg^{2+} cations depending on the salt composition. This is also the case in sulphate melts where SO_4^{2-} anions are present, while a fluoaluminate ion $(AlF_6)^{3-}$ is present in the AlF_3–NaF fused salt system.

REFERENCES

1 Moore, J.J. 1980 *Fundamentals of Foundry Technology*, chap. 4, ed. P.D. Webster. Portcullis Press.
2 Meadows, D.L. *et al.* 1972 *Limits to Growth*. London: Earthband.
3 Kellog, H.H. 1976 *J. Metall.*, 28, (12), 29–32.
4 Moore, J.J. 1978 *Int. Metall. Rev.*, 5, review 234, pp. 241–64.

FURTHER READING

1 Aitchinson, L. and Kondic, V. 1953 *The Casting of Non-Ferrous Metals*. London: Macdonald and Evans.

2 BCIRA 1965 *Cupola Operation and Control.* Birmingham: BCIRA.
3 Flinn, R.A. 1963 *Fundamentals of Metal Casting.* Reading, Mass.: Addison Wesley.
4 Heine, R.W., Loper, C.R. and Rosenthal, P.C. 1967 *Principles of Metal Casting.* New York: McGraw-Hill.
5 Kondic, V. 1968 *Metallurgical Principles of Founding.* London: Edward Arnold.
6 Strauss, K. (ed.) 1970 *Applied Science in the Casting of Metals.* Oxford: Pergamon Press.

For self-test questions on this chapter, see Appendix 5.

9

Corrosion of Metals

There are many definitions of metallic corrosion but in essence they state that it is the reaction between a metal and its environment which invariably leads to a degradation in the metal's mechanical and physical properties. Thus, to allow corrosion to occur unabated can be a costly error and, therefore, some measure of corrosion protection and a component replacement programme are well advised procedures.

Applying an inflation factor to the 1971 Hoar report[1] figures for the annual costs of corrosion in the UK, the current (1979) cost would be approximately £4000 million. In the USA, recent figures[2] indicate an annual corrosion cost of £35 000 million. It has been estimated that application of available corrosion-prevention technology could lead to savings of up to 25%.

In this chapter the effects of aqueous corrosion are outlined first with some deeper considerations of stress effects. The theory of aqueous corrosion and methods of protection then follow. Finally 'dry' corrosion is discussed, i.e. oxidation in the absence of water, its causes and how it may be controlled.

9.1   Aqueous corrosion

9.1.1   The basic corrosion cell

Aqueous corrosion is an electrochemical process, i.e. a chemical effect which induces the passage of an electric current. Because metals are good conductors they are particularly susceptible to this type of behaviour. The overall effect of corrosion is that a metal undergoes an electrochemical reaction as a result of its environment to form a compound similar to the one from which it was originally won.

In order to produce a continuous flow of current a complete cell must be set up comprising two half-cells as described in Section 6.3. The components of such a cell are: (i) an anode, the site at which a metal is corroded (dissolved); (ii) a cathode, which is not consumed in the corrosion process; (iii) a conductor, if the anode and cathode are not in contact; and (iv) an electrolyte, which permits flow of ions between electrodes. Such a cell must be electrochemical (*galvanic* or *current producing*) in character rather than electrolytic (*current consuming* via an external source). The basic electrochemical cell is shown in *Fig. 9.1*. Electrochemical cells are considered in Section 6.3 while electrolytic cells are discussed in Section 6.4. Electrons are produced at the anode by the reaction,

$$M \longrightarrow M^{n+} + ne \qquad (9.1)$$

In an electrochemical cell this reaction makes the polarity of the anode negative. In an electrolytic cell the anode is still the electrode at which electrons are produced but the anodic polarity in this case is positive. The electrons are, in effect, 'pumped away' by the applied electrical source (see Section 6.3.5 and Appendix 3). The fundamental cathodic process is the consumption of electrons (i.e. reduction), but this may be

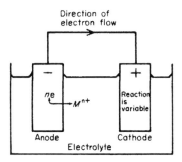

Figure 9.1 Principle of electrochemical corrosion cell

achieved in a number of ways depending on the environment. A common cathodic reaction is the hydrogen evolution reaction (sometimes abbreviated to HER):

$$2H^+ + 2e \longrightarrow H_2 \tag{9.II}$$

in acid solutions. The HER in neutral or alkaline solutions is

$$H_2O + e \longrightarrow \tfrac{1}{2}H_2 + OH^- \tag{9.III}$$

Alternatively, oxygen reduction may occur in acid solution as:

$$\tfrac{1}{2}O_2 + 2H^+ + 2e \longrightarrow H_2O \tag{9.IV}$$

or in neutral or alkaline solutions by:

$$\tfrac{1}{2}O_2 + H_2O + 2e \longrightarrow 2OH^- \tag{9.V}$$

Other possible reactions are metal deposition:

$$M^{n+} + ne \longrightarrow M \tag{9.VI}$$

metal ion reduction, e.g.

$$M^{3+} + e \longrightarrow M^{2+} \tag{9.VII}$$

or anion reduction, e.g. for nitric acid,

$$2NO_3^- + 4H^+ + 2e \longrightarrow N_2O_4 + 2H_2O \tag{9.VIII}$$

In practice, of course, anodes and cathodes are not as shown in *Fig. 9.1*, except in bimetallic contact. In the simplest case they may arise from two different metals contacting each other, the surface of one metal becoming anodic with respect to the

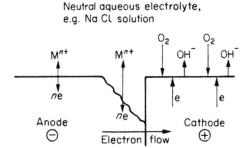

Figure 9.2 Principle of electrochemical corrosion cell existing between two different metals in contact with each other

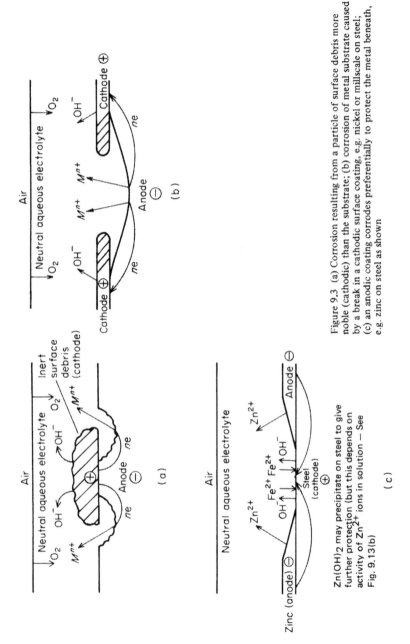

Figure 9.3 (a) Corrosion resulting from a particle of surface debris more noble (cathodic) than the substrate; (b) corrosion of metal substrate caused by a break in a cathodic surface coating, e.g. nickel or millscale on steel; (c) an anodic coating corrodes preferentially to protect the metal beneath, e.g. zinc on steel as shown

$Zn(OH)_2$ may precipitate on steel to give further protection (but this depends on activity of Zn^{2+} ions in solution — See Fig. 9.13(b)

other (*Fig. 9.2*). More usually corrosion cells are very small and numerous and occur at different points on the same metal surface. In this case anodes and cathodes arise from local differences in the metal structure or environment. Frequent examples are breaks in metal coatings, the presence of impurities or different alloy phases at the surface and local differences in oxygen concentration in the electrolyte. The corrosion process is markedly affected by the nature of the electrolyte. Total immersion or condensed droplets produce very different corrosion effects. Also, since ionic movement between electrodes is required, electrical conductivity of the electrolyte is a significant factor in determining corrosion rates.

The relative areas of anodes and cathodes are extremely important in this respect since corrosion rates are increased when the cathode area is large compared with the anode area.

9.1.2 Classification of electrochemical corrosion cells

There are considered to be three types of cell: (i) dissimilar electrode (galvanic) cells; (ii) concentration cells; and (iii) differential temperature cells. Each will be considered in turn.

9.1.2.1 Dissimilar electrode cell

A standard example of this type of cell is that of a copper pipe connected to an iron pipe. The principle is again shown by *Fig. 9.2*. Assuming standard conditions, consideration of the electrochemical series (Sections 6.3.5 and Appendix 2) shows that copper has a more positive redox potential and, therefore, becomes cathodic to the iron, which dissolves. Advantage is taken of this effect in the cathodic protection (Section 9.1.10.3) of iron by zinc (galvanising): zinc is anodic to iron and, therefore, dissolves in preference to the iron which is, therefore, protected. This situation is shown schematically in *Fig. 9.3(c)*. Note that the corrosion rate of zinc is approximately 0.03 times that of iron when these metals are not in bimetallic contact, i.e. zinc corrodes much more slowly than iron. Other examples of this type of cell are: (i) a bronze propellor contacting a steel hull; (ii) surface debris (*Fig. 9.3(a)*); (iii) imperfections in surface coatings (*Figs 9.3(b)*, *(c)*); (iv) the presence of different phases at the surface (*Fig. 9.4*);

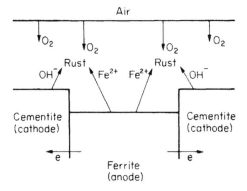

Figure 9.4 Preferential corrosion of ferrite in pearlite

and (v) local stresses as a result of cold worked grains in contact with annealed grains or simply the difference in stress between grain boundary metal and grains (the region of highest stress becomes the anode). Galvanic effects often result in intense localised attack called pitting, examples of which are shown by the deep recesses in *Figs 9.3(a)*, *(b)* and *(c)*. Pitting will be considered more generally in Section 9.1.3.2, pitting.

Whether a metal becomes an anode, and corrodes, depends not only on the other electrode but also on the environment. For example, according to the electrochemical series tin is cathodic to iron. Therefore, in a galvanic cell comprising tin plated mild steel the steel corrodes producing toxic Fe^{2+} ions. However, in anaerobic (oxygen-free) conditions in a 'tin' can organic preservatives in the food form soluble complex ions of tin. This reduces the activity of Sn^{2+} ions such that E_{Sn} becomes less than E_{Fe}. Therefore, iron becomes cathodic to tin and is protected with the added advantage that Sn^{2+} ions are non-toxic (see Section 6.3.12). Another example of *polarity reversal* is shown by metals which form protective oxide coatings. Thus aluminium and titanium both behave as more noble metals when coupled (galvanically) with zinc; aluminium (or more correctly Al_2O_3) acts as a cathode and zinc dissolves. By taking pairs of metals in common environments a new series may be developed to facilitate polarity (i.e. corrosion) prediction. Thus there are *galvanic* series for metals in sea water and soil (see Appendix 4). When a protective oxide film is broken and cannot be reformed in, for example, sea water, the metal or alloy may corrode and become *active*. Hence the *passive* (protection) and *active* (corrosion) positions of stainless steel in the galvanic series for sea water.

9.1.2.2 Concentration cells

These cells comprise two identical electrodes in contact with solutions of different component concentrations.

Salt concentration (metal ion) cell

Fig. 9.5(a) shows the principle of such a cell and an example of how such a cell could occur is given in *Fig. 9.5(b)*. The electrode in the more dilute solution tends to dissolve.

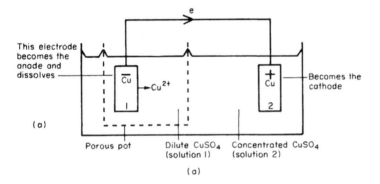

(a)

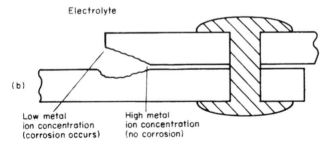

Figure 9.5 (a) Principle of the salt concentration corrosion cell; (b) example of corrosion between rivetted plates caused by a salt concentration cell

Equation (6.21) gives the cell emf and Section 6.3.14 explains the theory of concentration cells. Note that *Figs 9.5(a)* and (*b*) refer to a metal in a solution of its own ions. Such a situation is rarely found in practice. Consider a buried steel pipe passing through two layers of soil, one layer having a higher concentration of ions (Al^{3+}, Ca^{2+}, SO_4^{2-}, Cl^-, etc) than the other. Anodic sites are established on the section of pipe in the high concentration layer.

Differential aeration cell

This cell is due to differences in oxygen concentration and is more common than the metal ion cell. *Fig. 9.6* shows the principle and instances of occurrence. This theory has also been considered in Section 6.3.14. Since the corrosion of iron is of major importance rusting, as depicted in *Fig. 9.6(d)*, will be considered in more detail. This

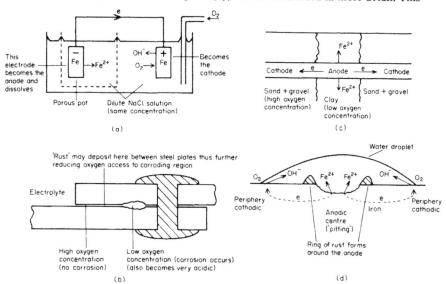

Figure 9.6 (a) Principle of the differential aeration cell; (b) example of corrosion between rivetted plates caused by a differential aeration cell; (c) corrosion of steel pipeline in soil caused by differential aeration effects of different soils; (d) corrosion pitting of iron caused by differential aeration effects of a drop of water

cell is set up because of the differences in oxygen concentration between the centre and the periphery of the drop of water. Thus at the centre where the concentration of available oxygen is least the anodic reaction predominates, i.e.

$$Fe \longrightarrow Fe^{2+} + 2e \qquad (9.IX)$$

At the peripheral regions the oxygen supply from the atmosphere maintains the cathodic reaction of reaction (9.V). These two equations may be combined:

$$2Fe + 2H_2O + O_2 \longrightarrow 2Fe(OH)_2 \qquad (9.X)$$

Iron (II) hydroxide is somewhat unstable in aerated solution and is oxidised to form brown 'rust'—hydrated iron (III) oxide, $Fe_2O_3.H_2O$. When the oxygen supply is limited black magnetite, Fe_3O_4, forms.

In general, the concentration of oxygen in solution decreases as depth increases. Therefore, the deepest sections of immersed structures will be anodic to those nearer the water line. *Fig. 9.7(a)* shows the relationship between differential aeration and depth of immersion. Where the volume of liquid is limited, e.g. a small container, the oxygen supply at the lower regions becomes used up. The anode area then spreads upwards. A particularly intense form of corrosion may occur at the water line of steel vessels which contain aqueous solutions. This 'water line attack' is shown in *Fig. 9.7(b)*. Again differential aeration causes corrosion which is a maximum just below the meniscus. Any corrosion precipitate tends to stick to the water–air interface leaving the

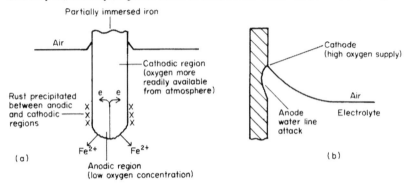

Figure 9.7   (a) Corrosion of partially immersed iron plate caused by differential aeartion; (b) water-line attack caused by differential aeration

anode reaction unobstructed. This form of attack is particularly prevalent when insufficient inhibitor (Section 9.1.10.1) has been added. In plain sodium chloride solution, sodium hydroxide forms at the water line which is itself an inhibitor and water line attack does not occur.

Both types of concentration cell can exist simultaneously and may oppose or support each other.

9.1.2.3   *Differential temperature cell*

This type of cell can occur when electrodes of the same material are immersed in an electrolyte of the same initial composition but each is at a different temperature. Such a situation occurs in heat exchanger boilers or immersion heaters. At first sight the Nernst equation indicates that as T increases E becomes more negative (the activity of metal ions in solution is less than unity therefore its logarithmic value is negative). This supports 'hot spot' corrosion observed in practice. However, temperature influences $E^\ominus$, solubility of corrosion product and ionic conductivity so that prediction of polarity with certainty is difficult. In practice the hot electrode is usually anodic but long-term tests may be needed to establish polarity and whether polarity changes with time.

9.1.3   Types of corrosion damage

There are two categories of corrosion damage, uniform attack and local attack.

9.1.3.1   *Uniform attack*

This is the most common form of corrosion. It causes the greatest destruction of metal on a weight basis but it enables the service life of equipment to be accurately estimated. This form of corrosion may be expressed in terms of depth of penetration of corrosion or of weight loss. The units often quoted are ipy, inches penetration per year; mpy, mil penetration per year (1 mil = 0.001 in); ipmo, inches penetration per month; $g\ m^{-2}\ d^{-1}$, g per m^2 per day; and mdd, mg per dm^2 per day. However, the SI unit is mm yr^{-1}, millimetres penetration per year. This unit should not be confused with mpy. (Sometimes millimetres per year are also written mpy so great care must be taken when interpreting data). Conversion factors are available, e.g.

$$mdd = 10 \times g \, m^{-2} \, d^{-1} \quad \text{and} \quad mm \, y^{-1} = \frac{36.52}{\rho} \times mdd \tag{9.1}$$

where ρ is the density of the metal (kg m^{-3}) and

$$mdd = \frac{\rho}{1.44} \times ipy \tag{9.2}$$

For service conditions, less than about 0.1 mm yr^{-1} is considered satisfactory and more than about 1 mm yr^{-1} unsatisfactory. The use of materials with intermediate corrosion resistance depends on the application.

9.1.3.2 Local attack

This type of damage is much more difficult to predict than uniform attack. It causes premature failure and, therefore, is of much greater technical importance than uniform corrosion.

Crevice corrosion

Crevice corrosion, shown schematically in *Fig. 9.6(b)* is rather more complex than the brief consideration in Section 9.1.2.2 implies. Differential aeration is present but it is not the main corrosive effect. Consider the case of overlapping iron plates in sea water. The solution in the crevice gradually becomes depleted in oxygen by the normal corrosion processes until virtually all the oxygen is removed and hence differential aeration conditions exist. Metal dissolution continues in the crevice, the corresponding cathodic reaction of oxygen reduction being transferred to local regions outside the crevice. The result is a build-up of positive charge within the crevice which attracts mobile Cl$^-$ ions. Hydrolysis occurs according to:

$$FeCl_2 + H_2O \rightarrow Fe(OH)_2 + 2HCl$$
$$\downarrow$$
'rust'

Precipitation of rust (Section 9.1.2.2) promotes this reaction and conditions in the crevice become increasingly acidic. This argument is supported by analysis of crevice solutions which shows chloride concentrations of up to ten times greater than the bulk solution and pH values of between 2 and 3 in neutral bulk solutions. Increasing acidity promotes further metal dissolution and increases cation formation. The reaction becomes self-propagating and is said to be autocatalytic. As the corrosion rate in the crevice increases so the cathodic oxygen reduction reaction on the nearby external surfaces increases, thereby cathodically protecting these surfaces. It is important to realise that there is an incubation period, perhaps of several months, before crevice corrosion is established; the process then accelerates.

Pitting corrosion

This phenomenon occurs mainly on the surfaces of metals and alloys which form passive oxide films where local film breakdown creates a small anode–large cathode combination. For example, stainless steel shows pitting in static chloride solutions. Local variations in solution composition or alloy surface structure help initiate pits but not all pits continue to grow. Once a pit has become established it grows by the same accelerating, autocatalytic mechanism as crevice corrosion. In fact, pitting corrosion may be regarded as a form of crevice corrosion which creates its own crevice. There is a link between pitting and crevice corrosion in that metals and alloys which show pitting also suffer crevice corrosion, although the reverse is not always true.

Pitting is a particularly insidious form of attack because the small visible surface blemish gives no indication of the considerable undercutting of the surface which usually occurs. The pit itself may pass unnoticed if corrosion products are not completely removed, resulting in unexpected perforation of a pipe or tank.

Dealloying

This is essentially the removal of a more base component of an alloy from a more noble one by corrosion. This effect is exemplified by dezincification of brasses. The two phase 60% copper–40% zinc brasses are particularly susceptible to this type of attack, but the α brasses, containing less than 30% zinc, are by no means immune. Depending on solution conditions, dezincification may occur by either dissolution of brass followed by precipitation of copper, or simply selective dissolution of zinc. Either way the result is a porous, mechanically-weak structure of copper.

Dezincification of brasses may be significantly reduced by addition of 1% tin (Admiralty brass) or, even better, by addition of 1% tin and 0.04% arsenic (arsenical Admiralty brass). Similar additions to 60–40 brasses are not nearly as effective. Where dezincification continues to be a problem it may be necessary to use cupronickels instead.

A further example of dealloying is provided by graphitisation of grey cast irons in which the anodic iron-rich ferrite slowly dissolves to leave a mechanically-weak, porous network of graphite. If a coarse graphite network or free graphite is absent graphitisation cannot occur. Thus, white cast irons and spheroidal graphite (SG) irons are immune. Where the problem occurs it is often overcome by using an austenitic cast iron, 'Ni-resist' (nominally Fe–20% Ni–5% Cr).

Intergranular corrosion

As the name suggests corrosion of this kind occurs by localised attack at grain boundaries which behave as anodes to the larger surrounding cathodic grains. Improperly heat treated austenitic stainless steels and duralumin-type alloys are prone to this type of attack although non-electrochemical attack can give the same effect, e.g. sulphur penetration of the grain boundaries in nickel at high temperatures. The intergranular attack of carbon steels in the presence of nitrates and carbonates is particularly important.

Weld decay: During the welding of an austenitic stainless steel a local effect may occur in the heat-affected zone which heat treats the steel at an undesirable, *sensitising*, temperature. The critical temperature range is 400–900 °C. Within this range carbon diffuses rapidly to the grain boundaries where it combines with chromium to precipitate chromium carbide denuding the area adjacent to the grain boundary of chromium. The effects are summarised in *Fig. 9.8(a)*. Anodic attack occurs adjacent to the grain boundaries which is exacerbated by the small anode–cathode area ratio, and the diffusion of chromium at normal temperatures is too slow to compensate for the loss. This type of attack may be prevented by having a low carbon content (less than 0.02%) to reduce chromium carbide formation, or rapidly cooling the steel to below 400 °C, or *stabilising* the steel by addition of titanium or niobium. These latter elements form more stable carbides than chromium and hence they inhibit carbon diffusion to grain boundaries.

Knife line attack (KLA):   Intergranular attack on niobium and titanium stabilised stainless steels may occur at sensitising heat treatment temperatures achieved during welding. This effect is called knife line attack and occurs in a very narrow region in the metal adjacent to the weld whereas weld decay occurs at a greater distance from the weld

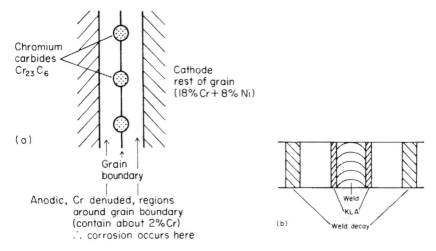

Figure 9.8 (a) Schematic representation of cause of weld decay; (b) schematic comparison of location of zones of weld decay and knife-line attack (KLA)

(Fig. 9.8(b)). When heated to temperatures in excess of 1100 °C titanium and niobium carbides go into solution. On rapid cooling these elements remain in solution and are not re-precipitated. Subsequent heat treatment *sensitises* the material, i.e. precipitates chromium carbide as if no titanium or niobium were present. The remedy is to heat treat stabilised stainless steels after welding to a temperature (~1050 °C) which permits solution of chromium carbide and precipitation of titanium or niobium carbide. The subsequent cooling rate then has little significance.

Hydrogen damage

Atomic hydrogen, released as the cathode product during corrosion (reaction (9.II)), may not all form molecular hydrogen (H_2) bubbles at the surface but some may diffuse into the metal in atomic form. Here, locked at voids or dislocations for example, molecular hydrogen can form to build up very high internal pressures. This effect produces two types of damage, hydrogen blistering and hydrogen embrittlement. The latter condition can also occur by formation of unstable hydrides. Blistering and embrittlement by hydrogen are problems often detected in electroplating and electro-polishing processes as well as corrosion. Hydrogen embrittlement may also be a cause of stress corrosion cracking (p. 344).

Two other forms of hydrogen damage are decarburisation and hydrogen attack. In general these occur at high temperatures in the presence of gaseous, normally moist, hydrogen. Each of these effects leads to a reduction of mechanical properties. Decarburisation removes carbon from steels as methane either as a surface effect or within the metal. Hydrogen attack refers to the chemical reduction of oxides in some oxide dispersion-strengthened materials.

Erosion–corrosion

The effect of this form of corrosion is shown schematically in *Fig. 9.9*. Nearly all corrosive media are able to bring about erosion-corrosion whether they are gaseous or

liquid and nearly all metals and alloys are susceptible to this form of corrosion.
However, metals and alloys which are capable of forming hard, dense, adherent continuous surface films, notably titanium, aluminium and stainless steels, have a much

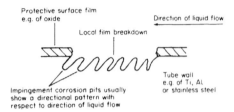

Figure 9.9 Schematic diagram representing impingement corrosion erosion-corrosion

greater resistance to erosion-corrosion. Thus titanium, which forms protective TiO_2,
is very resistant to erosion-corrosion by sea water, chloride solutions and nitric acid.
Brasses are more resistant to erosion-corrosion in chloride solutions than copper
because of formation of a CuO film which is more adherent than the $CuCl_2$ film formed
on copper. Often, the protective films are oxides and are *self-healing* under oxidising
conditions. Where reducing media are used such a film cannot reform once damaged
and erosion-corrosion (and any other corrosion process) occurs.

The nature and composition of surface films often change with pH in aqueous solutions (see Pourbaix diagrams, Section 9.1.4). Carbon steel subjected to flowing distilled
water at 50 °C shows marked erosion-corrosion below pH6 and also at pH8 due to
formation of granular Fe_3O_4, which becomes increasingly soluble as the pH decreases
from 6. Minimum erosion-corrosion occurs at pH6 and pH10 due to hydroxide formation which presumably confers protection by reducing oxygen and cation diffusion
(see Section 9.2).

The velocity of the corroding medium is very important particularly when solid
particles are present to increase the erosion factor. For some materials there is a critical electrolyte velocity above which erosion-corrosion occurs rapidly. However,
electrolyte flow may have adverse or beneficial effects. For example, a high velocity
increases the supply of oxygen or other gases at the metal surface which may *depolarise*
(Section 9.1.5.5) the cathodic reaction, i.e. increasing the ease with which a cathodic
reaction takes place and consequently increasing the corrosion rate. On the other hand
increasing the oxygen content at iron or zinc surfaces promotes protective film repair
and a consequent decrease in corrosion rate (cf. Evans[3], 'eccentric whirler' experiment).
Higher velocities also promote inhibitor (Section 9.1.10.1) effectiveness, e.g. less
sodium nitrite is required to protect steel from high velocity tap water than under slow
flowing conditions. In addition, the possibility of pitting by differential aeration cells
set up by deposition of silt or similar materials, especially at pipe bends, is considerably
reduced by high velocity electrolytes. In general, however, high velocities mean an overall increase in erosion-corrosion. This indicates the importance of design. Bends in
piping, where fluids strike the metal directly, produce a particularly rapid form of
erosion-corrosion called impingement attack. Obviously all the bends in a pipe system
cannot be removed but where unavoidable erosion-corrosion is likely to be a problem
suitable design can make replacement of pipe sections easier. High temperatures also
aggravate this type of damage as it does most forms of corrosion.

Cavitation erosion: When an object such as a propeller rotates in water the pressure on the trailing surface of the blade fluctuates continually. At some points very low pressures are produced which create tensile forces high enough to exceed the interatomic bonding forces of the liquid. The result is that at these points atoms are torn apart to create voids or cavities (or vacuum 'bubbles'). Once formed, these voids subsequently collapse in about 1 μs. When the void walls collide a tremendous shock wave is generated which exerts a pressure of up to 1.5 GN m^{-2} on a small adjacent area of metal surface. This pressure plastically deforms the metal. It has been estimated that two million cavities may form and collapse in one second on one small area. The resultant hammering effect gives rise to the typical closely pitted (honeycomb) appearance of cavitation erosion. The damage produced is caused by combination of a mechanical effect which creates pits and breaks down any protective surface films and a chemical effect, which is the reaction between the liquid and the metal in the pits.

Because of the nature of the process the ability of a material to work harden and thereby become more resistant to mechanical attack is advantageous as are high corrosion fatigue life, tensile strength, ductility and hardness. Hence stainless steels are particularly resistant to cavitation erosion.

Aerated liquids show a lesser tendency to produce cavitation than de-aerated liquids. Oxygen enters the vacuum bubbles thereby decreasing the vacuum slightly. As a consequence the damage caused is reduced. Air bubble injection over a metal surface subjected to cavitation conditions cushions the pressure waves caused by cavity collapse and reduces cavitation damage. Hydrogen liberated by cathodic protection (Section 9.1.10.3) also has this cushioning effect.

Fretting corrosion: This form of corrosion is a combination of mechanical wear and atmospheric oxidation and frequently occurs between close fitting metal components. The situation is shown schematically in *Fig. 9.10*. The surfaces are usually under load and subject to slight relative movement (less than 25 μm amplitude). The result is damage to the contact surfaces and formation of an oxide debris which, for iron, is usually Fe_3O_4. Initially, under the action of the oscillatory motion, the air-formed oxide film present on the metal surface breaks down. This permits metal-to-metal

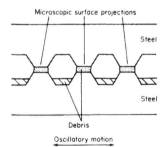

Figure 9.10 Schematic representation of fretting corrosion

contact at surface projections which bond strongly (weld) together. This theory is supported by friction studies which show that the initially low coefficient of friction rises to a high value, and electrical resistance studies which record a marked decrease in electrical resistivity. The welded junctions are subsequently subjected to plastic deformation and work harden until they break under the combined action of shear forces and the corrosive action of the atmosphere (because of the vibratory motion a corrosion fatigue process may be envisaged–p. 345)–again supported by electrical measurements. In dry air resistivity rises to a very high value indicating not only the loss of intermetallic contact but also the presence of an insulating medium, i.e. metal oxide, between the contact surfaces. The forward sliding motion of the broken welds exposes fresh (reactive) metal to the oxidising atmosphere. The oxide is then partially cleaned off by the reverse stroke. This results in a build-up of oxide debris between the surfaces, increasing resistivity and abrasive wear. The presence of oxygen must exert an effect on the wear process since fretting damage is reduced in vacuo or inert atmospheres. Moist air also reduces the wear rate. It has been suggested that hydrated oxides form which have a lubricating action because they are soft.

Fretting damage also decreases as temperature increases. Above a certain minimum temperature the formation of loose oxide debris is replaced by an adherent oxide *glaze* which has a low coefficient of friction. This occurs above about $180\,^{\circ}C$ for mild steel. The oscillatory motion also produces a fatigue effect. The severe local plastic deformation at welds combined with a local temperature rise nucleates small cracks at the points of contact. The growth of these cracks may then lead to failure by fretting fatigue. However, metals which have high resistance to fatigue have not been found to be more resistant to fretting fatigue–in fact alloys of lower fatigue strength often have better resistance. The reason for this is that fretting takes place under conditions of high strain. The fatigue performance of a material under these conditions is not necessarily the same as that under normal testing conditions.

Inserting soft, slip absorbing materials or lubricants between the surfaces and using a corrosion free environment or corrosion resistant material will help to prevent fretting corrosion. Applying large loads to two materials to prevent movement may be a dangerous practice since any slight movement at this higher load will increase debris formation and work hardening. Fretting fatigue can be controlled by inducing hard surfaces by shot-peening, nitriding, etc., but general work hardening, e.g. of steels, decreases the fretting fatigue life.

Stress corrosion cracking (SCC)

This phenomenon is the premature failure due to the combined action of a static tensile stress and a corrosive environment. It occurs most commonly in those corrosion resistant alloys which are capable of producing a protective oxide layer. The theory is not fully understood and some proposals will be considered later.

The appearance of fracture surfaces after SCC is similar to that after brittle fracture but, of course, corrosion processes are the cause. Cracking may be intergranular or transgranular, branched or multibranched. There are no general rules which apply to one particular material or one specific environment since the same material may show quite different modes of fracture in different environments. However, most metals will exhibit SCC when stressed in specific environments, e.g. stainless steels in chloride solutions, mild steels in nitrates and carbonates, or brasses in ammoniacal solutions. The important variables that determine length of time before failure are level of stress, environment, temperature, metal composition and metal structure. There is general agreement that the initiation of stress corrosion cracking takes place at a surface pit or notch. A similar defect may occur if there is a break in a surface film permitting local corrosion to take place. However, there are several theories of propagation of the root of the crack each of which may apply in different metal–environment systems. In each case the crack proceeds by a series of 'jumps'.

One theory of transgranular cracking which explains SCC in many alloys suggests plastic deformation takes place at the crack tip (see *Fig. 9.11*). The combination of an applied stress on the very small area at the crack tip results in the yield stress being exceeded in this local region. Thus, the material at A' deforms plastically opening the tip slightly. This process would fracture any protective films present. As the tip 'yawns' open fresh solution is drawn in to dissolve the highly anodic metal in the tip at favourable sites such as emergent dislocations. Concentration polarisation (Section 9.1.5.4) which would reduce the corrosion rate at the crack tip, is prevented to a

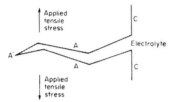

Figure 9.11 Transgranular crack propagation during stress corrosion cracking. A' = anode of yielding metal, current density $\simeq 10^3$ A m^{-2}; A = anodic sides of crack, current density $\simeq 10^{-1}$ A m^{-2}; C = cathodic surface at which oxygen is reduced

large extent due to the constant dilution of solution by the 'yawning' action. The sides of the crack (A) take little part in the process and are only weakly anodic; the current density values in *Fig. 9.11* indicate the lower corrosion rates prevalent at the sides of the crack. Transgranular SCC has also been explained in terms of preferential anodic dissolution at stacking faults (i.e. mistakes in the atomic stacking sequence) or at piled up dislocations. Alloys such as austenitic stainless steels and α-brasses are particularly susceptible to transgranular SCC and have low stacking fault energies whereas Cu–P, Cu–Si and Cu–Al alloys have high stacking fault energies and fail by an intergranular SCC mechanism. The repeated fracture and formation of a film initially at a stress raiser such as a surface slip step has also been proposed. Applied stress fractures the film locally, anodic dissolution occurs and the film reforms. This cycle is repeated until failure occurs. Absorption of hydrogen or solute from the corroding medium has also been suggested. Atomic hydrogen may be absorbed at the tip and diffuses into the metal faster than it can be evolved as hydrogen gas, producing hydrogen embrittlement. On the other hand solute atoms absorbed at the crack tip could weaken the metallic bonding at this point and promote cracking (*stress sorption*).

Film fracture and reforming has also been proposed as a mechanism for intergranular SCC. The applied stress fractures the film at a grain boundary where it reforms, the crack becoming deeper with each cycle. Selective anodic dissolution may also be a possibility, the function of the applied stress being merely to maintain a gap. Thus, an alloy solute or solute depleted zone produced by precipitation may provide anodic areas leading to local dissolution of the metal. One factor that is known about SCC is that for many metals if cathodic polarisation is increased by such methods as deoxygenation, inhibitors or cathodic protection SCC may be delayed or eliminated. However, in the case of austenitic stainless steels although the cathodic reaction is the reduction of oxygen (reaction 9.V)) hydrogen is also formed. Therefore, an incorrect cathode potential may produce hydrogen embrittlement since cathodic polarisation promotes hydrogen formation.

Corrosion fatigue

This term implies reduction in fatigue resistance under conditions of simultaneous corrosion and cyclic stresses. This phenomenon also is not fully understood but occurs, in general, by transgranular cracking with little branching and is most common in those environments which cause pitting. Under corrosion fatigue conditions S–N curves for ferrous materials no longer show a fatigue limit but resemble the S–N curves of non-ferrous materials. However, the final fracture surface bears some similarity to normal fatigue fractures, i.e. a large area of corrosion products and a small roughened area indicating the final, brittle fracture zone.

The damage ratio (corrosion fatigue strength divided by air fatigue strength) is often quoted for comparison, e.g. 0.4 for aluminium alloys, 0.5 for stainless steels, 0.2 for plain carbon steels and 1.0 for copper.

The production of so-called intrusions and extrusions by which normal fatigue is believed to take place produces irregularities, slip steps and slip band cracks at the surface. It is possible that the fresh surfaces produced by these mechanisms provide active anodic sites in aqueous environments. This approach has experimental support since increased anodic polarisation decreases corrosion fatigue life whereas increased cathodic polarisation increases it (thereby protecting the metal), unless hydrogen embrittlement is influential in the fatigue process.

Although anodic inhibition should provide a cure, in practice it is limited because incomplete inhibition occurs at the crack tip.

Fatigue life in corrosive media is usually increased by cladding, painting or plating and, ideally, the final surface should be notch free. Compressive surface stresses may be induced by shot-peening or surface alloying but these only help prevent the nucleation of notches. Where a pit nucleates and propagates to a depth greater than the stressed region the beneficial effects of surface stresses are lost. Cathodic protection (Section 9.1.10.3) has also been used successfully except in conditions of low pH. In this case the hydrogen produced by the cathodic reaction may lead to embrittlement. Maintaining a protective oxide film by anodic protection has been applied successfully to plain carbon and stainless steels.

9.1.4   Pourbaix diagrams

An introductory discussion of Pourbaix diagrams has been given earlier (Section 6.3.17). Pourbaix investigated electrochemical equilibria between metals and pure water over the entire pH scale, producing pictorial summaries in the form of an electrode potential versus pH graph in the same way that predominance area diagrams were constructed (Sections 7.2.2, 7.3.2.1). Before considering metal systems the E–pH behaviour of water must be examined. The Nernst equation (Section 6.3.12) for dilute solutions at $25\,^{\circ}$C after converting from ln to $\log_{10}$ becomes,

$$E = E^{\ominus} + \frac{0.059}{n}\ \log_{10}\left(\frac{a_{\text{reacts}}}{a_{\text{prods}}}\right) \tag{9.3}$$

and applied to reaction (9.II) gives

$$E_{\text{H}} = \frac{0.059}{2}\ \log_{10}\left(\frac{a_{\text{H}^+}^2}{p_{\text{H}_2}}\right)$$

Since pH $= -\log_{10}a_{\text{H}^+}$, we have

$$E_{\text{H}} = -0.059\text{pH} - 0.030\ \log_{10}p_{\text{H}_2} \tag{9.4}$$

Therefore, the electrode potential of this equilibrium depends on both pH and hydrogen partial pressure. Similarly, for reaction (9.IV)

$$E_{\text{O}_2} = +1.23 + \frac{0.059}{2}\ (\tfrac{1}{2}\log_{10}p_{\text{O}_2} + 2\ \log_{10}a_{\text{H}^+})$$

$$= +1.23 + 0.015\ \log_{10}p_{\text{O}_2} - 0.059\ \text{pH} \tag{9.5}$$

Thus, a dependence on both pH and p_{O_2} is shown. The same equation is derived from reaction (9.V).

Equations (9.4) and (9.5) represent the reversible* electrode potentials for the hydrogen electrode and oxygen electrode respectively. If a standard pressure of 1 atm is taken then $p_{O_2} = 1$ atm $= p_{H_2}$. Graphs of electrode potential against pH in each case are linear (*Fig. 9.12*) and at any pH value the lines are separated by 1.23 V. It must be remembered that lines AB and CD on *Fig. 9.12* represent *equilibria* and at a given pH a point on one of these lines represents the reversible electrode potential, E_r.

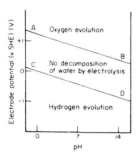

Figure 9.12 Pourbaix diagram for water.
(SHE = Standard Hydrogen Electrode)

Considering equation (9.4), if, at a given pH, E_H decreases p_{H_2} must increase. In other words, decreasing the electrode potential from its equilibrium value makes the reaction proceed in the direction of reduction of H^+ ions, i.e. hydrogen is formed. Similarly at potentials between AB and CD, i.e. $E_H > E_{r,H}$, the reaction proceeds in the direction of oxidation of H_2 gas, i.e. formation of H^+ ions. For the reversible oxygen electrode reaction (9.IV) (i.e. line AB), equation (9.5) shows that increasing E_{O_2} above its equilibrium value at a constant pH favours formation of water (or OH^- ions if reaction (9.V) is considered). The Pourbaix diagram makes no distinction between reactions (9.IV) and (9.V). In general, when $E > E_r$ oxidation takes place, when $E < E_r$ reduction is favoured.

The resulting diagram has three distinct zones. In the middle zone, between lines AB and CD, neither hydrogen nor oxygen are formed and, therefore, water is thermodynamically stable. The significance of these lines in corrosion will be considered later.

Consider an electrolytic cell comprising two platinum electrodes dipping into water (usually acidified to improve conductivity). When the applied voltage reaches 1.23 V the electrode potential of the anode is +0.817 V and that of the cathode is −0.413 V (note the polarity is opposite to that in an electrochemical cell, see Section 6.4 and Appendix 3) assuming pH 7, and water begins to be decomposed. Hydrogen is evolved at the cathode which becomes, in effect, a hydrogen electrode and the anode becomes an oxygen electrode. Increasing the applied voltage further increases the reaction rate. In practice electrode potentials rather more positive than +0.817 V and more negative than −0.413 V are required to decompose water electrolytically. This extra potential is called *overpotential* and will be discussed in Section 9.1.5.3.

Metal–water equilibria may be superimposed onto the E–pH diagram for water; see *Fig. 9.13(a)* for zinc–water equilibria. The equilibria represented by the lines 1 to 5 are:

1   $Zn^{2+} + 2e \rightleftharpoons Zn$ $E = -0.763 + 0.030 \log a_{Zn^{2+}}$

2   $Zn(OH)_2 + 2H^+ + 2e \rightleftharpoons Zn + 2H_2O$ $E = +0.419 - 0.059$ pH

3   $ZnO_2^{2-} + 2H_2O + 2e \rightleftharpoons Zn + 4OH^-$ $E = +0.436 - 0.118$ pH $+ 0.030 \log a_{ZnO_2^{2-}}$

4   $Zn^{2+} + 2H_2O \rightleftharpoons Zn(OH)_2 + 2H^+$ $\log K = -2\text{pH} - \log a_{Zn^{2+}}$

5   $Zn(OH)_2 \rightleftharpoons ZnO_2^{2-} + 2H^+$ $\log K = -2\text{pH} + \log a_{ZnO_2^{2-}}$

---

*i.e. *thermodynamically* reversible, see Section 2.2.1.1.

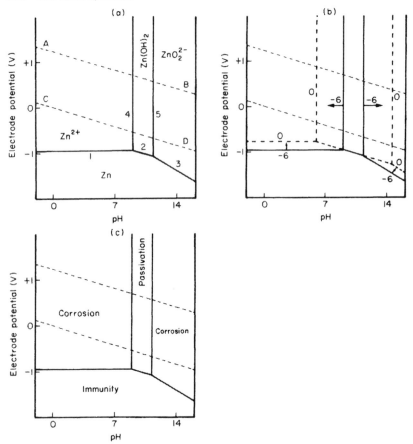

Figure 9.13 Pourbaix diagrams for zinc–water: (a) $a_{Zn^{2+}} = 10^{-6}$ g ions l^{-1}; (b) shows the effect of increasing $a_{Zn^{2+}}$ from 10^{-6} to 1 g ions l^{-1}; (c) shows domains of corrosion, passivation and immunity (cf. part (a))

Line 1 is independent of pH since it does not involve H^+ ions and is, therefore, parallel to the pH axis. The point of intersection of this curve with the vertical axis is at -0.943 V not at the equilibrium potential of -0.763 V. This is because Pourbaix diagrams are drawn using an equilibrium activity for metal cations or anions of 10^{-6} g ions l^{-1} which corresponds to dissolution of 0.065 mg of zinc (0.056 mg of iron, etc.) per dm^3 of solution. This arbitrary value was selected by Pourbaix as being so small that the dissolution of metal necessary to obtain it could then be ignored. In other words, zinc, for example, is regarded as uncorrodable (immune) or corrodable in an aqueous solution which initially contains no zinc ions if the quantity of zinc which can be dissolved by the solution is lower or higher respectively than 0.065 mg dm^{-3}. The effect of increasing activity to unity (10^0) as used to determine the electrochemical series is shown in *Fig. 9.13(b)*. Lines 2 and 3 are pH dependent and slope accordingly. No E–pH relationship is produced for lines 4 and 5 because, as no electrons are involved, both reactions are independent of electrode potential and depend only on pH and solubility product. Hence they are parallel to the E-axis. The effect of changing ion activity in this case must be examined by considering the equilibrium constants for the reactions. By determining equilibrium constants for each of the reactions, using appropriate thermodynamic data, the pH limits of the different equilibria may be calculated (see Sections 7.2.2 and 7.3.2).

Fig. 9.13(c) shows how the Pourbaix diagram is divided into zones, or domains, of corrosion, passivation and immunity. The domains represent the areas where these reactions are thermodynamically feasible (cf. predominance areas in *Figs 7.5* and *7.26*). Thus, in the corrosion domains of *Fig. 9.13(c)* zinc ions are formed in solutions up to pH of 8.5 (at activity 10^{-6} g ions l^{-1}) and zincate ions in highly alkaline solutions above pH 12. Similarly passivation is induced by formation (precipitation) of zinc hydroxide. However, in practice, where a hydroxide precipitate is formed it will only protect the metal surface if it has the necessary mechanical and physical properties. If the electrode potential is within the immunity domain any ions, (Zn^{2+} or ZnO_2^{2-} depending on the pH), or $Zn(OH)_2$ present form zinc metal, i.e. solid zinc is thermodynamically more stable than its anions or cations. The significance of lines AB and CD can now be

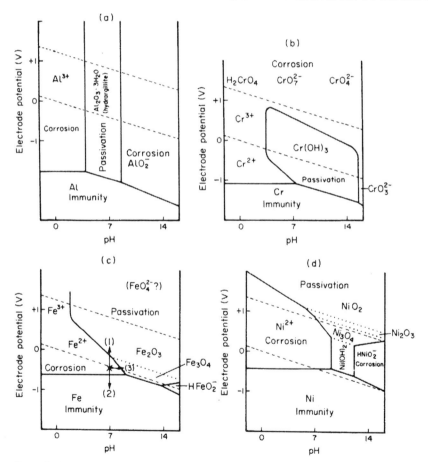

Figure 9.14 Pourbaix diagrams for: (a) aluminium–water; (b) chromium–water (note the different soluble ions in the corrosion domain possible at suitable combinations of E and pH); (c) iron–water (at high potentials especially when pH > 1.73 for $a_{M^{n+}} = 10^{-6}$ g ions l^{-1} iron is no longer passive; it is possible that soluble FeO_4^{2-} ions form but this has not been verified experimentally—see *Fig. 9.38*); (d) nickel–water

seen. At pH 7 and a zinc potential of -0.7 V for example, zinc corrodes, water is reduced to hydrogen ($p_{H_2} > 1$ atm) and any dissolved oxygen is reduced to OH^- ($p_{O_2} < 1$ atm). If $E < -0.943$ V zinc does not corrode (immunity) but the other reactions are the same. At potentials between -0.413 V and $+0.817$ V pure water is stable. Above $+0.817$ V zinc corrodes with oxygen evolution ($p_{O_2} > 1$ atm). A similar approach may be adopted for other pH values.

Fig. 9.14 shows schematic E–pH diagrams for other metals. The large regions of passivity shown by chromium, iron and nickel have special significance when corrosion protection is considered. The diagram for aluminium shows corrosion domains at low and high pH values similar to zinc. These zones are also shown by lead and tin and are a feature of amphoteric metals which produce cations in acidic solutions and anions in alkaline solutions.

Pourbaix diagrams are used to predict: the spontaneous direction of a reaction when the environmental conditions are changed; the composition of corrosion products at a particular potential or pH; conditions which are likely to increase or decrease corrosion. They are also used to establish mechanisms of corrosion reactions from experimental data of pH and electrode potential.

E–pH diagrams suffer severe limitations since they are based on thermodynamic equilibria and do not consider the kinetics of corrosion. However, Pourbaix diagrams and kinetic data considered together have considerable practical importance. Another limitation is that the only reactants considered are hydrogen and hydroxyl ions, metal and water, which is unlikely in practice. Other ions affect the equilibria and diagrams of the type M–H_2O–X are now available where X is another anion, e.g. SO_4^{2-}, CN^-, Cl^- at a specified concentration. The effect of these extra anions is to widen the corrosion domain if soluble compounds are formed and to widen the passivity domain if insoluble compounds are produced. Passivity in M–H_2O systems is associated with hydroxide precipitation but, as stated above, this precipitate may not in fact be protective especially if precipitation does not occur on the metal surface. In addition, environmental changes such as change in flow rate may break down an otherwise protective film. Many equilibria produce hydrogen or hydroxyl ions and, therefore, there may be local surface pH changes which are not detected by bulk pH measurements. These local changes of pH could have considerable influence on the ability to predict corrosion behaviour in a given environment. Finally, there is a temperature limitation since Pourbaix diagrams are based on thermodynamic data at 25 °C. However, there are now techniques which permit diagrams for high temperature aqueous corrosion to be constructed from data at 25 °C.

9.1.5   Electrode kinetics

9.1.5.1   The electrical double layer

When a metal dissolves (reaction (9.I)) it may become increasingly negatively charged if positively charged metal ions pass into solution. This continues until a dynamic equilibrium is reached (Section 6.3.1) such that for any cation formed a metal atom must also be formed simultaneously by the reverse reaction. At this point the excess negative charge at the surface of the metal just balances out the excess positive charge due to cations in solution adjacent to the metal. (Under standard conditions the potential difference thus created between metal and solution would be the standard electrode potential.) Therefore, a separation of charge exists and it was first compared to the parallel plates of a capacitor by Helmholtz. The *Helmholtz double layer* of charge separation is shown schematically in *Fig. 9.15*. The distance between the layers is about 0.1–0.5 nm (about 1 ion diameter). The potential difference across the layers is only about 1 V but combined with the small distance of separation a very high field strength of about 10^9 V m^{-1} is produced. A field strength of this magnitude is high enough to pull ions across the double layer to maintain equilibrium.

The simple approach given above is sufficient for many needs but when adsorption is considered (Section 6.7) a more detailed model is required. The electrolyte itself contributes to the potential difference between the layers. If an aqueous solution is used molecules of water are not simply neutral, covalently-bonded molecules. An oxygen atom has a greater attraction for the shared pairs of electrons with the result that each water molecule forms a dipole (*Fig. 6.1*). Oxygen forms the negatively charged 'end', hydrogen the positively charged 'end' and a potential difference exists across the molecule. The presence of dipoles enables water molecules to be electrostatically attracted to an electrode and thereby make a contribution to the potential difference across the double layer. The orientation of the dipoles depends on the sign of the charge at the metal surface. In addition, dipoles are attracted to ions in solution to form a hydration or solvation sheath around each one. Thus, the OHP in *Fig. 9.15*

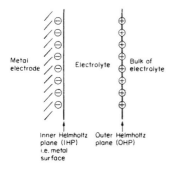

Inner Helmholtz
plane (IHP)
i.e. metal
surface

Outer Helmholtz
plane (OHP)

Figure 9.15 Schematic representation of the Helmholtz double layer at an electrode which has an excess negative charge in its surface layer

would be more correctly considered in terms of hydrated cations or 'aquo'-cations, M^{n+}(aq). Where a metal surface has only a slight excess charge dipoles of *both* orientations may be adsorbed. Therefore, the potential difference across the double layer may be considered to be due to two factors not one as was proposed in the simple model above: first, the size of the charges at each of the layers and secondly the layer of adsorbed dipoles at the metal surface.

So far adsorption has been explained in terms of electrostatic forces only. This approach cannot explain conclusions that a charged surface can adsorb ions of the same charge. This effect is clarified by using the concept of contact (or specific) adsorption. Electrostatic forces are long-range forces but there are also short-range chemical forces. These chemical forces induce nearby anions for example to contact a negatively charged metal surface and be chemisorbed (Section 6.7.2). This concept is important when inhibitors are discussed (Section 9.1.10.1).

The Helmholtz double layer model in which a layer of ions in aqueous solution is held firmly to a charge metal surface is applicable to concentrated solutions only. As the solution becomes more dilute a diffuse mobile charge layer (the *Gouy–Chapman layer*) is proposed which extends up to 1 μm into the bulk of the liquid from the OHP. The net charge of this diffuse layer is equal and opposite to that of the Helmholtz double layer. In very dilute solutions the contribution of the Helmholtz double layer becomes insignificant. The relationship between excess surface charge and surface tension has been studied by measuring the height of a capillary of charged mercury in different solutions. The results are sometimes referred to as electrocapillary effects. Normally the atoms in the surface of a metal are under tension due to the interatomic forces of attraction in the same way that atoms in the surface of a liquid are in tension (Section 4.2). The presence of an excess charge (positive or negative) in the metal surface provides a repulsive, surface-expanding effect which counteracts surface tension effects. Consider a positively charged metal surface polarised by an increasing negative potential, e.g. a battery applying a negative charge to the surface. The increasing potential cancels out the positive charges until a stage is reached when there is no charge on the surface. At this point there are no repulsive forces in the surface and the surface tension is a maximum. Increasing the potential further causes the negative charge to build up and the surface tension decreases. The point at which maximum surface tension is achieved is called the potential of zero charge, PZC (or the potential of the electrocapillary maximum). Clearly the charge on the OHP of the Helmholtz double layer must reverse on passing through the PZC.

The PZC has particular relevance to adsorption at metal surfaces. Thus anions or the O^{2-} ends of water dipoles are adsorbed at the IHP on the positive side of the PZC and cations or the H^+ ends of water dipoles on the negative side of the PZC. Adsorption of these ions increases surface tension at a given potential by reducing the effects of the surface charges of opposite sign. At the PZC anions, cations and molecules can be adsorbed.

9.1.5.2   Exchange current density

Current, I, is a measure of rate of transfer of charge since one ampere is the same as one coulomb per second. In electrode kinetics the area of the electrode must also be considered. This may be done by using current density, i, which is current divided by electrode area (A m^{-2} or A cm^{-2}). Therefore, current density is directly proportional to rate of charge transfer and may be used as a measure of rate of reaction of an electrode process. In practical corrosion terms a current density of 10 mA m^{-2} (1 μA cm^{-2}) for example does not appear to be a very meaningful measure of corrosion rate. However, the following equation is useful in this respect:

$$i = \frac{3.0575 n \rho}{W} \times 10^{-3} \times mm\ yr^{-1}$$

where i is the current density in A m^{-2}, ρ is the density in kg m^{-3}, W is the relative atomic mass and n has its usual meaning. Therefore, for iron for example, $\rho = 7880$ kg m^{-3}, $W = 55.85$, $n = 2$ (from Fe $\rightarrow$ Fe^{2+} + 2e) and 10 mA m^{-2} = 0.01 A m^{-2}. Hence a current density of 10 mA m^{-2} is equivalent to 0.0116 mm yr^{-1} (or 2.5 mdd from equation (9.1)).

Consider a metal immersed in a solution of its own ions. Let $\vec{i}$ be the rate of the anodic, dissolution process, $M \rightarrow M^{n+} + ne$, and let $\overleftarrow{i}$ be the rate of the cathodic, discharge process, $M^{n+} + ne \rightarrow M$. Initially, an immediate (and momentary) transport of charge takes place during which time $\vec{i} \neq \overleftarrow{i}$. The reaction which is energetically more favourable predominates and determines the electrode potential (i.e. the potential of the electrode with respect to the solution at equilibrium). For instance, if the anodic process is favoured the electrode dissolves. An excess of electrons is left in the metal which results in a negative electrode potential. At equilibrium, the rates of the anodic and cathodic processes are equal, i.e. there is no net transfer of charge and

$$\vec{i} = \overleftarrow{i} = i_0$$

where i_0 is called the exchange current density.

Each reversible electrode reaction has its own exchange current density. Furthermore an electrode will not achieve its reversible (equilibrium) electrode potential (for the reaction $M^{n+} + ne \rightleftharpoons M$) unless its i_0 value is much greater than the i_0 values of any other reversible reactions in the system. (Remember a high value for i_0 denotes a rapid rate of reaction.) Hence in a system in which the reactions (9.1), (9.II) and (9.IV) are all possible the final equilibrium potential of the metal is decided by reaction (9.1) since the i_0 values for the other (slower) reactions are usually smaller.

The main factors affecting i_0 for a particular metal are temperature, composition of electrolyte (SO_4^{2-}, Cl^-, etc.) and concentration of metal ions. Surface roughness also affects i_0 because of its dependence on surface area. As will be discussed in Section 9.1.5.6, i_0 is a very important parameter in determining corrosion rates.

9.1.5.3   Thermodynamic irreversibility and polarisation

Fig. 9.16(a) represents the free activation energy relationship between a metal in equilibrium with a solution of its own ions. The free activation energy wells for the anodic and cathodic reactions are at the same level and, therefore, the dissolution and discharge reactions need the same free energy of activation, ΔG^*, to proceed and are thermodynamically reversible. There is no net reaction since $\overrightarrow{i} - \overleftarrow{i} = 0$ and the

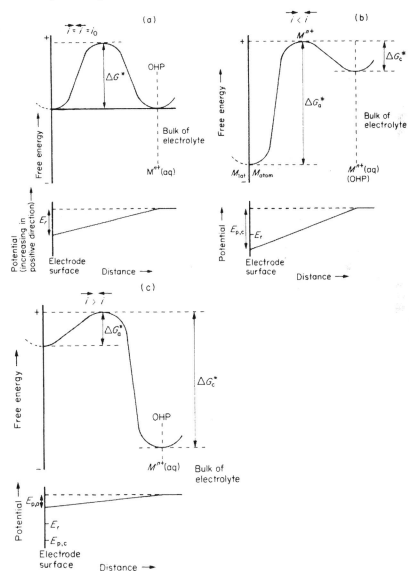

Figure 9.16   Free energy and electrical potential plotted against distance for: (a) an electrode at equilibrium; (b) an electrode made cathodic by polarisation; (c) an electrode made anodic by polarisation

electrode potential assumes its equilibrium value, E_r. If the electrode at equilibrium is now connected to the negative pole of an external source of current or galvanically coupled to a more anodic, base, material, electrons are supplied to the electrode. Thus the potential of the electrode with respect to the solution is made more negative than its equilibrium value and the potential difference across the double layer increases. Consequently the field strength increases and ions are pulled out of solution across the double layer in an attempt to cancel out the effect of the extra electrons and achieve a new equilibrium. Provided the applied source (or the galvanic coupling) makes the electrode negative enough no new equilibrium is obtained and a net deposition (cathodic) current, $i_c = \overleftarrow{i} - \overrightarrow{i}$, flows. The new free energy relationship is shown in *Fig. 9.16(b)*. Clearly, the activation energy for the cathodic process, ΔG_c^*, is made more favourable than the anodic process and the rate of transfer of charge by the cathodic process $\overleftarrow{i}$ is faster. Equilibrium conditions are no longer present and the situation is one of thermodynamic irreversibility: the electrode is said to be polarised. The difference between the reversible electrode potential, E_r, and the new polarised electrode potential, E_p, is called the overpotential, η:

$$\eta = E_p - E_r \tag{9.6}$$

Therefore, for a net cathodic process,

$$\eta_c = E_{p,c} - E_r$$

and since $E_{p,c} < E_r$ then η_c must be negative, i.e. the electrode potential has moved to a more negative value as a result of cathodic polarisation.

If the original electrode, initially at its equilibrium potential is connected to the positive pole of an external source or galvanically coupled to a more cathodic, noble, material, the external source acts as an electron sink, electrons are lost from the electrode and its potential, with respect to the solution, becomes more positive. The free energy relationship is shown in *Fig. 9.16(c)*. The anodic dissolution process is favoured in order to create more electrons in the electrode in an attempt to cancel out the effects of the electron sink. The result is that a net anodic current, $i_a = \overrightarrow{i} - \overleftarrow{i}$, flows and the electrode is polarised. The overpotential, η_a, is defined by equation (9.6) as,

$$\eta_a = E_{p,a} - E_r$$

Since $E_{p,a} > E_r$ then η_a is positive and polarisation in this case has shifted the electrode potential in a positive direction. Putting the above in a different way, for a metal at equilibrium to become a cathode a potential, the overpotential, must be applied in such a way that the initially reversible equilibrium electrode potential is made more negative. Similarly, for a metal to behave as an anode an electrode potential more positive than the equilibrium value is required.

9.1.5.4   Types of polarisation

Activation polarisation

A specific activation overpotential, η^A, must be applied to an electrode at equilibrium in order to achieve an appreciable anodic or cathodic rate. If the free energy of activation (activation energy), ΔG^*, in *Fig. 9.16(a)* is small, the overpotential required for a reactant to overcome this energy barrier is also small. Thus the displacements of the free energy curve shown in *Figs 9.16(b)* and (c) will be correspondingly small. Silver, for example, requires only a small displacement from its equilibrium potential to facilitate dissolution or deposition of metal. If ΔG^* is high, e.g. for iron, a considerable activation overpotential must be applied before the free energy wells are displaced enough for the anodic and cathodic reactions to proceed at a significant rate.

Electrode processes usually proceed in a series of steps each step having its own activation energy. The step which has the highest activation energy proceeds at the slowest rate and it is this rate which determines the speed of the reaction (Section 3.6.1). Consider the hydrogen evolution reaction (HER) given by reaction (9.II). Initially, hydrogen ions diffuse or migrate to the metal surface where they are discharged and adsorbed (discharge step A):

$$H^+ + e \longrightarrow H(ads)$$

In iron and iron alloys this discharge step may be accompanied by simultaneous absorption of hydrogen into the metal producing hydrogen embrittlement. For other metals, the H(ads) form hydrogen molecules by a chemical desorption step, B:

$$H(ads) + H(ads) \longrightarrow H_2$$

or by an electrochemical desorption step, C:

$$H(ads) + H^+ + e \longrightarrow H_2$$

Which of these three steps is in fact rate controlling depends on the metal at which the hydrogen is discharged. For nickel it is step A, for platinum it is step B and for niobium it is step C. In 1M HCl solution at 20 °C the corresponding values of η^A are -0.31 V, 0.00 V and -0.40 V respectively. The ability of a metal to act as a catalyst affects this hydrogen overpotential (or hydrogen overvoltage) and platinum is a better catalyst than either niobium or nickel. A very poor catalyst such as lead requires an activation overpotential of -1.16 V before hydrogen is evolved at the surface; in this case step A is again rate controlling. The most important parameter is the exchange current density for the HER at a metal surface: platinum has a high i_0 value ($\sim 10^{-2}$ A cm^{-2}), whereas lead has a very low i_0 value of about 10^{-13} A cm^{-2}. Thus, the HER at a platinum surface is more easily polarised (lower η_c^A) than the HER at a lead surface (see Sections 9.1.5.2, 9.1.5.6).

The oxygen reduction reaction, reactions (9.IV) and (9.V) (oxygen overpotential), also proceeds in a series of steps but these steps are not completely understood at present.

Activation overpotential has been defined by Tafel as

$$\eta_a^A = b_a \log \left(\frac{i_a}{i_0} \right) \quad \text{and} \quad \eta_c^A = b_c \log \left(\frac{i_c}{i_0} \right)$$

for anodic and cathodic processes respectively. b_a and b_c are temperature-dependent constants and the equations hold for overpotentials greater than 0.052 V. The other parameters have been defined in Section 9.1.5.3. These equations may be combined to give a general Tafel equation:

$$\eta = a + b \log i \tag{9.7}$$

For a given reaction i_0 is constant, therefore, a and b are constants. For an anodic process a is negative and b is positive, for a cathodic process a is positive and b is negative. The Tafel equation clearly shows a linear relationship between overpotential and $\log i$. For overpotentials less than 0.052 V this relationship is no longer linear; in this case overpotential and i become linearly related. *Figs 9.17(a)* and (b) show plots of electrode potential against $\log i$ for net anodic and cathodic reactions respectively at a single electrode. At some activation overpotential η_a^A or η_c^A, the electrode potential of the polarised electrode is $E_{p,a}$ or $E_{p,c}$ respectively. The extrapolated Tafel lines meet at $\log i_0$. Clearly the reactions shown ignore the presence of the other electrode reaction which must also be present. The most important factor is the significance of i_0 on the rate of the electrode process. Consider *Fig. 9.17(a)*. As i_0 decreases, the Tafel curve effectively moves to the left. Therefore, at some arbitrary polarisation potential

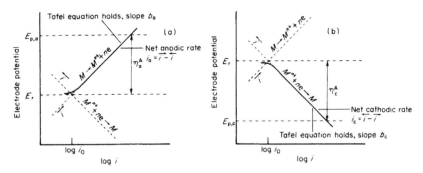

Figure 9.17 Electrode potential against log (current density) for a single exchange process $M^{n+} + ne \rightleftharpoons M$. In (a) the anodic process $M \rightarrow M^{n+} + ne$, predominates; in (b) the reverse, cathodic, process predominates.

Note. The continuous curves are simulated experimental E–log i (polarisation) curves in which i_a and i_c are the measurable *applied* current densities. Extrapolation of the Tafel region to E_r reveals i_0 for the $M^{n+} + ne \rightleftharpoons M$ equilibrium. In a practical situation the initial potential of an electrode would be its reversible equilibrium value, E_r, and $\overleftarrow{i}$ and $\overrightarrow{i}$ values prior to i_0 are not measurable

(such as $E_{p,a}$) decreasing i_0 decreases the current density, i.e. the rate of dissolution decreases, at which $E_{p,a}$ is reached.

Thus, where an electrode process involves an appreciable activation energy, i_0 is immensely important. This conclusion is of great value when corrosion resistance is considered.

Concentration polarisation

Under irreversible conditions, when for example the cathodic process $M^{n+} + ne \rightarrow M$ occurs very rapidly, ions from the bulk of the solution are unable to diffuse across the double layer quickly enough to replace the discharged ions. Therefore, a concentration gradient is set up across the double layer and a limiting rate of reaction, i.e. a limiting current density i_L, which cannot be exceeded is achieved. Similarly, for a very rapid anodic process, ions cannot solvate and diffuse away quickly and a build-up of ions occurs at the electrode surface. Once again a limiting current density is reached. The expression for this limiting current density is:

$$i_L = \frac{DnFc}{\delta(1-t)}$$

where D is the diffusion coefficient of the metal ion in solution, t is the transport number (Section 6.2.5) and c is the metal ion concentration in the bulk of the solution; n and F have their usual meanings. δ is sometimes quoted as the thickness of the Helmholtz double layer. However, in practice, electrode concentration gradients extend beyond the double layer to 0.1–0.5 mm from the electrode surface. Therefore it is more usual to use δ to represent the thickness of this broader *diffusion layer*.

At a cathode concentration overpotential η_c^c is expressed as

$$\eta_c^c = 2.3 \frac{RT}{nF} \log \left(1 - \frac{i}{i_L}\right)$$

As the applied current density approaches i_L so η_c^c tends towards infinitely large values producing a deviation away from Tafel linearity (see *Fig. 9.18*). A similar effect but of opposite sign occurs at the anode. At sufficiently high potentials alternative electrode processes, e.g. hydrogen evolution, may occur. Only then can current density exceed i_L, the new reaction having its own limiting current density.

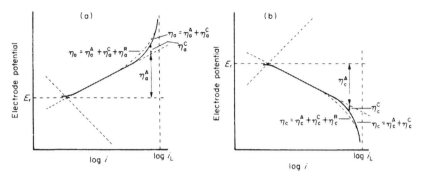

Figure 9.18 Experimental polarisation curves for (a) anodic and (b) cathodic processes showing the effect of concentration and resistance polarisation on the curves of *Fig. 9.17* (*i* in this case represents the *applied* current density)

Any factor which can reduce the local concentrations at an electrode can increase i_L, i.e. at a given current density (rate) η^C is reduced. Thus, stirring, and increasing the temperature, which increases the diffusion rate, can both reduce η^C since both reduce δ. The action of gravity, by using a horizontal rather than a vertical electrode, can also reduce η^C. The nature of the cation also has a marked effect on η^C because different cations have different diffusion coefficients.

Resistance polarisation

When a conducting film of oxide or adsorbed layer of gas is present on the surface of an electrode the electrical resistance of the electrolyte-to-electrode path increases. Under equilibrium conditions there is no net flow of current. Therefore, E_r is unaffected by the film since it depends only on the initial and final states (activities) of the metal ion in the $M^{n+} + ne \rightleftharpoons M$ equilibrium. However, under non-equilibrium, polarising conditions a net current flows and a potential drop is produced across a surface film. This effect gives rise to the need for an extra overpotential, the resistance overpotential, η^R (sometimes called filming overpotential or IR drop).

In practice η^R cannot be determined absolutely. The measuring probe used always measures the resistance of a minute amount of electrolyte as well as the resistance of the film. Therefore, the resistance overpotential should be defined as:

$$\eta^R = I \text{ (resistance of surface film}$$
$$+ \text{ resistance of small 'length' of electrolyte)}$$

I is the current flowing. The total overpotential at an electrode is given by

$$\eta = \eta^A + \eta^C + \eta^R \qquad (9.8)$$

If η_a and η_c are the total anodic and cathodic overpotentials respectively the effect of combined polarisation is shown in *Fig. 9.18*.

In many corrosion processes η^C and η^R are often insignificant compared with η^A, although, as will be seen later, the presence of oxygen may make η^C very important. In electrolytic cells used for plating, extraction or refining ionic diffusion rates and high current densities increase the relative importance of η^C and η^R considerably.

9.1.5.5 Potential–current (E–I) diagrams (Evans diagrams)

When a single metal corrodes tiny anodic and cathodic areas are formed on the surface (Section 9.1.1). A simple example is the corrosion (dissolution) of zinc in acid solution. The cathodic reaction in this case is hydrogen evolution (reaction (9.II)) and the anodic and cathodic reactions mutually polarise each other. This corrosion situation is analogous to a $Zn|Zn^{2+}||H^+|H_2|Pt$ cell in which parameters such as current and individual electrode potentials may be measured as an external resistance is altered. Obviously in corrosion these parameters cannot be measured (except, possibly, in galvanic, two metal cells). A graph of potential against current for such a cell is shown in *Fig. 9.19.* Polarisation of anode and cathode proceeds to $E_{p,a}$ and $E_{p,c}$ such that

$$E_{p,c} - E_{p,a} = I_p R$$

$(E_{p,c} - E_{p,a})$ may be called the working emf of the cell, R is the total resistance of the cell due to metallic connections and electrolyte and surface films. In a corrosion process such as that of zinc in acid R may be regarded as being negligibly small since the electrolyte has a very high conductivity, the anodes and cathodes are very close

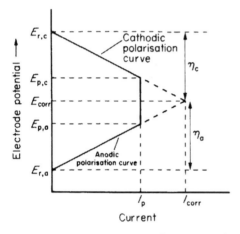

Cur rent   Figure 9.19 Mixed corrosion control

together and the contribution from any surface films, in this case, is zero. Thus, $E_{p,c}$ and $E_{p,a}$ tend towards the same value. The intersection of the anodic and cathodic polarisation curves gives the current, I_{corr}, and the corrosion potential, E_{corr} (sometimes called the mixed or compromise potential). An interesting point to note from *Fig. 9.19* is

$$E_{r,c} - E_{r,a} = E_{r, cell} = \eta_a + |\eta_c|$$

where $|\eta_c|$ represents the magnitude but not the negative sign of η_c. This equation implies that the corrosion rate is dependent on both thermodynamic and kinetic considerations. In an Evans diagram I_{corr} is taken as a measure of corrosion rate, although, strictly, current density should be considered. However, if the areas of the anode and cathode are taken to be equal I may be replaced by current density. In the case of zinc dissolution in acid the reaction occurs uniformly over the surface (Section 9.1.3.1) and, therefore, the anodic and cathodic areas are equal. However, this is not often the case in practice. No account is taken of exchange current density. The polarisation curves are normally drawn as straight lines implying Tafel behaviour, i.e. activation polarisation only. Even bearing these restrictions in mind Evans diagrams represent a very convenient, simple way of pictorially presenting corrosion rates. However, some of these restrictions are overcome by E–log i diagrams (Section 9.1.5.6).

The corrosion situation represented by *Fig. 9.19* is brought about by approximately equal contributions to polarisation by both anode and cathode. Such a system is said to be under mixed control. When polarisation occurs mainly at the cathode (*Fig. 9.20(a)*), such as when iron corrodes in water, the system is under cathodic control. The corrosion of zinc in reducing acids (e.g. dilute H_2SO_4) is of this type since the reaction is controlled by the high activation energy of the HER at the cathode. When magnesium corrodes in natural waters (rather than pure water) anodic polarisation predominates to give anodic control (*Fig. 9.20(b)*). Sometimes an insulating surface film is formed which prevents metal–electrolyte contact, for example, the cathodic deposition of calcium carbonate on iron in hard waters. Thus, the resistance overpotential increases and the situation is under resistance control (*Fig. 9.20(c)*).

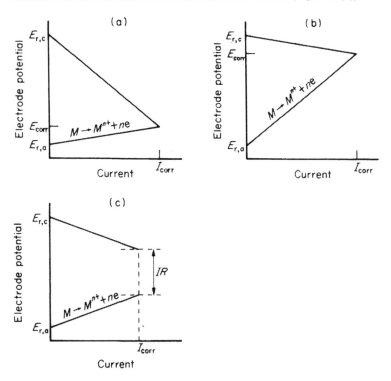

Figure 9.20 Types of corrosion control: (a) cathodic; (b) anodic; (c) resistance

The influence of hydrogen overpotential on corrosion rates may be easily seen using Evans diagrams. Consider iron and zinc corroding in acid solutions, i.e. two separate metal–hydrogen cells. $E_{Fe}^{\ominus} = -0.440$ V, $E_{Zn}^{\ominus} = -0.763$ V and $E_{H}^{\ominus} = 0.00$ V, therefore, using equation (6.20), $E_{cell}^{\ominus}$ values are $+0.440$ V and $+0.763$ V for the iron and zinc cells respectively assuming standard thermodynamic conditions in each cell. Under non-standard thermodynamic conditions the Nernst equation can be used and $E_{r,\,cell}$ values compared in each case. Using equation (6.15a) it may be seen that the iron has a lower tendency to dissolve (corrode) than zinc since its ΔG_{cell} value is less negative. Now consider the E-I diagram of *Fig. 9.21*. Clearly, the iron dissolves at a faster rate than zinc since I_{corr} (Fe) is greater than I_{corr} (Zn). The reason for this is that a higher activation energy is required to produce hydrogen evolution at a zinc surface than at an iron surface, i.e. the activation overpotential for the HER on zinc is greater than that for iron, e.g. in 1M HCl at 20 °C η_c^A is -1.13 V for zinc and -0.45 V for iron. In fact the HER is controlled by the exchange current density and this will be discussed in the next section.

The presence of an alternative easier cathodic reaction changes the original cathodic reaction, i.e. *depolarises* it, to make the corrosion rate more rapid. For example in

Fig. 9.22(a) the presence of impurities or more noble alloying elements in zinc makes the HER much easier, i.e. a smaller η_c^A for hydrogen evolution is required at a given value for I, and the corrosion rate increases. Similarly reactions which are controlled by oxygen diffusion to the surface, i.e. concentration polarisation, result in a limiting reaction rate (*Figs 9.22(b), (c)* and *(d)*). In *Fig. 9.22(b)* increasing the oxygen content increases $E_{r,c}$ (equation (9.5)) and consequently the corrosion rate increases.

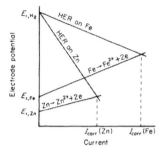

Figure 9.21 Evans diagram showing that iron corrodes faster than zinc in acid solution even though zinc has a more negative (anodic) reversible electrode potential

In *Fig. 9.22(c)* the presence of H_2O_2 provides a ready supply of oxygen and depolarises the initial diffusion controlled reaction. If nitric acid is present the ready reduction of the nitrate ion (reaction (9.VIII)) provides an even easier cathodic reaction and the corrosion rate increases still further. *Fig. 9.22(d)* shows the effect of increasing agitation in aerated solutions. The supply of oxygen to the cathode increasingly depolarises the original reaction. Above a certain velocity, v_2, there is no further

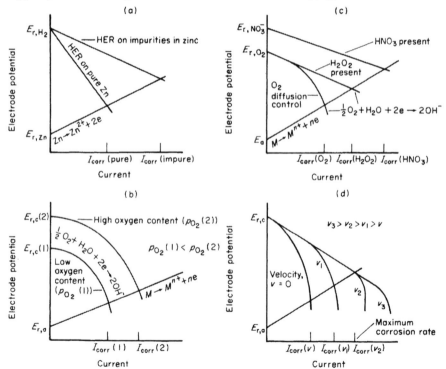

Figure 9.22 Evans diagrams representing different corrosion situations

increase in corrosion rate because the rate of oxygen supply no longer controls the corrosion reaction.

Where anode and cathode areas are considerably different little useful information may be obtained from a simple potential–current diagram. However, if an Evans potential–current density curve is drawn the significance of relative anodic and cathodic areas may be immediately seen. *Fig. 9.23(a)* represents the situation where

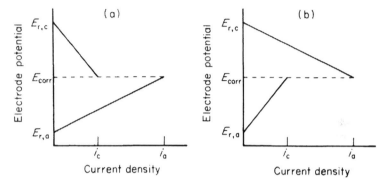

Figure 9.23 Evans diagrams showing the effect of different anode/cathode areas on corrosion rate: (a) cathode area > anode area; (b) anode area > cathode area. $i = I$/electrode area

the cathode area is larger than the anode area. Therefore, for the same current, i_a is greater than i_c and marked dissolution of the anode takes place. If the anode area is large compared with the cathode area (*Fig. 9.23(b)*) corrosion of the anode is virtually unaffected.

9.1.5.6 Potential-current density (E-log i) diagrams (Stern diagrams)

Modification to the Evans approach, notably by the inclusion of exchange current density, has led to a more accurate representation of the corrosion of a metal, e.g. iron in acid represented by *Fig. 9.24(a)* or schematically as in *9.24(b)*. These diagrams show the two exchange processes which take place. The reverse reactions of hydrogen

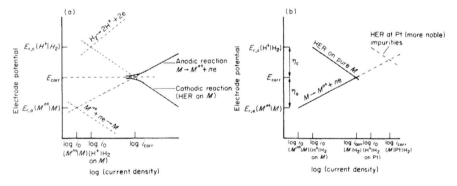

Figure 9.24 Stern diagrams: (a) actual; (b) schematic; where two exchange processes take place.

Note. The full curves in (a) simulate polarisation curves derived experimentally from a simple corrosion system, e.g. iron in dilute sulphuric acid. Clearly the initial potential in such a case is E_{corr}. Extrapolation of the Tafel regions back to E_{corr} gives i_{corr}

ion and metal atom formation may be ignored at high overpotentials. Nevertheless, these diagrams do show that a metal undergoes both anodic and cathodic reactions one of which tends to predominate.

Although the polarisation curves have been drawn to imply Tafel behaviour, i.e. activation polarisation is rate controlling, obviously the influence of concentration and resistive effects at higher current densities may be included. The influence of i_0 on the corrosion may be clearly seen by reference to *Fig. 9.24(b)*. As implied in the previous section, i_0 for platinum is higher than for other metals. Therefore, if a platinum salt were added to the system platinum would be deposited on the metal surface. The effect of this would be to shift i_0 ($H^+|H_2$) to the right as shown and thereby increase the corrosion rate. Thus, the original cathodic reaction is depolarised by platinum. A similar argument may be used to explain the fact that pure zinc has virtually no reaction in reducing acids whereas impure zinc readily evolves hydrogen (Section 9.1.5.5); $i_0(H^+|H_2)$ on pure zinc is much lower than $i_0(H^+|H_2)$ at impurities in zinc. The greater corrosion rate of iron compared with zinc (Section 9.1.5.5) is due to the fact that $i_0(H^+|H_2)$ on zinc is 10^{-7} A m^{-2} whereas $i_0(H^+|H_2)$ on iron is much larger at 10^{-2} A m^{-2}. Hence, interpretation of E–log i diagrams is carried out in much the same way as Evans diagrams, but more experimental data are required to construct the former.

9.1.5.7 Wagner-Traud diagrams

In a Wagner and Traud potential–current diagram the reaction rate per unit area of electrode is given diagrammatically (*Fig. 9.25(a)*). The anodic polarisation curve is taken to be positive and drawn on the right-hand side of the potential axis, the opposite convention is used for the cathode curve. The position of the horizontal axis is not critical and is placed at a convenient potential which does not obstruct the E–I

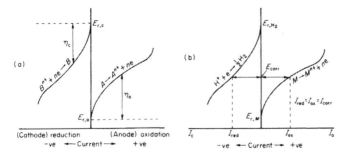

Figure 9.25 Wagner–Traud polarisation diagrams: (a) general; (b) showing balance point

curves under consideration. Wagner and Traud's basic argument was that the total rate of oxidation in a corrosion process equals the total rate of reduction. Since current is a measure of rate of reaction then at the oxidation–reduction 'balance' point, $I_{red} = I_{ox} (= I_{corr})$, see *Fig. 9.25(b)*. Sometimes these curves are plotted as potential against current density. If a balance point is cited in this case equal reaction areas are implied which, as mentioned earlier, is unlikely in practice.

9.1.5.8 Passivity and anodic E-log i curves

When the potential of a metal such as chromium in aqueous solution pH 5 is increased from its equilibrium potential a protective film about 1 to 10 nm thick forms on the metal surface. This film makes the metal *passive*, i.e. unreactive, even though a metal–environment reaction is thermodynamically feasible. The potential enters the passivity domain of the appropriate Pourbaix diagram (*Fig. 9.14*). Passivity occurs at a potential above the equilibrium potential for the reaction between the metal and one of its oxides, M_xO_y:

$$M_xO_y + 2yH^+ + 2ye \rightleftharpoons xM + yH_2O$$

As the potential increases beyond $E_r(M_xO_y/M)$ the linear Tafel relationship is no longer obeyed (*Fig. 9.26*). A limiting current density, the critical passivating current density, i_{crit}, is achieved. At a higher potential, the (primary) passivating potential, E_{pass}, a passivating film forms and the current density decreases to a minimum value, i_{pass}. This current is maintained as the electrode potential is raised until at some potential the current density begins to increase beyond i_{pass}. This point marks the onset of transpassivity and is due to the breakdown of the passivating film. This

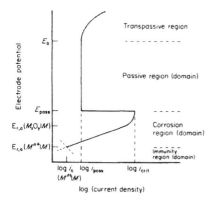

Figure 9.26 Anodic polarisation curve for a metal showing passivity (cf. with Pourbaix diagrams, of chromium, and also *Fig. 9.38*)

region is associated with gas evolution or metal corrosion or both simultaneously. The gas formed depends on the nature and concentration of the anions present, e.g. oxygen by OH^- oxidation (the reverse of reaction (9.V)). Chromium and chromium steels in acid solutions at high potentials form soluble anions such as chromate ($Cr_2O_7^{2-}$) rather than continue to form insoluble, passivating Cr_2O_3. Similarly, iron in sulphuric acid initially forms oxygen but on increasing the potential further oxygen and Fe^{3+} ions are produced together. In alkaline solutions transpassive iron dissolves as ferroate, FeO_4^{2-}, ions.

The passive film is produced by chemisorption of oxygen to form a monatomic layer. Oxygen molecules are then weakly chemisorbed onto this layer. Cation migration into the monatomic oxygen layer subsequently produces a monolayer of some form of oxide, the thickness of which increases with time to a maximum value at a particular electrode potential. The chemisorption process involves chemical bonding between metal and oxygen and, therefore, the metal requires spare electrons to facilitate this process. Most metals which show passivation are transition metals which have unpaired electrons in the d-shells and are, therefore, capable of achieving chemisorption.

The current density, i_{pass}, is associated with flow of ions through the film. Ionic current in an oxide film requires a high field strength of about 10^8 V m^{-1}, which may be achieved at the prevailing potential of about 1 V by the presence of a very thin film. The value of i_{pass} remains fairly constant at potentials producing the passive film and is a measure of (oxidation) dissolution rate. For the system to remain passive and for the field strength across the film to remain constant the thickness of the film must also be constant at a given potential. The effect of raising the potential is to increase the film thickness which thus maintains a constant field strength across the film. Film growth is achieved by combination of O^{2-} and OH^- at the oxide–solution interface with extra cations which have diffused through the film by the action of the initially higher electric field. Clearly the film produced must have little chemical solubility in the surrounding electrolyte as well as a low dissolution rate (i_{pass}). The film must also be thermodynamically stable. However, protection cannot be provided unless factors such as adhesion and mechanical strength are also satisfied.

Sometimes the primary passivation potential is called the Flade potential, E_F. Strictly E_{pass} and E_F are not the same. The difference is due to the fact that Flade measured the change in passivation potential as the already passivated metal began to lose its passivity on decreasing the potential of the metal. E_{pass} on the other hand is the potential at which passivity begins (increasing E_M) and takes account of the IR drop of the initially formed film. However, the resulting value for E_{pass} is only greater than E_F by between 0.001 and 0.003 V which for most purposes is small enough for the two potentials to be regarded as equal.

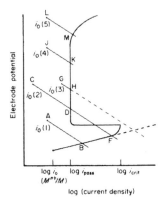

Figure 9.27 E–log i curves for a passivatable metal showing the effects of different cathodic reactions

Fig. 9.27 shows the variety of cathodic reactions, represented by different i_0 values, which can occur. $i_0(1)$ could be the HER represented by reaction (9.II) when a metal such as iron corrodes in dilute acid. The resulting corrosion rate is given by point B. CDF represents the unstable passivating condition, e.g. iron initially passivated in fuming nitric acid. It should be noted that fuming nitric acid is used because it contains a higher concentration of passivating NO_2^- ions (as HNO_2) than concentrated nitric acid. If the protective film is undamaged iron remains passive but if damaged the film cannot be repaired and corrosion takes place at a rate given by point F. Curve JK indicates spontaneous or self-passivation. Curve LM indicates the situation where corrosion occurs in the transpassive region at rate M, such as for chromium in oxidising acids (in non-oxidising acids chromium passivates according to curve JK). GH represents the effect of noble alloying additions on the corrosion rate in acids. Noble elements make the HER easier (Section 9.1.5.4) and depolarise the cathodic reaction increasing the corrosion rate. Where a metal may be passivated this effect may be used to advantage since the corrosion rate is initially increased beyond i_{crit} allowing the metal to become passive. For example, titanium corrodes in boiling 10% hydrochloric acid but an alloying addition of 0.1% palladium reduces the corrosion rate by two orders of magnitude.

Alloying with metals which passivate more readily than the base metal can reduce i_{crit} to such an extent that self-passivity may be induced. Thus elements such as chromium and nickel, which have lower i_{crit} and E_{pass} than iron, reduce i_{crit} for iron to very low levels. Additions of up to 18% chromium to iron reduces i_{crit} even more, which accounts for the passive nature of stainless steels in many environments. Similarly, additions of more than 70% nickel to copper reduces i_{crit} and i_{pass}. The reason for the success of these alloying elements is believed to be due to the presence of unfilled d-orbitals in the alloying, usually transition, element. The filled d-orbitals of a non-transition metal, e.g. copper, supply electrons to the d-band of, for example, nickel and the alloy assumes transition metal properties. As long as some d-orbitals are unfilled these properties remain. The transition alloying element must be present to a certain minimum amount to provide the resultant alloy with sufficient unfilled d-orbitals in order to produce passive film formation. For copper–nickel and iron–chromium alloys passivity occurs above minimum concentrations of about 30% nickel and 12% chromium respectively. Studies of d-bands have shown fair agreement with these minima and the minima at which d-orbitals become vacant. The effects of adding these elements on E–log i curves is shown in *Fig. 9.28* together with passivity reducing factors.

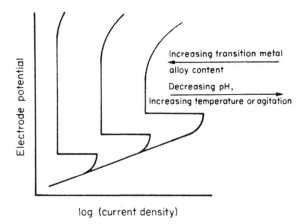

Figure 9.28 Schematic representation of the effects shown on passivation

   Passivity can also be induced by purely chemical means. For example, in the presence of sulphate ions insoluble lead sulphate is precipitated on lead as a result of solubility product considerations rather than by cation transport across a thin film. The lead sulphate formed protects the lead underneath but because the precipitate is coarse and porous the effect can only be regarded as partial passivation. If passivation were defined in terms of E_{pass} and i_{crit} then clearly lead sulphate formation cannot be classed as passivity. However, a more general approach defines passivity as resistance to further corrosion by formation of a protective barrier between metal and solution in an environment in which the metal is thermodynamically unstable.

   Under certain conditions breakdown of the film can occur. This may proceed by uniform dissolution over the metal surface or by local dissolution to leave the metal unprotected. Local breakdown may be a result of the condition of the metal itself producing weak spots in the film. Surface discontinuities, such as inclusions, phase changes, grain boundaries, sharp changes in shape and surface finish, could lead to pitting or undermining of the film in a suitable electrolyte. Mechanical effects such as abrasion or bending and poor heat treatment may achieve the same deleterious effects. Electrochemical breakdown is also possible and occurs by so-called reductive dissolution, i.e. the production of metal cations by a cathode reaction together with vacancy formation in the film due to loss of oxygen anions, i.e.

$$2Fe^{3+}(film) + O^{2-}(film) + Fe + 2H^+(aq) \rightarrow 2Fe^{2+}(film)$$
$$+ Fe^{2+}(aq) + O^{2-}(vacancy) + H_2O$$

In some cases pits may be plugged by corrosion products and thereby reduce corrosion. However, it is unlikely that pits can 'self-heal' because repair of an oxide film in a pit means the use of OH^- ions which in turn means the local H^+ ion concentration increases. Therefore, within a pit the electrolyte may be strongly acidic even if the bulk electrolyte is neutral. The possibility of corrosion problems produced by pits is aggravated by the presence of 'aggressive' ions such as Cl^- for iron, chromium and stainless steels. Chloride ions are strongly adsorbed and displace water dipoles from the double layer. Thus self-healing of a damaged film is prevented and intense local attack can occur especially in a pit. Where a film is not damaged adsorption of chloride ions in sufficient numbers is proposed such that repulsive forces between them crack the protective film. Metals and alloys which may be passivated and which are prone to pitting show a characteristic breakdown potential E_b on the E–log i curve. At this point pitting begins and the current density begins to rise. *Fig. 9.29* is a schematic

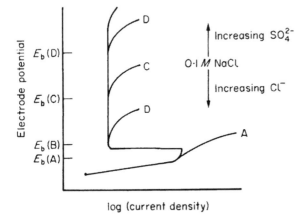

Figure 9.29 Schematic representation of the effect of aggressive ions (Cl^-) and inhibitive ions (SO_4^{2-}) on passivation of stainless steel

representation of the effects of increasing chloride and sulphate ions on the E–log i curve of an 18-8 (Cr–Ni) stainless steel in sodium chloride solution. Addition of an inhibitor such as sulphate ion (or alloying) raises E_b or may suppress it completely, whereas increase in chloride ion concentration moves the onset of pitting to lower potentials.

9.1.6   Determination of E-log i curves

There are two types of instrument which may be used for determining polarisation curves and these are shown diagrammatically in *Fig. 9.30*. The *galvanostat* produces a constant current and enables the steady state potential of the test electrode to be taken with respect to a reference electrode. The *potentiostat* maintains the potential of a test (working) electrode constant with respect to a standard reference electrode and permits the current density to be measured.

Both instruments may be used for investigating polarising behaviour except when passivating materials are studied. The reason may be seen by reference to *Fig. 9.31*. Because the galvanostat only controls current, at i_{crit} there is a sudden jump in anode potential from the active to the transpassive region. Therefore, this instrument cannot give any information concerning the passive region. Thus, potentiostatic techniques should be used in studying passivating materials.

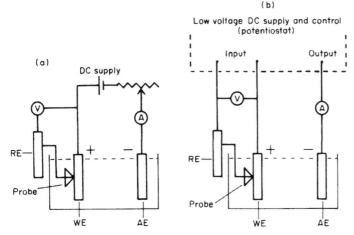

Figure 9.30 Schematic representation of circuits for: (a) galvanostatic and (b) potentiostatic studies (RE = reference electrode, WE = working electrode being measured, AE = auxiliary electrode)

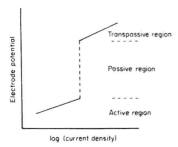

Figure 9.31 Galvanostatic anodic polarisation curve for a passivatable metal

9.1.7 Anodising

The metals aluminium, niobium, tantalum, titanium and zirconium do not show transpassive behaviour (unless in halide solutions). When these metals are anodically polarised the natural passive oxide film of 1–10 μm thickness grows, provided the potential gradient across the film exceeds approximately 1 MV mm^{-1}. Aluminium is the metal most commonly anodised commercially and a range of oxide thicknesses of up to 0.1 mm is produced for architectural and engineering purposes. The oxide film is complex, usually comprising two layers, an inner barrier layer a few nanometres thick and a much thicker porous outer layer. Applied voltages are usually less than 100 V. Other operating conditions, such as the electrolyte and its concentration and temperature, need to be chosen to ensure an oxide thickness appropriate to the component application.

9.1.8 Electropolishing and electrochemical machining

There are distinct similarities between the shape of the E–log i curves for passivation and potential–current curves for electropolishing. *Fig. 9.32* shows a schematic set of curves; curve 1 relates to passivation, curves 2 and 3 represent electropolishing conditions. Under electropolishing conditions the AB portion of the curve is characterised by etching and microscopic smoothing of peaks and ridges occurs. CD and C$'$D$'$

are the *polishing plateaux* and DF and D'F represent current densities at which gas evolution and pitting occur. Curves 1, 2 and 3 all involve the formation of a surface film but 2 and 3 show a much higher anodic dissolution rate (i_{pol} and i'_{pol}) than true passivity, i_{pass} (e.g. 5×10^4 A m^{-2} compared with 10^{-1} A m^{-2}). In CD and C'D' macroscopic levelling occurs but the reason for its occurrence is not clear. It is known, however, that in the regions BC and BC' the concentration of cations builds up at the surface to form the so-called Jacquet layer. Cation diffusion through this layer results in a limiting current density as shown in *Fig. 9.32*. At CD a thin, rapidly dissolving film forms on the Jacquet layer whereas in C'D' only a monolayer or no

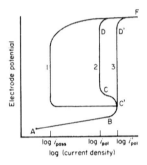

Figure 9.32 Comparison of schematic *E*-log *i* curves for passivation and electropolishing

layer at all may be present. The correct combination of electrolyte and diffusion rate through the Jacquet layer with or without film formation and dissolution results in a macroscopically level surface.

Materials which have ultra-high strength or very high oxidation resistance are required for more exacting purposes, e.g. turbine blades. These materials are very hard or tough and cannot be machined economically by conventional means. Electrochemical machining (ECM) makes these difficult (passivatable) materials the anode of an electrolytic cell, the cathode approximates to the mirror image of the required shape. Operating voltages are in the transpassive region. The dissolution rate is very high, about 200 mm^3 s^{-1}, at current densities of about 100–300 kA m^{-2} (10–30 A cm^{-2}). These extremely high current densities are achieved by using a rapidly flowing electrolyte (about 10 km s^{-1}) between the cathode and workpiece with a very small distance (about 20 μm) between the two electrodes. Any anodic products such as metal hydroxides are swept away by the flowing electrolyte which also dissipates heat. One major problem with the method is selection of a suitable electrolyte for a given alloy which will produce a sharp change from the transpassive to the passive condition. Ideally dissolution of the workpiece should occur at selected (transpassive) sites and there should be no dissolution at adjacent (passive) sites where the potential is lower. In other words the throwing power (Section 6.4.10) of the electrolyte should be low. Closely associated with this problem is the equally important problem of tool (cathode) design and machining tolerances. Thus the cathode is not the exact mirror image of the required shape particularly when the shape is complex. Where suitable combinations of electrolyte, workpiece and tool geometry have been found excellent results have been achieved.

The high flow rate of the electrolyte increases the current density at which the electropolishing plateau occurs and this factor together with the rapid dissolution rates results in the bonus of an electropolished workpiece (e.g. stainless steel). Some materials, e.g. titanium, can be obtained with a 'satin' finish, anodic dissolution taking place by a process of etching rather than by passive film formation on a polishing plateau.

9.1.9 Environmental aspects of aqueous corrosion

9.1.9.1 Corrosion in the atmosphere

Here the term atmosphere is used to imply air and its contaminants and is not used in its most general sense. Air contains both solid and gaseous impurities, e.g. carbon compounds, dust, water vapour, NO_2, H_2S, NH_3, chlorides and metal oxides, depending on whether the geographical area is rural, marine, urban or industrial. For example, the air in marine areas would contain more chlorides than other regions whereas the air in an industrial region contains a relatively high proportion of SO_2, C and carbon compounds. Additional variations are provided by climate with variables such as temperature, rainfall, relative humidity and wind affecting corrosion rates. Moisture content is particularly important and corrosion is higher where the relative humidity is higher (e.g. tropical climates). However, moisture in the form of rainfall may have beneficial effects since it may wash corrosive substances away. This aspect of rainfall is important since it may reduce corrosion in exposed positions whereas in sheltered, almost rain-free sites corrosion may continue virtually unchecked. Solid airborne contaminants are important since they may be relatively inert and set up galvanic or differential aeration cells. Alternatively they may absorb reactive gases to form an electrolyte with condensed water vapour. Of the gaseous contaminants SO_2 probably exerts the greatest effect since it is converted to H_2SO_4 in moist or humid air. This is the main problem when restoring corroded car bodywork. Abrasion and repainting do not remove H_2SO_4 present in pits and corrosion continues after painting.

9.1.9.2 Corrosion in water

Water may be classified as rainwater, freshwater or sea water, i.e. natural waters, potable water (domestic drinking water), and distilled or purified waters.

Rainwater has the same impurities as listed under atmospheric corrosion and will not be considered further.

Freshwater in the form of lakes, underground streams, etc., may contain dissolved gases such as O_2, CO_2, N_2, NH_3, H_2S and SO_2 and dissolved mineral salts such as calcium and magnesium carbonates, sulphates and hydrogen carbonates which determine the hardness of the water. Organic matter and bacteria may also be present. Of these, oxygen is probably the most important constituent since it is a very effective cathodic depolariser and is present in all natural waters. An additional variable is pH, which for natural waters is between 4.5 and 8.5. For steels this range of pH has little effect on corrosion rate but for pH sensitive metals such as copper it may be more important. An increase in flow rate and temperature generally increases corrosion since both increase oxygen diffusion to cathodic sites.

Sea water contains a relatively large proportion of NaCl (about 3%) together with Mg, Ca, K and trace amounts of nearly all existing elements. Nevertheless, the proportions of these elements are virtually constant anywhere in the world. The presence of such high quantities of dissolved salts gives sea water a conductivity about 400 times greater than that of freshwater. Oxygen content decreases with increasing depth and increasing temperature and in general corrosion rates increase in both cases. However, this depth dependence is complicated by the fact that deep cold currents in Atlantic regions may contain more oxygen than the warmer water above it. Sea water temperature varies round the globe, the general range being $-2\,^{\circ}$C to $35\,^{\circ}$C although higher temperatures may be reached in hot shallow regions. The pH of sea water remains almost constant within the range 8.1 to 8.3 although pH may drop to 7 where stagnant regions allow bacterial activity. The effects of surface water movement, i.e. waves, exerts considerable effect on corrosion rates of partly immersed structures (see *Fig. 9.33*). The well aerated regions in the splash zone become cathodic to areas just below the water surface. In 'flat calm' conditions such rapid corrosion would not

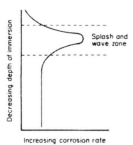

Figure 9.33 Corrosion rate of a partially immersed structure in sea water

occur. The galvanic series for metals in sea water is given in Appendix 4 and considered in Section 9.1.2.1.

9.1.9.3   Corrosion in soils

The principal variables are soil composition and soil particle size which control the important components of the corrosion process, namely moisture content, oxygen content, electrical conductivity, pH and bacterial content. The moisture content of soils can vary between two extremes which may be regarded as atmospheric and total immersion. In general a drier soil will produce less general corrosion than a waterlogged soil. The ability to retain water and oxygen is governed by particle size. Thus, a coarse gravel is likely to retain less oxygen than a clay soil and even a dry soil will retain some water. Therefore, a steel structure buried in clay will corrode to a greater extent than if it were buried in loam.

It should be noted, however, that, initially, pitting occurs to a greater extent in well aerated soils than in poorly aerated soils. In well aerated soils, a corrosion product, such as rust, may form which reduces the corrosion rate. In poorly aerated soils oxygen is hindered in gaining access to the metal surface and protective film formation is prevented. The presence of different soils can produce differential aeration cells as the metal pipe passes through the different regions. The anodic and cathodic regions created may be several kilometres apart and the effect is called *longline corrosion* (see *Fig. 9.6(c)*). However, burial in cinders, which should give well aerated conditions, produces a much higher rate of corrosion of steel, copper and lead than poorly aerated soils. This effect is caused by the excellent cathodic nature of cinders (see Appendix 4).

The nature and type of compounds present will govern the conductivity and pH of a soil. In general, the higher the conductivity the more corrosive the soil while dissolved CO_2 may form H_2CO_3, carbonic acid, to produce acid conditions. The pH of most soils falls within the range 5 to 8. Bacterial attack can also affect pH. Some thiobacillus bacteria form sulphuric acid by oxidation of S or S^{2-} thereby inducing corrosion not only of metals but of non-metallic materials such as concrete. One particularly virulent form of bacterial attack affects iron and steel in anaerobic neutral, waterlogged soil.

The cathodic reaction is hydrogen evolution which occurs slowly. However, where sulphate ions are present the bacteria *Desulphovibria desulphuricans* depolarises the cathodic reaction by:

$$SO_4^{2-} + 4H_2 \xrightarrow{\text{bacteria}} S^{2-} + 4H_2O$$

The result is very rapid attack of the iron and the formation of a non-protective FeS scale.

Where buried metal pipes pass under electrical installations, such as electric railways or cathodic protective systems, there is the possibility of 'stray' electrical currents.

These may give rise to a form of longline corrosion called *stray current corrosion* (*Fig. 9.34*). Stray currents which leak to earth are offered a path of relatively low resistance by a metal pipeline since soils generally have a high resistivity. The pipeline becomes cathodic where the stray (positive) current enters and anodic where the current leaves. The source of current may be alternating or direct. Where neutral or alkaline solutions exist around the pipeline the pH will be increased at the cathodic point since the cathodic reaction is the production of OH^- ions (reaction (9.V)). If alternative currents are present polarity reversals will occur. For a steel pipeline in alkaline conditions the

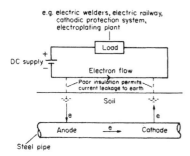

Figure 9.34 The principle of stray current corrosion

reversal of the cathode to an anode may lead to local passivation (see $Fe–H_2O$ Pourbaix diagram. Section 9.1.4). However, for lead pipes anodic attack will occur and is enhanced in alkaline conditions.

9.1.10 Prevention of aqueous corrosion

9.1.10.1 Modification of the environment

In aqueous environments metal–water reactions may be investigated thermodynamically by means of Pourbaix diagrams (Section 9.1.4). The point X in the iron–water system (*Fig. 9.14(c)*) represents the situation where iron is corroding in a neutral aqueous solution. Directions 1 and 2 represent the effects of anodic and cathodic protection (Section 9.1.10.3) whereas direction 3 implies that increase in pH may be used to reduce the corrosion of iron by passivation. However, if the pH were substantially increased corrosion would resume for metals such as zinc and aluminium.

The two main methods of controlling corrosion in aqueous environments are oxygen removal and the use of inhibitors. However, in large tracts of natural waters neither of these methods is possible and cathodic protection is often used. CO_2 is often dissolved in natural waters forming carbonic acid and lowering the pH. Dissolved gases may be removed by physical or chemical treatments, e.g. increasing the temperature (lowering oxygen solubility) or holding at low pressure (Sievert's law) and flushing with an inert scavenging gas such as nitrogen. Chemical de-aeration is achieved by using hydrazine (N_2H_4) for neutral or acid solutions (although it is the HER which becomes increasingly important at low pH values) and sodium sulphite for alkaline solutions:

$$N_2H_4 + \tfrac{5}{2}O_2 \longrightarrow 2H^+ + 2NO_2^- + H_2O$$
$$SO_3^{2-} + \tfrac{1}{2}O_2 \longrightarrow SO_4^{2-}$$

This chemical method is used to protect steel water boilers. In this case the oxygen cannot escape into the atmosphere since the system is 'closed'. Elevated temperatures increase the diffusion of oxygen to the cathode which is the rate controlling reaction of the corrosion process. However, the oxygen solubility (concentration) decreases and a balance occurs which maintains an approximately constant corrosion rate up to the boiling point. However, where a passive oxide film produces protection oxygen must be present for continuous film replacement and repair. Oxygen removal in this case is clearly undesirable. Therefore, oxygen can behave as both a cathodic depolarising agent which increases the corrosion rate (*Fig. 9.22(b)*) and as a corrosion inhibitor.

Inhibitors are substances which, when added to a corrosion system, decrease or eliminate anodic dissolution. This may be achieved by lowering anodic or cathodic reaction rates or both simultaneously. One of the most simple classifications of inhibitor identifies anodic and cathodic types which act upon anodes and cathodes respectively to increase the corresponding polarisation (*Fig. 9.35*) and thereby reduce the corrosion rate from I_{corr} to I'_{corr}. This classification, although convenient, conceals some of the complex mechanisms involved in corrosion inhibition.

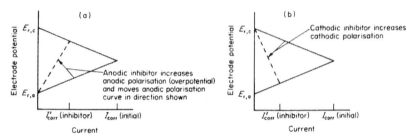

Figure 9.35 Evans diagrams showing the effects of: (a) anodic inhibition and (b) cathodic inhibition

Inhibitors may be considered as two fundamental types: (a) those which form a protective barrier film on anodes or cathodes by a reaction between the metal and the environment; and (b) those which are initially adsorbed directly onto the metal surface by interaction between surface charge and ionic and/or molecular dipole charges. This division of inhibitor types results principally from the pH of the solution in which they operate.

Type (a) inhibitors function in neutral or, in some cases, alkaline solutions in which the main cathodic reaction is oxygen reduction (reaction (9.V)) and a corroding metal surface is covered by a film of oxide or hydroxide, for example. Therefore, the purpose of this type of inhibitor is either to produce a protective film or stabilise an already existing one. Some inhibitors of this type react specifically with the cathodic reaction product (OH⁻ ions in neutral aerated solutions) to precipitate a film at cathodic regions. This film not only prevents direct metal—oxygen interaction but is generally a poor conductor. Thus, the rate of oxygen reduction at the film—electrolyte interface is reduced because the availability of electrons at this interface is reduced and the cathodic polarisation increases. Hence these inhibitors are regarded as cathodic inhibitors. Examples of this sort of inhibitor are:

(i) soluble salts of metals such as Mg, Mn, Ni and Zn which form insoluble hydroxides in cathodic regions:

$$Ni^{2+}(aq) + 2OH^-(aq) \longrightarrow Ni(OH)_2(s)$$

(ii) soluble calcium hydrogen carbonate, $Ca(HCO_3)_2$ (as in temporarily hard water) may produce insoluble calcium carbonate in alkaline conditions:

$$HCO_3^- + OH^- \longrightarrow CO_3^{2-} + H_2O$$

(iii) polyphosphates such as sodium hexametaphosphate $(NaPO_3)_6$ ('calgon') also form a protective film but of a much more complex composition than in (i) or (ii).

A second class of type (i) inhibitors could be called electrochemical passivators. They stabilise an existing, air-formed oxide film by initially increasing the current density of a passivatable metal beyond i_{crit} to produce the typically thin, passive, oxide film. This effect may be achieved by oxidising anions such as chromate, CrO_4^{2-}, and nitrite, NO_2^-. Because these inhibitors affect the anodic reaction they are regarded as anodic inhibitors. Their primary task is to reduce the rate of dissolution of the oxide film by adsorption and then possibly effect complex ion formation. The ions facilitate film repair by producing local increases in i_{corr} beyond i_{crit} and plugging pores in the oxide film by producing insoluble reaction products with dissolved metal cations. These ions may also reduce the adsorption of aggressive anions which could promote breakdown.

Oxidising agents such as those above supply the oxygen to produce protective γ-Fe_2O_3 on iron, for example, since oxygen itself acts as the inhibitor. However, there are also non-oxidising anodic inhibitors which, because of the previous statement, only function in aerated solutions. Such inhibitors for mild steel are sodium hydroxide, carbonate, silicate, and orthophosphate which produce alkaline solutions necessary for Fe_2O_3 formation (see Pourbaix diagrams).

Another inhibitor frequently used is sodium benzoate but this is only effective when the potential of iron is positive to the PZC (Section 9.1.5.1). In fact, although benzoates and also silicates and orthophosphates, are initially adsorbed at anodic sites they subsequently promote oxide formation at both anodic and cathodic sites.

Type (ii) adsorption inhibitors function in acid solutions in which the main cathodic reaction is hydrogen evolution (reaction (9.II)) and a corroding metal surface is free of any film. Adsorption has been used to explain the action of some type (i) inhibitors but this was an integral part of the formation of a reaction product. Adsorption inhibitors, on the other hand, are adsorbed at local anodic and/or cathodic sites and themselves provide the barrier between the metal surface and the environment. As a consequence it is not always necessary to have a complete monolayer in order to produce a significant decrease in corrosion rate.

Typical examples of this type of inhibitor are provided by pickling inhibitors (restrainers) used for mild steel. Once the mill scale has been removed by the sulphuric or hydrochloric acid pickle the bare metal is exposed. Pickling inhibitors, such as those based on thiourea, $(NH_2)_2CS$, are adsorbed at the clean metal surfaces and tend to inhibit both anodic and cathodic reactions. The mechanism of adsorption is complex but relies on the metal surface charge relative to the PZC (Section 9.1.5.1). Thus, iron in hydrochloric acid is positive with respect to the PZC facilitating anion adsorption whereas in sulphuric acid iron becomes negatively charged with respect to the PZC and cations may be adsorbed. In general adsorption inhibitors are organic compounds and bond to the metal surface by a coordinate type of linkage. This type of bonding requires the presence of lone pairs of electrons as provided by elements in groups V and VI of the periodic table, e.g. nitrogen, oxygen and sulphur. Therefore, many adsorption inhibitors contain such elements in thiourea and substituted thioureas, aromatic amines and aldehydes, in such a structure to enable the 'active' atoms to form the positive (nitrogen) or negative (sulphur and oxygen) ends of a dipole. Inhibition by adsorption is also provided by organic compounds containing unsaturated bonds, very large molecules, e.g. proteins and also simple halides. However, some quaternary ammonium compounds do not have lone pairs but are still adsorbed.

Once a type (ii) inhibitor has been adsorbed if may prevent or reduce corrosion simply by blocking anodic or cathodic sites as stated before. However, it may only be weakly inhibitive in which case it may be reduced to produce a secondary, stronger inhibitor. If a reaction, such as anodic dissolution, occurs in a stepwise sequence with the formation of intermediate products an inhibitor may interfere with one of the steps, e.g. by complex ion formation. Thus that step and, therefore, the whole reaction may be made slower. The size, shape and charge on the ion or dipole significantly influences the inhibitive properties. For example, if a molecule is very large, e.g.

protein-based, a physical barrier may be set up between the metal and the solution. On the other hand long molecular chains may, if present in sufficient concentrations, repel one another because of similar dipole charges and produce only weak adsorption. Conversely, attractive forces between dipoles would produce strong adsorption but this requires the presence of two inhibitors and adsorption of two oppositely charged species of ion or dipole (see PZC, Section 9.1.5.1). The charge on the ion, etc., can also alter the electrical double layer. Adsorption of anions, for example, could facilitate the HER (reaction (9.II)) since more electrons would be available.

Inhibitors must be present in a minimum concentration for them to be fully effective. When an anodic inhibitor concentration falls below the specified minimum, small anodic areas are exposed resulting in very high local current densities and intense pitting. This effect has led to inhibitors being classified as safe or dangerous. Anodic inhibitors are regarded as dangerous but may be 100% efficient above the minimum inhibitor concentration. Anodic inhibitors such as benzoates, silicates and ortho-phosphates for iron mentioned earlier, protect both anodic and cathodic areas once they have been adsorbed at anodic sites and are not regarded as dangerous. Cathodic inhibitors produce no intense attack by incomplete filming of the cathode and are considered safe, but at best they are only about 90% efficient. In this context efficiency is used as a measure of anode protection and may be defined as

$$\text{Protective efficiency} = \frac{R_0 - R}{R_0} \times 100\%$$

where R and R_0 are the corrosion rates with and without inhibitor respectively. Sometimes the effect of slowing a reaction is quoted as the

$$\text{Retardation efficiency} = R_0/R$$

In general addition of excessive amounts of inhibitor is unnecessary not only on the grounds of economy and waste disposal but simply because a limiting corrosion rate is achieved at high inhibitor concentrations. In some cases high concentrations may be positively undesirable. For example, although some pickling inhibitors show a minimum corrosion rate at a specific concentration, if this concentration (about 2%) is exceeded corrosion increases again. The reason for this effect is not clear. A similar effect is shown by polyphosphates in which case the corrosion increase is due to soluble complex formation.

In practice it is unusual for a single inhibitor to be used alone. Protection of multi-metal or alloy systems may necessitate the presence of a number of different inhibitors. In addition some mixtures of inhibitors, particularly mixtures of anodic and cathodic types, produce inhibitive properties which are far superior to those of any one of the inhibitors used on its own. This effect is called *synergism*. Often synergism in a two-inhibitor system is achieved by very small amounts of the second compound. Examples of synergistic inhibitors are chromate + polyphosphate + zinc ions in potable waters and benzoate + chromate ions in neutral solutions. Such inhibitors should prevent pitting.

As stated previously, type (i) and type (ii) inhibitors function at different pH values. In some cases a higher temperature may simply require more inhibitor, but often inhibitors which function at lower temperatures become useless at high temperatures and vice versa. For example, polyphosphates are hydrolysed at elevated temperature and their inhibitive properties are lost. Partial immersion conditions may also present problems especially where passivating inhibitors are used. The water line is particularly vulnerable and whereas inhibitors such as phosphates protect under conditions of complete immersion they may intensify attack at the water line. Inhibitors may also produce deleterious effects by the actual process of protection. For example, thiourea reduces the evolution of hydrogen at a high tensile steel surface in acid solution and thereby reduces the corrosion rate. However, atomic hydrogen builds up on the surface to such an extent that it diffuses into the alloy resulting in hydrogen embrittlement.

The prevention or control of atmospheric corrosion depends on the space in which the atmosphere is contained. There are three distinct classes of 'space', outdoor; indoor; and small closed containers. The levels of gaseous contaminants, in particular SO_2, for outdoor conditions cannot be removed but only controlled to within broad limits. Therefore, alloying, use of a protective coating or even a completely different material are the main methods of controlling corrosion. Indoors it is possible to use air conditioning to give a clean, dust-free atmosphere with a low relative humidity. In large box sections, e.g. of bridges, desiccants have been used with limited success. In small enclosed spaces a vapour phase inhibitor (VPI), or volatile corrosion inhibitor (VCI) may be used. A VPI is a soluble film-forming inhibitor which has a low vapour pressure. In general a VPI comprises an inhibitive anion of nitrite, carbonate or benzoate attached to a large organic cation which acts as a 'parachute' for the anion. Correct combinations of these two components produce compounds with vapour pressures of between 0.1 mm and 1 mm of mercury at ambient temperatures. This ensures that the rate of inhibitor evaporation is neither so rapid that it is quickly lost nor so slow that it does not inhibit effectively. Two commonly used VPI's are dicyclohexylamine nitrite and cyclohexylamine carbonate. In an enclosed space they quickly volatilise and dissolve in any condensed moisture present to protect in the usual way. VPI's may be used as powders or impregnated in wrapping paper. The choice of a suitable VPI is complicated by the fact that VPI's vary in volatility, pH of the aqueous solution formed and ability to protect some metals and attack others.

Corrosion in soil is complex. Not only are there local environmental conditions to consider but also bacterial attack, longline and stray current effects. Sometimes pipe lines and storage tanks are buried in a 'back fill' of high-resistance sand (see Section 9.1.10.3). Often, however, for steel, cathodic protection which is expensive if used alone, is combined with a suitable coating, e.g. bitumen or pitch around which PVC sheeting is wrapped followed by another layer of bitumen. Cathodic protection controls longline and stray current effects and, because alkaline conditions are created, bacterial attack is also halted. If bacterial attack were a major problem biocides could be added to the back filling material but regular dosing would be required which could conflict with pollution regulations.

9.1.10.2 *Modification of the metal*

The corrosion properties of many metals may be enhanced by alloying, e.g. Ti–Pd (*Fig. 9.27*). Chromium is passive in non-oxidising acids and transpassive in oxidising acids such as nitric acid where the cathodic reaction is reduction of NO_3^- ion. Alloying with a noble metal in this latter case increases $i_0(5)$ (*Fig. 9.27*) and, therefore, the corrosion rate (M) merely moves to a higher value in the transpassive region. In theory, in acid conditions a noble metal such as gold or platinum provides the best corrosion resistance. In practice of course cost and mechanical strength must also be considered. Therefore, metals or, more usually, alloys are used which simulate noble metal behaviour by passive oxide film formation. Hence aluminium, titanium, chromium and nickel–chromium alloys and stainless steels may be used for specific environments. Reducing conditions may lead to oxide breakdown but aerated solutions facilitate oxide self-healing. Nickel–copper alloys on the other hand, which are only suitable for non-oxidising acids, show best corrosion resistance in de-aerated acid solutions.

Acid resistance may be achieved by alloying with an element which can produce an acidic oxide. For example, cast iron containing 14% silicon produces a protective oxide of SiO_2 while 2% molybdenum in stainless steels improves acid resistance possibly by the influence of acidic MoO_3 in the complex protective oxide film. If an adherent insoluble film can be formed on the metal surface by a metal–acid reaction corrosion resistance may also be achieved, e.g. magnesium in hydrofluoric acid forms MgF_2 and lead in sulphuric acid forms $PbSO_4$.

Magnesium and nickel and their alloys are among materials most resistant to alkaline attack. Metals such as iron and zinc are subject to attack at high pH by ferrate and zincate formation (see Pourbaix diagrams, *Fig. 9.14*) respectively.

Mild steel is the most common structural material. Its resistance to atmospheric attack may be improved by alloying additions of phosphorus, nickel, chromium and copper. As little as 0.2% copper to mild steel produces improved weathering resistance by making the oxide film more protective.

9.1.10.3   Electrical methods

There are three methods which may be discussed under this heading: cathodic protection, anodic protection and electrical insulation.

Consider the corrosion system shown in *Fig. 9.36(a)*. The anodic reaction $M \rightarrow M^{n+} + ne$ is as normal and the cathodic reaction is $C^{n+} + ne \rightarrow C$ where C represents one of the cathodic reactants given in Section 9.1.1. If electrons (current) are supplied to the entire system, e.g. coupling the metal to the negative terminal of a

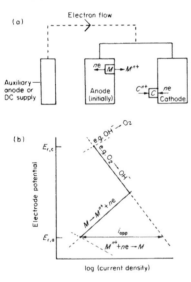

Figure 9.36 Cathodic protection:
(a) in principle; (b) Stern diagram

battery, the rate of the anodic reaction decreases and the cathodic reaction rate increases. In other words anodic dissolution is reduced and M has become protected. Because the concentration of electrons in the anode increases the residual charge and hence the electrode potential becomes more negative, i.e. the overpotential of the anode is reduced because anodic polarisation decreases (see *Fig. 9.36(b)*). Similarly, the cathode becomes polarised to an increasing extent. Therefore, if the cathode potential can be increased such that the anode potential reaches $E_{r,a}$ net anodic dissolution stops (see Section 9.1.5.2). In effect the entire metal surface now acts as a cathode with respect to the applied source and the metal is cathodically protected. The supply of current necessary may be achieved in two ways. The metal to be protected may be made the cathode of an electrolytic cell (impressed current) or it may be connected to a more reactive metal (sacrificial anode) and made the cathode of a galvanic cell. The latter may be achieved by plating, e.g. zinc on iron (galvanising). The determination of $E_{r,a}$ is often difficult in practice. However, it may not be necessary to achieve $E_{r,a}$ to provide adequate protection. This produces a cost saving since the applied current (density), i_{app}, need not be so high. A high degree of overprotection, i.e. supplying enough current to make the anodic potential much more negative than $E_{r,a}$, is wasteful. It may in some cases lead to anodic dissolution by metal–hydrogen reactions because of the increased volume of cathodic product. In theory, increasing i_{app} should merely promote the cathodic reaction $M^{n+} + ne \rightarrow M$ and move the electrode potential into the immunity domain of the Pourbaix diagram. *Fig. 9.37* shows examples of cathodic protection of underground tanks and pipes. Often coatings are used together with

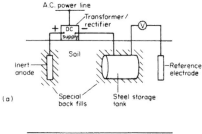

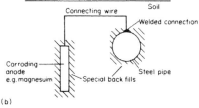

Figure 9.37 Cathodic protection using:
(a) impressed current; (b) sacrificial anode

special backfills to improve corrosion resistance. A protective coating on the pipework necessitates a lower impressed current for cathodic protection. In the case of pipelines sacrificial protection is very convenient to install but generally the output current is uncontrolled. Sacrificial anodes are often scrap steel or graphite and need to be replaced at regular intervals which may be from six months up to more than five years. The use of inert anodes in impressed current systems enables the output from them to be controlled at some prerequisite level relative to a reference electrode.

Both methods may be used to protect the steel hulls of ships. Impressed current methods use an inert anode of platinum-clad titanium. Sacrificial methods use magnesium anodes rivetted to the hull but in freshwater magnesium passivates by forming $Mg(OH)_2$ and aluminium–calcium–zinc anodes are then used. Anodic protection relies on the ability to passivate a metal surface and is, therefore, limited to metals and alloys which form a passive film. *Fig. 9.38* relates the Pourbaix diagram for iron to the E–log i passivation curve. An external current is applied in the opposite direction to cathodic protection, i.e. electrons are removed from the metal to be protected, e.g. by

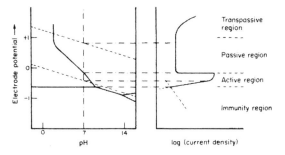

Figure 9.38 Comparison between the Pourbaix diagram and E–log i curve for iron at pH 7

coupling the metal to the positive terminal of a battery. Therefore, anodic polarisation increases up to i_{crit} at which point passivation begins. A potential between E_{pass} and E_b, the breakdown potential is used to maintain passivity. Thus, the current required to produce passivity is high but that required to maintain it is very low. Potential control is maintained by means of a potentiostat (*Fig. 9.39*). Obviously metals which have low values of E_{pass} are more easily passivated, e.g. E_{pass} values for Fe, Cr, Ti are 0.58 V, −0.22 V and −0.24 V respectively. When compared with cathodic protection anodic protection has a number of disadvantages but also some very important advantages. Anodic protection is used for example to protect vessels of about ten million

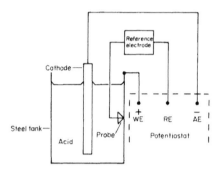

Figure 9.39 Anodic protection of a steel tank

litres capacity which contain highly corrosive acids. It has a high throwing power (Section 6.4.10) i.e. plant remote from the cathode is passivated. The cost of installing anodic protection is high because of the potentiostat and reference electrodes required. The high current densities necessary to achieve passivity also imply high costs. However, these costs are often justified since cheaper materials may be used, e.g mild steel instead of stainless steel or aluminium. The operating costs are low because the current densities needed to maintain anodic protection (e.g. 10^{-1} A m^{-2}) are about an order of magnitude lower than those needed for cathodic protection.

However, current density increases if the liquid is stirred or depleted of oxygen. The applied current itself is often a direct measure of corrosion rate and, therefore, of protection. The current applied in cathodic protection gives no such indication. However, anodic control requires constant monitoring of plant since if local film breakdown were to occur and the film did not or could not reform, dissolution at this point would be very rapid. The operating conditions for anodic protection can be accurately determined by laboratory measurements whereas cathodic protection usually requires some form of empirical testing.

Electrical insulation methods have been mentioned in previous sections. The principle involves that of increasing the resistivity of some part of the electrochemical corrosion cell such that part of the energy available to drive the cell must be used in overcoming this resistance. The result is that the system becomes resistance controlled (*Fig. 9.20(c)*). This effect may be achieved in practice by special high resistance backfills around pipelines for example. Longline effects may be reduced by inserting insulating discs between short lengths of pipe. Where dissimilar metals are bolted together use of insulating washers or high resistivity jointing compounds can prevent galvanic cell formation.

9.1.10.4   Protective coatings

Coatings may be metallic or non-metallic and protect either by exclusion, i.e. preventing direct metal–atmosphere contact, or by sacrificial cathodic protection. Vitreous enamels, lacquers and some paints act by exclusion as do coatings of more noble metals, e.g. nickel or chromium on steels. However, such coatings must be pore-free otherwise accelerated attack of the more base (anodic) metal substrate occurs. In sacrificial protection breaks in the coating cause preferential dissolution of the coating which protects the metal substrate (*Fig. 9.3(c)*). The baser coating often dissolves in any case and this must be taken into account when layer thickness and service conditions are linked. In all cases proper surface preparation treatments prior to coating are vital; methods used include abrasive cleaning, acid and alkaline cleaning solvents and vapour degreasing.

Metallic coatings

Electrodeposition (Section 6.4.10) may be used to apply gold, silver, copper, tin, nickel, cadmium, chromium and zinc as thin, relatively pore-free coatings, although gold and silver should never be used for corrosion control because they can set up intense bimetallic corrosion. The thickness of such coatings may be readily controlled.

Hot dipping immerses the metal to be protected in a bath of molten zinc, tin, lead or aluminium, i.e. low melting point metals. Good adherence is achieved due to the formation of an alloy layer at the coating–metal interface. The process gives a good coating (greater than 0.01 mm) but thickness control is poor.

In *cladding*, the metal to be protected is sandwiched between layers of a more oxidation resistant metal and then hot or cold rolled. Both coating and substrate must have similar deformation characteristics. A subsequent annealing treatment (especially after cold rolling) may be required to produce alloying necessary for bonding at the metal–coating interfaces. Cladding may be used to protect mild steel by aluminium, copper and nickel. Some aluminium alloys and stainless steels may be clad by aluminium.

In the *cementation* process a corrosion resistant metal is allowed to diffuse into the surface of a component. Thus zinc (sherardising), aluminium (calorising) or chromium (chromising) may be introduced into the metal surface by heating the component in a suitable powder (aluminium, zinc) or gaseous atmosphere (chromium). Chromising, for example, may be used to produce a corrosion resistant case on a nitrided steel surface. This is not to be confused with cementation precipitation discussed in Section 7.3.2.3.

Vacuum deposition produces very thin films of metals such as aluminium, zirconium, chromium and copper on metallic or non-metallic substrates. A subsequent anneal is usually needed to promote bonding. Although primarily used for producing decorative finishes the possibility of applying it to the continuous coating of steel strip is being investigated.

Spraying converts most metals to a high pressure jet of molten droplets. The initial keying action of the coating is provided by a rough substrate surface. Subsequent annealing is required to promote adhesion by alloying. Heat treatment also reduces the brittleness of the coating. Spraying is most commonly used for applying zinc and aluminium to mild steel, but it has been used to apply tin, lead, molybdenum, stainless steel, silver and nickel for specialised applications. Methods which involve alloying as a means of producing an adherent coating are often classified as *diffusion coatings*.

Non-metallic coatings

Anodising was discussed in Section 9.1.7. The porous, anodically-thickened oxide film may be painted or dyed before the pores are finally sealed by, for example, boiling water immersion.

Phosphate conversion coating or *phosphating* is a chemical passivation technique in which a metal, usually steel but also aluminium, tin and cadmium, reacts with phosphoric acid. A phosphate coating provides limited corrosion protection but it does provide an excellent pretreatment before painting and has good retention properties for enamels, paints and oils, e.g. retaining lubricant oils prior to deformation processes.

Chromate conversion coating or *chromating* provides a metal oxide–chromium oxide passivating film on the surfaces of metals such as aluminium, magnesium, tin, zinc, cadmium and sometimes copper and silver. Unlike phosphating, chromating is a complex redox reaction brought about by immersion in a sodium dichromate–sulphuric acid bath. Chromating may also be achieved electrolytically although this method is less usual, its main purpose being to obtain a film which is more resistant to abrasion than chemical chromating.

Inorganic coatings include materials such as cement, zinc silicates and vitreous enamels, i.e. brittle coatings used as environment excluders. Protection is only provided for as long as the coating remains undamaged. Important advantages of cement coatings are that they are cheap, readily applied and readily repaired.

Organic coatings include oil-based paints, pitch, tar, bitumen, plastics and rubbers. The most widely used of any coating is paint. Paints are porous to oxygen and water and, therefore, many paints contain inhibitors to overcome this problem. Inhibitors, e.g. red lead (Pb_3O_4) and zinc chromate, are usually contained in the paint used as the *primer*, although nowadays the use of lead compounds is decreasing. Some paints for steels contain about 95% metallic zinc to provide further protection by sacrificial protection ('cold galvanising').

9.1.10.5   Other factors

The above methods have all considered corrosion to be the most important parameter. In practice a suitable choice of protective method depends on an interrelationship between many factors dependent in turn on the practical use to which the material is to be put. Therefore, other factors to be considered include structural design, mechanical properties, physical properties, ease of fabrication, weldability, availability and cost. It is not surprising then that the most common basic material is mild steel. When combined with an appropriate coating mild steel has a satisfactory combination of properties for a variety of environments and is relatively cheap. In some cases it is not necessary to use metals at all particularly where elevated temperatures or changes in temperature do not take place or where high mechanical strength and ductility are not required. Thus, materials such as stone, rubber, plastics, glass, graphite and porcelain may be used in a limited number of applications.

9.2   Oxidation–dry corrosion

Oxidation or 'dry' corrosion implies the reaction between a metal and any oxidising gas, e.g. carbon dioxide, oxygen, sulphur or a mixture of oxidising gases at ambient and elevated temperatures. However, the most common form of metal oxidation is brought about by oxygen and, therefore, the following discussions are concerned principally with oxide formation and growth.

Although the $\Delta G^{\ominus} - T$ diagram indicates that the standard free energy of formation of most oxides increases (becomes more positive) and, therefore, stability decreases as the temperature rises (see p. 85), oxide films on metal surfaces often thicken at higher temperatures indicating the importance of kinetics. The rate of oxide formation cannot be defined simply since it depends on a number of factors such as temperature, time, surface finish of metal substrate, continuous or discontinuous nature of oxide, internal stress in oxide and volatility of oxide. In general, once a continuous oxide layer has become thick enough one of a number of mathematical relationships between oxide thickness and time may be obeyed at a particular temperature.

Although there is no clear dividing line between an oxide film and an oxide scale an oxide layer of up to 1 μm thick is regarded as a film. Thus, a thin film is less than about 200 nm thick and refers to the early stages of oxide nucleation.

An oxide layer above 1 μm thick is considered to be a scale. Sometimes the term scale is used to denote a thick visible oxide layer formed under highly oxidising conditions. Therefore, on this basis a film refers to the invisible layer preceding scale formation. However, classification in terms of visibility is not exclusive since thin films are often visible because of the interference colours they produce. Tarnishing is not a further class of oxide formation but refers to the discolouration or dulling of a metal surface due to the presence of an oxide film.

9.2.1   Formation of the oxide layer

The first stage in the oxidation of a perfectly clean metal surface is chemisorption of oxygen. Oxygen molecules, adsorbed at the metal surface, dissociate to form atoms which then share electrons with adjacent metal atoms. As this rapid process occurs some metal atoms move into the plane of adsorbed atoms. The result is an extremely stable, adherent, continuous monolayer about 1 nm thick. Further film growth proceeds by a combination of chemisorption of oxygen and diffusion of electrons, metal and oxygen ions. At this stage, however, growth is confined to energetically favourable nucleation sites where local epitaxial accumulations of oxide form. The nuclei then grow laterally by surface diffusion of metal and oxygen to produce, ideally, a continuous layer of uniform thickness.

The lateral growth of different nuclei results in a definite grain structure, the grain boundaries providing easier diffusion paths for ions and hence a faster growth rate. The number of nuclei is strongly affected by temperature (an increase leads to a reduction in nuclei formation) and crystallographic orientation of the metal substrate, e.g. a maximum number of nuclei for FeO formation is found on the (100) planes of iron. The orientation of the film is similar to that of the grain beneath (epitaxy) and rate of growth varies with orientation. Thus a polycrystalline material showing no preferred orientation exhibits grains of oxide of different thicknesses and orientations. The similarity of orientation between oxide and metal tends to become less marked as the oxide thickens.

The formation of a protective oxide necessitates a continuous oxide film which contains minimum defects and which is adherent and coherent with the metal substrate and is relatively plastic in order that it can withstand thermal and mechanical stresses.

9.2.2   The rate laws for oxide growth

Some graphical relationships between film thickness and time are given in *Fig. 9.40*. The interpretation of y as film thickness is an ideal approach. In practice it is more convenient to measure y as a weight loss or gain per unit area. Clearly the two parameters are directly related. Hence the units of y may be given as kg m^{-2} rather than

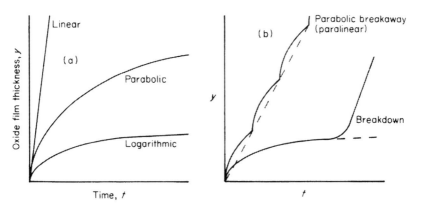

Figure 9.40 Oxidation rates: (a) different rate laws of oxidation; (b) oxidation behaviour as a result of oxide discontinuity

simply a unit of length. The logarithmic law and corresponding growth rate (dy/dt) may be expressed as:

$$y = K_1 \log_{10}\left(\frac{t}{K_2} + 1\right) \quad \text{and} \quad \frac{dy}{dt} \propto \frac{1}{t}$$

K_1 and K_2 (and the K terms below) are temperature dependent constants. This law applies to thin films and films formed at low temperatures. The law is also obeyed where oxidation involves void formation within the film (see oxidation of nickel, Section 9.2.4). The parabolic law is given by

$$y^2 = K_3 t + K_4 \quad \text{and} \quad \frac{dy}{dt} \propto \frac{1}{y}$$

This law is obeyed by many metals at higher temperatures and, therefore, relates to thicker films than the logarithmic law. From the graph it may be deduced that both parabolic and logarithmic growth afford protection to the metal after a time.

The linear law,

$$y = K_5 t + K_6 \quad \text{and} \quad \frac{dy}{dt} = \text{constant}$$

applies to the initial stages of oxidation before the film is thick enough to be protective. In addition this law may be followed where the film is non-protective, porous or cracked or where a phase change in the film, such as oxide volatilisation, prevents protection. The condition of parabolic breakaway or intermittent oxidation may occur when local cracking or spalling of the film occurs at intervals due to large internal stresses in the film. The effect of repeatedly building up and breaking down the protective nature of the film prevents continued parabolic growth taking place and the result is an effectively linear growth rate. This combination of parabolic 'jumps' and overall linear growth rate is sometimes called paralinear behaviour. Breakdown occurs when the film becomes completely non-protective due, for example, to a change in the environment from an oxidising to a reducing nature. The laws given above are by no means the only thickness–time relationships. At low temperatures and for thin oxide films inverse logarithmic and asymptotic laws have been identified:

$$\frac{1}{y} = K_7 - K_8 \log_{10} t$$

and

$$y = K_9[1 - \exp(-K_{10}t)]$$

A cubic law,

$$y^3 = K_{11} t + K_{12}$$

intermediate between the logarithmic and parabolic laws also exists for some metals.

It is important to note that a metal may obey different rate laws over different ranges of temperatures and in different atmospheres. The oxidation of iron in air provides one such example showing a growth rate which is logarithmic below 200 °C and parabolic between 200 °C and 900 °C. Another example is the oxidation of copper in air which shows paralinear behaviour at about 500 °C but when the oxide film becomes more plastic at about 800 °C a true parabolic growth rate is observed.

The tendency for a film to become unprotective by cracking or spalling was originally predicted by Pilling and Bedworth who derived the following relationship:

$$\text{Pilling–Bedworth ratio} = \frac{\text{volume of 1 mol. oxide } (V_{ox})}{\text{volume of metal to produce } V_{ox}}$$

Therefore, if this ratio is less than unity, e.g. MgO (0.81), Na_2O (0.55), the oxide is formed in tension and should not be protective or continuous. A linear growth rate would be expected. If the Pilling–Bedworth ratio is approximately unity, e.g. Al_2O_3 (1.24), a stress-free protective film is possible. If the ratio is much greater than one, e.g. Cr_2O_3 (2.0), WO_3 (3.35), the oxide should be continuous but may be too bulky and tend to spall due to compressive stresses set up in the oxide film.

In practice these rules are not necessarily followed. For example, magnesium shows linear behaviour above about 400 °C but below this temperature the oxide film is continuous, protective and follows a logarithmic growth law. Aluminium forms a very protective oxide but so does chromium. The problem lies in considering oxidation to be controlled only by the chemical reaction between a metal and oxygen. In fact oxide formation is much more complex and other parameters such as diffusion of ions and electrons through the oxide layer must be considered also. Therefore, the Pilling–Bedworth ratio cannot be used to predict the protective nature of films for all metals at all temperatures. However, it may be used to give an approximate indication of internal stresses provided the molecular formula of the oxide is known.

9.2.3 The structure of oxides

Oxides, like metals, contain defects. These defects may be ion vacancies or interstitials and the concentration of each is not usually equal. This inequality gives rise to non-stoichiometry. However, even though these defects are present the arrangement of ions and defects in the oxide lattice must maintain electrical neutrality. In fact the presence of defects results in metal oxides having semiconductor properties. Therefore, oxides may be classed as n-type or p-type.

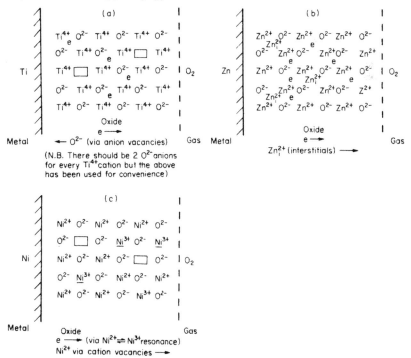

Figure 9.41 Schematic representation of the structure of non-stoichiometric oxides: (a) TiO_2; (b) ZnO; (c) NiO

The n-type semiconductors contain excess metal and conduct by means of negative charge carriers, namely, electrons. These electrons are present to maintain electrical neutrality in the presence of excess metal cations. This metal excess may be achieved in two ways: (a) by anion vacancies as in Al_2O_3, Fe_2O_3, TiO_2 (non-stoichiometric formula TiO_{2-x} for example), and (b) by interstitial cations as in ZnO ($Zn_{1+x}O$). Similarly, p-type semiconductor oxides have excess anions in the lattice (metal deficient) and conduct by means of *positive holes* which are, essentially, cations which have an extra positive charge due to missing electrons, e.g. Ni^{3+}. Anion excess may also be achieved in two ways: (a) by interstitial anions, of which there appear to be no examples, and (b) by cation vacancies, e.g. Cu_2O, FeO, NiO ($Cu_{2-x}O$ for example). *Figs 9.41(a)*, *(b)* and *(c)* show schematic representations of the structures of n- and p-type oxides. Each lattice is drawn electrically neutral with respect to the defects present. Excess electrons are shown for the n-type lattices and positive holes (Ni^{3+} ions, i.e. Ni^{2+} − e) for the p-type oxide. Clearly there should be twice as many oxygen ions on the nominally TiO_2 lattice as there are shown but the lattice has been drawn in this way for convenience.

Electrical conductivity within an oxide is controlled by electron flow whilst oxidation rate is controlled by flow of ions (ionic conduction). Hence, any factor which increases ionic (anion or cation) conductivity also increases the oxidation rate.

Since the oxide film must always be electrically neutral in order to remain stable ionic and electronic conductivity must be considered together. Thus in p-type semiconductor oxides electron flow occurs via positive holes (electron deficient sites) while ionic conductivity (and hence oxidation) occurs by diffusion of cations via vacant cation sites. In n-type semiconductors electron flow occurs by movement of free electrons while ionic conductivity (and hence oxidation) occurs by diffusion of O^{2-} ions via vacant anion sites or by diffusion of interstitial cations. Therefore, any factor which increases cation vacancies in p-type oxides or O^{2-} anions or interstitial cations in n-type oxides also increases the oxidation rate. This explanation of oxide growth by means of a defective lattice mechanism is applied to oxides growing according to a parabolic rate law (Section 9.2.4). Thin films, which grow according to a logarithmic law, have few defects. In this case a very high electric field is considered to exist across the oxide such that ions are pulled across the film to make it grow (Section 9.2.5).

9.2.4   The Wagner theory of parabolic growth

At one time it was believed that all oxidation took place by a process of oxygen diffusion through an oxide layer to the metal surface. Although this is true for some metals 'marker' experiments using a thin, discontinuous layer of platinum or Cr_2O_3 on an initially clean metal surface have shown different results for zinc, iron, copper and nickel. For example, after oxidising iron above 200 °C, in the parabolic growth rate region, Cr_2O_3 markers were observed on the metal—oxide boundary. This effect can only be explained by diffusion of metal cations outwards rather than oxygen anion diffusion inwards. Inward diffusion of oxygen anions would have resulted in the markers being found on the oxide—atmosphere interface.

Wagner predicted the above behaviour and was the first to propose that parabolic oxide growth was in fact caused by anion and cation diffusion under the action of a concentration gradient across the oxide film. Diffusion (migration) of electrons also occurs under the action of a potential gradient. Therefore, since oxide growth supposedly occurs by a diffusion process it should be possible to derive the parabolic growth law from basic diffusion laws. Hence, considering Fick's first law of diffusion:

$$J_x = -D_x \, dc/dx$$

J_x is the flux (or diffusive flow) of an atomic or ionic species, x, in a single direction, i.e. the number of moles (gram atoms or atoms) flowing through unit area in unit time. D_x is the diffusion coefficient or diffusivity (units $m^2 s^{-1}$ for example), dc is the change in concentration of diffusing species across a distance dx in moles (gram atoms or atoms) per unit volume. In other words diffusion is proportional to concentration gradient. The negative sign indicates flow from a high to a low concentration.

Consider a mass, dm, of the diffusing species passing through area, A, in time, dt. Fick's law then becomes (ignoring sign):

$$\frac{dm}{dt} = DA \frac{dc}{dx}$$

Integrating:

$$m = DA \frac{dc}{dx} t + \text{constant} \tag{9.9}$$

Since $m = 0$ when $t = 0$, then the constant is zero.

Now consider the diffusing species to have relative atomic mass A_m and to form an oxide of molar mass M, density ρ and that b atoms of the diffusing species are required per mole of oxide. The thickness of the oxide film is y and diffusion across a cross-sectional area A occurs.

$$\text{Volume of oxide} = yA \qquad \text{Mass of oxide} = yA\rho$$

Therefore,

$$\text{No. of moles of oxide} = \frac{yA\rho}{M}$$

The mass of the diffusing species in 1 mole of oxide is bA_m and so the total mass of the diffusing species, m, is given by

$$m = \frac{yA\rho}{M} bA_m \tag{9.10}$$

Combining equations (9.9) and (9.10) and substituting $dx = y$, we get

$$\frac{yA\rho}{M} bA_m = DA \frac{dc}{y} t$$

Therefore,

$$y^2 = Kt$$

Thus diffusion appears to be closely associated with oxide growth in the parabolic region.

If oxidation occurs by a process of ion and electron transport the process may be considered purely as a flow of current. The rate constants, k, calculated by this approach are in close agreement with those derived experimentally for many metal-oxide systems. Parabolic oxide growth is a thermally activated process and hence,

$$k = A \exp\left(\frac{-E_A}{RT}\right) \qquad \text{(Section 3.5.1)}$$

should apply where A is a constant and E_A is the activation energy (J mol^{-1}) for diffusion. In a system in which only one ion species diffuses experimental values of E_A correspond closely to the activation energy for self-diffusion.

Clearly there is considerable evidence to support the Wagner theory. As well as ion and electron diffusion Wagner also assumes that the oxide is adherent, continuous, electrically neutral and non-stoichiometric. Another important assumption is that the oxide film is thick enough to be able to ignore the so-called space-charge regions at the metal–oxide and oxide–atmosphere interfaces. The significance of these regions will be discussed in Section 9.2.5.

Applying the Wagner approach to the growth of ZnO (above 400 °C) the excess cations and an equivalent number of electrons to maintain electrical neutrality are located on interstitial sites (*Fig. 9.41(b)*). Oxygen at the oxide–gas interface is chemisorbed into the oxide by

$$\tfrac{1}{2}O_2 + 2\,\text{\textcircled{e}} \longrightarrow O^{2-} \tag{9.XI}$$

where \text{\textcircled{e}} represents free electrons. It may be reasonably assumed that the O^{2-} ions are chemisorbed onto the anion sub-lattice which builds up locally leaving gaps, i.e. vacancies, in the zinc sub-lattice. The zinc interstitial ions, Zn_i^{2+}, diffusing under the driving force of a concentration gradient across the oxide, then fill up the cation vacancies. Thus, the interstitial defects are annihilated to create a 'perfect' lattice in the region of the oxide–gas interface, i.e. Zn_i^{2+}, and O^{2-} ions combine to form stoichiometric ZnO:

$$Zn_i^{2+} + O^{2-} \rightleftharpoons ZnO$$

Combining these two equations results in an expression for dissociation of zinc oxide:

$$\tfrac{1}{2}O_2 + Zn_i^{2+} + 2\,\text{\textcircled{e}} \rightleftharpoons ZnO \tag{9.XII}$$

A constant concentration gradient is assumed to exist across the oxide and interstitial zinc ions and electrons are replaced at the metal–oxide interface by

$$Zn \longrightarrow Zn_i^{2+} + 2\,\text{\textcircled{e}}$$

Therefore, the overall effect is thickening of the oxide by reaction at the oxide–gas interface.

A similar approach may be used for the anion deficient n-type oxide, TiO_2, in the temperature range (550–850 °C) for parabolic growth (*Fig. 9.41(a)*). Oxygen is chemisorbed at the metal–gas interface by reaction (9.XI). It is worth noting that the 'free' electrons in TiO_2 (or ZnO) may be thought of in terms of Ti^{3+} (or Zn^+) ions. Since vacancies are present in the anion sub-lattice the oxygen ions diffuse inwards towards the metal–oxide interface. In effect the anions created at the surface cancel out the vacancy defects in this region. The process is continuous but may be regarded as creating a local stoichiometric lattice at any instant of time, therefore:

$$2O^{2-}\square + 2O^{2-} + Ti^{4+} \rightleftharpoons TiO_2$$

where $O^{2-}\square$ is an anion vacancy. Combining this reaction with (9.XI) gives:

$$2O^{2-}\square + O_2 + 4\,\text{\textcircled{e}} + Ti^{4+} \rightleftharpoons TiO_2 \tag{9.XIII}$$

To enable oxidation to continue anion vacancies must be created at the metal–oxide interface. These defects would be formed if anions were to give up their electrons to form atomic oxygen which are then dissolved in the metal. Experimental work has verified this proposal since oxygen is very soluble in titanium at these temperatures and a solid solution of oxygen has been identified beneath the titanium surface. At the same time as vacancies are created titanium cations are created by $Ti \rightarrow Ti^{4+} + 4\,\text{\textcircled{e}}$. Because the oxide is constantly growing into the metal (markers remain at the oxide–gas interface) stresses are set up within the oxide. However, breakaway has not been observed in the parabolic region. Paralinear behaviour is observed above 850 °C but the nature of the scale changes to become layers of TiO, Ti_2O_3 and TiO_2. In any case the Wagner mechanism no longer applies outside the parabolic growth region.

For a p-type oxide such as NiO in the parabolic growth region (above 1000 °C) chemisorption of oxygen proceeds by a different mechanism to that in n-type oxides. There are no free electrons in p-type oxides so in order to form oxygen ions electrons are taken from neighbouring Ni^{2+} ions to leave them as Ni^{3+} ions. Since these latter ions have a higher positive charge than Ni^{2+} ions they are called positive holes. To preserve electrical neutrality within the oxide the number of positive charges must be reduced. This can only be achieved by simultaneously creating cation vacancies at the oxide–gas interface as oxygen is chemisorbed. Therefore, the formation of one O^{2-} ion produces two positive holes and one nickel cation vacancy, $Ni^{2+}\square$, and may be represented by

$$\tfrac{1}{2}O_2 \longrightarrow O^{2-} + 2 \oplus + Ni^{2+}\square$$

where $\oplus$ represents a positive hole. At an instant of time surface vacancies may be regarded as being filled by outward diffusing cations and a local stoichiometric lattice formed, i.e.

$$Ni^{2+} + O^{2-} \rightleftharpoons NiO$$

Combining these two reactions:

$$Ni^{2+} + \tfrac{1}{2}O_2 \rightleftharpoons NiO + 2 \oplus + Ni^{2+}\square \qquad (9.XIV)$$

Thus oxidation proceeds by inward diffusion of vacancies (outward diffusion of cations). Electrons migrate to the oxide–gas interface by positive hole conduction. In effect Ni^{3+} ions migrate towards the metal–oxide interface and electrons in the opposite direction by resonance between Ni^{2+} and Ni^{3+} ions, i.e. $Ni^{2+} \rightleftharpoons Ni^{3+} + e$. Electrons and cations are created at the metal–oxide interface by $Ni \rightarrow Ni^{2+} + 2e$. A cation enters a vacancy and, therefore, its original atomic site becomes a vacancy. Because of this 'inward diffusion of vacancies' it would be reasonable to predict that under suitable conditions the vacancies could coalesce to form holes or voids at the metal–oxide interface. Experimental observations have in fact identified voids (called Kirkendall voids) in a number of p-type oxides including NiO. The presence of voids may alter the growth rate law in operation. An initially rapid parabolic oxidation rate may become logarithmic, i.e. become slower, because internal voids prevent diffusion of ions etc. across the oxide. This effect means that the conductivity of the film is reduced and hence the growth rate is reduced since conductivity and growth rate are directly related. This explains why the logarithmic law is obeyed by films which are much thicker than would normally be expected.

The Wagner theory has been successfully applied to explain the effects of oxygen pressure on oxidation. Applying the law of mass action (Section 2.2.3.1) to reaction (9.XII)

$$[Zn_i^{2+}]\,[\oplus]^2[O_2]^{\tfrac{1}{2}} = \text{constant } [ZnO]$$

where [ ] denotes concentration. Since the concentration of non-interstitial Zn^{2+} ions and O^{2-} ions may be regarded as virtually constant, [ZnO] may be incorporated into the constant term. $[O_2]$ may be replaced by p_{O_2}, oxygen partial pressure. Thus, the equation becomes

$$[Zn_i^{2+}]\,[\oplus]^2 p_{O_2}^{\tfrac{1}{2}} = \text{constant} \qquad (9.11)$$

Furthermore since every interstitial cation is associated with two free electrons then $[Zn_i^{2+}] = 2[\,e\,]$ and equation (9.11) becomes

$$[\oplus] = \text{constant } p_{O_2}^{-1/6} \qquad (9.12)$$

The electrical conductivity of semiconductors is predominantly electronic. Therefore, in n-type oxides conductivity, which is dependent on electron concentration, should be proportional to $p_{O_2}^{-1/6}$. In practice values of between $p_{O_2}^{-1/4}$ and $p_{O_2}^{-1/6}$ are obtained.

TiO_2 may be considered in a similar way to give

$$[O^{2-}\square]^2[\oplus]^4 p_{O_2} = \text{constant} \qquad (9.13)$$

Since $[O^{2-}] = 2[\,e\,]$ this equation reduces to (9.12) again. This relationship has also been verified experimentally. In general equation (9.12) only applies over certain pressure ranges. At only moderate pressures in the parabolic region the concentration of defects at the oxide–gas interface may be regarded as negligibly small since the annihilation of surface defects is very rapid. Therefore, because the concentration gradient across the oxide effectively determines the growth rate any subsequent increase in pressure has little effect on oxide growth.

Experimental observations on ZnO, TiO_2, Al_2O_3, ZrO_2 and other n-type oxides have shown that the growth rate of each of these oxides is independent of oxygen above a certain level. Thus, for example, equation (9.11) becomes

$$[Zn_i^{2+}][\circledcirc]^2 = \text{constant}$$

At relatively low pressures the Wagner mechanism no longer applies. The concentration of defects at the oxide–gas interface increases. The rate controlling process becomes the arrival of oxygen molecules at the oxide surface and their subsequent chemisorption. For a p-type oxide such as NiO application of the law of mass action to reaction (9.XIV) gives:

$$[Ni^{2+}\square][\oplus]^2 = \text{constant } p_{O_2}^{\frac{1}{2}} \tag{9.14}$$

Conduction is by positive holes and since for every $Ni^{2+}\square$ there must be two $\oplus$ then

$$[\oplus] = \text{constant } p_{O_2}^{1/6}$$

Therefore, conductivity should be proportional to $p_{O_2}^{1/6}$, which has been found in practice. Similarly for Cu_2O:

$$2Cu^+ + \tfrac{1}{2}O_2 \rightleftharpoons Cu_2O + 2\oplus + 2Cu^+\square$$

and conductivity is proportional to $p_{O_2}^{1/8}$. In this case experimental results vary between $p_{O_2}^{1/6}$ to $p_{O_2}^{1/7}$.

Unlike n-type oxides, the growth of p-type oxides must show continued oxygen dependence at high pressures if the Wagner mechanism holds since the concentration of defects at the oxide–gas interface should increase as the pressure increases. This dependence of growth rate on oxygen pressure has been found in practice.

The Wagner (–Hauffe) theory may also be used to explain the effects of small alloying additions, *dopants*, on oxidation rates. The general rule is that for n-type oxides lower valency additions increase the oxidation rate whereas addition of higher valency metals reduce the oxidation rate. For p-type oxides the opposite effects would be expected. Although this rule is not totally reliable it does apply in a majority of cases. Typical examples are given below.

If small additions of Li_2O are made to ZnO, Li^+ ions will take up positions on the cation lattice effectively replacing some Zn^{2+} ions. To preserve electrical neutrality fewer free electrons are needed. Therefore, from equation (9.11), the concentration of interstitial zinc ions increases and hence the oxidation rate increases.

Similarly, addition of Al^{3+} ions would be expected to increase the number of electrons required for electrical neutrality. Hence the interstitial defects and growth rate should decrease. Experimental determination of the parabolic rate constant (kg^2 m^{-4} s^{-1}) for pure zinc, zinc + 0.1 mol% Li and Zn + 0.1 mol% Al at about 400 °C and 1 atm pressure gives values 10^{-4}, 10^{-3} and 10^{-5} respectively, which supports the Wagner approach.

When Cr^{3+} ions are added to TiO_2 the number of vacant anion sites increases because fewer electrons are needed to maintain electrical neutrality (equation (9.13)). Inward oxygen diffusion and hence growth rate increase. Addition of W^{6+} ions reduces the number of vacant sites by a similar argument and the oxidation rate decreases. The addition of higher valency elements to TiO_2 is complicated by the fact that these elements have a lower affinity in general for oxygen and do not dissolve to any great extent in the oxide. The beneficial effects of these additions are only short term and at longer times the scales become porous and the oxidation rate increases. The addition of Li^+ ions to a p-type oxide such as NiO replaces some Ni^{2+} ions. Each Li^+ ion may be regarded as a negative charge in the lattice by comparison with the Ni^{2+} ion it replaces and, therefore, a positive hole is required to maintain electrical neutrality. By equation (9.14) this implies that the concentration of cation vacancies is reduced and, therefore, the oxidation rate is reduced. By a similar argument addition of higher valency ions such as Cr^{3+} reduces the number of positive holes increasing the cation vacancy concentration and hence the oxidation rate increases. The result is a nickel-chromium alloy which has a *higher* oxidation rate than pure nickel. This result appears to conflict with the fact that Nichrome (nickel–20% chromium) is extremely resistant to oxidation, much more so than pure nickel. This disparity is only an apparent one and is caused by the different concentrations of chromium in each case (see Section 9.2.7).

9.2.5 Growth of thin oxide films

Electrons needed to chemisorb oxygen and create O^{2-} ions at an oxide–gas interface are supplied from a narrow region at the surface which therefore suffers electron depletion. Thus a negative surface charge is associated with a layer containing positive 'space charges'. Electrical neutrality does not exist within this layer. The effect of the electric field created tends to counteract the chemisorption process which therefore becomes slower. Ions move across this layer under the action of a potential gradient rather than by diffusion under a concentration gradient. Similarly a negative space charge layer may be considered at a metal–oxide interface. Between the layers an electrically neutral 'Wagner zone' exists across which diffusion of defects can occur.

At high temperatures and relatively thick scales the effect of the electric field transport regions may be ignored and the Wagner diffusion mechanism discussed previously is rate-controlling. For thin films of less than 1 μm at intermediate or lower temperatures the retardation effects of space charge layers (hundreds of nm wide) on ion movement must be considered to explain logarithmic growth. For very thin films at room temperature (up to ~10 nm) the effects of space charges are considered to be negligible compared with surface charges. The voltage measured across the oxide is about 1 V. Therefore for an oxide thickness of 1 nm the field strength is 10^9 V m^{-1} which is strong enough to pull ions across the oxide. As the oxide thickness increases the field strength decreases and it would be reasonable to predict a reduction in oxidation rate. In practice growth of air-formed films at room temperature follows a logarithmic law and a maximum thickness of 1 to 5 nm is achieved after about four weeks.

9.2.6 Scales on multivalent metals

So far an oxide on the surface of a pure metal has been considered to be of one composition. However, for metals which show more than one valency state, notably the transition metals, overall oxidation rate is difficult to predict because oxides of different compositions are formed in layers on the metal surface. A typical example is shown by iron above 600 °C (*Fig. 9.42*). The situation shown is an ideal state and in fact some oxygen can diffuse through the p-type FeO and Fe_3O_4 layers. Below 570 °C FeO is unstable and is virtually absent from scales. Between 200 and 570 °C, the parabolic region, oxygen diffuses through the Fe_2O_3 layer and to some extent through the Fe_3O_4 layer and may even penetrate the metal surface. Below 200 °C the oxidation rate changes from parabolic to logarithmic. A thin film of Fe_3O_4 alone is now present and the mechanism outlined in Section 9.2.5 applies. Other examples of multilayer scales are TiO, Ti_2O_3 and TiO_2 on titanium and MnO, Mn_3O_4 and Mn_2O_3 on manganese.

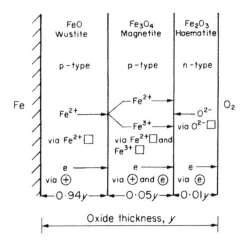

Oxide thickness, *y*

Figure 9.42 Oxide scale on iron
above 600 °C

9.2.7 Oxidation of alloys

There are two possibilities to be considered when a single-phase binary alloy is oxidised.
Either one component may be preferentially oxidised to produce a single oxide film
or both components may be oxidised to produce relatively complex oxide arrangements. The first possibility includes doping which has been discussed in Section 9.2.4.
Complications can arise however if the solute is volatile. In this case alloying solute
atoms may diffuse through the oxide to the oxide–gas interface and evaporate. Since
doping involves small percentages of alloying elements selective oxidation may be
placed in this category. Thus if an alloy contains a less noble constituent with a high
affinity for oxygen it is possible to oxidise only that one constituent. This may be
achieved by oxidising at a partial pressure of oxygen which is less than the dissociation
pressure of the more noble alloying element. For example, at a suitably low pressure
at about 800 °C a protective, pure alumina (Al_2O_3) film may be produced on Cu–Al
alloys; copper oxide being unable to form at this pressure. The selective oxidation
principle is used in the production of tarnish-resistant silver (92.5% Ag, 6.5% Cu,
1% Al). The main disadvantage of the process is that the protective oxide film is not
self-healing in any atmosphere other than low oxygen pressure. Therefore, once the
film is damaged in atmospheric use normal oxidation conditions apply and the break
in the film is filled by non-protective oxide. The second possibility is the most common
and applies to most alloys. A number of alternatives are possible within this category.
Suppose a binary alloy consists of solvent A and solute B. On oxidation, AO and BO
may form simultaneously (at different rates). Thus AO is likely to be the major component and BO could be present as discrete particles. Consequently further reaction
at BO and AO interfaces is possible to produce a compound oxide of A and B. The
most common compound oxides are the spinels. If A were divalent and B trivalent
the formula of the spinel would be $AO.B_2O_3$. Spinels also show p- or n-type behaviour;
like simple oxides the fewer the defects the more protective the spinel. $MgO.Al_2O_3$
on aluminium magnesium alloys is an example of a protective n-type spinel. Fe_3O_4 is
in fact a p-type spinel, $FeO.Fe_2O_3$, and during the oxidation of iron Fe^{2+} and
Fe^{3+} cations diffuse via tetrahedral and octahedral cation vacancies in the spinel.
A further alternative concerning the presence of AO and BO together is that BO could
form a continuous protective oxide layer at the metal–oxide interface. BO could then
reduce or prevent diffusion of cations outwards and O^{2-} anion diffusion inwards. The
scale on Cu–10% Al below 800 °C comprises Cu_2O with islands of Al_2O_3. Above
800 °C faster diffusion rates of Cu^+ and Al^{3+} ions permit rapid formation of Cu_2O

and, eventually, of a continuous layer of Al_2O_3 between the metal and Cu_2O. Once formed, the Al_2O_3 layer prevents outward diffusion of Cu^+ ions. A similar effect is achieved by alloying aluminium with silicon or iron with chromium.

Small additions of chromium increase the oxidation rate of nickel up to a maximum at about 5% chromium (Wagner-Hauffe theory, Section 9.2.4). Above this composition the oxidation rate falls because chromium forms particles of protective Cr_2O_3 and $NiO.Cr_2O_3$ spinel rather than enter the NiO lattice. At about 20% chromium (Nichrome) the oxidation rate of the alloy decreases to below that of pure nickel. At lower temperatures (800 °C), thin films (100 nm) consist mainly of Cr_2O_3 with only a trace amount of spinel whereas thin films at higher temperatures (1000 °C) consist mainly of spinel. Scales of about 10 μm at higher temperatures contain both Cr_2O_3 and $NiO.Cr_2O_3$. When chromium is added to iron, chromium oxidises at the alloy–oxide interface. Cr_2O_3 has little solubility in FeO and, therefore, islands of spinel $FeO.Cr_2O_3$ form in a matrix of FeO. At long times chromium can diffuse into the other oxides of iron present to form complex iron–chromium spinels. When the chromium content is sufficiently high a virtually continuous layer of Cr_2O_3 forms at the alloy–oxide interface. At temperatures below 1000 °C protection is achieved by adding more than 12% chromium while above 1000 °C more than 17% chromium is required. An important feature of the Cr_2O_3 layer is that it is self-healing.

Under suitable conditions internal oxidation of alloys can occur by oxygen diffusing through the oxide to the alloy faster than solute atoms can diffuse in the other direction. The solute is oxidised at the alloy–oxide interface to produce a layer of oxide particles, a subscale, below the scale on the alloy surface. Clearly, internal oxidation is favoured by a surface oxide which is permeable to oxygen, high oxygen solubility of the alloy, slow diffusion of the solute element and low concentration of solute in the alloy. For example, copper–10% aluminium forms an external scale but when the aluminium content is less than 0.1% there is insufficient aluminium to form a continuous Al_2O_3 layer at the alloy–oxide interface. Internal oxidation then takes place since oxygen ingress is facilitated. Subscale formation is also observed on copper–0.1% silicon alloys and in iron alloys containing small amounts of silicon, aluminium and chromium at high temperatures. Internal oxidation does not always involve the formation of a subscale. Penetration of oxygen along grain boundaries may lead to oxide formation at these sites. Other energetically favourable sites for oxide formation may be provided on suitable crystallographic planes or at points within grains. Cast iron shows a rather different form of internal oxidation at about 650 °C when a phase change from α to γ produces slight expansion of the iron matrix compared with the graphite flakes. Under oxidising conditions oxygen penetrates along the graphite flakes to produce CO and residual O_2. On cooling a volume change is observed and the effect is called *growth* of cast iron. Growth may be overcome by alloying with nickel (to stabilise the γ phase), chromium (to improve oxidation resistance) or silicon (to raise the α to γ transition temperature) or reducing the size of the carbon particles to prevent a network of graphite being present. In general, internal oxidation produces deleterious effects but it may be used to produce oxide dispersions in, for example, nickel–aluminium alloys (Al_2O_3 formation) with a view to enhancing high temperature creep strength.

Alloys of high molybdenum content show extremely rapid oxidation rates under certain conditions. This rapid oxidation is called catastrophic oxidation or *breakaway* (*Fig. 9.40(b)*). For example, iron-20% molybdenum at 1000 °C is not oxidised very rapidly in air but when alloyed with nickel or chromium the oxidation rate increases dramatically. The exact reason for this behaviour has not been established but it is known that it takes place at the metal–oxide interface. Catastrophic oxidation is also strongly associated with the formation of a partly liquid oxide phase. The formation of this phase may be assisted by the low melting point of MoO_3 (795 °C) since eutectic combination of oxides would reduce the melting point further. This is supported by the fact that copper–8% aluminium oxidises catastrophically in an

MoO_3 atmosphere. Contact with V_2O_5 (melting point 675 °C) also produces catastrophic oxidation in some heat-resisting alloys. Even when present in small quantities, such as in fuel ash containing more than 1% V_2O_5, the oxidation rate of steels is increased considerably by the presence of V_2O_5.

9.2.8   Oxidising atmospheres

The most common oxidising atmosphere is, of course, air. Air contains 80% nitrogen which, although generally unimportant, may affect the oxidation rate of metals, such as chromium, aluminium and titanium, which dissolve nitrogen and form stable nitrides. Contamination of air by small amounts of water vapour and oxides of nitrogen and sulphur exert a noticeable, but unpredictable, effect on oxidation. Yet another factor is the velocity of air. An increase in flow rate (wind speed) produces an increased oxidation rate (compare corrosion in static and running water, Section 9.1.3.2, erosion-corrosion). The products of combustion of fuels form a variety of gas mixtures which significantly affect oxidation rates of many metals. Owing to the organic nature of nearly all fuels complete combustion produces CO_2, water vapour and residual nitrogen. Incomplete combustion results in a more complex mixture including oxygen and residual fuel hydrocarbons with reducing gases such as hydrogen and CO. In addition SO_2, SO_3 and H_2S are often produced. Clearly the situation is complicated. The composition of the atmosphere may vary. The presence of reducing gases may alter the composition of a scale and, therefore, its protective nature. In general oxidation rates are increased by any contaminating constituent but particularly where flue gases are present. In these conditions the protective, continuous film produced in air is not formed, instead the normally protective oxide forms islands, resulting in linear growth rates.

Obviously, reactions other than oxide formation can take place. For example, nickel-based alloys in the presence of sulphur compounds are considerably weakened by the formation of intergranular, low melting point (645 °C) sulphide eutectics. On the other hand sulphide scales are formed on the surfaces of many metals. The protective nature of such scales is dependent on their defect structure but even so most sulphide scales are more defective than oxide scales.

Where hydrocarbons are present carburisation of the metal is possible. Under these conditions nickel–chromium and nickel–iron–chromium alloys at 1000 °C undergo a severe embrittling form of attack by a combination of carburisation and oxidation. Carbon in the atmosphere diffuses into the alloy to cause depletion of chromium by formation of chromium carbide. Hence, less chromium is available to maintain the protective Cr_2O_3 layer which, therefore, becomes discontinuous and non-protective. When mechanically tested the fracture surfaces of these alloys have a green colour and the condition is called *green-rot*. The presence of hydrogen rarely produces surface films since hydrides tend to be unstable. Instead the gas is dissolved because it is very soluble in many metals. It may react in a number of ways to reduce mechanical properties. Hydrogen may diffuse into the metal in atomic form then recombine at suitable sites to form molecular hydrogen thereby producing hydrogen embrittlement. It may also embrittle by reducing oxides, for example intergranular oxides in tough pitch copper (see section 9.1.3.2, hydrogen damage).

One product of combustion not mentioned so far is ash. The composition of ash is complex and it may contain all or some of the following depending on the fuel: Al_2O_3, CaO, Cl_2, Fe_2O_3, MgO, Na_2O, P_2O_5, SiO_2, TiO_2, V_2O_5. The deposition of ash on a metal may remove any protective oxides initially present by acting as a flux. If the temperature is such that the ash melts improved metal–ash contact is made and rapid oxidation can occur. The ingredient, V_2O_5, is particularly important in this respect (see Section 9.2.7, catastrophic oxidation). However, an ash need not necessarily be harmful since it may act as a surface barrier between the alloy and harmful constituents of the atmosphere. On the other hand, if such an ash were porous, it could permit

selective diffusion of harmful gases through it thereby increasing the oxidation rate.
Where high velocity gases are involved, e.g. in gas turbines, the major problem caused
by ash is erosion.

9.2.9 Oxidation protection

Methods of protection from 'dry' corrosion are concerned principally with applying a
suitable coating or modifying the metal or the environment. The major requirement
of a protective oxide is that it should be adherent and highly resistant to diffusion of
ions and electrons, i.e. in effect an oxide should be a good electrical insulator. Some
metals, such as aluminium, form oxides of this type. Alloying elements may be added
in very small quantities to reduce the growth rate of the oxide layer (Wagner–Hauffe
theory, Section 9.2.4). Often larger proportions of elements are added which have a
greater affinity for oxygen than the base metal. For this reason aluminium is added to
copper and iron. Additions of noble elements do not have the same advantageous
effects here as they sometimes have in aqueous corrosion (Section 9.1.5.8). Unless the
alloy contains more than 50% noble element the oxidation rate remains virtually
unchanged. In some cases alloying produces a protective mixed oxide, a spinel of
general formula $MO.M_2'O_3$ where M is a divalent metal and M' is a trivalent metal.
Examples of spinels formed by alloying occur in aluminium–magnesium, nickel–
chromium and iron–chromium alloys and many others. More details on alloying
effects are given in Section 9.2.7. It is worth noting here that protection by oxides
and spinels is not confined to oxidation alone but is an integral part of protection
from aqueous corrosion.

In the main the alloys considered so far have only had two components. However,
ternary and quaternary alloys have been developed, the extra alloying additions having
a 'synergistic' (Section 9.1.9.1) or reinforcing effect on oxidation resistance. For
example, iron–24% chromium has a maximum operating temperature of about 1000 °C,
but additions of 5.5% Al and 2% Co (producing Kanthal) increase this temperature to
1300 °C. Other examples of combining the oxidation resistance of a number of elements are given by Alumel (see below) and stainless steels, for example, Fe–18% Cr,
8% Ni, 2% Si.

Once again, as in Section 9.1.10.5, there are factors to be considered other than
oxidation, e.g. mechanical strength, temperature, spalling, ductility, ease of fabrication, cost, etc. For example, where iron–chromium alloys are used under thermal
cycling conditions there is a tendency for the oxide to spall. This effect is reduced by
addition of a rare earth, e.g. 1% yttrium. High strength at high temperatures necessitates the use of chromium–nickel steels. Iron–8% aluminium has good oxidation resistance but is brittle. Nichrome (80% nickel–20% chromium) is very resistant to oxidation but difficult to fabricate. Inconel (nickel–16% chromium–7% iron) is easier than
Nichrome to work and weld but has a lower oxidation resistance than Nichrome, but
can still be used in air up to 1100 °C.

The accuracy of a thermocouple must not be affected by oxidation and must also
possess reasonable mechanical strength. At temperatures up to 1100 °C Chromel
(nickel–10% chromium)/Alumel (nickel–2% aluminium–2% manganese–1% silicon)
thermocouples satisfy these requirements. For higher temperatures (but with some loss
of sensitivity) platinum/platinum–10% rhodium thermocouples are used. For critical
applications at high temperatures the cost aspect becomes relatively less important.
Protection is also obtained by coating with metals or ceramic materials such as carbides, nitrides or oxides. The main properties of a coating are that it should be adherent
and capable of withstanding thermal shock. The methods available for applying metallic
coatings are essentially the same as those outlined in Section 9.1.0.4. Cathodic deposition (from molten electrolytes), displacement (electroless) reactions using molten
salts, spraying, vapour phase deposition, sintering, cladding, dipping and diffusion from
powders have all been used. Ceramic coatings may be applied by spraying, sintering
and dipping. It may be noted that most metals protect by formation of an oxide which

may itself be classed as a ceramic. In general ceramic coatings, which include Cr_2O_3 and Al_2O_3, need to be thin otherwise there is a tendency to crack and spall.

Further reading

Birks, N. and Meier, G.H. 1983 *Introduction to High Temperature Oxidation of Metals*. London: Edward Arnold.

Committee for Industrial Technologies *Controlling Corrosion*, Vols 1–6 and *Guides to Practice in Corrosion Control*, Vols 1–16. London: Dept of Industry and Central Office of Information (Available from National Corrosion Service (NPL) England).

Department of Trade and Industry 1971 *Report of the Committee on Corrosion and Protection* (Hoar Report). London: HMSO.

Evans, U.R. 1981 *Introduction to Metallic Corrosion*. London: Edward Arnold.

Fontana, M.G. and Greene, N.D. 1978 *Corrosion Engineering*. New York: McGraw-Hill.

Gray, A.G. 1978 What corrosion really costs, *Metal Prog*. June, 29.

Hauffe, K. 1965 *Oxidation of Metals*. New York: Plenum Press.

Inst. Metall. Refresher Course Proc.: Corrosion and Protection of Metals, April, 1974. London: Iliffe Books.

Jones, D.A. 1971 The application of electrode kinetics to the theory and practice of cathodic protection, *Corr. Sci.* **2**, No. 6, 439–51.

Kubaschewski, O. and Hopkins, B.E. 1962 *Oxidation of Metals and Alloys*. London: Butterworths.

National Association of Corrosion Engineers 1980 Basic Corrosion Course. Houston: NACE.

Parker, R.H. 1977 *An Introduction to Chemical Metallurgy*. Oxford: Pergamon Press.

Parkins, R.N. 1982 *Corrosion Processes*. London: Elsevier Applied Science Publishers.

Pludek, V.R. 1977 *Design and Corrosion Control*. Basingstoke: Macmillan.

Pourbaix, M. 1966 *Atlas of Electrochemical Equilibria in Aqueous Solutions*. Oxford: Pergamon Press Cebelcor.

Riggs, O.L. Jr and Locke, C.E. 1981 *Anodic Protection*. New York: Plenum Press.

Scully, J.C. 1975 *The Fundamentals of Corrosion*. Oxford: Pergamon Press.

Shreir, L.L. (ed) 1976 *Corrosion*, Vols 1 and 2, 2nd edn. London: Newnes-Butterworths.

Uhlig, H.H. 1985 *Corrosion and Corrosion Control*. New York: Wiley.

Walker, R. and Ward, A. 1969 The theory and practice of anodic protection, *Metall. Rev. No. 137 Metals & Mat.*, 143–51.

Waterhouse, R.B. 1971 *Fretting Corrosion*. Oxford: Pergamon Press.

West, J.M. 1986 *Basic Corrosion and Oxidation*. Chichester: Ellis Horwood.

For self-test questions on this chapter, see Appendix 5.

Appendix 1

Electron Configuration of the Elements

Z	Symbol		Electron configuration		
1	H			$1s^1$	
2	He			$1s^2$	
3	Li	He		$2s^1$	
4	Be	He		$2s^2$	
5	B	He		$2s^2$	$2p^1$
6	C	He		$2s^2$	$2p^2$
7	N	He		$2s^2$	$2p^3$
8	O	He		$2s^2$	$2p^4$
9	F	He		$2s^2$	$2p^5$
10	Ne	He		$2s^2$	$2p^6$
11	Na	Ne		$3s^1$	
12	Mg	Ne		$3s^2$	
13	Al	Ne		$3s^2$	$3p^1$
14	Si	Ne		$3s^2$	$3p^2$
15	P	Ne		$3s^2$	$3p^3$
16	S	Ne		$3s^2$	$3p^4$
17	Cl	Ne		$3s^2$	$3p^5$
18	Ar	Ne		$3s^2$	$3p^6$
19	K	Ar		$4s^1$	
20	Ca	Ar		$4s^2$	
21	Sc	Ar	$3d^1$	$4s^2$	
22	Ti	Ar	$3d^2$	$4s^2$	
23	V	Ar	$3d^3$	$4s^2$	
24	Cr	Ar	$3d^5$	$4s^1$	
25	Mn	Ar	$3d^5$	$4s^2$	
26	Fe	Ar	$3d^6$	$4s^2$	
27	Co	Ar	$3d^7$	$4s^2$	
28	Ni	Ar	$3d^8$	$4s^2$	
29	Cu	Ar	$3d^{10}$	$4s^1$	
30	Zn	Ar	$3d^{10}$	$4s^2$	

Z	Symbol		Electron configuration			
31	Ga	Ar		$3d^{10}$	$4s^2$	$4p^1$
32	Ge	Ar		$3d^{10}$	$4s^2$	$4p^2$
33	As	Ar		$3d^{10}$	$4s^2$	$4p^3$
34	Se	Ar		$3d^{10}$	$4s^2$	$4p^4$
35	Br	Ar		$3d^{10}$	$4s^2$	$4p^5$
36	Kr	Ar		$3d^{10}$	$4s^2$	$4p^6$
37	Rb	Kr			$5s^1$	
38	Sr	Kr			$5s^2$	
39	Y	Kr		$4d^1$	$5s^2$	
40	Zr	Kr		$4d^2$	$5s^2$	
41	Nb	Kr		$4d^4$	$5s^1$	
42	Mo	Kr		$4d^5$	$5s^1$	
43	Tc	Kr		$4d^6$	$5s^1$	
44	Ru	Kr		$4d^7$	$5s^1$	
45	Rh	Kr		$4d^8$	$5s^1$	
46	Pd	Kr		$4d^{10}$	$5s^0$	
47	Ag	Kr		$4d^{10}$	$5s^1$	
48	Cd	Kr		$4d^{10}$	$5s^2$	
49	In	Kr		$4d^{10}$	$5s^2$	$5p^1$
50	Sn	Kr		$4d^{10}$	$5s^2$	$5p^2$
51	Sb	Kr		$4d^{10}$	$5s^2$	$5p^3$
52	Te	Kr		$4d^{10}$	$5s^2$	$5p^4$
53	I	Kr		$4d^{10}$	$5s^2$	$5p^5$
54	Xe	Kr		$4d^{10}$	$5s^2$	$5p^6$
55	Cs	Xe			$6s^1$	
56	Ba	Xe			$6s^2$	
57	La	Xe		$5d^1$	$6s^2$	
58	Ce	Xe	$4f^2$		$6s^2$	
59	Pr	Xe	$4f^3$		$6s^2$	
60	Nd	Xe	$4f^4$		$6s^2$	
61	Pm	Xe	$4f^5$		$6s^2$	
62	Sm	Xe	$4f^6$		$6s^2$	
63	Eu	Xe	$4f^7$		$6s^2$	
64	Gd	Xe	$4f^7$	$5d^1$	$6s^2$	
65	Tb	Xe	$4f^9$		$6s^2$	
66	Dy	Xe	$4f^{10}$		$6s^2$	
67	Ho	Xe	$4f^{11}$		$6s^2$	
68	Er	Xe	$4f^{12}$		$6s^2$	
69	Tm	Xe	$4f^{13}$		$6s^2$	
70	Yb	Xe	$4f^{14}$		$6s^2$	
71	Lu	Xe	$4f^{14}$	$5d^1$	$6s^2$	
72	Hf	Xe	$4f^{14}$	$5d^2$	$6s^2$	
73	Ta	Xe	$4f^{14}$	$5d^3$	$6s^2$	
74	W	Xe	$4f^{14}$	$5d^4$	$6s^2$	
75	Re	Xe	$4f^{14}$	$5d^5$	$6s^2$	
76	Os	Xe	$4f^{14}$	$5d^6$	$6s^2$	
77	Ir	Xe	$4f^{14}$	$5d^9$	$6s^0$	
78	Pt	Xe	$4f^{14}$	$5d^9$	$6s^1$	

Z	Symbol		Electron configuration			
79	Au	Xe	$4f^{14}$	$5d^{10}$	$6s^1$	
80	Hg	Xe	$4f^{14}$	$5d^{10}$	$6s^2$	
81	Tl	Xe	$4f^{14}$	$5d^{10}$	$6s^2$	$6p^1$
82	Pb	Xe	$4f^{14}$	$5d^{10}$	$6s^2$	$6p^2$
83	Bi	Xe	$4f^{14}$	$5d^{10}$	$6s^2$	$6p^3$
84	Po	Xe	$4f^{14}$	$5d^{10}$	$6s^2$	$6p^4$
85	At	Xe	$4f^{14}$	$5d^{10}$	$6s^2$	$6p^5$
86	Rn	Xe	$4f^{14}$	$5d^{10}$	$6s^2$	$6p^6$
87	Fr	Rn			$7s^1$	
88	Ra	Rn			$7s^2$	
89	Ac	Rn		$6d^{1\,1}$	$7s^2$	
90	Th	Rn		$6d^2$	$7s^2$	
91	Pa	Rn	$5f^2$	$6d^1$	$7s^2$	
92	U	Rn	$5f^3$	$6d^1$	$7s^2$	
93	Np	Rn	$5f^4$	$6d^1$	$7s^2$	
94	Pu	Rn	$5f^6$		$7s^2$	
95	Am	Rn	$5f^7$		$7s^2$	
96	Cm	Rn	$5f^7$	$6d^1$	$7s^2$	
97	Bk	Rn	$5f^8$	$6d^1$	$7s^2$	
98	Cf	Rn	$5f^{10}$		$7s^2$	
99	Es	Rn	$5f^{11}$		$7s^2$	
100	Fm	Rn	$5f^{12}$		$7s^2$	
101	Md	Rn	$5f^{13}$		$7s^2$	
102	No	Rn	$5f^{14}$		$7s^2$	
103	Lw	Rn	$5f^{14}$	$6d^1$	$7s^2$	

Appendix 2

The Electrochemical (Redox) Series (Standard electrode (reduction) potentials at 25 °C)

(a) **Aqueous acid solutions**

Electrode	reaction	E^0 (V)
$Li^+ + e \rightleftharpoons Li$		-3.05
$K^+ + e \rightleftharpoons K$		-2.93
$Cs^+ + e \rightleftharpoons Cs$		-2.92
$Ca^{2+} + 2e \rightleftharpoons Ca$		-2.87
$Na^+ + e \rightleftharpoons Na$		-2.71
$Mg^{2+} + 2e \rightleftharpoons Mg$		-2.37
$Al^{3+} + 3e \rightleftharpoons Al$		-1.66
$Ti^{2+} + 2e \rightleftharpoons Ti$		-1.63
$Zn^{2+} + 2e \rightleftharpoons Zn$		-0.76
$Cr^{3+} + 3e \rightleftharpoons Cr$		-0.74
$Fe^{2+} + 2e \rightleftharpoons Fe$		-0.44
$Cd^{2+} + 2e \rightleftharpoons Cd$		-0.40
$Ni^{2+} + 2e \rightleftharpoons Ni$		-0.25
$Sn^{2+} + 2e \rightleftharpoons Sn$		-0.14
$Pb^{2+} + 2e \rightleftharpoons Pb$		-0.13
$2H^+ + 2e \rightleftharpoons H_2$		0.00
$Cu^{2+} + e \rightleftharpoons Cu^+$		$+0.15$
$Cu^{2+} + 2e \rightleftharpoons Cu$		$+0.34$
$\frac{1}{2}O_2 + H_2O + 2e \rightleftharpoons 2OH$		$+0.40$
$Fe^{3+} + e \rightleftharpoons Fe^{2+}$		$+0.77$
$Ag^+ + e \rightleftharpoons Ag$		$+0.80$
$Hg_2^{2+} + 2e \rightleftharpoons 2Hg$		$+0.85$
$NO_3^- + 3H^+ + 2e \rightleftharpoons HNO_2 + H_2O$		$+0.94$
$Pd^{2+} + 2e \rightleftharpoons Pd$		$+0.99$
$Pt^{2+} + 2e \rightleftharpoons Pt$		$+1.20$ (approx.)
$\frac{1}{2}O_2 + 4H^+ + 4e \rightleftharpoons H_2O$		$+1.23$
$Cl_2 + 2e \rightleftharpoons 2Cl^-$		$+1.36$
$Au^{3+} + 3e \rightleftharpoons Au$		$+1.50$
$Co^{3+} + e \rightleftharpoons Co^{2+}$		$+1.82$
$S_2O_8^{2-} + 2e \rightleftharpoons 2SO_4^{2-}$		$+2.01$

(b) Aqueous basic solutions

Electrode	reaction	$E^{\ominus}$ (V)
$Mg(OH)_2 + 2e \rightleftharpoons Mg + 2OH^-$		-2.69
$Cr(OH)_3 + 3e \rightleftharpoons Cr + 3OH^-$		$-1:30$
$Zn(OH)_2 + 2e \rightleftharpoons Zn + 2OH^-$		-1.25
$ZnO_2^{2-} + 2H_2O + 2e \rightleftharpoons Zn + 4OH^-$		-1.22
$Fe(OH)_2 + 2e \rightleftharpoons Fe + 2OH^-$		-0.88
$H_2O + e \rightleftharpoons \frac{1}{2}H_2 + OH^-$		-0.83
$Ni(OH)_2 + 2e \rightleftharpoons Ni + 2OH^-$		-0.72
$S + 2e \rightleftharpoons S^{2-}$		-0.48
$O_2 + 2H_2O + 4e \rightleftharpoons 4OH^-$		$+0.40$

Appendix 3

Electrode Polarity and Ion Discharge in Electrochemical and Electrolytic Cells

Electrode polarity

An explanation of electrode polarity reversal of electrochemical cells compared with electrolytic cells is somewhat complex and is explained in some textbooks by the fact that electrochemical cells produce an electrical potential from a chemical reaction while electrolytic cells produce a chemical reaction from an externally applied electrical potential. The following more detailed discussion is offered in an effort to explain this phenomenon.

Consider the electrolytic and electrochemical cells shown in *Fig. A3.1.* The reactions in part (a) are:

At the anode	$Cu \longrightarrow Cu^{2+} + 2e$	(A3.I)
At the cathode	$2H^+ + 2e \longrightarrow H_2$	(A3.II)

The reactions in part (b) are:

At the anode	$Fe \longrightarrow Fe^{2+} + 2e$	(A3.III)
At the cathode	$\frac{1}{2}O_2 + H_2O + 2e \longrightarrow 2OH^-$	(A3.IV)

From these reactions it may be seen that the anodic and cathodic reactions are of the

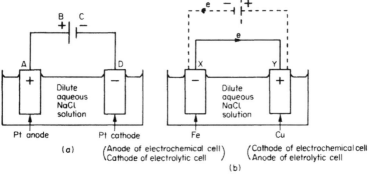

Figure A3.1 A comparison of anodic and cathodic polarity in: (a) an electrolytic cell and (b) an electrochemical cell. The latter also shows the change in electrode nature when the connecting wire (full line) is replaced by a circuit including a battery (broken line). Note that the polarity of each electrode remains the same and electrons still flow from anode to cathode

same type in each case, i.e. the anodic reaction is oxidation (production of electrons) and the cathodic reaction is reduction (electron consumption). The direction of electron flow is from anode to cathode in each case. However, the sign of electrode polarity in each case is reversed. This polarity reversal is further exemplified by considering the inclusion of a battery in the electrochemical circuit of *Fig. A3.1(b)*. With the positive terminal of the battery connected to the copper electrode and the negative terminal to the iron, the polarity, according to part (b), remains the same, but the nature of the electrode changes. Copper becomes anodic and dissolves according to reaction (A3.I), and hydrogen is produced (reaction (A3.II)) on iron which is therefore cathodic.

The reason for this apparent reversal of polarity lies in the source of the reaction. In the electrochemical cell the driving force is dissolution of iron ($E^{\ominus} = -0.44$ V)* which is more readily achieved than dissolution of copper ($E^{\ominus} = +0.34$ V)*. Clearly, the iron electrode must have a more negative potential at X than the copper electrode at Y for electrons to flow in the direction indicated. In the electrolytic cell the driving force is the electrochemical reaction at C. Hence the polarity of D is *positive* with respect to C and the polarity of A is *negative* with respect to B.

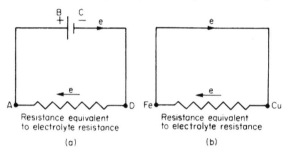

Figure A3.2 Comparison of equivalent circuits for: (a) an electrolytic cell and (b) an electrochemical cell

However, if the reactions (A3.I) and (A3.II) are to proceed, the polarity of D must be negative with respect to A across the solution in order to attract the necessary ions. Consideration of an equivalent circuit *Fig. A3.2(a)* shows this to be the case.

Similarly, in the electrochemical cell (*Fig. A3.2(b)*), the copper electrode is in fact more negative than the iron electrode across the electrolyte. Therefore it would be expected that positively charged ions should be attracted to the copper electrode rather than to the iron electrode. This requirement is supported by the fact that hydrogen forms at the copper electrode in a so-called simple cell comprising a zinc–copper couple in dilute sulphuric acid. In addition, in the iron–copper cell shown, rust (basically $Fe(OH)_2$) is precipitated between the two electrodes. This observation indicates diffusion of Fe^{2+} ions from the iron electrode to the copper electrode and diffusion of OH^- ions in the opposite direction. In the situation given by reaction (A3.IV) it is possible that water dipoles are attracted to the copper surface where a reaction occurs with the surface oxide (i.e. oxygen) present.

The idea that a cathode in an electrochemical cell is more negative than the anode across the solution finds support from *absolute electrode potential* considerations (Section 6.3.5). Thus, if

$$E_M = \phi_{H_2} - \phi_M$$

for E_M to be positive (e.g. copper), the absolute electrode potential ϕ_M must be more

---

*It must be remembered that $E^{\ominus}$ values relate to equilibrium conditions between an electrode and its ions in solution. Equilibrium does not initially exist in *Fig. A3.1(b)* and $E^{\ominus}$ values are used here to indicate relative dissolution tendencies of the two electrode materials.

negative or less positive than ϕ_{H_2}. For E_M to be negative ϕ_M must be less negative (more positive) than ϕ_{H_2}. Therefore the cathode of an electrochemical cell will be more negative or less positive than the anode in terms of absolute electrode potentials. This situation may be represented by *Fig. A3.3*.

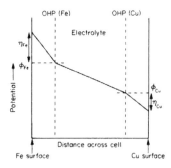

Figure A3.3 Schematic representation of potential gradients in an electrochemical cell

This concept of absolute electrode potentials upsets the principle of drawing $E-I$ or $E-\log i$ curves for corrosion reactions but only insofar as the polarities are reversed. The overall E_{corr} and I_{corr} (i_{corr}) values are unchanged. In any case ϕ values are unobtainable. Another important factor about considering electrode potentials at all is that an initial equilibrium state between electrode atoms and ions of the same material in solution is assumed. This is rarely the case in corrosion reactions but is true for cells such as the Daniell cell.

Ion discharge

A topic which has been implied in the electrolysis reactions quoted is the preferential discharge of ions. In dilute aqueous sodium chloride there are four species of ions derived from the ionisation reactions,

$$NaCl \rightleftharpoons Na^+ + Cl^- \quad \text{and} \quad H_2O \rightleftharpoons H^+ + OH^-$$

Cl^- and OH^- ions, being negatively charged, are attracted to the (positive) anode. It should be noted that there is not an immediate 'rush' of anions through the electrolyte to the anode. Ions move by a 'drifting' process as a result of the potential gradient created across the bulk solution. Once ions have reached the outer Helmholtz plane (OHP) automatic transfer across to the metal surface does not necessarily occur. The potential across the double layer (due to polarisation) must be high enough to create conditions which are energetically favourable for increasing the probability of ion transfer.

The relevant standard electrode potentials (IUPAC) or discharge potentials, are:

$$H_2O + \tfrac{1}{2}O_2 + 2e \longrightarrow 2OH^- \qquad E^\ominus = +0.40 \text{ V} \qquad (A3.1)$$

$$Cl_2 + 2e \longrightarrow 2Cl^- \qquad E^\ominus = +1.36 \text{ V} \qquad (A3.2)$$

Considering equation (A3.1), for a neutral solution, activity of OH^- ions is 10^{-7}. Therefore, by the Nernst equation, assuming $p_{O_2} = 1$ atm:

$$E = 0.4 + \frac{0.059}{2} \log_{10}\left(\frac{1}{(10^{-7})^2}\right)$$

$$= +0.81 \text{ V}$$

This is the theoretical (reversible) discharge potential. In practice a potential more positive than this would be required in order to overcome polarisation effects. However, little error will be introduced here by ignoring polarisation effects and so only reversible potentials will be considered.

The reaction which takes place preferentially at an electrode during electrolysis will be that which has the most negative ΔG value, i.e. the most positive electrode potential for that specific reaction ($\Delta G = -nFE$). For example, to evolve Cl_2 gas at the anode, the *anodic* potential is -1.36 V under standard thermodynamic conditions, i.e. the *reverse* of equation (A3.2), and the potential *applied* to the anode must be in excess of 1.36 V. Similarly to evolve O_2 at the anode a potential in excess of 0.81 V must be *applied* to the anode ($a_{OH^-} = 10^{-7}$, $p_{O_2} = 1$ atm). Therefore, when a potential is applied and the anode is made progressively more positive, it would be expected that oxygen evolution would occur first. This is achieved when inert electrodes are considered. In this case increasing the operating anode potential further, to a value greater than the reversible potential of chlorine, produces simultaneous evolution of oxygen and chlorine. At higher concentrations of sodium chloride (i.e. Cl^- ions) chlorine is evolved before oxygen. Extending this argument, if SO_4^{2-} ions were also present, then because

$$S_2O_8^{2-} + 2e \longrightarrow 2SO_4^{2-} \qquad E^{\ominus} = +2.01 \text{ V}$$

SO_4^{2-} ions have a much lower tendency to be oxidised at an anode than chloride or hydroxyl ions.

When an anode such as copper is involved another reaction becomes possible:

$$Cu^{2+} + 2e \longrightarrow Cu \qquad E^{\ominus} = +0.34 \text{ V}$$

Applying the Nernst equation:

$$E = 0.34 + \frac{0.059}{2} \log_{10} \left(\frac{a_{Cu^{2+}}}{1} \right)$$

Initially, in *Fig. A3.1(a)* the concentration of copper ions in solution is virtually zero and E assumes an infinitely large, negative value ($a_{Cu^{2+}} = 10^{-x}$, where x is very large). Clearly, application of a potential to the anode strongly favours copper dissolution above any other reaction because the discharge potential for copper is far more negative than the discharge potential for oxygen or chlorine evolution.

Therefore, the simple conclusion to be drawn from this discussion so far is that the electrode reaction which has the most *negative* (or least positive) reversible potential (IUPAC) is likely to take place at an anode.

A similar reasoning can be used to determine the reaction which takes place preferentially at the cathode. The cathode attracts Na^+ and H^+ (strictly, hydrated H_3O^+) ions:

$$Na^+ + e \longrightarrow Na \qquad E^{\ominus} = -2.71 \text{ V} \qquad \text{(A3.3)}$$

$$2H^+ + 2e \longrightarrow H_2 \qquad E^{\ominus} = 0.00 \text{ V} \qquad \text{(A3.4)}$$

Applying the Nernst equation to equation (A3.4):

$$E = 0.00 + \frac{0.059}{2} \log_{10} a_{H^+}^2$$

Assuming in neutral solution $a_{H^+} = 10^{-7}$ then,

$$E = -0.41 \text{ V}$$

This is the reversible or *discharge* potential required to produce hydrogen at 1 atm pressure. Therefore, as the cathode becomes more negative the first process to occur will be hydrogen evolution. In theory, increasing the cathodic potential to a sufficiently negative value should produce sodium deposition also. In practice, however, sodium cannot be deposited from aqueous solution because the overpotential is infinitely high. Therefore, in general, the electrode reaction which has the *most positive* (least negative) discharge potential is likely to take place at the *cathode*.

It should be noted that in an electrolytic cell which uses *inert* electrodes the anode becomes in effect an oxygen electrode, and the cathode a hydrogen electrode. Thus, from electrochemical cell considerations, a cell is set up, the emf of which opposes the applied voltage since the oxygen electrode is cathodic to the hydrogen electrode. Therefore, for small applied currents, the anodic and cathodic reactions continue until the emf of the galvanic cell is equal and opposite to the applied potential. Thus the voltage–current curve (*Fig. 6.18*) should show zero (or positive) current for all applied voltages up to the point at which the applied voltage is equal to the maximum emf of the cell (the decomposition voltage, E_D). For applied voltages greater than E_D free electrolysis occurs. From *Fig. 6.18* it is clear that a small residual current is present before E_D is reached. This is caused by diffusion of electrode products away from the electrodes thereby reducing the back emf.

Appendix 4

Galvanic Series

In sea water

When potentials are quoted in this series it should be noted that they are only valid as anodic potentials. The cathode reaction is invariably the oxygen reduction reaction on the various metals, which implies constant potential but variation in i_0 values.

          Base, active or anodic end
               Magnesium
               Zinc
               Aluminium (commercial purity)
               Cadmium
               Duralumin
               Steel
               Cast iron
               High nickel cast iron (Ni-resist)
               18/8 stainless steel (active)
               Lead–tin solders
               Lead
               Tin
               Nickel (active)
               Brasses
               Copper
               Bronzes
               Silver solder
               Nickel (passive)
               18/8 stainless steel (passive)
               Silver
               Titanium
               Graphite
               Gold
               Platinum
          Noble or cathodic end

Note. pH range of natural sea water is 8.1 to 8.3.

In soil*

> Base, active or anodic end
> > Magnesium
> > Zinc
> > Aluminium
> > Clean mild steel
> > Rusted mild steel
> > Cast iron
> > Lead
> > Mild steel (in concrete)
> > Copper, brasses and bronzes
> > High silicon cast iron
> > Carbon, coke, graphite
> Noble or cathodic end

Note. General pH range of soil is 5–8.

*After von Fraunhofer, J.A. 1974 *Concise Corrosion Science*. London: Portcullis Press.

Appendix 5

Questions

Questions on chapter 1

(1) Write the electron configuration for: (i) neon (atomic no. 10), (ii) calcium (atomic no. 20), (iii) iron (atomic no. 26), (iv) bismith (atomic no. 83).

(2) Determine whether the bond formed between the following sets of two elements is predominantly covalent or ionic (electrovalent): (i) Ba and O, (ii) Ti and C, (iii) Cs and F, (iv) Cs and I.

(3) Determine the crystallographic structures of the following compounds given their cation/anion radii ratios (r^+/r^-) in nm as: (i) MgO (0.066:0.140), (ii) SiO_2 (0.042:0.140), (iii) CsI (0.167:0.220).

(4) State how the atomic size varies: (i) across a period and (ii) down a group in the periodic table.

(5) Explain the trend in *Tables 1.5, 1.9, 1.10, 1.12* of increasing atomic number on ionisation elements in s-, d- and p-block elements.

Questions on chapter 3

Initial rate method

(1) When titanium alloys are in contact with molten aluminium at 800 °C, a reaction takes place and the titanium decreases in thickness. From the following data find the apparent order of the reaction:

Atomic % of Ti	Initial rate (cm s^{-1})
100.00	4.201×10^{-6}
95.35	3.527×10^{-6}
93.86	2.944×10^{-6}
90.14	2.500×10^{-6}
88.50	2.222×10^{-6}

First order kinetics

(2) A sample of iron containing sulphur is in contact with a slag at 1570 °C. After 17 minutes, 34.5% of the sulphur has been removed from the iron. Calculate:

(a) the percentage of sulphur remaining in the iron after 60 minutes, (b) the time required for 60% of the sulphur to be removed. The process is first-order.

(3) Using similar conditions to those described in question (2), but for a different slag at 1398 $^\circ$C, the sulphur content of the slag changed as follows with time of refining:

Time (min)	0	18	63	100	173
[S] (%)	100	82.5	26.8	13.1	2.6

Show that the process is first order and find the rate constant.

(4) Thermal decomposition of nickel carbonyl in accordance with the equation:

$$Ni(CO)_4(l) \rightarrow Ni(s) + 4CO(g)$$

where (l), (s) and (g) denote liquid, solid and gaseous states respectively, was followed by measurement of the rate of increase of carbon monoxide volume, with the following results:

Time (s)	0	25	50	75	100	150	200	250
Volume CO (cm^3)	6.3	11.6	16.3	20.2	26.1	30.4	33.5	41.5

Determine the rate constant and half-life from the appropriate first-order plot.

Second order kinetics

(5) When a sample of ϵ-iron nitride was heated in vacuo at 450 $^\circ$C the following results were obtained for denitriding:

Time (min)	[N]/(atoms per 100 Fe atoms)
20	14.3
30	9.1
50	5.4
60	4.3
80	3.2

Show that the reaction is second order and find the rate constants.

Arrhenius equation

(6) Using the Arrhenius equation for chemical reaction rates:

(a) For a reaction, $A = 1 \times 10^{11}$ s^{-1} and $E_A = 100$ kJ mol^{-1}. Calculate k at 298 K and 310 K.
(b) For a reaction, $k = 2 \times 10^8$ s^{-1} at 300 K. Calculate E_A if $A = 1 \times 10^{10}$ s^{-1}.

(7) The rate constant for the pyrolysis of trimethyl bismuth is 0.113 s^{-1} at 370 $^\circ$C and 0.525 s^{-1} at 400 $^\circ$C. Deduce E_A, A and the rate constant at 380 $^\circ$C.

(8) The thermal isomerisation of cyclopropane to give propene obeys a first order law and the following table gives the rate constant, k, at three temperatures:

Temperature (°C)	470	500	519
k (s^{-1})	1.13×10^{-4}	5.95×10^{-4}	16.7×10^{-4}

Show graphically that these data conform to the Arrhenius equation and derive the activation energy, E_A, and the A factor.

Collision theory

(9) For the reaction

$$2HI(g) \rightarrow H_2(g) + I_2(g)$$

calculate the collision frequency per m^3 at 700 K and 1011 325 N m^{-2} pressure. Under these conditions the number of molecules $= 1.05 \times 10^{25}$ m^{-3}. ($\sigma = 350$ pm.) Hence find the rate if $E_A = 184$ kJ mol^{-1}.

Mechanism

(10) $2NO(g) \underset{k_2}{\overset{k_1}{\rightleftharpoons}} N_2O_2(g)$

where both k_1 and k_2 are high.

$$O_2(g) + N_2O_2(g) \xrightarrow{k_3} 2NO_2(g)$$

where k_3 is low. Show that for the above reaction:

$$\text{Rate} = k[NO]^2 [O_2]$$

Questions on chapter 4

(1) The mole fractions of the three components, 1, 2, 3, in a ternary liquid solution are given as:

$$N_1 = 0.18$$
$$N_2 = 0.30$$
$$N_3 = 0.52$$

Calculate:

(i) The integral molar free energy of mixing, ΔG^m, assuming ideal mixing.
(ii) The integral molar free energy of mixing, ΔG^m, assuming non-ideal mixing where:

$$\log \gamma_3 = -1.1$$

determined from Gibbs–Duhem equation (ternary)

$$\log \gamma_2 = -0.7$$

and $a_1 = 0.45$

determined by an electrochemical sensor.
(iii) The integral excess free energy of mixing.
(iv) Establish whether these slag components mix exothermically or endothermically, assuming a regular solution.

(2)   The following values have been determined for the activity of mercury in liquid mercury–bismuth alloys at $320\,°C$ from vapour pressure measurements. Calculate the activity of bismuth in a 20 atom% Bi alloy at this temperature.

N_{Hg}	0.949	0.893	0.851	0.753	0.653	0.537	0.437	0.330	0.207	0.063
a_{Hg}	0.961	0.929	0.908	0.840	0.765	0.650	0.542	0.432	0.278	0.092

(3)   Liquid iron at $1600\,°C$ is in equilibrium with hydrogen gas at 0.01 atmosphere pressure. Calculate the wt% of hydrogen dissolved in the liquid for the following three cases:

   (i)   an ideal solution of hydrogen in liquid iron
   (ii)   the Henrian activity coefficient of hydrogen in iron is 0.2
   (iii) 0.5 wt% carbon, 0.01 wt% oxygen and 0.05 wt% titanium are dissolved in the liquid iron with the following interaction coefficients, f_H^X, of these elements in the Fe–H solution:

$$f_H^C = 0.07;\ f_H^O = 5.00;\ f_H^{Ti} = -0.06$$

   (iv)   explain why some interaction coefficients are positive while others are negative.

Questions on chapter 5

(1)   Briefly outline the functions and properties required in a slag.
(2)   A slag is composed of three oxides at $1600\,°C$.

$$N_{MnO} = 0.20;\ N_{CaO} = 0.50;\ N_{SiO_2} = 0.30$$

where N = mole fraction. Calculate the basicity ratio of the slag using three different relationships.

(3)   Discuss the effect of slag temperature and basicity on the following slag–metal reactions:

   (a) $MnO_{(slag)} + C = Mn_{(Fe)} + CO;\ \Delta G^\theta = 290\,300 - 173.22T$ J
   (b) $SiO_{2(slag)} + 2C = Si_{(Fe)} + 2CO;\ \log K = -31\,000/T + 20.65$
   (c) $Ti_{(Fe)} + 2CO = TiO_{2(slag)} + 2C;\ \Delta G^\theta = -165\,000 + 83.68T$ J

a_{TiO_2} increases slowly as basicity increases.

(4)   On examination of the isoviscosity lines plotted in *Fig. 5.18*, explain the effect of adding Al_2O_3 to both basic and acidic $CaO–SiO_2$ slags.

(5)   Fully explain the trends for sulphur removal as presented in *Fig. A5.1*. Note that the term 'excess base' has a similar meaning to excess number of O^{2-} ions, i.e. $n_{O^{2-}}$.

Other questions that would provide a useful learning exercise include the following.

(6)   Discuss the effect of:
   (i)   cation and anion size
   (ii)   ionic bond fraction
   (iii) coordination number
   on the basic/acid characteristics of an oxide.

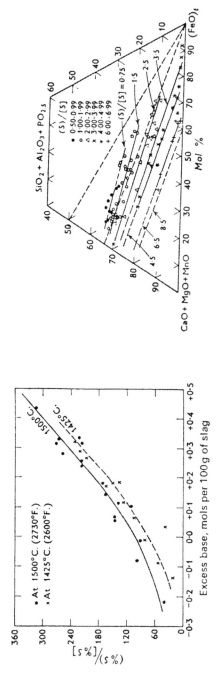

Figure A5.1 Sulphur partitioning between the slag and metal, i.e. (%S)/[%S]: (a) from Hatch and Chipman[22]; (b) from Bishop *et al.*[23]

(7) Explain the sequence of structural changes that occur when increasing additions of CaO are added to SiO_2 in the liquid state.

(8) With reference to *Figs 5.6, 5.8, 5.9* and *5.10*, explain the reason for immiscibility in liquid oxide systems.

(9) Compare the ionic slag theories of Temkin, Flood and Masson with the molecular slag theory. Special emphasis should be placed on the assumptions made and the applications of each to practical slag systems.

Questions on chapter 6

(1) Calculate the pH of:

   (a) 3×10^{-3} mol dm^{-3} HNO_3
   (b) 3×10^{-3} mol dm^{-3} NaOH

   (assuming $Y_{\pm} = 1$).

(2) The conductivity of a 1×10^{-2} mol dm^{-3} solution of KCl is 0.000 141 S cm^{-1} at 25 °C. Calculate the molar conductivity in

   (a)   S cm^2 mol^{-1}
   (b)   S m^2 mol^{-1}

(3) Calculate the electrode potential of the cadmium half-cell if the emf of the following cell is 0.7 V:

$$Cd(s)|\ Cd^{2+}(aq)||\ calomel(satd)$$

$E_{cal,\ satd} = 0.242$ V

(4) Calculate the electrode potential of the Fe^{3+}/Fe^{2+} electrode at 298 K when $a_{Fe^{3+}}/a_{Fe^{2+}} = 0.01$, $E^0_{Fe^{2+}}/E^0_{Fe^{3+}} = 0.77$ V.

(5) Electrorefining of copper using acidic copper sulphate solution is carried out at a current of 15 000 A and 0.4 V. If the output per 24 hour day is 408.5 kg on a total cathode area of 10 m^2 calculate:

   (a)   current efficiency,   (b)   energy efficiency,
   (c)   current density and   (d)   thickness of deposit

   (Relative atomic mass of copper, 64; density of copper, 8930 kg m^{-3}.

(6) 10 circular plates each 300 mm in diameter are to be electroplated with nickel on one side only to a thickness of 25 μm. If the cathode current density is 3 A dm^{-2} and the cathode current efficiency is 94% calculate the plating time. Assume that the 10 components are electroplated together, not singly, in a nickel sulphate bath. Relative atomic mass of nickel, 58.71; density of nickel, 8900 kg m^{-3}; Faraday constant, 96 500 C.)

Questions on chapters 7 and 8

(1) A mineral concentrate contains both MnS and NiS. Using *Fig. 7.6* determine a suitable temperature for the selective sulphating roasting of MnO to $MnSO_4$ while retaining the Ni as NiO.

(2) A concentrate contains FeS, MnS, Cu_2S and NiS. Using data from *Fig. 7.6* determine the relationship between P_{O_2} and P_{SO_2} which provides a partial roasting process to be conducted in this concentrate at $600\,°C$ in which the Cu and Ni remain as sulphides and the Mn and Fe are dead roasted. Assume unit activity for all oxides and sulphides.

(3) The calcine product from question (2) is subsequently smelted and converted, and fire refining is used to preferentially oxidise Ni from the Cu–Ni impure liquid metal. Using data from *Fig. 2.11* determine the limiting ratio of a_{Ni}/a_{Cu} beyond which further fire refining at $1200\,°C$ becomes useless. Assume unit activity for both oxides.

(4) Determine the partial pressure of Cu in Cu-containing steel which is held in a ladle at $1600\,°C$, given that the boiling point of Cu is $2560\,°C$, the latent heat of evaporation for Cu is 302 kJ mol^{-1}, 0.6wt% Cu is present in the steel and the Henrian activity coefficient for Cu in the steel is -0.02.

(5) From examination of *Fig. 2.11*, determine the effect of increasing temperature on the equilibrium p_{CO}/p_{CO_2} ratio required for the reduction of metal oxides with CO whose reaction lines $(2M + O_2 = 2MO)$ lie: (i) above and (ii) below the $2CO + O_2 = 2CO_2$ reaction line.

(6) Electrowinning of copper from acidified copper sulphate solution is conducted at a pH of 1 and a p_{O_2} of 0.2 atm. From the data given in Appendix 2 determine the electrode reactions, the overall cell reaction and the theoretical cell voltage required to extract pure Cu at the cathode.

(7) Calculate the required partial pressure of CO (p_{CO}) in the AOD process at $1600\,°C$ for the production of a 0.02%C, 18%Cr steel given the following data:

$$2Cr_{(Fe)} + 3O_{(Fe)} = Cr_2O_{3\,(slag)}; \; \Delta G = -887\,400 + 383T \text{ J}$$

$$C_{(Fe)} + O_{(Fe)} = CO; \; \Delta G^{\ominus} = -110\,600 - 110.7T \text{ J}$$

f_c (Henrian activity coefficient of C in the steel) $= -0.12$, γ_{Cr} activity coefficient of Cr in the steel) $= -4.0$. Assume $a_{Cr_2O_3\,(slag)} = 1$.

(8) Cu^{2+} and Ni^{2+} ions are present in a leached solution through which is bubbled hydrogen gas at 1 atm pressure. If the activities of Cu^{2+} and N^{2+} in solution are 0.2 and 0.3 respectively, determine the values of pH required to first precipitate Cu and subsequently Ni from the solution at 298 K.

(9) A cupola operator is encountering difficulties with high silicon contents in the cast iron. Given the following relationship what would you advise regarding suitable cupola operation practice?

$$SiO_{2\,(slag)} + 2C = Si_{(Fe)} + 2CO; \; \Delta G^{\ominus} = +593\,570 - 396.16T \text{ J}$$

(10) The same cupola as in question (9) is now operating under your advised practice. Using the net diagram *Fig. 8.4*, determine suitable operating conditions for the iron/coke ratio, blast volume and molten iron temperature to provide efficient melting practice for the 550 mm internal diameter cupola.

Questions on chapter 9

(1) Calculate the electrode potential for the equilibrium

$$\tfrac{1}{2}O_2 + H_2O + 2e = 2OH^-$$

in air-saturated water. ($E^{\ominus} = +0.40$ V; assume p_{O_2} is 0.2 atm and pH is 7.)

(2) Calculate the electrode potential for iron in equilibrium in aqueous solution. ($a_{Fe^{2+}}$ is 10^{-6} and $E^{\ominus}$ is -0.44 V.)

(3) Using the data from questions (1) and (2) calculate:

(a) the corrosion rate of iron in air-saturated water in A m^{-2} and mm yr^{-1}
(b) the corrosion potential of iron in this environment. What assumption is implicit in this approach?

Tafel coefficient for the anodic reaction, $b_a = 0.065$ V dec^{-1}. Exchange current density for Fe^{2+} + 2e = Fe = 10^{-4} A m^{-2}. Exchange current density for O_2 reduction reaction on iron = 10^{-10} A m^{-2}.

(4) Determine the corrosion rate of iron in air-saturated sea water when corrosion is controlled by concentration polarisation. (Diffusivity for oxygen in water is 10^{-9} m^2 s^{-1}, diffusion layer thickness is 100 μm, transport number is zero and n is 2. Assume sea water contains 0.3 mol m^{-3} oxygen.)

(5) Calculate the corrosion rate of iron in aerated water from the following data: p_{O_2} is 0.2; η_c^c is -0.06 V; b_c^A is -0.15 V dec^{-1}. i_0 for oxygen reduction on iron is 10^{-10} A m^{-2}; E_{corr} is -0.45 V.

(6) Aerated reducing acid flows along a stainless steel pipe. Determine whether the steel is likely to corrode: (a) in service and (b) during shutdown periods. (Concentration of dissolved oxygen is 10^{-3} mol dm^{-3}; i_{crit} for stainless steel in the acid is 2 A m^{-2}; δ for the flowing acid is 0.5 μm and for static acid is 0.5 mm; diffusion coefficient for dissolved oxygen is 10^{-3} mm^2 s^{-1}; transport number is zero.)

(7) Calculate the corrosion rate of iron in oxygen-free acid, pH 2, if the reaction is:

$$Fe + 2H^+ \rightarrow Fe^{2+} + H_2$$

$E^{\ominus}$ (Fe/Fe^{2+}) = -0.44 V; $E^{\ominus}$ (2H$^+$/H$_2$) = 0.00 V; $a_{Fe^{2+}}$ = 10^{-6}

i_0 (Fe/Fe^{2+}) = 10^{-4} A m^{-2}; i_0 (2H$^+$/H$_2$ on Fe) = 10^{-2} A m^{-2};

b_a = 0.06 V dec^{-1}; b_c = -0.12 V dec^{-1}; p_{H_2} = 1.0.

(8) Determine the weight loss per week from an iron pipe when the corrosion current is 10 mA. (Relative atomic mass of iron, 55.85.)

(9) (a) Calculate the theoretical capacity, A h kg^{-1}, and weight loss, kg (A yr)$^{-1}$, of a sacrificial magnesium anode.
(b) Calculate the life span of a 55 kg magnesium anode: (i) in soil and (ii) in sea water, if the current outputs in these environments are 0.3 A and 3.5 A respectively.

(Relative atomic mass of magnesium is 24.3; assume that, in practice, magnesium dissolves twice as fast as theory predicts.)

(10)   Steel pilings of immersed area 5500 m^2 are used to support a pier in sea water. If the current density required to provide cathodic protection is 80 mA m^{-2}, calculate the minimum number of sacrificial zinc anodes, each of 110 kg, which would be needed to protect this structure for 8 years. (Relative atomic mass of zinc is 65.4; assume that, in practice, zinc dissolves 1.1 times faster than theory predicts.)

Appendix 6

Solutions to Questions

Note: Where logs are involved $\ln x = 2.303 \log_{10} x$ has been used throughout, so slight differences may occur where $\ln x$ has been used directly.

Solutions to questions on chapter 1

(1) (i)   Ne: $1s^2 ; 2s^2 ; 2p^6$
    (ii) Ca: $1s^2 ; 2s^2 ; 2p^6 ; 3s^2 ; 3p^6 ; 4s^2$
    (iii) Fe; $1s^2 ; 2s^2 ; 2p^6 ; 3s^2 ; 3p^6 ; 3d^6 ; 4s^2$
    (iv) Bi: $1s^2 ; 2s^2 ; 2p^6 ; 3s^2 ; 3p^6 ; 3d^{10} ; 4s^2 ; 4p^6 ; 4d^{14} ; 5s^2 ; 5p^6 ; 5d^{10} ; 6s^2 ; 6p^3$.

(2) From electronegativity differences between the two elements: (i) ionic, (ii) covalent, (iii) covalent, (iv) ionic.

(3) (i) octahedral; coordination no. of 6 for Mg and O, (ii) tetrahedral; coordination no. of 4 for Si and 2 for Si, (iii) body central cubic; coordination no. of 8 for Cs and I.

(4) (i) Atomic size decreases with increasing atomic no. across a period, (ii) Atomic size increases with increasing atomic no. down each group.

(5) Explanation based on atomic size, nuclear charge, etc., such that the following general points can be stated: ionisation energies will decrease on increasing atomic size (atomic no.) since electrons are less tightly held. Ionisation energies decrease in order s, p, d, f since s-electrons are nearer to the nucleus and therefore more tightly held compared with f-electrons which are furthest from the nucleus. Second ionisation energies are higher than first since once the first electron is removed the remainder are held more tightly to the nucleus.

Solutions to questions on chapter 2

Section 2.1.2.1 (enthalpy changes)

(1) $\Delta H^{\ominus}_{1523} = +161.1 \text{ kJ}$

(2) (a) $\Delta H^{\ominus}_{298} = +625 \text{ kJ (using } H^{\ominus}_{298}[Al_2O_3(s)] = -1669 \text{ kJ mol}^{-1};$
        $H^{\ominus}_{298}[ZnO(s)] = -348 \text{ kJ mol}^{-1})$
    (b) $\Delta H^{\ominus}_{298} = +117 \text{ kJ (using } H^{\ominus}_{298}[MgCO_3(s)] = -1113 \text{ kJ mol}^{-1};$
        $H^{\ominus}_{298}[MgO(s)] = -602 \text{ kJ mol}^{-1}; H^{\ominus}_{298}[CO_2(g)] = -394 \text{ kJ mol}^{-1})$

(3) $\Delta H^{\ominus}_{298} = -87 \text{ kJ mol}^{-1}$

(4) $\Delta H^{\ominus}_{\text{trans}}$ for $H_2O(l) \rightarrow H_2O(g)$

416

(5) $H_{298}^{\ominus}[C_2H_5OH(l)] = -275 \text{ kJ mol}^{-1}$ (using $H^{\ominus}[H_2O(l)] = -286 \text{ kJ mol}^{-1}$; $H^{\ominus}[CO_2(g)] = -395 \text{ kJ mol}^{-1}$)

(6) $\Delta H_{298}^{\ominus} = -460 \text{ kJ mol}^{-1}$

Section 2.1.2.2 (Hess's law)

(1) (a) $\Delta H_{f,298}^{\ominus} = +110 \text{ kJ mol}^{-1}$

    (b) $\Delta H_{f,298}^{\ominus} = -338.1 \text{ kJ mol}^{-1}$

    (c) $\Delta H_{f,298}^{\ominus} = -277.6 \text{ kJ mol}^{-1}$

(2) $\Delta H_{f,298}^{\ominus} = -141 \text{ kJ mol}^{-1}$

(3) $\Delta H_{1873}^{\ominus} = -361.8 \text{ kJ}$

Section 2.1.4.2 (Kirchoff's equation)

(1) (a) $\Delta H^{\ominus}$ becomes more negative with increase in temperature

    (b) slope negative and of small magnitude

    (c) $\Delta H_{598}^{\ominus} = +177.6 \text{ kJ}$

(2) $\Delta H_{1373}^{\ominus} = +369.44 \text{ kJ}$

(3) $\Delta H_{298}^{\ominus} = -283.21 \text{ kJ}$

Section 2.2.1.1 (First law of thermodynamics)

(1) (a) $\Delta nRT = 3716 \text{ J}$

    (b) $\Delta nRT = -1239 \text{ J}$

(2) $\Delta H_{298}^{\ominus} = -3059.2 \text{ kJ}$

(3) $\Delta H_{300}^{\ominus} = -1199 \text{ kJ}$

Section 2.2.1.2 (Entropy changes)

(1) $\Delta S_{353}^{\ominus} = 87.25 \text{ J K}^{-1} \text{ mol}^{-1}$

(2) (a) $\Delta S_{298}^{\ominus} = -327 \text{ J K}^{-1} \text{ mol}^{-1}$

    (b) $\Delta S_{298}^{\ominus} = -156 \text{ J K}^{-1} \text{ mol}^{-1}$

    (c) $\Delta S_{298}^{\ominus} = -167 \text{ J K}^{-1} \text{ mol}^{-1}$

(3) $S_{1800}^{\ominus}[Cu(l)] = 86.96 \text{ J K}^{-1} \text{ mol}^{-1}$

(4) (a) $\Delta S_{298}^{\ominus} = -35.2 \text{ J K}^{-1} \text{ mol}^{-1}$

    (b) $\Delta S_{798}^{\ominus} = 350.0 \text{ J K}^{-1} \text{ mol}^{-1}$

Section 2.2.2.4 (Free energy changes)

(1) (a) $\Delta G_{298}^{\ominus} = +175.1 \text{ kJ mol}^{-1}$

    (b) $\Delta G_{1373}^{\ominus} = -41.4 \text{ kJ mol}^{-1}$

    (c) $\Delta G_{298}^{\ominus} = -244.2 \text{ kJ mol}^{-1}$

(2) (a) $\Delta G_{298}^{\ominus} = +208 \text{ kJ mol}^{-1}$

    (b) $\Delta G_{298}^{\ominus} = +21 \text{ kJ mol}^{-1}$

    (c) $\Delta G_{298}^{\ominus} = -835 \text{ kJ}$

    (d) $\Delta G^{\ominus}_{298} = +78$ kJ

(3)   (a)   (i) $\Delta H^{\ominus}_{500} = -202$ kJ, (ii) $\Delta S^{\ominus}_{500} = +18$ J K^{-1} mol^{-1},
             (iii) $\Delta G^{\ominus}_{500} = -211$ kJ

       (b)   (i) $\Delta H^{\ominus}_{500} = +236$ kJ, (ii) $\Delta S^{\ominus}_{500} = +217$ J K^{-1} mol^{-1},
             (iii) $\Delta G^{\ominus}_{500} = +127.5$ kJ. Reversal temperature = 1088 K.

(4) $\Delta G^{\ominus}_{1200} = -128.2$ kJ

Section 2.2.3.5 (Equilibria)

(1) $K = 0.06$

(2) $K_{1120} = 11.64$; $p_{CO_2} = 8.59 \times 10^{-10}$ atm

(3)   (i)   (a)   no effect   (ii)   (a)   decreased yield
           (b)   no effect   (b)   increased yield
           (c)   increased yield   (c)   decreased yield
           (d)   decreased yield   (d)   increased yield

Section 2.2.3.6 (Relations between free energy and equilibrium constant)

(1) $K_{1200} = 3.81 \times 10^5$

(2)   (a) $K_{1173} = 4.54$

       (b) $p_{H_2} = 0.0095$ atm (9.5×10^{-3} atm)

(3) $K_{298} = 7.52$

(4) $\Delta G_{1773} = -155.5$ kJ (i.e. reaction is feasible)

Section 2.2.3.7 (Dissociation pressures)

(1)   (a) $p_{O_2} = 9.52 \times 10^{-19}$ atm

       (b) $p_{O_2} = 9.93 \times 10^{-9}$ atm

(2) $\Delta G^{\ominus}_{1373} = -683.5$ kJ mol^{-1}

Section 2.2.3.9 (Variation of equilibrium constant and vapour pressure with temperature)

(1) $K_{600} = 9.35 \times 10^5$

(2) $\Delta H^{\ominus} = -116$ kJ

(3) $\overset{\circ}{p}_{1873} = 0.0483$ atm (4.83×10^{-2} atm)

Section 2.2.3.10 (Free energy–temperature diagrams)

Most answers are approximate as the diagrams do not allow great accuracy

(1)   (a)   above 650 $^\circ$C

       (b)   above 1230 $^\circ$C

       (c)   above 1550 $^\circ$C

(2)   600 $^\circ$C (however, temperatures in excess of this, around 1000 $^\circ$C in fact, required as at this higher temperature there is less tendency for the C to oxidise to CO_2)

(3) $200\,^{\circ}C$

(4)   above $610\,^{\circ}C$

(5) $\Delta G^{\ominus}_{1273} = -100\,kJ$

(7)   (i) Cu_2O reduction by $H_2 = \Delta G^{\ominus}_{1200} = -180\,kJ$

     (ii) Cu_2O reduction by $C = \Delta G^{\ominus}_{1200} = -260\,kJ$

(9) $1400\,^{\circ}C$

(10) $p_{CO}/p_{CO_2} = 4.4 \times 10^4$ at $1100\,^{\circ}C$

(11) $520\,^{\circ}C$

Solutions to questions on chapter 3

(1) $n \approx 6$, indicating (possibly) that nucleation is the RDS.

(2)   (a) $0.0249 \times 100 = \ln(100/[S])$. $[S] = 22.4\%$. (b) $t = 36.8\,min$.

(3)   A plot of $\ln[S]$ versus time gives a straight line of slope
$0.053\,min^{-1}$. Thus $k = 0.053\,min^{-1} = 8.8 \times 10^{-4}\,s^{-1}$.

(4)   A plot of $\ln(41.5 - vol.\ CO)$ versus time gives $k = 6.6 \times 10^{-3}\,s^{-1}$.
$t^{1/2} = \ln 2/k = 105\,s$.

(5)   A plot of $1/[N]$ versus time gives a straight line. Slope $= k = 8.52 \times 10^{-5}$
(N atoms per 100 Fe atoms)$^{-1}$ min^{-1} or 1.42×10^{-6} (N atoms per 100 Fe
atoms)$^{-1}$ s^{-1}.

(6)   (a) $k = 1 \times 10^{11} \exp(-100\,000/8.314 \times 298)\,s^{-1} = 2.96 \times 10^{-7}\,s^{-1}$
    (b) $k = 1.41 \times 10^{-6}\,s^{-1}$.

(7) $E_A = 184.2\,kJ\,mol^{-1}$
$A = 1.0 \times 10^{14}\,s^{-1}$.
$k_{633} = 0.184\,s^{-1}$.

(8)   A plot of $\ln k$ versus $1/T$ (after conversion to kelvin) gives a straight line,
slope $= -A = 3.248 \times 10^4 = 1E_A/R$. Thus $E_A = 270\,kJ\,mol^{-1}$.

(9)   Collision frequency, Z, is $1.02 \times 10^{34}\,m^{-3}\,s^{-1}$. Rate $= 3.75 \times 10^{20}\,m^{-3}\,s^{-1}$.
To convert to more usual units, divide by the Avogadro number, 6.02×10^{23}
mol^{-1}. Rate $= 6.23 \times 10^{-7}\,mol\,dm^{-3}\,s^{-1}$.

(10) For the rate determining step:

$$Rate = k_3\,[O_2]\,[N_2O_2] \qquad (1)$$

For the fast pre-equilibrium

$$k = k_1/k_2 = [N_2O_2]/[NO]^2$$

$$\therefore \qquad [N_2O_2] = k\,[NO]^2 \qquad (2)$$

Substituting in (1) from (2)

$$Rate = k_3 k\,[O_2]\,[NO]^2$$

Solutions to questions on chapter 4

(1)   (i) $\Delta G^m_{(ideal)} = -15\,730\,\text{J mol}^{-1}$

     (ii) $\Delta G^m_{(real)} = -41\,450\,\text{J mol}^{-1}$

     (iii) $\Delta G^E = -25\,720\,\text{J mol}^{-1}$

     (iv) $\Delta H^m_{(regular)} = -25\,402\,\text{J mol}^{-1}$

    $\therefore$ exothermic mixing.

(2) $a_{Bi} = 0.28$.

(3)   (i)   wt%H = 0.0001 wt%

     (ii)   wt%H = 0.0005 wt%

     (iii)   wt%H = 0.0004 wt%

     (iv)   A strong reaction between the third element and hydrogen results in a negative interaction coefficient, e.g. Ti forms halides with hydrogen. If the additional third element reacts more strongly with iron than with hydrogen, e.g. carbon or oxygen, a positive interaction coefficient is produced, see Section 4.3.7.

Solutions to questions on chapter 5

(1)   Discussion of following required. Functions: protection; insulation; acceptance of unwanted species; control of refining media. Properties: low melting point (high fluidity); low specific gravity; basicity composition needed to perform required refining processes, e.g. low activity in slag of unwanted species.

(2)

$$B = \frac{N_{MnO} + N_{CaO}}{N_{SiO_2}} = \frac{0.20 + 0.50}{0.30} = 2.33$$

$$V = \frac{N_{CaO}}{N_{SiO_2}} = \frac{0.50}{0.30} = 1.67$$

$$n_{O^{2-}} = n_{CaO} + n_{MnO} - 2n_{0.30} = 0.1$$

(3)   Discussion based on following points:

    (a)   Increase $T \to \Delta G^\theta$ more negative, i.e. MnO reduction favoured.
Increase $B \to$ increases a_{MnO} and MnO reduction favoured.

    (b)   Increase $T \to K$ larger therefore SiO_2 reduction favoured.
Increase $B \to$ decreases a_{SiO_2} and SiO_2 reduction not favoured.

    (c)   Increase $T \to \Delta G^\theta$ more positive TiO_2 reduction favoured.
Increase $B \to$ increases a_{TiO_2} and TiO_2 reduction favoured.

(4)   Explanation based on following points: Al_2O_3 –amphoteric. (a) In basic composition $Al_2O_3 + 3O^{2-} \to 2AlO_3^3$, increases viscosity owing to development of network of AlO_3^3. (b) In acidic composition $Al_2O_3 \to 2Al^{3+} + 3O^{2-}$; decreases viscosity on account of increased ionisation, O^{2-}.

(5)   Explanation based on following points:

$$S_{Fe} + O^{2-}_{(slag)} = S^{2-}_{(slag)} + O_{(Fe)}; \quad K = \frac{a_{S^{2-}_{(slag)}} \times a_{O_{(Fe)}}}{a_{S_{(Fe)}} \times a_{O^{2-}_{(slag)}}}$$

and $\Delta G^{\ominus} = +A - BT$ (form)

so S removal requires high $a_{O^{2-}}$, low $a_{O_{(Fe)}}$, high temperature if not oxidising conditions.

Increase $n_{O^{2-}_{(slag)}}$ increases $a_{O^{2-}_{(slag)}} \rightarrow$ increases (S)/[S]

Increase $T \rightarrow$ increase (s)/[S] for reducing conditions
Increase CaO, etc. $\rightarrow$ increases (S)/[S] on account of increased $n_{O^{2-}_{(slag)}}$

Addition of weaker basic oxide FeO compared with CaO decreases (S)/[S] for highly basic slags, i.e. dilution of CaO with FeO.

(6), (7), (8), (9) No formal answers will be provided for these questions. The reader should consult Section 5.4.

Solutions to questions on chapter 6

(1)   (a)   pH $= 2.52$
      (b)   pH $= 11.48$

(2)   (a) $141\,S\,cm^2\,mol^{-1}$
      (b) $1.41 \times 10^{-2}\,S\,m^2\,mol^{-1}$

(3) $E_{Cd} = -0.458\,V$

(4) $E = 0.65\,V$

(5)   (a) 95%, (b) $0.35\,kW\,h\,kg^{-1}$, (c) $1500\,A\,m^{-2}$, (d) 4.6 mm

(6)   43.2 min

Solutions to questions on chapters 7 and 8

(1)   Between $1000\,^{\circ}C$ and $1400\,^{\circ}C$

(2) $2.21 \times 10^{-5}\,p_{SO_2}^{2/3} > p_{O_2} > 4 \times 10^{-6}\,p_{SO_2}^{2/3}$

(3) $a_{Cu}^4/a_{Ni}^2 \leqslant 7.77$ or, $a_{Ni} \geqslant (1.29 \times 10^{-7})\,a_{Cu}^2$

(4) $p_{Cu,1873} = 4.15 \times 10^{-6}$ atm

(5)   For reaction lines of $2M + O_2 = 2MO$ which are below the $2CO + O_2 = 2CO_2$ reaction line, the reduction of MO with CO requires lower p_{CO}/p_{CO_2} ratios to reduce MO as the temperature increases, i.e. it becomes easier to reduce the oxide with CO on increasing temperature. The reverse is true for oxide formation lines which lie above the $2CO + O_2 = 2CO_2$ reaction line.

(6)   Cathodic reaction:

$$Cu^{2+} + 2e = Cu; \quad E_c^{\ominus} = +0.34 \, V$$

Anodic reaction:

$$2OH = \tfrac{1}{2}O_2 + H_2O + 2e; \quad E_a^{\ominus} = -0.4 \, V$$

Overall cell reaction:

$$Cu^{2+} + 2OH^- = Cu + \tfrac{1}{2}O_2 + H_2O$$

$E_{cell} = 0.82 \, V$

(7) $p_{CO} = 5.7 \times 10^{-2} \, atm$

(8)   Cu will be precipitated at all pH values. To precipitate Ni pH > 4.44.

(9)   Operate the cupola at lower hearth temperatures and cold blast and use a basic slag practice since increasing temperature will increase Si in Fe and a basic slag will lower the value of $a_{SiO_2 \, (slag)}$.

(10)   An efficient cupola melting rate of $8.2 \, t \, m^{-2} \, hr^{-1}$ is produced using an Fe/coke ratio of 18.1, a blast volume of $15 \, m^3 \, min^{-1}$ (at STP) producing a low molten iron temperature of about $1370 \, ^{\circ}C$.

Solutions to questions on chapter 9

(1)   From equation (9.5), $E_{r,O_2} = 0.83 \, V$

(2) $E_{r,Fe} = -0.62 \, V$

(3)   (a) $0.27 \, A \, m^{-2} = 0.31 \, mm \, yr^{-1}$, i.e. $A \, m^{-2} \simeq mm \, yr^{-1}$. However, this relationship only holds for Fe, Ni, Al and Cu which have similar values for n/W. It does not apply to Zn, Ti, Mg or to most other metals.
       (b) $E_{corr} = -0.36 \, V$
       The calculation assumes Tafel behaviour, i.e. charge-transfer control. In aerated solutions the oxygen reduction reaction at the cathode will promote concentration polarisation, the polarisation curves being similar to those shown in *Fig. 9.22(d)*.

(4) $i_L = 0.6 \, A \, m^{-2}$
       Under these conditions this represents a maximum value of corrosion rate and will only change if temperature rises and/or degree of turbulence changes thereby affecting D, c and δ.

(5) $i_{corr} = 0.014 \, A \, m^{-2}$

(6) $i_L(flowing) = 385 \, A \, m^{-2}$
       $i_L(static) = 0.385 \, A \, m^{-2}$
i.e. $i_L(flowing) > i_{crit}$; no corrosion
       $i_L(static) < i_{crit}$; corrosion probable so the pipes must be drained, washed and dried during extended shutdown periods to reduce the likelihood of pitting corrosion.

(7) $i_{corr} = 1.28 \, A \, m^{-2}$ (i.e. $1.28 \, mm \, yr^{-1}$ approx—see answer to Question (3)).

(8)   Weight loss = 91.3 g
      If corrosion were uniform over a large structure this weight loss would be
      insignificant, but if corrosion were local then this weight loss could result in
      perforation.

(9)   (a) $m/It = 3.97 \, \text{kg} \, (\text{A yr})^{-1}$
      (b)   (i) 55 kg anode lasts 46.2 years in theory or 23.1 years in practice.
            (ii) in sea water, a 55 kg magnesium anode would last almost 2 years in
                 practice.

(10) 377 (rounding up) anodes needed.

Index

18051227R00252

Printed in Great Britain
by Amazon